Sparse Sensing and Sparsity Sensed in Multi-sensor Array Applications

Xiangrong Wang · Xianghua Wang · Weitong Zhai ·
Kaiquan Cai

Sparse Sensing and Sparsity Sensed in Multi-sensor Array Applications

Springer

Xiangrong Wang
Beihang University
Beijing, China

Weitong Zhai
Beihang University
Beijing, China

Xianghua Wang
Beijing University of Posts
and Telecommunications
Beijing, China

Kaiquan Cai
Beihang University
Beijing, China

ISBN 978-981-99-9560-8 ISBN 978-981-99-9558-5 (eBook)
https://doi.org/10.1007/978-981-99-9558-5

This Springer imprint is published by the registered company Springer Nature Singapore Pte Ltd.
The registered company address is: 152 Beach Road, #21-01/04 Gateway East, Singapore 189721, Singapore

This book is dedicated to our hard-working parents, who have devoted all their love to their children, and all for once.

Preface

Sensing with a single sensor cannot provide the wealth of information for the processing of signals in modern-day applications. Multi-sensor arrays use a set of sensors to sample the electromagnetic waveform in the spatial domain, and thus find extensive use in different sensing modalities, such as radar, telescope, acoustic, ultrasound, etc. While uniform arrays provide the full complement of information and are straightforward to deal with, they inherently contain significant redundancy, which opens the door to save on their size, weight, and power (SWaP). Sparse sensor arrays are under-sampled such that several sensors are removed from their uniform counterparts, thus ensuring SWaP savings. The resulting configuration provides what we refer to as sparsity in sensing. On the other hand, sparsity in the sensed refers to sparsity that is imposed on the collected signal, such as in spectral domain or in the sensed scene. Exploitation of the sparsity in the sensed can significantly enhance the performance of signal processing systems.

Sensing can be generally divided into active and passive sensing. Active sensing is achieved by transmitting probing signals and measuring the backscattered waveforms. Passive sensing, on the other hand, fulfills its mission using signals of opportunity. Regardless of active or passive sensing, sensor array configurations define the structure of sensing systems and affect the underlying sensing functionality. For the exploration of sparsity in the sensing, the key is to optimize the placement of a given number of sensors along the physical aperture to satisfy some specific criteria. In the case of sparsity in the sensed, the objective is two-fold: the first is, to develop signal processing algorithms to compensate for the missing information caused by the imposed sparsity; the second is to fully leverage the underlying sparsity of signals in some transformed domain in order to improve the sensing performance.

Sparse arrays achieve sparsity in sensing the spatial domain and have found applications in RF health monitoring, automotive industry, satellite navigations, sonar, acoustics, and ultrasound. Sparse arrays have also attracted much interest in the defense industry due to their effectiveness and efficiency compared to uniform linear or planar arrays that house the same number of antennas. In essence, there is a strong interest from both defense and civilian sectors. Driven by the importance of sparsity, research into sparse sensing and processing techniques continues unabated. The vigorousness of research activities is especially evident from the number of conferences, special sessions, M.S. theses, and Ph.D. dissertations dedicated to sparse array design.

The intent of this book is to provide a summary of the authors' research in sparse array processing, especially in radar applications. In particular, our focus is bringing together a number of contributions on both sparse sensing design (including both active and passive sensing) and sparsity-sensed processing techniques, which are established as state-of-the-art sparsity exploration techniques. This book provides an in-depth study on sparse sensing for several major multi-sensor array applications, such as beampattern synthesis, adaptive beamforming, target detection, direction of arrival estimation, and dual-functional radar communications. Our objective is to document the progress made in this area of research and provide a handy reference for researchers interested in this field. In the process, we hope that this book will stimulate additional work in this area and lead to further advances in the state of the art.

This book is organized as follows: Part I comprises Chaps. 1 and 2. Chapter 1 provides a review on the array processing fundamentals from the viewpoint of mathematical duality between arrays and filters. Chapter 2 briefly introduces four main applications of multi-sensor arrays, those are adaptive beamforming, adaptive sidelobe canceller, direction of arrival estimation, and target detection.

Part II of this book focuses on the sparse sensing design via antenna selection for different array applications. In Chap. 3, we consider the thinned array beampattern synthesis by iterative soft-thresholding-based optimization algorithms, followed by the sparse quiescent beamformer design combining adaptive and deterministic constraints. In Chap. 4, we consider sparse array design for adaptive beamforming, which is a more effective method for interference suppression compared to deterministic beamforming in Chap. 3. A new parameter, referred to as spatial correlation coefficient (SCC), is proposed to characterize the spatial separation between the desired signal and the interference subspace. The squared absolute value of the SCC is directly related to the output SINR, and thus maximizing the output SINR is equivalent to minimizing the SCC. Optimum sparse array design methods are then proposed in terms of minimizing the SCC.

The sparse array design for maximum output SINR in Chap. 4 assumes priori knowledge of source and interference DOAs, which may significantly limit its practical applications. In Chap. 5, we introduce the cognitive-driven optimization of sparse sensing, which can fulfill the task of environmental sensing automatically and adaptively. We first propose the adaptive beamformer design by regularized complementary antenna switching, which first collects the full array data via a set of deterministic complementary sparse arrays and then reconfigures the sparse array as the optimum configuration for adaptive beamforming. Cognitive sparse beamformer design via regularized switching network is then proposed, which is capable of swiftly adapting the beamformer geometry via antenna removement and addition according to the environmental dynamics through a "perception-action" cycle.

In Chap. 6, we investigate the sparse MIMO array transceiver design for enhanced adaptive beamforming. A simple scenario of known environmental conditions is first considered. The cognitive-driven optimization of sparse array transceiver for MIMO radar beamforming is then introduced, which further eliminates the prerequisite of prior information. In Chap. 7, we proceed to examine the problem of slow target detection in heterogeneous clutter. In order to address the issue of limited training data, we propose a novel thinned STAP via selecting an optimum subset of antenna-pulse pairs in the metric of maximum output signal-to-clutter-plus-noise ratio (SCNR). The proposed thinned STAP strategy defines a new parameter, named spatial spectrum correlation coefficient (S^2C^2), to analytically characterize the effect of space-time configuration on STAP performance and reduce the dimensionality of traditional STAP. Three algorithms are proposed to solve the antenna-pulse pair selection problem. The effectiveness of the proposed strategy is confirmed by extensive simulation results, especially by employing the MCARM data set.

As a final chapter in Part II, we introduce the sparse sensing for dual-functional radar communications (DFRC). DFRC systems have recently been proposed to enable the coexistence of radar and wireless communications, alleviating the radio frequency spectral congestion. The nominal array configuration for existing DFRC systems is uniform and of fixed-structured, however, it is not necessarily optimum in every sense, and ignores the additional degrees of freedom (DoFs) provided by the flexibility in configuring the antenna array. In Chap. 8, we first explore the sparse array design for DFRC by antenna selection where same or different antennas are assigned to different functions. The array configuration is utilized as an additional spatial degree of freedom (DoF) to suppress the cross-interference and facilitate the cohabitation of the two system functions. We then consider the sparse array optimization for spatial index modulation, where sparse antenna array configurations are utilized to embed communication information.

Part III of this book discusses the sparsity in the sensed for enhanced multi-sensor array processing. In Chap. 9, we propose the array thinning via antenna selection for enhanced DOA estimation in both single-source and multi-source scenarios. In Sect. 9.2, we considered the single-source case. Problem formulation and optimal antenna selection in terms of minimizing the CRB for both isotropic and directional arrays were provided. We proposed a Dinklebach-type algorithm and convex relaxation to solve the combinational optimization in polynomial time. In Sect. 9.3, we investigated the optimal sparse MIMO transceiver design for enhanced DOA estimation in the multi-source scenario. A reweighted l_1-norm minimization with convex relaxation was adopted to promote the binary sparsity.

In Chap. 10, we studied the sparsity sensed for enhanced DOA estimation in the coarray domain employing both fully augmentable arrays and partially augmentable arrays. The utility of difference coarray could increase the number of estimated signals, which was far more than the number of physical antennas. In Sect. 10.2, two novel DOA estimation approaches based on the MOP were proposed, those are MOP interference spectrum and polynomial rooting. In Sect. 10.3, we utilized a probabilistic Bayesian inference method for DOA estimation for both fully and partially augmentable arrays. The proposed method can overcome the shortcomings of the sparse signal reconstruction method and MUSIC, such as spurious peaks and inaccurate power estimation.

I would like to take this opportunity to thank my two Ph.D. supervisors, Prof. Elias Aboutanios from the University of New South Wales in Australia and Prof. Moeness Amin from Villanova University in the USA, who introduced such a great research topic to me. The authors in this book have been working on this topic for more than 10 years and this completion of this book is based on the summary of the authors' research on sparse array processing. Finally, we would like to acknowledge the enthusiasm and support from our lovely students, our friends, and our family.

Beijing, China
June 2023

Xiangrong Wang

Contents

Part II Sparse Sensing via Antenna Selection

Acronyms

Abbreviations

1-D	One-Dimensional
2-D	Two-Dimensional
A/D	Analogue to Digital
ADM	Alternate Direction Method
AMF	Adaptive Matched Filter
AOA	Angle Of Arrival
ASO	Amplitude Sparse Optimization
BCS	Bayesian Compressive Sensing
BFN	Beamforming Network
BQP	Boolean Quadratic Programming
CCM	Clutter plus noise Covariance Matrix
CCP	Convex Concave Procedure
CFAR	Constant False Alarm Rate
CM	Correlation Measurement
CNR	Clutter-to-Noise Ratio
CPI	Coherent Processing Interval
CRB	Cramer Rao Bound
CRPA	Controlled Radiation Pattern Antenna
CUT	Cell Under Test
CW	Continuous Wave
DCS	Difference of Convex Sets
DFT	Discrete Fourier Transform
DMI	Direct Matrix Inversion
DOA	Direction Of Arrival
DoFs	Degrees Of Freedom
DRR	Dynamic Range Ratio
EMI	ElectroMagnetic Interference
FAR	False Alarm Rate
FFT	Fast Fourier Transform

FIM	Fisher Information Matrix
FISTA	Fast Iterative Soft-Thresholding Algorithm
GLRT	Generalized Likelihood Ratio Test
GNSS	Global Navigation Satellite Systems
GSO	Group Sparse Optimization
IF	Intermediate Frequency
IID	Independent and Identically Distributed
INR	Interference-to-Noise Ratio
KKT	Karush-Kuhn-Tucker
LMI	Linear Matrix Inequality
LMS	Least Mean Square
LP	Linear Programming
MBPE	Model-Based Parameter Estimation
MCARM	Multi-Channel Airborne Radar Measurements
MCM	Modified Correlation Measurement
MDV	Minimum Detectable Velocity
MHA	Minimum Hole Array
MIMO	Multiple Input Multiple Output
min	Minimization
ML	Maximum Likelihood
MMV	Multiple Measurement Vector
MOP	Minimum Output Power
MRA	Minimum Redundancy Array
MSE	Mean Squared Error
MTD	Moving Target Detection
MTI	Moving Target Indication
MUSIC	MUltiple-SIgnal Characterization
MVE	Minimum Variance Estimation
PCA	Principle Component Analysis
PDF	Probability Density Function
PRF	Pulse Repetition Frequency
PRI	Pulse Repetition Interval
PRN	Pseudo-Random Noise
PSL	Peak Sidelobe Level
RAAA	Reconfigurable Adaptive Antenna Array
RF	Radio Frequency
RLS	Recursive Least Square
RVM	Relevance Vector Machine
s.t.	Subject To
SCC	Spatial Correlation Coefficients
SCNR	Signal-to-Clutter-plus-Noise Ratio
SDP	Semi-Definite Programming
SINR	Signal-to-Interference-plus-Noise Ratio
SLA	Standard Linear Array
SMI	Sample Matrix Inversion

SMV	Single Measurement vector
SNR	Signal-to-Noise Ratio
SOCP	Second-Order Cone Programming
SSCC	Spatial Spectral Correlation Coefficients
SSR	Sparse Signal Reconstruction
STAP	Space-Time Adaptive Processing
SVD	Singular Value Decomposition
TOA	Time Of Arrival
UCA	Uniform Circular Array
ULA	Uniform Linear Array

Symbols

λ	The wavelength of corresponding frequency
$\mathcal{D}_a$	Array aperture length
$\mathcal{R}_a$	Distance between the array and the transmitted source
$\mathbf{P} = [\mathbf{p}_1, \ldots, \mathbf{p}_N]$	Position matrix of an N-antenna array
$\mathbf{p}_i$	Position vector of the ith antenna
θ	Elevation angle
ϕ	Azimuth angle
c	The speed of light
$k_0 = 2\pi/\lambda$	The wavenumber
$\mathbf{u} = \sin\theta$	The electral angle
$\mathbf{x} \in \mathbb{R}^N$	A vector with dimension N in real domain
$\mathbf{x} \in \mathbb{C}^N$	A vector with dimension N in complex domain
$\mathbf{x}(i)$	The ith entry of the vector $\mathbf{x}$
$\mathbf{x}_{i,c}$	The ith column vector of the corresponding matrix $\mathbf{X}$
$\mathbf{x}_{i,r}$	The ith row vector of the corresponding matrix $\mathbf{X}$
$\mathbf{x}^{(i)}$	Vector $\mathbf{x}$ in the ith iteration
$\mathbf{X} \in \mathbb{R}^{N\times M}$	A matrix with dimension $N \times M$ in real domain
$\mathbf{X} \in \mathbb{C}^{N\times M}$	A matrix with dimension $N \times M$ in complex domain
$\mathbf{X}_{ij}$	The entry of $\mathbf{X}$ in the ith row and jth column
$\|\|\mathbf{x}\|\|_2 = \mathbf{x}^H\mathbf{x}$	l_2-norm of the vector $\mathbf{x}$
$\|\|\mathbf{x}\|\|_0 = C(\mathbf{x})$	l_0-norm of $\mathbf{x}$, i.e., the number of non-zero entries of $\mathbf{x}$
$\|\|\mathbf{x}\|\|_1 = \Sigma_i \|\mathbf{x}(i)\|$	l_1-norm of the vector $\mathbf{x}$
$\|\|\mathbf{x}\|\|_p = (\Sigma_i \|\mathbf{x}(i)\|^p)^{\frac{1}{p}}$	l_p-norm of the vector $\mathbf{x}$
$\|\|\mathbf{x}\|\|_{\mathbf{M}} = \mathbf{x}^H\mathbf{M}\mathbf{x}$	Weighted l_2-norm of the vector $\mathbf{x}$
$\|\|\mathbf{X}\|\|_{2,1} = \Sigma_i \|\|\mathbf{x}_{i,r}\|\|_2$	The $l_{2,1}$-norm for promoting group sparsity in each row
$\|\mathbf{x}\|$	The absolute value of the vector $\mathbf{x}$ element-wise
$\lambda_{\max}(\mathbf{M})$	The maximum eigenvalue of the matrix $\mathbf{M}$
$\lambda_{\min}(\mathbf{M})$	The minimum eigenvalue of the matrix $\mathbf{M}$

$\mathrm{sign}(\cdot)$	The sign function, taking value 1 is $\cdot > 0$, value -1 if $\cdot < 0$, and value 0 if $\cdot = 0$
$\mathrm{SIGN}\,(\mathbf{x})$	The subgradient function of the l_1-norm $\|\mathbf{x}\|_1$
$\log(x)$	Natural logarithm of x
$\frac{\partial f}{\partial x}$	The first derivative of the function f with respect to x
$\frac{\partial^2 f}{\partial x^2}$	The second derivative of the function f with respect to x
$E\{x\}$	Expectation of the variable x
$x \sim \mathcal{N}(\mu, \sigma^2)$	x obeys Gaussian distribution with mean μ and variance σ^2
$x \sim \mathcal{CN}(\mu, \sigma^2)$	x obeys complex Gaussian distribution with mean μ and variance σ^2
$\mathbf{x} \sim \mathcal{N}(\boldsymbol{\mu}, \sigma^2\mathbf{I})$	$\mathbf{x}$ obeys Gaussian distribution with mean $\boldsymbol{\mu}$ and covariance matrix $\sigma^2\mathbf{I}$
^{H}	The transpose conjugate operation
^{T}	The transpose operation
*	The conjugate operation
$\mathcal{R}\{\bullet\}$	Real part of $\bullet$
$\mathcal{I}\{\bullet\}$	Imaginary part of $\bullet$
$\mathcal{C}\{\bullet\}$	Column space of the matrix $\bullet$
$\mathrm{d}(\bullet)$	Diagonal matrix with the vector $\bullet$ along the diagonal
$\mathrm{vec}(\bullet)$	Obtain a vector by stacking each column of the matrix $\bullet$
$\mathrm{tr}(\bullet)$	Trace of the matrix $\bullet$
$\mathbf{1}^N$	A vector with all N entries being one
$\max\{a, b\}$	The maximum value between a and b
$\max_x f(x)$	Maximize the function f with respect to the variable x
$\min_x f(x)$	Minimize the function f with respect to the variable x
$\mathrm{mean}\,(\mathbf{x})$	The mean value of vector $\mathbf{x}$
$a(\mathrm{mod}\ b)$	The residual of a divided by b
$\int_a^b f(x)dx$	Integration of $f(x)$ with respect to variable x from a to b
$f_\beta(x, a, b)$	Type I beta distribution with parameters a and b
$n! = 1 \times 2 \times \cdots n$	Factorial of integer n
C_m^n	Combination of selecting n elements from m candidates
$\mathbf{X} \succeq 0$	Matrix $\mathbf{X}$ is positive semidefinite
$\mathbf{X} \succ 0$	Matrix $\mathbf{X}$ is positive definite
$x \in \{0, 1\}$	Variable x is either 0 or 1
$x \in [0, 1]$	Variable x is within the convex interval [0, 1]
$\approx$	Approximately equal
$\odot$	Hadamard product, i.e., element-wise product
$\otimes$	Kronecker product
$\oslash$	Element-wise division

Part I
Foundation

Chapter 1
Array Processing Fundamentals

Signal processing is ubiquitous, ranging from our human body processing light and acoustic radiations to electronic devices, such as cell phones and televisions. The goal of signal processing is to extract as much information as possible from the environment. Generally, the received signal is characterized and processed in two domains, temporal domain and spectral domain, and sometimes in the joint-domain, such as time-frequency analysis. Array signal processing is a prominent and specialized branch of signal processing that processes the signal in the spatial domain. Antenna arrays play an important role in diverse application areas, such as radar, sonar systems and wireless communications [1] to list a few. Such an array consists of a set of N sensors that are spatially distributed at known locations with reference to a common reference point. These sensors collect signals from sources in their field of view. Depending on the distance of the sources from the array, the sources can be characterized as near-field or far-field. For an antenna array of maximum aperture length $\mathcal{D}_a$, the far-field region occurs when the following two conditions are met: [2]

$$\mathcal{R}_a > \frac{2\mathcal{D}_a^2}{\lambda}, \tag{1.1}$$

$$\mathcal{R}_a \gg \lambda, \tag{1.2}$$

where $\mathcal{R}_a$ denotes the distance between the antenna array and the source and λ is the wavelength corresponding to the operating frequency. Due to the source energy distribution, the source can be considered as a point source if it is sufficiently far away from the receiver, or as a distributed source if it is close to the receiver. Array processing can further be characterized as being wideband or narrowband, depending on whether the received signals are considered wideband or narrowband signals. In this book, we focus on the narrowband, far-field point source cases. Extensive research to general cases, such as distributed sources, can be conducted further based on the work presented in this book.

X. Wang et al., *Sparse Sensing and Sparsity Sensed in Multi-sensor Array Applications*, https://doi.org/10.1007/978-981-99-9558-5_1

Similar to Analogue to Digital (A/D) converters in the temporal domain, we sample a waveform in the spatial domain with an array of sensors. Sensing with a single sensor cannot provide the wealth of information for the processing of signals in modern day applications. Thus, multi-sensor arrays find extensive use in different sensing modalities, such as radar, telescope, acoustic and ultrasound, etc. While uniform arrays provide the full complement of information and are straightforward to deal with, they inherently contain significant redundancy, which opens the door to save on their size, weight, and power (SWaP). Sparse sensor arrays are under-sampled such that several sensors are removed from their uniform counterparts, thus ensuring SWaP savings. The resulting configuration provides what we refer to as sparsity in sensing. On the other hand, sparsity in the sensed refers to sparsity in the sensed scene, particularly referring to the sensed spatial spectrum. Exploitation of the sparsity in the sensed can significantly enhance the performance of signal processing systems. Driven by the importance of sparsity, research into sparse sensing and processing techniques continues unabated.

Most of the array processing applications fall into one of the two categories: detection or estimation of signals. The former determines whether the interested signals exist or not, and the latter estimates the parameters of signals, such as direction of arrival (DOA), power levels and the signal waveforms from the noisy outputs of multiple sensors. In practical scenarios, the received signals are inevitably corrupted by various kinds of interferences, jammers and thermal noise. Antenna arrays are capable of spatial filtering, which makes it possible to receive the desired signal from a particular direction while simultaneously blocking interferences from other directions. Regardless of applications, spatial filtering retains as a latent first step followed by either the detection or estimation of signals. The spatial filtering performance of beamformers is jointly determined by two factors, array configuration and excitation weights. Especially, array configurations are important spatial degrees of freedom (DoFs) for enhanced spatial filtering. For the exploration of sparsity in the sensing, the key is to optimize the placement of a given number of sensors along the physical aperture, that is referred to as array configuration, to satisfy some specific criteria. This book brings together a number of contributions on both sparse sensing design and sparsity-sensed processing techniques for different detection and estimation applications, which are established as state-of-the-art sparsity exploration techniques.

1.1 Arrays and Spatial Filters

Antenna arrays use a set of antennas to sample signals in the spatial domain from the sources in their field of view (FoV). To illustrate a simple array beamforming operation, consider an N-antenna array as shown in Fig. 1.1. A Cartesian coordinate system is utilized in the entire book without special notice and we assume that the array is placed on the x-y plane. The positions of the N antennas are denoted as $\mathbf{p}_n = [x_n, y_n, z_n]^T, n = 1, \ldots, N$ and $\mathbf{P} = [\mathbf{p}_1, \mathbf{p}_2, \ldots, \mathbf{p}_N]$. In the far-field scenaio,

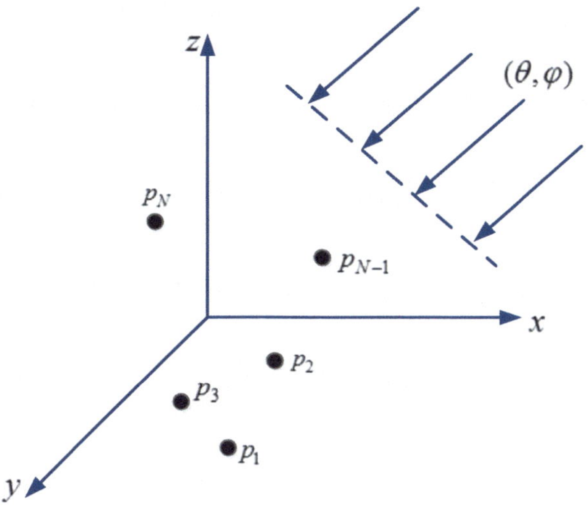

Fig. 1.1 An N-antenna array placed on the x-y plane

the input signal is a plane wave coming from the direction (θ, ϕ) for elevation and azimuth respectively with a frequency f, i.e. the corresponding wavelength is $\lambda = c/f$. The electrical angle of arrival is then defined as,

$$\mathbf{u} = [u_x, u_y, u_z]^T = [\cos\theta\cos\phi, \cos\theta\sin\phi, \sin\theta]^T, \tag{1.3}$$

where ^{T} stands for transpose operation. The received signal of the array exhibits different time delays corresponding to the time of arrival (TOA) at the various sensors. Denote the source position as $\mathbf{p}_o = [x_o, y_o, z_o]^T$ and then the TOA of the ith antenna can be expressed as $t_i = (\mathbf{p}_i - \mathbf{p}_o)^T\mathbf{u}/c$, and the TOA of the jth antenna is similarly $t_j = (\mathbf{p}_j - \mathbf{p}_o)^T\mathbf{u}/c$. The time delay between the two antennas is then $\tau_{i,j} = (\mathbf{p}_i - \mathbf{p}_j)^T\mathbf{u}/c$ with c denoting the propogation speed of light. In the case of uniform linear array (ULA) placing along the z-axis, the delay can be simplied as $\tau = d\sin\theta/c$ with d denoting the unit inter-element spacing. The time delay of the reception time between the nth antenna and the first one is then $(n-1)\tau$. The signal model of ULA is illustrated in Fig. 1.2. The different TOA causes differed phase shift of the received signal. For example, the signal received by the ith antenna is phase shifed by,

$$\Psi_i = 2\pi f t_i = 2\pi f(\mathbf{p}_i - \mathbf{p}_o)^T\mathbf{u}/c. \tag{1.4}$$

Putting all the phase shifts into one vector yields the fundamental concept in the subject of array signal processing, that is array steering vector, i.e.,

$$\mathbf{a}(\mathbf{u}) = [e^{jk_0(\mathbf{p}_1-\mathbf{p}_o)^T\mathbf{u}}, e^{jk_0(\mathbf{p}_2-\mathbf{p}_o)^T\mathbf{u}}, \ldots, e^{jk_0(\mathbf{p}_N-\mathbf{p}_o)^T\mathbf{u}}]^T, \tag{1.5}$$

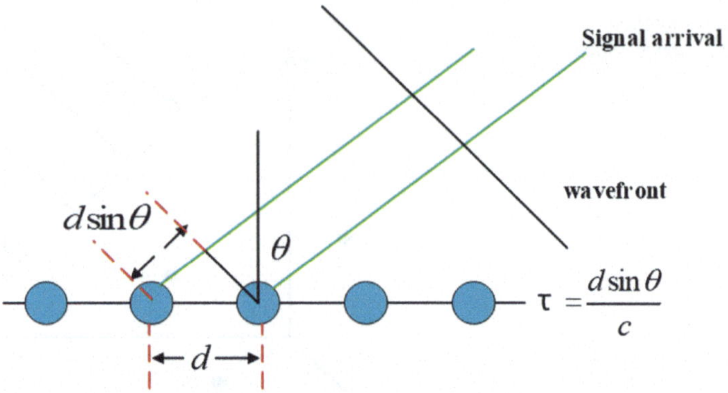

Fig. 1.2 The signal model of a uniform linear array

where the wavenumber is defined by $k_0 = 2\pi/\lambda = 2\pi f/c$. For ULA, the array steering vector in Eq. (1.5) is simplified as,

$$\begin{aligned}\mathbf{a}(\mathbf{u}) &= [1, e^{j2\pi f\tau}, \ldots, e^{j2\pi f(N-1)\tau}]^T, \\ &= [1, e^{jk_0 d\sin\theta}, \ldots, e^{jk_0(N-1)d\sin\theta}]^T.\end{aligned} \tag{1.6}$$

Here, we take the signal received by the first antenna as the phase reference. Analogous to the digital frequency $\omega = 2\pi f T_s$ with T_s denoting the sampling interval in the frequency domain, we define the spatial frequency as,

$$\omega_s = k_0 d \sin\theta. \tag{1.7}$$

In the framework of Nyquist Sampling Thereom, the digital frequency falls in the range $\omega \in [-\pi, \pi]$. Spatial frequency, as a mathematical dual of digital frequency, has the same range $\omega_s \in [-\pi, \pi]$ for a standard ULA, the inter-element spacing of which is $d = \lambda/2$, as clearly observed from Eq. (1.7). Based on the definition of the spatial frequency ω_s, the array steering vector of a ULA in Eq. (1.6) can be rewritten as,

$$\mathbf{a}(\omega_s) = [1, e^{j\omega_s}, \ldots, e^{j\omega_s(N-1)}]^T. \tag{1.8}$$

Obviously, the steering vector $\mathbf{a}(\omega_s)$ exhibits the same form of discrete Fourier transform vector, which, again, validates the mathematical dual relationship between time (frequency) domain and spatial domain. *In a nutshell, the array signal processing can be interpreted via an analogy with time-domain signal processing that we have already been familiar with. This feature is one of the specialities of this book.*

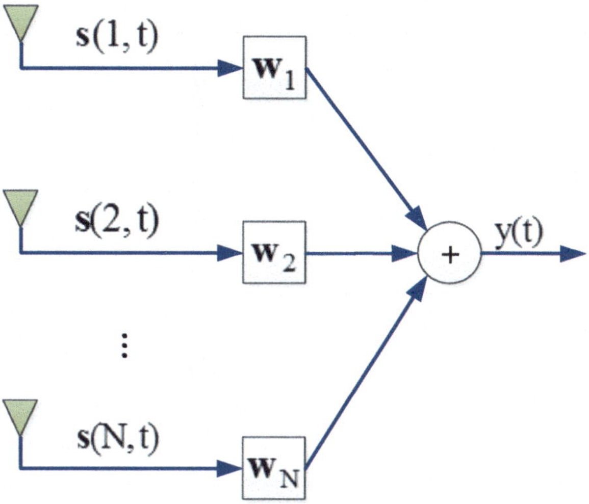

Fig. 1.3 The structure of a beamformer

If $s(t)$ is the desired signal that could be received at the origin of the coordinate system, then the received signal vector of the N-antenna array is expressed in terms of the array steering vector as,

$$\mathbf{s}(t) = s(t)\mathbf{a}(\mathbf{u}) = s(t)[e^{jk_0(\mathbf{p}_1-\mathbf{p}_o)^T\mathbf{u}}, e^{jk_0(\mathbf{p}_2-\mathbf{p}_o)^T\mathbf{u}}, \ldots, e^{jk_0(\mathbf{p}_N-\mathbf{p}_o)^T\mathbf{u}}]^T. \tag{1.9}$$

The beamformer involves adjusting the amplitude and phase of the output of each sensor and sum them together, as shown in Fig. 1.3. Then the output of the beamformer at the time instant t can be expressed as,

$$y(t) = \mathbf{w}^H\mathbf{s}(t), \tag{1.10}$$

where $\mathbf{w} \in \mathbb{C}^N$ is the complex weight vector and ^{H} denotes the conjugate transpose operation. In the case of the ULA, the beamformer is tantamount to a finite impulse response (FIR) filter, as shown in Fig. 1.4, where z^{-1} denotes one unit delay of τ, and the beamforming weights are equivalent to filter coefficents.

Analogous to frequency response used to describe the temporal filter, beampattern is an important metric for characterizing the performance of antenna arrays, which is defined as,

$$B(\theta, \phi) = |\mathbf{w}^H\mathbf{a}(\theta, \phi)|. \tag{1.11}$$

Beampattern characterizes the specifics of a spatial filter formed by an antenna array, which is introduced in the next section.

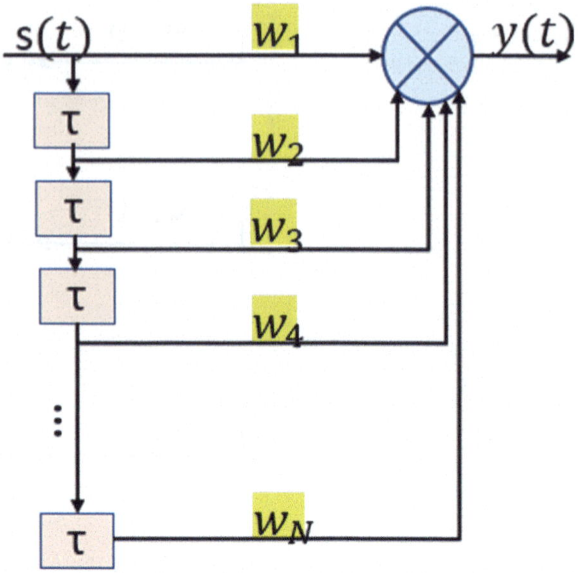

Fig. 1.4 The equivalent FIR structure of a ULA beamformer

1.2 Beampattern Characteristics

We take a ULA with N antenna elements located on the z-axis with a uniform inter-element spacing d (d is set to half-wavelength for standard ULA) as an example to introduce some important metrics of array beampattern. For ULA, the positions of N antennas are simplified as $p_n = (n-1)d, n = 1, \ldots, N$. Note that linear arrays exibit a cone of estimation ambiguity, referred to as ambiguity cone. That implies linear arrays can only resolve one angular direction (which is elevation here) and have no resolution capability in the azimuth ϕ direction. The ambiguity cone is illustrated in Fig. 1.5.

And then the electrical angle is defined simply in terms of elevation angle for linear arrays, i.e. $u = \sin\theta$ and the beampattern becomes,

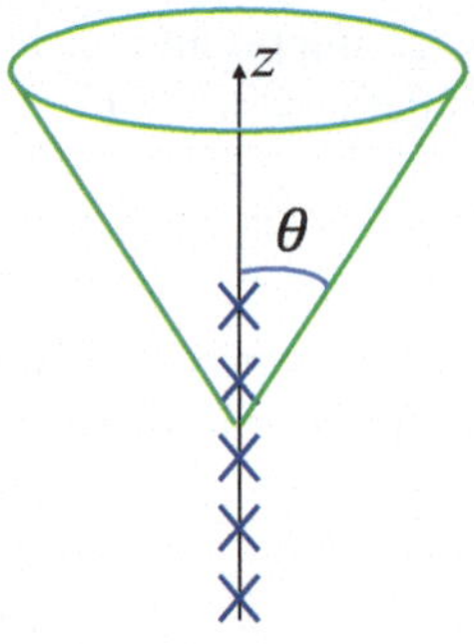

Fig. 1.5 Ambiguity cone of a linear array, which cannot resolve the azimuth angle

$$B(\theta) = |\mathbf{w}^H \mathbf{a}(\theta)|. \tag{1.12}$$

We take uniform weighting as an example, i.e.,

$$w_n = \frac{1}{N}, n = 1, \ldots, N, \tag{1.13}$$

where w_n is the nth entry of the weight vector $\mathbf{w}$. Substituting Eq. (1.13) into Eq. (1.12), the beampattern is given by,

$$\begin{aligned} B(\theta) &= \frac{1}{N} \left| \sum_{n=1}^{N} e^{jk_0(n-1)d \sin\theta} \right|, \\ &= \frac{1}{N} \left| \frac{1 - e^{jk_0 N d \sin\theta}}{1 - e^{jk_0 d \sin\theta}} \right|, \\ &= \frac{1}{N} \left| \frac{\sin(\frac{k_0 d}{2} N \sin\theta)}{\sin(\frac{k_0 d}{2} \sin\theta)} \right|, -\pi \leq \theta \leq \pi. \end{aligned} \tag{1.14}$$

We plot the beampattern of a 10-antenna ULA with $d = \lambda/2$ in Fig. 1.6. Substituting $d = \lambda/2$ into Eq. (1.14) simplies to,

$$|B(\theta)| = \frac{1}{N} \left| \frac{\sin(\frac{\pi}{2} N \sin\theta)}{\sin(\frac{\pi}{2} \sin\theta)} \right|, -\pi \leq \theta \leq \pi. \tag{1.15}$$

The first important metric characterizing a beampattern is 3-dB beamwidth (or half-power beamwidth), which is defined as the angular distance between the two points where $|B(\theta)|^2 = 1/2$ or equivalently $|B(\theta)| = 1/\sqrt{2}$.

The second important metric of the beampattern is peak sidelobe level (PSL), which represents the robustness of the array against noise and interferences.

Now we increase the inter-element spacing to $d = 2\lambda$, the beampattern of the uniform weighting is modified in Fig. 1.7. This figure illustrates the important concept of "grating lobe", that is a lobe of the same hight as the main lobe. Grating lobes can result in estimation ambiguity, that can be resolved only when there is some prior information about the DOA of the source. Generally, we find that if the array is required to steer at $0 \leq \theta \leq \pi$, then the inter-element spacing should satisfy,

$$d \leq \lambda/2, \tag{1.16}$$

to avoid the grating lobes. Also note that the decrease of inter-element space d may incur an increased mutual coupling effect among neighbouring antennas [2]. The problem of grating lobes in the spatial domain is identical to the aliasing phenomenon in time series analysis, which occurs when we sample the waveform in time domain with a frequency below the Nyquist requirement [3]. We refer to a ULA with $d = \lambda/2$ as a standard linear array (SLA).

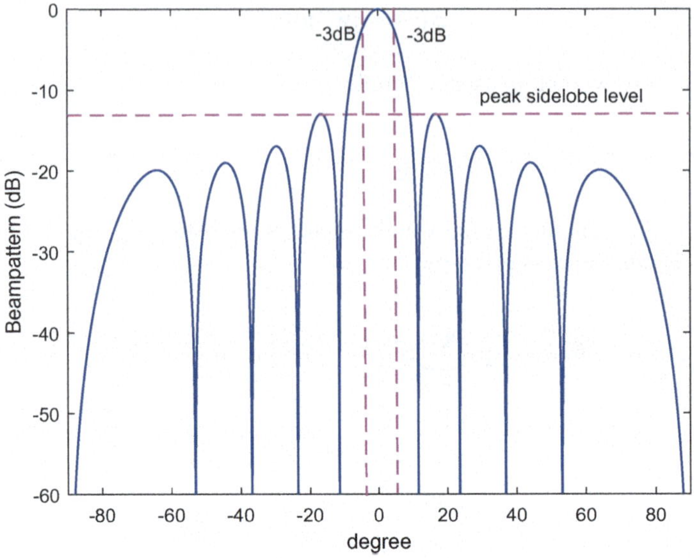

Fig. 1.6 The beampattern of uniform weighting on a 10-antenna ULA. The inter-element spacing is $d = \lambda/2$

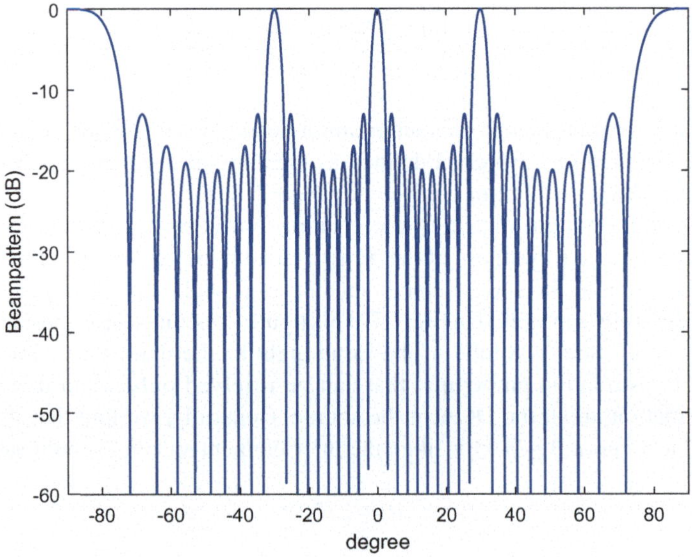

Fig. 1.7 The illustration of grating lobes. The inter-element spacing is $d = 2\lambda$

1.3 Synthesis of Antenna Arrays

An important subset of beamforming techniques for spatial filtering are the so-called null-steering algorithms [4–7]. It is well known that by proper weighting of the array element (prior to summation), it is possible to create nulls in prescribed directions to suppress the directional interferences. In other words, the energy received from plane waves impinging on the array from certain directions cancels out at the array output. This property can be used to receive the desired signals in the presence of multiple sources of interference. The weights of the linear combiner of the array elements are designed to steer nulls in the directions of interferers, while ensuring that signals received from the desired directions pass undistorted and unattenuated. Since the directions of the interferers and the desired signals are usually not known a prior and are often time-varying (due to the motion of the array or the sources or due to propagation effects), null-steering algorithms need to be adaptive.

Beamforming algorithms can be broadly classified into open-loop and closed-loop techniques [8]. Examples of closed-loop null steering algorithms can be found in [4–12]. These algorithms adjust the weights of the linear combiner based on the error signal at the combiner output obtained from the received data. In other words, the closed-loop algorithms are data-dependent and adaptive. Open-loop null-steering algorithms involve a two-step procedure: first, we estimate the DOAs of all the plane waves impinging on the array, using some direction-finding techniques. Having estimated the DOAs, we compute a set of complex weights for the linear combiner that places nulls in the estimated interferer directions. These weights are then used by the linear combiner to compute the array output. This is an open-loop procedure in the sense that the received data is not required in adjusting the weights. Open-loop null-steering algorithms are also referred to as deterministic array synthesis.

Closed-loop algorithms are potentially more robust than open-loop algorithms since they can adjust to uncertainties in the phase-gain characteristics of the array elements and do not require array calibration. Open-loop algorithms do require array calibration and more detailed knowledge of the gain-phase response of the array elements. However, they are less sensitive to convergence problems and usually do not suffer from signal cancellation [13]. In this book, we discuss both open-loop array synthesis and adaptive closed-loop beamforming.

1.3.1 Synthesis of Uniform Arrays

The first sidelobe of a uniform array is only about 13.5dB down from the main lobe level as shown in Fig. 1.6. This is undesirable for many directive applications, such as in radar and in direction finding. Array synthesis is designating the weight of each sensor output in order to obtain a beampattern with desirable properties. It is analogous to the design of filter frequency response due to the mathematical

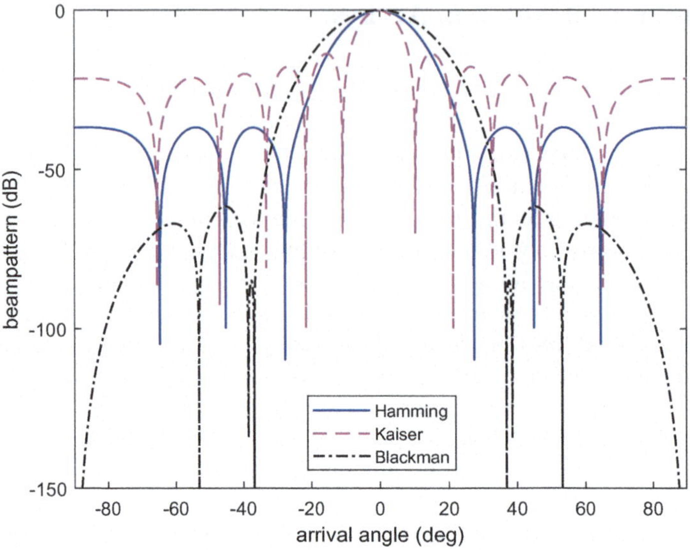

Fig. 1.8 Beampatterns of different spectral weighting methods. The number of antennas is 11

duality between the temporal and spatial domain. There are a number of different classic approaches for synthesizing ULAs, such as spectral weighting (or window), array polynomials, least square and minmax design to list a few [1]. In Fig. 1.8, we present three beampatterns of Hamming, Blackman and Kaiser windows. It is noted that there is always a trade-off between the mainlobe width and the PSL. Usually, we synthesize an array with the minimum possible mainlobe width for the specified PSL, such as two classical synthesis techniques, called Dolph-Chebychev [14] and Taylor weighting [15, 16], which result in constant and decaying sidelobes respectively. It should be noted that all the aforementioned weighting methods are only applicable for ULAs.

Synthesizing a beampttern with constant sidelobe levels tacitly assumes that the interferences are equally likely to arrive anywhere in the sidelobe region. If we have some prior knowledge about the locations of the interferences, we could synthesize the antenna array with nulls steering at the interfering signals. We assume that there is a desired beampattern, $\mathbf{f}_{\mathrm{d}}$, that can be synthesized by a discrete array, i.e.,

$$\mathbf{f}_{\mathrm{d}} = \mathbf{A}^H \mathbf{w}_{\mathrm{d}}, \tag{1.17}$$

where $\mathbf{A}$ is the steering matrix with each column being the steering vector corresponding to a sampled arrival angle θ. Then a least square minimization can be utilized to approximate a second beampattern subject to a set of null constraints, i.e.,

$$\begin{aligned} &\min \|\mathbf{w}_{\mathrm{d}} - \mathbf{w}\|_2, \\ &\text{s.t. } \mathbf{w}^H\mathbf{C} = \mathbf{0}, \end{aligned} \tag{1.18}$$

where the matrix $\mathbf{C}$ comprises of the steering vectors of L interferences, i.e., $\mathbf{C} = [\mathbf{a}(\theta_1), \ldots, \mathbf{a}(\theta_L)]$. Here the l_2-norm of a vector $\|\mathbf{x}\|_2$ stands for $\|\mathbf{x}\|_2 = \mathbf{x}^H\mathbf{x}$. There exists an analytical solution to the least square minimization in Eq. (1.18) that can be obtained through Lagrangian multipliers,

$$\mathbf{w} = (\mathbf{I} - \mathbf{C}(\mathbf{C}^H\mathbf{C})^{-1}\mathbf{C}^H)\mathbf{w}_{\mathrm{d}}, \tag{1.19}$$

where $\mathbf{I}$ is an identity matrix with the corresponding rank. We can observe that $\mathbf{w}$ is the component of $\mathbf{w}_{\mathrm{d}}$ in the subspace orthogonal to the constraint subspace. Let us take an 11-antenna ULA as an example. Suppose the desired signal is arriving from boresight (that is 0°) and there are two interferences coming from 30° and −60°, respectively. The desired beampattern is a Chebychev pattern with −100 dB sidelobes. We place two nulls steering at the interferences and the synthesized beampattern is shown in Fig. 1.9. Note that the PSL becomes −44 dB due to the consumption of DoFs on the null formation.

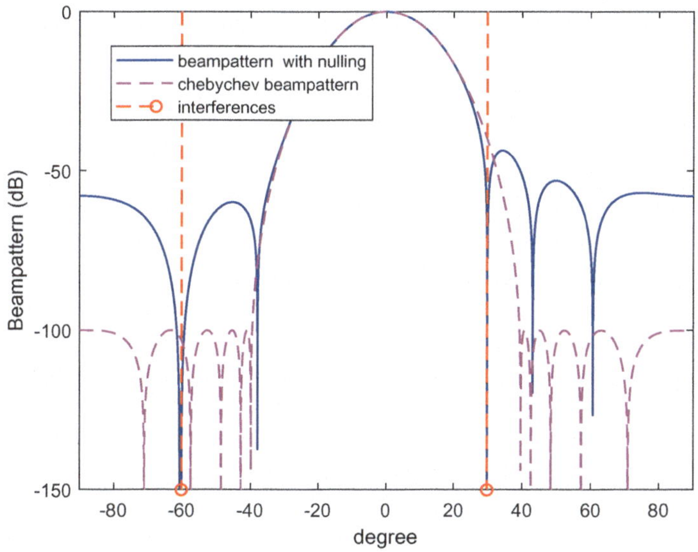

Fig. 1.9 Synthesized Chebychev beampattern with nulls

1.3.2 Synthesis of Non-uniform Arrays

Classical antenna array synthesis techniques, such as Windowing, Dolph-Chebyshev and Taylor synthesis, efficiently taper the amplitude of the excitation of the elements for equally spaced arrays to generate a desired far-field radiation pattern. Thus the least square method in Eq. (1.19) for placing nulls towards the undesired interferences is only amenable to ULAs. The analysis of unequally spaced antenna arrays originated from the work of Unz [17], who developed a matrix formulation to obtain the current distribution necessary to generate a prescribed radiation pattern from an unequally spaced linear array. Subsequent to the initial concept of Unz, recent design techniques focus on two categories of non-uniform arrays: arrays with randomly spaced elements and thinned arrays, that are derived by selectively discarding some elements of an initial equally spaced array.

In the first category, Harrington [18] developed a perturbational method to reduce sidelobe levels of uniformly excited N-element linear arrays by employing non-uniform spacing. Furthermore, he demonstrated that the close-in sidelobes can be reduced in height to approximately $2/N$ times the main-lobe field intensity level. Andreasan [19] exploited the use of emerging digital computation techniques to develop empirical results on unequally spaced arrays. Two important conclusions resulted from his work: (1) the 3-dB beamwidth of the mainlobe depends primarily on the length of the array and (2) the sidelobe level depends primarily on the number of elements in the array and to a minimal extent on the average spacing of the array when the latter exceeds about two wavelength. Kumar [20] described an approach based on Legendre transformation and an inversion algorithm to obtain the unequal element spacings from prescribed far-zone electric field and current distribution. He further generalized a unified mathematical approach to non-linear optimization of multi-dimensional array geometries in [21]. Zhou [22] proposed an adaptive array method for arbitrary array synthesis which optimizes a weighted l_2 norm between desired and achieved patterns. Ismail [23] presented a technique for null steering by minimizing the sum of the squares of the total element position perturbations based on a uniformly spaced array.

Lebret [24] showed that a variety of antenna array pattern synthesis problems can be expressed as convex optimization problems, that can be numerically solved with great efficiency by interior-point methods. The synthesis problems involve arrays with arbitrary geometry and element directivity constraints on far and near-field patterns over narrow or broad frequency bandwidth and some important robustness constraints. Based on Lebret's work, Wang [25] proposed an algorithm for array pattern synthesis with robustness and power constraints using Semi-Definite Programming (SDP). Yu [26] also formulated a robust adaptive beamformer as a SDP with new constraints on the magnitude response, the beamwidth and response ripple of the robust response region. Fuchs [27] transformed the narrow beam low sidelobe synthesis for arbitrary arrays into either a Linear Programming (LP) or a Second Order Cone Programming (SOCP). Moreover, Comisso [28] introduced an iterative Auxiliary Phase algorithm for arbitrary power pattern synthesis, including multi-beam synthesis and radiation suppression within large angular regions. Yang [29] introduced a fast pencil beam pattern synthesis method for large unequally spaced antenna arrays, which is based on an interpolation in a least square sense and iterative Fast Fourier Transform (FFT). There are also different kinds of heuristic approaches proposed in the literature for synthesizing non-uniformly spaced antenna arrays, such as Genetic Algorithm [30, 31], Particle Swarm Optimization [32] and Differential Evolution [33] and so on. Isernia [34] discussed a hybrid approach to the synthesis of excitations and locations of non-uniformly spaced arrays, which takes definite advantage from the convexity of the problem with respect to excitation variables, and exploits a Simulated Annealing procedure as fas as location variables are concerned.

The second category of thinned arrays aims at synthesizing a desired beampattern with the smallest number of antennas using sparsity-promoting algorithms, such as reweighted l_1-norm [35], Bayesian inference [36], soft-thresholding shrinkage [37], and other compressive sensing methods [38]. Our previous work in [39, 40] inspected the common deterministic sparse array design for multiple switched beams in both cases of fully and partially-connected radio frequency (RF) switch network. Fully-connected switch network can facilitate the inter-connection of any input port to all output ports [41]. While partially-connected switch network constrains regularized antenna switching, which divides the full array into contiguous groups and precisely one antenna is selected in each group to compose the sparse array. The regularization of switched antennas is more practical in terms of circuit routing, connectivity and array calibration [42] and furthermore restricts the maximum inter-element spacing, which, in turn, reduces the unwanted sidelobes. Other types of deterministic sparse arrays, such as minimum redundancy arrays (MRAs) and nested arrays [43], are designed to enable the estimation of more sources than physical sensors. However, they might exhibit significantly low array gain and are not suitable for interference mitigation [44, 45]. Nevertheless, very few work has examined the optimal design of thinned array for closed-loop adaptive beamforming and having at the same time well-controlled quiescent patterns. Needless to say the cognitive-driven optimization of sparse beamfomrers for different array applications, eliminating the requirement of prior environmental knowledge and the full array covariance. The solution to this problem constitutes an important part of this book.

1.4 Multi-input Multi-output Array

Multi-Input Multi-Output (MIMO) radar was first proposed at the international radar annual conference in the early 21st century as a new radar system. The concept of MIMO was initially applied in wireless communication systems. The transmitter and receiver of the communication system use multiple antennas, which can effectively improve the transmission capacity and robust performance [46–48]. Capitalizing on the idea of MIMO communication, MIMO radar adopts transmit waveform diversity technology to improve the DoF of the radar system and thus obtain higher detection performance. We first introduce the concept and basic working principles of MIMO array, and then we elaborate on the signal model of colocated MIMO array.

1.4.1 Introduction of MIMO Array

Different antennas of MIMO arrays can transmit different waveforms, which can be either completely uncorrelated or partially correlated. This is referred to as waveform diversity. Compared with the traditional phased array, the characteristics of waveform diversity make MIMO arrays have the following specific advantages:

- Flexible transmit waveform. Traditional phased arrays realize transmit beamforming by controlling the phase of each transmit array element. Usually, only a single transmit beam can be formed in a single pulse, and the shape of the beam is not easy to control. However, each antenna of the transmit array of MIMO radar can transmit different signals, which provides more degrees of freedom for transmit beamforming. This can not only search targets with wide beams, but also track different targets with multiple narrow beams. Therefore, transmit beamforming is a research focus of MIMO array. In addition, facing the increasingly complex working environment, MIMO radar can design the transmission waveform according to the working requirements and prior information, so as to improve the performance of system parameter estimation, target detection and tracking [49–52].
- Improved parameter estimation. Benefit from the additional DoFs brought by waveform diversity, MIMO radar can separate the echo signals of different transmission paths at the receiver, form a larger virtual aperture, and obtain higher spatial angle resolution and parameter recognition. Therefore, taking full advantage of waveform diversity, MIMO radar can obtain higher parameter estimation accuracy and quantity than phased array radar with only receiving degrees of freedom [53–55].
- Improved interference suppression. On one hand, benefiting from the additional DoFs obtained by waveform diversity, MIMO radar can effectively suppress clutter and interference; On the other hand, for the multipath clutter generated by low altitude/ultra-low altitude targets, MIMO radar can reduce the impact of multipath clutter by transmit beamforming according to the working environment [56–58].

In order to give full play to the above advantages of MIMO radar, it is necessary to start from the two aspects of the radar's transmit beampattern and transmit waveform, so as to make full use of the DoFs at the transmit end [59, 60]. For the transmit beampattern, the traditional phased array adopts the working mode of single narrow beam scanning, which can not meet the actual needs. For example, all antennas of phased array transmit the same waveform and control the beam direction by applying different phases to the phase center. Usually, when performing the task of multi-target tracking, only single beam switching can be used, which will limit the radiation time to each target and thus hinder the high-precision estimation and tracking. While for the MIMO radar, due to the convenience and flexibility of its beam control mode, it can acquire the environmental information by designing omni-directional or specific beampatterns according to actual needs. If the problem of multi-target tracking is considered, we can optimize a multi-beam beampattern for the transmit end. Compared with the traditional narrow beam switching mode of the phased array, the multi-beam beampattern can observe and detect multiple targets at the same time, which extends the dwell time of the signal on each target, thereby improving the accuracy of estimation and tracking.

As for the transmit waveform, that is, the selection of transmit signal, the traditional radar mostly adopts fixed waveforms (such as linear frequency modulation signal, two-phase code, Barker code, polyphase code and so on) at the transmit end, and selects the waveform according to some factors (such as the mainlobe width, sidelobe level, the shape of three-dimensional ambiguity function) after matched filtering. In this case, adaptive processing is only carried out at the receive end. With the increasingly complex and diverse scenes faced by radar, the working mode of fixed transmit waveform applied by traditional radar can not gaurantee satisfactory results. Therefore, according to the different tasks and working scenes, the radar needs to adjust adaptively at the transmit end through dynamic perception. Combined with the concept of cognitive radar, the overall performance of radar can be greatly improved by continuously online waveform optimization.

According to the distance between the antennas of the transmit array and the receive array, MIMO radars can be divided into two categories: distributed MIMO radars and colocated MIMO radars [61–63]. The schematic diagrams of these two kinds of MIMO radars are presented in Figs. 1.10 and 1.11 respectively. The distributed MIMO radar, different from the conventional multistatic radar system, enables each transmit antenna to detect the radar target from different directions by utilizing spatial diversity technology and increasing the inter-element spacing of the transmit array. And the received echo signal can be considered as the superposition of independent fading signals. Combining these echo signals from different independent observation paths can obtain an approximately constant radar cross section (RCS) of the target, so as to suppress the RCS flicker of the target and obtain a large spatial diversity gain. The colocated MIMO radar, the inter-element spacing is relatively small. And usually, the transmit array and receive array are closely

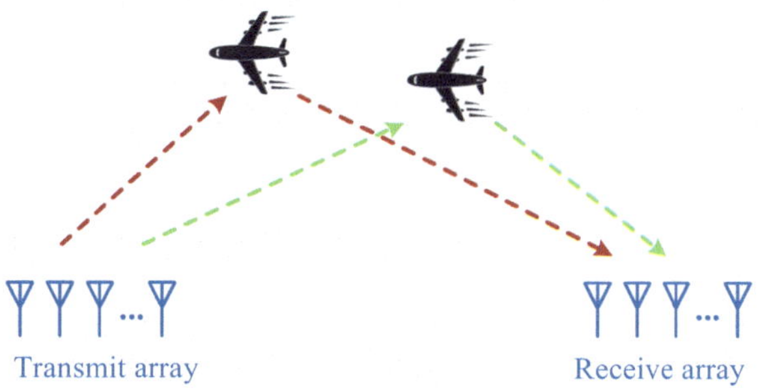

Fig. 1.10 Schematic diagram of distributed MIMO Radar

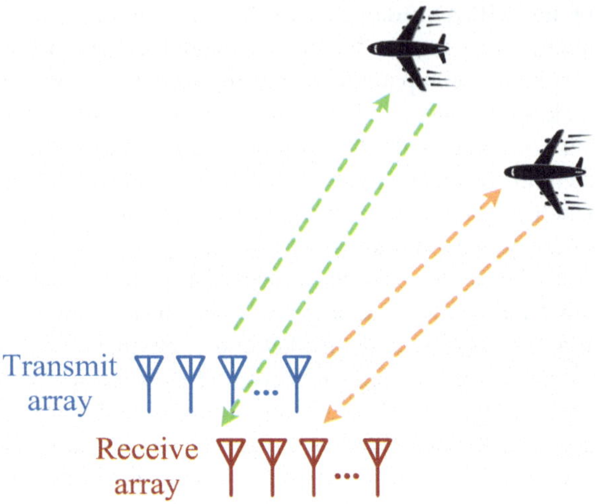

Fig. 1.11 Schematic diagram of colocated MIMO Radar

spaced or share the same antennas. Different from phased array radar system, the colocated MIMO array enables each transmit antenna to carry arbitrary waveform signals through waveform diversity technology. In particular, in terms of the form of transmit signals, phased array radar can be regarded as a special case of colocated MIMO radar.

1.4.2 Colocated MIMO Array

In this book, we mainly introduce the colocated MIMO array. The transmit waveform of the colocated MIMO array can be divided into two categories: orthogonal and correlated. When the transmit waveforms are orthogonal, the transmit beampattern is omni-directional. In this case, the virtual aperture expansion can be realized by signal separation at the receive end, so as to obtain high resolution. When the transmit waveforms are correlated, the transmit beampattern can be optimized to any desired shape. In this case, some complex transmit beampatterns can be designed so that the transmit energy can be reasonably allocated to the target area of interest.

We take the ULA as an example. Assuming that the transmitter of the MIMO radar is an ULA composed of N_t antennas and the receiver is another ULA composed of N_r antennas. The transmit signal of MIMO array can be expressed as,

$$y(t) = \mathbf{a}_T^T(\theta)\mathbf{s}(t), \tag{1.20}$$

where $\mathbf{a}_T(\theta)$ is the transmit steering vector and $\mathbf{s}(t)$ is the vector composed of the signal transmitted by each antenna. Then the transmit beampattern is given by,

$$
\begin{aligned}
B(\theta) &= E\{y(t)y^*(t)\}, \\
&= E\{\mathbf{a}_T^T(\theta)\mathbf{s}(t)\mathbf{s}^H(t)\mathbf{a}_T^*(\theta)\}, \\
&= \mathbf{a}_T^T(\theta)E\{\mathbf{s}(t)\mathbf{s}^H(t)\}\mathbf{a}_T^*(\theta),
\end{aligned} \tag{1.21}
$$

where $E\{.\}$ means taking expectation. We define the autocorrelation matrix of the transmit waveform as,

$$
\mathbf{R}_s = E\{\mathbf{s}(t)\mathbf{s}^H(t)\}. \tag{1.22}
$$

When the transmit signals are orthogonal to each other, $\mathbf{R}_s$ is an identity matrix, that is $\mathbf{R}_s = \mathbf{I}$. Substituting it into Eq. (1.21), we have,

$$
B(\theta) = \mathbf{a}_T^T(\theta)\mathbf{I}\mathbf{a}_T^*(\theta) = N_t, \tag{1.23}
$$

which indicates that the radiation power in any direction is the same, that is, omni-directional detection situation.

When the the transmit signals are correlated, matrix $\mathbf{R}_s$ is no longer an identity matrix. We can synthesize arbitrarily shaped transmit beampatterns by optimizing the matrix $\mathbf{R}_s$. A special case is the multi-carrier, that is, the transmit signals of each antenna can be regarded as a linear combination of a group of orthogonal signals. In this case, the transmit signal vector can be expressed as,

$$
\mathbf{s}(t) = \mathbf{W}\mathbf{s}_o(t), \tag{1.24}
$$

where $\mathbf{W}$ is the weighting matrix and $\mathbf{s}_o(t)$ is the vector of $N_{t'}$, $(N_{t'} \geq N_t)$ orthogonal signals. Substituting Eq. (1.24) into Eq. (1.21), the transmit beampattern is transformed into,

$$
\begin{aligned}
B(\theta) &= E\{\mathbf{a}_T^T(\theta)\mathbf{s}(t)\mathbf{s}^H(t)\mathbf{a}_T^*(\theta)\}, \\
&= \mathbf{a}_T^T(\theta)\mathbf{W}E\{\mathbf{s}_o(t)\mathbf{s}_o^*(t)\}\mathbf{W}^H\mathbf{a}_T^*(\theta), \\
&= \mathbf{a}_T^T(\theta)\mathbf{W}\mathbf{W}^H\mathbf{a}_T^*(\theta), \\
&= \mathbf{a}_T^T(\theta)\mathbf{R}_W\mathbf{a}_T^*(\theta),
\end{aligned} \tag{1.25}
$$

where $\mathbf{R}_W = \mathbf{W}\mathbf{W}^H$ is the beamforming weight autocorrelation matrix. According to Eq. (1.25), the optimization problem of the transmit beamforming can be transformed into the optimization of the beamforming matrix $\mathbf{W}$ or the autocorrelation matrix $\mathbf{R}_W$.

At the receive end, the received signal vector can be expressed as,

$$
\begin{aligned}
\mathbf{x}(t) &= \mathbf{a}_R(\theta)x(t), \\
&= \mathbf{a}_R(\theta)\mathbf{a}_T^T(\theta)\mathbf{s}(t),
\end{aligned} \tag{1.26}
$$

where $\mathbf{a}_R(\theta)$ is the receive steering vector. Applying the multi-carrier model, the received signal vector can be transformed into a matrix after matched filtering,

$$\begin{aligned}\mathbf{X} &= \int_T \mathbf{x}(t)\mathbf{s}_o^H(t)\mathrm{d}t, \\ &= \mathbf{a}_R(\theta)\mathbf{a}_T^T(\theta)\mathbf{W}.\end{aligned} \tag{1.27}$$

When $\mathbf{W} = \mathbf{I}$, the transmit signals are orthogonal, in this case, $\mathbf{X}$ can be regarded as the received signal of $N_t \times N_r$ virtual antennas derived from the transmit and receive arrays. When $\mathbf{W} \neq \mathbf{I}$, the transmit signals are correlated, in this case, $\mathbf{X}$ is the received signal of $N_{t'} \times N_r$ virtual antennas derived from the virtual transmit array (whose steering vector is $\bar{\mathbf{a}}(\theta) = \mathbf{W}^T\mathbf{a}_T(\theta)$) and receive array. Equation (1.27) indicates that the MIMO array can provide more DoFs.

Figure 1.12 depicts the generation principle of virtual antennas. As can be seen from the figure, the synthesis of transmit and receive arrays can generate a virtual array with larger aperture and more elements compared with the physical arrays, which greatly increases the available DoFs and enhances the spatial resolution. Figure 1.13 presents the generation of sum-coarray, which is a special case of MIMO array and a commonly-used array structure in automotive radar. Assuming that the inter-element spacing of the receive array is d and the inter-element spacing of the transmit array is $N_r d$. Then the transceiver can generate a virtual ULA consisting of $N_t N_r$ virtual antennas with an inter-element spacing of d.

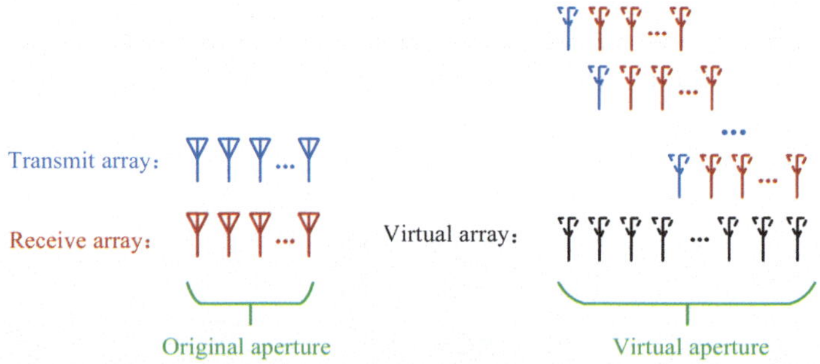

Fig. 1.12 Virtual array and virtual aperture of MIMO array

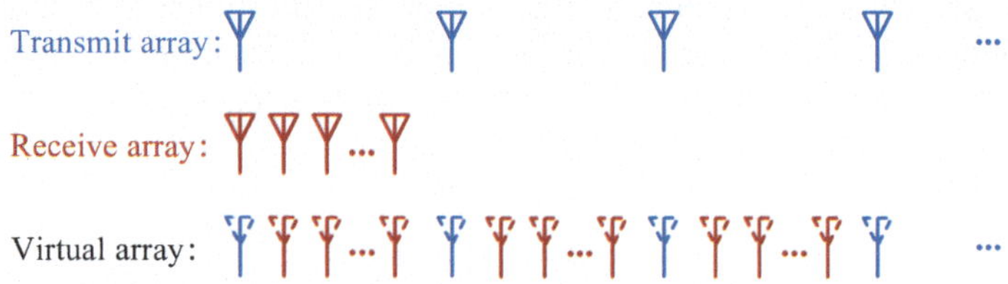

Fig. 1.13 A special MIMO virtual array: sum-coarray

1.5 Summary

In this chapter, we briefly reviewed the relevant array processing fundamentals, from the viewpoint of mathematical duality between arrays and filters. Antenna arrays use a set of antennas to sample the signal in the spatial domain from the sources in their FoV and then spatially filter the signal for further estimation or detection. Analogous to frequency response, beampattern is utilized to characterize the spatial filtering of antenna arrays, where different metrics are introduced to quantitatively measure the performance. Spectral windowing is a classical way to reduce the sidelobe level for uniform linear array. To further enhance the interference suppression, null forming is required to on the basis of spectral windowing. We then briefly reviewed the synthesis of unequally spaced arrays, where spectral windowing does not apply any more. The main focus of this book is sparse array synthesis and beamforming. Finally, MIMO arrays, which exhibit a few advantages over the phased array counterparts attributed to the waveform diversity, are introduced to complete this section. A large-apertured sum-coarray with more virtual antennas can be generated by the MIMO array, thus possessing more DoFs for performance enhancement. In a nutshell, antenna arrays sample and process the signals in the spatial domain, which present an analogy to the temporal and spectral processing, while exhibiting their own specialty and interesting aspects.

Chapter 2
Multi-sensor Array Applications

As mentioned in Chap. 1, an array samples and processes the waveform to create the output signal with the aim of extracting the information about the propagating waveform. Several sensors sampling the common field can be merged to produce more refined information about the waveform compared with single sensor. The goals of array processing are to cleverly combine the sensors' outputs such that

- the output signal-to-noise ratio (SNR) is maximally enhanced;
- the parameters of the waveform, such as the number of sources, the bearings of these sources and the waveforms they are emitting, are accuratly estimated;
- the sources of interest are reliably detected and tracked in space.

When the waveform is propogating in the space, it is unavoidably contaminated by noise and various interferences. In order to extract the contained information, spatial filtering techniques have been developed to amplify the desired signal while suppresing the unwanted noises over the several decades. To accomplish the task of spatial filtering, the signal has to be maximally separated from the noise in the subspace spanned by the array of sensors. To be specifically, the array should form a directional beam pointing towards the target while a deep null against the strong interference. The basic principle of spatial filtering is analogous to FIR filtering in the temporal domain as explained in Chap. 1. The difference is that array structure is also an important factor and designate DoF that can be utilized to enhance the spatial filtering performance, in addition to the filter coefficients. The focus of this book is to optimally design the array configuration in terms of maximizing the output SNR for enhanced spatial filtering.

One important parameter that is used to characterize the propagating waveform in a quantitative way is the direction of arrival (DOA) of the waveform. In this application, the received waveform can be adequately modeled as the sum of several plane waves plus white noise. Within this model, we can formulate an optimization problem to find the parameter values that most closely match the model to the waveform. One commonly used model is Fourier beamforming, that is measuring the average power

X. Wang et al., *Sparse Sensing and Sparsity Sensed in Multi-sensor Array Applications*, https://doi.org/10.1007/978-981-99-9558-5_2

in the beamformer output for each possible direction of propagation, which is then determined by the highest output power.

The detection problem is to determine whether a signal is present in a measurement contaminated by noise. In active sensing situations such as radar and sonar, a known waveform is generated which in turn propagates through a medium and is reflected by some target back to the point of origin. In addition to these target reflected signals, there may also be spurious returns such as clutter and jamming in the case of airborne radar. The ground clutter observed by an airborne platform is extended in both angls and range, and is spread in Doppler frequency because of the platform motion. Noiselike jamming signals are localized in angle and spread over all Doppler frequency. Interference cancellation requires multidimensional filtering over the spatial and temporal domains. Space-time adaptive processing (STAP) refers io the simultaneous processing of the spatial samples from an array antenna and the temporal samples provided by the echoes from multiple pulses of a radar coherent processing interval (CPI). STAP provides improved detection of targets obscured by mainlobe clutter, sidelobe clutter, and jamming.

It is interesting to note that, although these three goals correspond to different array applications, estimation and detection are essentially two distinct viewpoints of spatial filtering. Without sufficient output SNR, neither the parameters of the source can be estimated nor the target can be detected.

2.1 Adaptive Beamformers

In the open-loop array synthesis explained in Chap. 1, the arrival directions of desired signals and interferences are assumed as known parameters or can be estimated before the subsequent beampattern synthesis. However, these prior information may be difficult to obtain in practical senarios or the sensed electromagnetic environment is changing dynamically. The open-loop array synthesis fails in either of these cases. The second kind of beamformers automatically combine parameter (or approximate surrogates) estimation and interference nulling together. The resulting beamformers will adapt to the incoming data and are referred to as adaptive closed-loop beamformers. In this subsection, we develop the theory and practice of adaptive beamformers.

The adaptive beamformers can be divided into three general categories:

- Beamformers that estimate the noise covariance correlation matrix $\mathbf{R}_n \in \mathbb{C}^{N\times N}$ and use the estimate in an appropriate formula. This implementation requires the inversion of the sample covariance matrix and is frequently referred to as the sample matrix inversion (SMI) technique [64]. It is also referred to as the direct matrix inversion (DMI) technique [9] or the estimate-and-plug technique [11]. It is a block data processor.
- Beamformers that implement the inversion recursively are the second category. These recursive least square implementations potentially have performance that is similar to the SMI beamformer.

- A third approach is to adapt classical steepest descent algorithms to the optimization problem in order to find the optimum weight vector. This approach leads to the least mean square (LMS) algorithm. These algorithms require less computation, but converge slower to the optimum solution.

Since this book focuses on investigating the effect of array configuration, instead of different algorithms, on adaptive beamforming performance, we only consider the first category where the optimum filter weight vector is given by [64, 65],

$$\mathbf{w}_{\text{opt}} = \gamma \mathbf{R}_n^{-1} \mathbf{a}. \tag{2.1}$$

Here γ is an arbitrary constant and $\mathbf{a}$ is the steering vector of the desired signal (we omit the arrival angles (θ, ϕ) for simplicity). The optimal filtering problem can be easily understood as a combination of a whitening filter followed by a matched filter. Since $\mathbf{R}_n$ is positive semidefinite and Hermitian, it can be written, using the cholesky decomposition, in the form

$$\mathbf{R}_n = \mathbf{U}^H \mathbf{U}, \tag{2.2}$$

where the matrix $\mathbf{U}$ is an upper triangular matrix. Assume the received signal vector is $\mathbf{x}(t)$, that is the desired signal $\mathbf{s}(t)$ in Eq. (1.9) contaminated by strong interferences and white noise, $\mathbf{n}(t)$, i.e.,

$$\mathbf{x}(t) = \mathbf{s}(t) + \mathbf{n}(t). \tag{2.3}$$

Ignoring the constant γ and combing Eqs. (2.1) and (2.2), the output of the optimum array filter is,

$$\begin{aligned} y(t) &= \mathbf{w}_{\text{opt}}^H \mathbf{x}(t), \\ &= ((\mathbf{U}^H \mathbf{U})^{-1} \mathbf{a})^H \mathbf{x}(t), \\ &= ((\mathbf{U}^{-1})^H \mathbf{a})^H (\mathbf{U}^{-1})^H \mathbf{x}(t), \\ &= \bar{\mathbf{a}}^H \bar{\mathbf{x}}(t). \end{aligned} \tag{2.4}$$

This is easily seen as a matched filtering equation between a modified template signal $\bar{\mathbf{a}}$ and the filtered data vector $\bar{\mathbf{x}}(t)$. The transformed data vector, $\bar{\mathbf{x}}(t)$, becomes,

$$\begin{aligned} \bar{\mathbf{x}}(t) &= (\mathbf{U}^{-1})^H \mathbf{x}(t) \\ &= s(t)(\mathbf{U}^{-1})^H \mathbf{a} + (\mathbf{U}^{-1})^H \mathbf{n}(t) \\ &= s(t)\bar{\mathbf{a}} + \bar{\mathbf{n}}(t). \end{aligned} \tag{2.5}$$

Thus, $\bar{\mathbf{x}}(t)$ comprises of the modified template signal $s(t)\bar{\mathbf{a}}$ and a filtered noise component $\bar{\mathbf{n}}(t)$. The covariance matrix of the filtered noise component is,

$$
\begin{aligned}
\bar{\mathbf{R}}_n &= E\{\bar{\mathbf{n}}(t)\bar{\mathbf{n}}^H(t)\}, \\
&= (\mathbf{U}^{-1})^H E\{\mathbf{n}(t)\mathbf{n}^H(t)\}\mathbf{U}^{-1}, \\
&= (\mathbf{U}^{-1})^H \mathbf{R}_n \mathbf{U}^{-1}, \\
&= \mathbf{I}.
\end{aligned} \tag{2.6}
$$

Consequently, the transformed noise is now white. The covariance matrix $\mathbf{R}_n$ is not available in practice and must be estimated using the training data. Suppose there are in total L snapshots and the noise is much stronger than the desired signal in power, the maximum likelihood estimate of $\mathbf{R}_n$ is [1],

$$
\tilde{\mathbf{R}}_n = \frac{1}{L}\sum_{t=1}^{L}\mathbf{x}(t)\mathbf{x}^H(t). \tag{2.7}
$$

Based on Eq. (2.1), the Capon filter weights are then found to be, [66]

$$
\mathbf{w}_{\text{capon}} = \frac{\tilde{\mathbf{R}}_n^{-1}\mathbf{a}}{\mathbf{a}^H\tilde{\mathbf{R}}_n^{-1}\mathbf{a}}. \tag{2.8}
$$

Correspondingly, the Minimum Variance Estimation (MVE) spectrum based on the Capon beamformer of the direction (θ, ϕ) is,

$$
J(\theta, \phi) = \frac{1}{\mathbf{a}^H(\theta, \phi)\tilde{\mathbf{R}}_n^{-1}\mathbf{a}(\theta, \phi)}. \tag{2.9}
$$

Suppose there are two uncorrelated signals coming from 30° and 120° respectively with signal to noise ratio (SNR) being 20dB. We employ a standard linear array with 10 antennas and present both the MVE spectrum and the correlation spectrum, $J(\theta, \phi) = |\mathbf{a}^H(\theta, \phi)\mathbf{x}(t)|$, in Fig. 2.1. Clearly, the MVE spectrum exhibits much higher resolution than the correlation method that is equivalent to a Discrete Fourier Transform (DFT) of the received data vector in the spatial domain. Noted that Capon beamformer does not work in the case with coherent or highly correlated signals.

2.2 Adaptive Sidelobe Canceller

Adaptive array beamforming is an effective method of spatial filtering to eliminate unwanted interferences and improve the signal to interference-plus-noise ratio (SINR). However, the computational complexity of a fully adaptive beamformer is prohibitively high such that it is impossible to implement it in the practical applications. To enhance the anti-jamming performance of practical radar systems with a reduced complexity, a well-established and practical technique is adaptive sidelobe canceller (SLC) [67–71], which comprises one main channel and one auxiliary

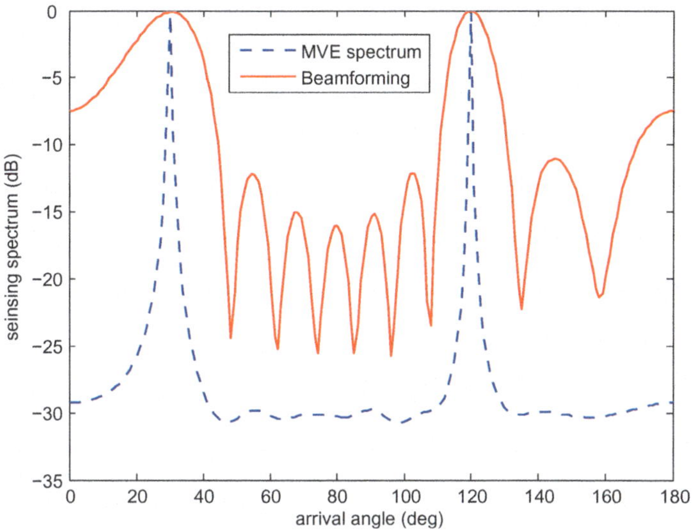

Fig. 2.1 The MVE spectrum and correlation spectrum

channel. The outputs of the main channel and the auxiliary channel are subtracted to produce the final output, in such a way that the interferences in the sidelobe region of the main channel are cancelled out [72–74]. As the adaptive part of SLCs only applies to the auxiliary channel, this greatly reduces the computational complexity.

Summarizing the existing SLC techniques in the literature, there are two commonly-used metrics for the SLC design. The first metric of minimum output power (MOP) is to minimize the total output power of the sidelobe canceller, equivalently to maximize the cancellation of interference components in the main channel. The other metric is beampattern matching (BPM), which constructs the auxiliary weights by matching the beampatterns of the auxiliary and the main channels towards the interferences' direction, and thus achieving the aim of interference cancellation when subtracting.

Generally, the SLC can be implemented in two different kinds of structures, antenna reuse and array splitting. In the first structure, the main channel is the full aperture array and the auxiliary channel shares some antennas with the main channel [75–77]. In the structure of array splitting, the available antennas are split between the main channel and the auxiliary channel in such a way that there are no overlapping antennas between two channels [78–80]. The main difference between the two structures lies in whether the noise of the main channel is correlated with that of the auxiliary channel. The two structures are plotted in Fig. 2.2, when the switches are on, the structure of antenna reuse is employed, while when the switches are off, the structure of array splitting is dictated.

2.2.1 Mathematical Model

Assume a full-aperture linear array with $MN(N = C \times M)$ antennas equally spaced with half-wavelength, which are divided into M non-overlapping subarrays and each subarray comprises C contiguous elements. When $C > 1$, SLC is composed on the basis of subarray. When $C = 1$, each subarray contains a single antenna, and the SLC degenerates to antenna basis. Clearly, a unified mathematical model can be established regardless of basis, thereby, we use the subarray basis to delineate the SLC as follows.

2.2.1.1 The Structure of Antenna Reuse

In the structure of antenna reuse, the main channel selects all M subarrays and the auxiliary channel selects L subarrays. As shown in Fig. 2.2, each subarray applies conventional beamforming to the received signal, and then the outputs of M subarrays are uniformly weighted to yield the output of the main channel. The beamforming weight of the auxiliary channel is designed according to different metrics in order to cancel the interferences present in the main channel.

The output of the ith subarray can be expressed as,

$$x_i(t) = \mathbf{a}_i^H \mathbf{n}_i(t) + \mathbf{a}_i^H \mathbf{H}_i \mathbf{j}(t), \tag{2.10}$$

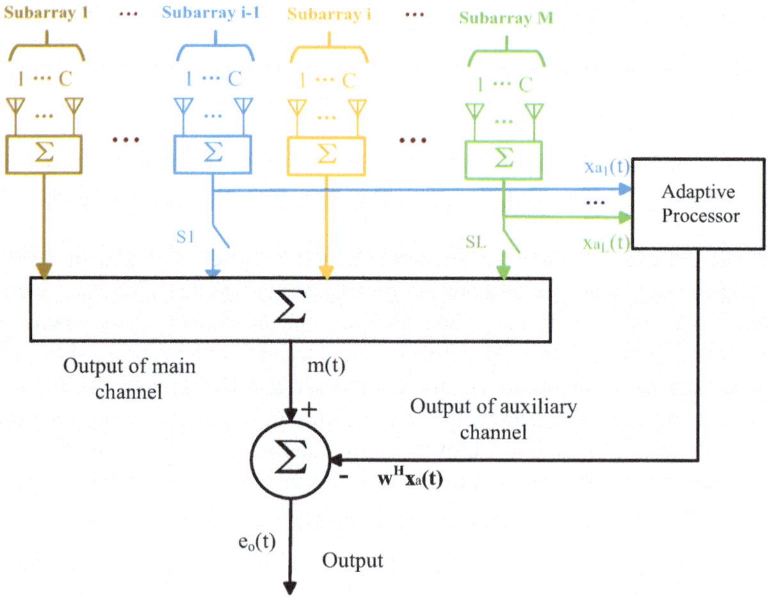

Fig. 2.2 Schematic diagram of adaptive sidelobe canceller

where $\mathbf{a}_i$ is the steering vector of the ith subarray pointing towards the desired signal, $\mathbf{j}(t)$ is a complex vector of $(K+1)$-dimension composed of one desired signal and K interferences, $\mathbf{n}_i(t)$ denotes Gaussian white noise, and $\mathbf{H}_i = [\mathbf{a}_i, \mathbf{h}_{i1}, \mathbf{h}_{i2}, ..., \mathbf{h}_{iK}]$ is composed of steering vectors associated with the ith subarray pointing towards the desired signal and K interferences.

Assume that L subarrays have been selected for the auxiliary channel, the signals received by the auxiliary channel can be expressed as,

$$\begin{aligned}
\mathbf{x}_a(t) &= [x_{(1)}(t), x_{(2)}(t), ..., x_{(L)}(t)]^T, \\
&= \begin{bmatrix} \mathbf{a}_{(1)}^H \mathbf{n}_{(1)}(t) + \mathbf{a}_{(1)}^H \mathbf{H}_{(1)} \mathbf{j}(t) \\ \mathbf{a}_{(2)}^H \mathbf{n}_{(2)}(t) + \mathbf{a}_{(2)}^H \mathbf{H}_{(2)} \mathbf{j}(t) \\ \cdots \\ \mathbf{a}_{(L)}^H \mathbf{n}_{(L)}(t) + \mathbf{a}_{(L)}^H \mathbf{H}_{(L)} \mathbf{j}(t) \end{bmatrix}, \\
&= \mathbf{S}_a \mathbf{N}_a(t) + \mathbf{S}_a \mathbf{M}_a \mathbf{j}(t), \\
&= \mathbf{S}_a \mathbf{N}_a(t) + \mathbf{V} \mathbf{j}(t).
\end{aligned} \tag{2.11}$$

For the convenience of representation, we define the following matrices,

$$\mathbf{S}_a = \begin{bmatrix} \mathbf{a}_{(1)}^H & \mathbf{0}_c^H & \ldots & \mathbf{0}_c^H \\ \mathbf{0}_c^H & \mathbf{a}_{(2)}^H & \ldots & \mathbf{0}_c^H \\ \mathbf{0}_c^H & \mathbf{0}_c^H & \ldots & \mathbf{0}_c^H \\ \mathbf{0}_c^H & \mathbf{0}_c^H & \ldots & \mathbf{a}_{(L)}^H \end{bmatrix}, \mathbf{N}_a(t) = \begin{bmatrix} \mathbf{n}_{(1)}(t) \\ \mathbf{n}_{(2)}(t) \\ \cdots \\ \mathbf{n}_{(L)}(t) \end{bmatrix}, \quad \mathbf{M}_a = \begin{bmatrix} \mathbf{H}_{(1)} \\ \mathbf{H}_{(2)} \\ \cdots \\ \mathbf{H}_{(L)} \end{bmatrix},$$

where the subscript "(i), $1 \leq i \leq L$" denotes the ith subarray of the auxiliary channel, $\mathbf{S}_a$ is a $L \times CL$-dimensional matrix composed of L steering vectors of the auxiliary channel pointing towards the desired signal and others are $C \times 1$-dimensional zero vectors $\mathbf{0}_c$, $\mathbf{M}_a$ is a $CL \times (K+1)$-dimensional matrix comprised of array manifold matrix $\mathbf{H}_i$ defined in Eq. (2.10), $\mathbf{N}_a(t)$ is a $CL \times 1$-dimensional Gaussian noise matrix, and $\mathbf{V} = \mathbf{S}_a \mathbf{M}_a$ is a $L \times (K+1)$-dimensional matrix. Here, the ith row of the matrix $\mathbf{V}$ corresponds to the conventional beamforming output of the desired signal and K interferences based on the ith subarray of the auxiliary channel. In other words, each row of $\mathbf{V}$ can be regarded as the beamspace steering vector.

As the main channel comprises all the N antennas, the output of the main channel can be expressed as,

$$m_r(t) = \sum_{i=1}^{M} x_i(t) = \mathbf{s}_{m_r}^H \mathbf{n}_{m_r}(t) + \mathbf{s}_{m_r}^H \mathbf{M_r} \mathbf{j}(t), \tag{2.12}$$

where $\mathbf{s}_{m_r}$ is the $N \times 1$-dimensional steering vector of the main channel pointing towards the target, $\mathbf{M_r} = [\mathbf{s}_{m_r}, \mathbf{m}_{r1}, \mathbf{m}_{r2}, ..., \mathbf{m}_{rK}]$ is the $N \times (K+1)$-dimensional array manifold matrix composed of $K+1$ steering vectors of the main channel pointing towards the desired signal and K interferences, and $\mathbf{n}_{m_r}(t)$ is the $N \times 1$-dimensional noise vector received by the main channel at time instant t, which can be modelled as the white Gaussian noise with power P_n. For clear representation, the subscript "r" indicates the structure of antenna reuse in the following subsections.

2.2.1.2 The Structure of Array Splitting

In the structure of array splitting, the main channel and the auxiliary channel select $M - L$ and L subarrays from M ones, respectively, and the two channels do not share any element. The signal processing procedure of auxiliary channel remains the same as that of antenna reuse structure. The output of the main channel is expressed as,

$$m_s(t) = \sum_{i=1}^{N-L} x_i(t) = \mathbf{s}_{m_s}^H \mathbf{n}_{m_s}(t) + \mathbf{s}_{m_s}^H \mathbf{M}_s \mathbf{j}(t). \tag{2.13}$$

where the subscript "s" indicates the structure of array splitting. The definitions of the matrix $\mathbf{M}_s$, vectors $\mathbf{s}_{m_s}$ and $\mathbf{n}_{m_s}$ are similar to $\mathbf{M}_r$, $\mathbf{s}_{m_r}$, $\mathbf{n}_{m_r}$ for antenna reuse, with elements composed of $M - L$ subarrays of the main channel.

By comparing Eqs. (2.12) and (2.13), we can find that for antenna reuse, the configuration of the main channel is fixed, while for array splitting, the configuration of the main channel changes with the variation of the auxiliary channel. So the mathematical models and solutions under the two structures are different. In order to distinguish these two structures, we use the subscript "r" to denote antenna reuse and "s" to denote array splitting in this section.

2.2.2 *Beamforming Weight Design for SLC*

Minimum output power (MOP) and beampattern matching (BPM) are two commonly-used metrics to design the beamforming weights for SLC auxiliary channels. However, both metrics are specific to the scenarios of small SNR, and will unavoidably cancel the desired signal when input SNR increases. In order to counteract this shortcoming of SLC, we resort to impose additional constraints, such as regularization penalty and zero forcing, to the two metrics for beamforming weight design, respectively. Since beamforming weight design applies to both structures of antenna reuse and array splitting, we ignore the subscripts "r" and "s" in this section for a clear explanation.

2.2.2.1 Minimum Output Power with Regularization Penalty (MOP-RP)

(a) Conventional MOP

According to Fig. 2.2, the output of SLC can be expressed as the subtraction between the main channel output and the auxiliary output. That is,

$$e_o(t) = m(t) - \mathbf{w}^H \mathbf{x}_a(t), \tag{2.14}$$

where $\mathbf{w}$ is the beamforming weight vector of the auxiliary channel, and $m(t)$ is the output of the main channel, which can be calculated by Eqs. (2.12) and (2.13) in the case of antenna reuse and array splitting, respectively. Furthermore, the output power of the SLC is,

$$\begin{aligned} P_o &= E\left\{|e_o(t)|^2\right\}, \\ &= E\left\{|m(t)|^2\right\} - \mathbf{x}_{a_m}^H \mathbf{w} - \mathbf{w}^H \mathbf{x}_{a_m} + \mathbf{w}^H \mathbf{R}_a \mathbf{w}. \end{aligned} \tag{2.15}$$

where $\mathbf{R}_a = E\{\mathbf{x}_a(t)\mathbf{x}_a^H(t)\}$ is the covariance matrix of the signal received by the auxiliary channel, and $\mathbf{x}_{a_m} = E\{\mathbf{x}_a(t)m^*(t)\}$ is the cross correlation vector of the signal received by the main channel and the auxiliary channel. The optimal weight vector $\mathbf{w}_o$ aims to minimize the output power of SLC per Eq. (2.15), based on which the optimal weight vector can be expressed as [81, 82],

$$\mathbf{w}_o = \mathbf{R}_a^{-1} \mathbf{a}_m. \tag{2.16}$$

(b) Regularization Penalty on Signal Cancellation

It is expected that the desired signal output of the auxiliary channel be nulled at any time instant, such that signal cancellation is avoided when subtracting the auxiliary from the main channel. That is,

$$\mathbf{w}^H \mathbf{x}_{a_d}(t) = \mathbf{w}^H \mathbf{V}\mathbf{D}\mathbf{j}(t) = 0, \tag{2.17}$$

where $\mathbf{x}_{a_d}(t)$ is the desired signal component received by the auxiliary channel, $\mathbf{D}$ is a selection matrix defined as $\mathbf{D} = \mathbf{f}\mathbf{f}^T$, with $\mathbf{f} = [1, \underbrace{0, ..., 0}_{K}]^T$. The purpose of the matrix $\mathbf{D}$ is to select the desired signal component from the data vector $\mathbf{j}(t)$. The output power of desired signal in the auxiliary channel is then,

$$E\left\{\left|\mathbf{w}^H \mathbf{x}_{a_d}(t)\right|^2\right\} = \mathbf{w}^H \mathbf{V}\mathbf{D}\mathbf{P}_j \mathbf{D}\mathbf{V}^H \mathbf{w} = 0, \tag{2.18}$$

where $\mathbf{P}_j$ is a diagonal matrix, with the power of the desired signal and K interferences distributing along its diagonal. When the output of the auxiliary channel contains partially the desired signal, there is unavoidably signal cancellation when subtracting the two channels.

(c) MOP-RP

The traditional MOP metric is modified here by adding the null constraint in Eq. (2.18), that is,

$$\begin{aligned}&\min_{\mathbf{w}} \; P_o \\ &\text{s.t.} \;\; \mathbf{w}^H \mathbf{V}\mathbf{D}\mathbf{P}_j \mathbf{D}\mathbf{V}^H \mathbf{w} = 0.\end{aligned} \tag{2.19}$$

The hard null constraint in Eq. (2.18) consumes L DoFs, we then resort to the soft-null formulation, that is,

$$\min_{\mathbf{w}} \; P_o + \beta \mathbf{w}^H \mathbf{V}\mathbf{D}\mathbf{P}_j \mathbf{D}\mathbf{V}^H \mathbf{w}. \tag{2.20}$$

The null constraint is added in the objective as a regularization penalty, β is a parameter used to emphasize between the output power and the penalty term.

By carefully observing Eq. (2.20), we can argue that the regularization penalty only acts on the desired signal attributed to the selection matrix $\mathbf{D}\mathbf{P}_j\mathbf{D}$. Therefore, we can rewrite Eq. (2.20) as follows,

$$\min_{\mathbf{w}} \; P_o + \widetilde{\beta} \mathbf{w}^H \mathbf{V}\mathbf{D}\mathbf{V}^H \mathbf{w}, \tag{2.21}$$

where $\widetilde{\beta} = \beta\sigma_s^2$ with σ_s^2 denoting the power of the desired signal. We refer to Eq. (2.21) as the MOP-RP metric, the optimal weight vector of which can be obtained by setting the complex gradient of Eq. (2.21) to zero. We can obtain that,

$$\mathbf{w}_{\text{mop}} = \mathbf{R}_a'^{-1} \mathbf{x}_{a_m}, \tag{2.22}$$

where $\mathbf{R}_a' = \mathbf{R}_a + \widetilde{\beta}\mathbf{V}\mathbf{D}\mathbf{V}^H$. Clearly that when $\widetilde{\beta} \rightarrow \infty$, the two formulations in Eqs. (2.19) and (2.21) are equivalent.

2.2.2.2 Beampattern Matching with Zero Forcing (BPM-ZF)

(a) Conventional BPM

We define the beampattern of the main channel as $F_m(\theta)$. It can be expressed as follows,

$$F_m(\theta) = \begin{cases} \mathbf{s}_{m_r}^H \mathbf{a}_r(\theta) & \text{antenna reuse} \\ \mathbf{s}_{m_s}^H \mathbf{a}_s(\theta) & \text{array splitting} \end{cases}, \tag{2.23}$$

where $\mathbf{a}_r(\theta)$ and $\mathbf{a}_s(\theta)$ are the steering vectors of the main channel in the case of antenna reuse and array splitting respectively.

For the auxiliary channel, we first apply conventional beamforming to each subarray and then combine the weighted output of all subarrays. According to Eq. (2.11), the auxiliary beampatterns in the case of antenna reuse and array splitting are same and given by,

$$F_a(\theta) = \mathbf{w}^H \mathbf{S}_a \mathbf{a}_a(\theta), \tag{2.24}$$

where $\mathbf{a}_a(\theta)$ is the steering vector of the auxiliary channel.

Thus, according to Eqs. (2.23) and (2.24), the beampattern of SLC can be expressed as,

$$F(\theta) = F_m(\theta) - F_a(\theta). \tag{2.25}$$

The conventional BPM metrics strive to match the main pattern with the auxiliary pattern in the directions of interferences [83]. As no constraints are imposed on the desired signal, the desired signal may get weakened significantly, especially when the interferences are spatially close to the desired signal. Thereby, we add a zero-forcing constraint to solve this problem. Specifically, the beampattern of the auxiliary channel matches that of the main channel so that the interferences of both channels cancel each other. In addition, the beampattern of the auxiliary channel should form a null in the direction of the desired signal, such that the interference cancellation will not weaken the desired signal.

(b) Zero-Forcing Constraints

In the direction of the desired signal, the auxiliary beampattern is forced to zero, that is,

$$F_a(\theta_0) = \mathbf{w}^H \mathbf{S}_a \mathbf{a}_a(\theta_0) = 0, \tag{2.26}$$

where θ_0 represents the DOA of the desired signal.

In the directions of interferences, the auxiliary beampattern and main beampattern should match with each other, that is,

$$\begin{aligned} F_a(\theta_i) &= \mathbf{w}^H \mathbf{S}_a \mathbf{a}_a(\theta_i) = F_m(\theta_i) \\ &= \mathrm{SLL}_i (i = 1, 2, 3, \ldots, K), \end{aligned} \tag{2.27}$$

where θ_i represents the DOA of the ith interference, K represents the total number of interferences, and SLL_i is the value of the main channel beampattern at the direction of the ith interference.

Compared with the matrix $\mathbf{M}_a$ defined in Eq. (2.11), it is not difficult to find that $\mathbf{a}_a(\theta_0)$ is the first column of matrix $\mathbf{M}_a$ and $\mathbf{a}_a(\theta_i)(i = 1, \ldots, K)$ is the second to $(K + 1)$th columns of matrix $\mathbf{M}_a$. Therefore, we can combine the above $K + 1$ constraints into the matrix-vector form as follows,

$$\mathbf{V}^H \mathbf{w} = \mathbf{g}, \tag{2.28}$$

where $\mathbf{g} = [0, \text{SLL}_1, \text{SLL}_2, ..., \text{SLL}_K]^T$ is a $K+1$-dimensional column vector.

(c) BPM-ZF

We design the optimal beamforming weights by minimizing the output power subject to beampattern matching constraints towards the interferences and zero-forcing constraints towards the desired signal concurrently. Mathematically, the problem can be described as follows:

$$\begin{aligned} \min_{\mathbf{w}} \quad & \mathbf{w}^H \mathbf{R}_a \mathbf{w} \\ \text{s.t.} \quad & \mathbf{V}^H \mathbf{w} = \mathbf{g}. \end{aligned} \tag{2.29}$$

From Eq. (2.11), we can obtain the theoretical covariance matrix of auxiliary channel, that is,

$$\mathbf{R}_a = E\{\mathbf{x}_a(t)\mathbf{x}_a^H(t)\} = P_n C \mathbf{I}_L + \mathbf{V}\mathbf{P}_j \mathbf{V}^H. \tag{2.30}$$

Substituting the covariance into Eq. (2.29) yields,

$$\begin{aligned} \mathbf{w}^H \mathbf{R}_a \mathbf{w} &= \mathbf{w}^H (P_n C \mathbf{I}_L + \mathbf{V}\mathbf{P}_j \mathbf{V}^H)\mathbf{w}, \\ &= P_n C \mathbf{w}^H \mathbf{w} + \mathbf{g}^H \mathbf{P}_j \mathbf{g}. \end{aligned} \tag{2.31}$$

Clearly, the term $\mathbf{g}^H \mathbf{P}_j \mathbf{g}$ is constant and irrelevant to beamforming weights, thus Eq. (2.29) can be simplified to minimize the noise power,

$$\begin{aligned} \min_{\mathbf{w}} \quad & \mathbf{w}^H \mathbf{w}, \\ \text{s.t.} \quad & \mathbf{V}^H \mathbf{w} = \mathbf{g}. \end{aligned} \tag{2.32}$$

Applying the Lagrange multiplier method, we can obtain the solution to the above problem as follows,

$$\mathbf{w}_{\text{bpm}} = \mathbf{V}(\mathbf{V}^H \mathbf{V})^{-1} \mathbf{g}. \tag{2.33}$$

2.3 Direction of Arrival Estimation

The problem of estimating the direction of arrival (DOA) of a plane wave is commonly referred to as the DOA estimation problem. It is important in radar, sonar, seismic systems and radio astronomy to list a few. For example, the two DOAs can be estimated from the two spectra in Fig. 2.1. Thus the Capon beamformer and correlation are considered as two commonly used DOA estimation approaches. In this subsection, we begin with determining the Cramer-Rao Bound (CRB), especially its dependence on the array configuration. We then look at some popular DOA estimation techniques: Maximum Likelihood (ML) [84, 85], MUSIC, MUSIC-root and ESPRIT algorithms [86].

2.3.1 CRB for DOA Estimation

The CRB has been analysed for array processing and studying the CRB helps us to gain an insight into the accuracy potential of any given estimation problem [87]. We derive the CRB of DOA estimation using 2-D planar arrays for a single source corrupted by white noise. The data model is therefore,

$$\mathbf{x}(t) = s(t)\mathbf{a} + \mathbf{n}(t), \tag{2.34}$$

where the desired signal, $s(t)$, may be random or deterministic. The model is called deterministic if $s(t)$ is assumed to be deterministic unknown and called random otherwise. The choice of one of the two models depends on applications. Since the CRBs of both data models have the same dependence on the array configuration [88], we consider the random waveform model in what follows. We assume both the desired signal $s(t)$ and white noise $\mathbf{n}(t)$ to be zero-mean Gaussian with covariance σ_s^2 and $\sigma_n^2\mathbf{I}$ respectively. Then the Probability Density Function (PDF) of the received data vector is,

$$f(\mathbf{x}(t); \theta, \phi) = \mathcal{N}(\mathbf{0}, \mathbf{\Sigma}), \tag{2.35}$$

where $\mathbf{\Sigma} = \sigma_s^2\mathbf{a}\mathbf{a}^H + \sigma^2\mathbf{I}$. Taking the logarithm to the PDF in Eq. (2.35) yields,

$$\mathcal{L}(\theta, \phi) = \log f(\mathbf{x}(t); \theta, \phi) = C - \mathbf{x}^H(t)\mathbf{\Sigma}^{-1}\mathbf{x}(t), \tag{2.36}$$

where C is an irrelevant constant to the DOA estimation. Define the SNR as $\rho = \sigma_s^2/\sigma_n^2$. According to the Sherman–Morrison formula, we have that

$$\mathbf{\Sigma}^{-1} = \sigma_n^{-2}(\mathbf{I} - \frac{\rho\mathbf{a}\mathbf{a}^H}{1 + N\rho}). \tag{2.37}$$

Since we are interested in taking derivatives of the expression in Eq. (2.36), we can ignore the constant terms. Now taking the second derivative of $\mathcal{L}(\theta, \phi)$ with respect to θ and then taking negative expectation yields the Fisher Information of θ,

$$\begin{aligned} J_{\theta\theta} &= -E\left\{\frac{\partial^2\mathcal{L}(\theta, \phi)}{\partial\theta^2}\right\} \\ &= \mathcal{G}\cos^2\theta\left\{\cos^2\phi\sum_{n=1}^{N}x_n^2 + \sin^2\phi\sum_{n=1}^{N}y_n^2 + \sin(2\phi)\sum_{n=1}^{N}x_n y_n\right\}, \end{aligned} \tag{2.38}$$

where $\mathcal{G} = 2N\rho^2 k_0^2/(1 + N\rho)$ is independent on the array geometry and x_n, y_n are coordinates of the nth antenna as defined in Sect. 1.1. Similarly, we can derive the Fisher Information of the azimuth angle ϕ as,

$$J_{\phi\phi} = \mathcal{G}\sin^2\theta\left\{\sin^2\phi\sum_{n=1}^{N}x_n^2 + \cos^2\phi\sum_{n=1}^{N}y_n^2 - \sin(2\phi)\sum_{n=1}^{N}x_n y_n\right\}. \tag{2.39}$$

The cross-terms of the Fisher Information Matrix (FIM) are,

$$J_{\phi\theta} = J_{\theta\phi} = \frac{\mathcal{G}}{4}\sin(2\theta)\left\{\sin(2\phi)(\sum_{n=1}^{N}x_n^2 - \sum_{n=1}^{N}y_n^2) - 2\cos(2\phi)\sum_{n=1}^{N}x_n y_n\right\}. \tag{2.40}$$

Therefore, the FIM can be written as

$$\mathbf{J} = \begin{bmatrix} J_{\theta\theta} & J_{\theta\phi} \\ J_{\phi\theta} & J_{\phi\phi} \end{bmatrix}. \tag{2.41}$$

Taking the inverse of the FIM, the CRBs of θ and ϕ are obtained as

$$C_{\theta\theta} = \frac{1}{\mathcal{G}\cos^2\theta}\cdot\frac{\sin^2\phi\sum_{n=1}^{N}x_n^2 + \cos^2\phi\sum_{n=1}^{N}y_n^2 - \sin(2\phi)\sum_{n=1}^{N}x_n y_n}{\sum_{n=1}^{N}x_n^2\sum_{n=1}^{N}y_n^2 - \sum_{n=1}^{N}x_n y_n}, \tag{2.42}$$

and

$$C_{\phi\phi} = \frac{1}{\mathcal{G}\sin^2\theta}\cdot\frac{\cos^2\phi\sum_{n=1}^{N}x_n^2 + \sin^2\phi\sum_{n=1}^{N}y_n^2 + \sin(2\phi)\sum_{n=1}^{N}x_n y_n}{\sum_{n=1}^{N}x_n^2\sum_{n=1}^{N}y_n^2 - \sum_{n=1}^{N}x_n y_n}, \tag{2.43}$$

respectively.

In the case of multiple sources (say the number of sources is Q), the received data $\mathbf{x}(t)$ can be modeled as,

$$\mathbf{x}(t) \sim \mathcal{CN}(\mathbf{m}(\alpha), \Gamma(\alpha)), \tag{2.44}$$

where α is a real-valued parameter vector that completely and uniquely specifies the distribution of the data $\mathbf{x}(t)$. Then, a general formula for the CRB on the covariance matrix of any unbiased estimate of α is,

$$[(\mathrm{CRB})^{-1}]_{ij} = \mathrm{tr}\left[\Gamma^{-1}(\alpha)\frac{\partial\Gamma(\alpha)}{\partial\alpha_i}\Gamma^{-1}(\alpha)\frac{\partial\Gamma(\alpha)}{\partial\alpha_j}\right] + 2\mathrm{Re}\left\{\frac{\partial m^*(\alpha)}{\partial\alpha_i}\Gamma^{-1}(\alpha)\frac{\partial m(\alpha)}{\partial\alpha_i},\right\} \tag{2.45}$$

where α_i denotes the ith component of α. In the case of random desired signal and $\alpha = \theta$, the CRB of a linear array can be expressed as,

$$\mathrm{CRB} = \frac{\sigma_n^2}{2J}\left\{\mathrm{Re}[\mathbf{D}^H(\mathbf{I} - \mathbf{A}(\mathbf{A}^H\mathbf{A})^{-1}\mathbf{A}^H)\mathbf{D}\odot\mathbf{P}^T]\right\}^{-1}, \tag{2.46}$$

where J is the total number of snapshots and $\mathbf{P}$ is the correlation matrix of the Q sources, σ_n^2 is the noise power, $\mathbf{A} = [\mathbf{a}_1, \mathbf{a}_2, \ldots, \mathbf{a}_Q]$ is the array manifold containing the steering vectors of the Q sources, and $\mathbf{D}$ is the derivative of $\mathbf{A}$ with respect to $\omega = (2\pi/\lambda)d \sin\theta$ with d denoting the inter-element spacing.

2.3.2 Maximum Likelihood

One way of estimating the DOA of an incoming signal is to maximize the likelihood that the signal is coming from that particular angle, which is referred to Maximum Likelihood (ML) [86]. The data model is shown in Eq. (2.34) and is assumed to be deterministic complex Gaussian, i.e.,

$$f(\mathbf{x}; \theta, s) = \mathcal{N}(s\mathbf{a}, \mathbf{R}_n), \tag{2.47}$$

where $\mathbf{R}_n = E\{\mathbf{n}\mathbf{n}^H\}$ is the noise covariance matrix. We have two unknown parameters, the magnitude s and the angle θ. The ML estimator is given by

$$\{\tilde{s}, \tilde{\theta}\} = \max_{s,\theta}\{f(\mathbf{x}; \theta, s)\}, \tag{2.48}$$

that is equivalent to

$$\begin{aligned}\{\tilde{s}, \tilde{\theta}\} &= \min_{s,\theta}\{(\mathbf{x} - s\mathbf{a})^H \mathbf{R}_n^{-1}(\mathbf{x} - s\mathbf{a})\}, \\ &= \min_{s,\theta}\{\mathbf{x}^H\mathbf{R}_n^{-1}\mathbf{x} - s\mathbf{x}^H\mathbf{R}_n^{-1}\mathbf{a} - s^*\mathbf{a}^H\mathbf{R}_n^{-1}\mathbf{x} + |s|^2\mathbf{a}^H\mathbf{R}_n^{-1}\mathbf{a}\}.\end{aligned} \tag{2.49}$$

We minimize Eq. (2.49) with respect to s^* first while treating s as an independent variable. We have that,

$$\tilde{s} = \frac{\mathbf{a}^H\mathbf{R}_n^{-1}\mathbf{x}}{\mathbf{a}^H\mathbf{R}_n^{-1}\mathbf{a}}. \tag{2.50}$$

Substituting $\tilde{s}$ into Eq. (2.49) yields the ML estimate of θ,

$$\tilde{\theta} = \max_{\theta}\left\{\frac{|\mathbf{a}^H\mathbf{R}_n^{-1}\mathbf{x}|^2}{\mathbf{a}^H\mathbf{R}_n^{-1}\mathbf{a}}\right\}. \tag{2.51}$$

An interesting aspect of this estimator is that when there is only one source and $\mathbf{R}_n = \sigma_n^2\mathbf{I}$, the ML estimator reduces to the correlation technique. This is expected because the correlation technique is equivalent to the matched filter, which is optimal in the single source case. The ML estimator is unbiased and asymptotically efficient (achieves the CRB with infinite number of samples). However, it is an impractical algorithm due to the assumption of prior knowledge of interference covariance matrix $\mathbf{R}_n$ and also the high computational cost.

2.3.3 MUSIC

A more efficient subspace-based algorithm, called MUSIC (MUltiple SIgnal Characterization) [84, 89], has received significant attention. Suppose there are M sources impinging on the N-antenna array ($N > M$). Applying the eigenvalue decomposition on the received data covariance matrix $\tilde{\mathbf{R}}_x$ in Eq. (2.7) yields,

$$\tilde{\mathbf{R}}_x = \mathbf{E}\mathbf{\Sigma}\mathbf{E}^H, \tag{2.52}$$

where $\mathbf{\Sigma} = \text{diag}\{\lambda_1, \ldots, \lambda_N\}$ is the eigenvalue matrix with descending eigenvalues $\lambda_1 \geq \lambda_2 \geq \cdots \geq \lambda_N$, and $\mathbf{E} = [\mathbf{e}_1, \ldots, \mathbf{e}_N]$ is the eigenvector matrix. There are M significant eigenvalues (and eigenvectors) corresponding to the estimated sources and the remaining $N - M$ eigenvectors are the basis for the noise subspace, i.e.,

$$\mathbf{E}_n = [\mathbf{e}_{M+1}, \ldots, \mathbf{e}_N]. \tag{2.53}$$

MUSIC plots the pseudo spatial spectrum,

$$P(\theta) = \frac{1}{\mathbf{a}(\theta)^H(\mathbf{E}_n\mathbf{E}_n^H)\mathbf{a}(\theta)} = \frac{1}{\|\mathbf{E}_n^H\mathbf{a}(\theta)\|_2^2}. \tag{2.54}$$

Note that since the signal steering vectors are orthogonal to the noise subspace, the denominator becomes zero when θ is a signal direction. Therefore, the estimated signal directions are the M largest peaks in the pseudo-spectrum.

2.3.4 Root-MUSIC

The problem of the MUSIC algorithm described above is that its accuracy is limited by the discretization at which the pseudo-spectrum $P(\theta)$ is evaluated. More importantly, it requires either human interaction or a comprehensive search algorithm to decide on the largest M peaks. This is an extremely computationally intensive process. Therefore, MUSIC is not very practical either. This is where Root-MUSIC [90] comes in.

Note that MUSIC is a technique that estimates the spectrum of the incoming data, i.e. it is a spectral estimation technique. Root-MUSIC, on the other hand, is an example of a model-based parameter estimation (MBPE) technique. A crucial aspect of Root-MUSIC is that the estimation technique is only valid for ULAs. For now, define $z = e^{jk_0 d\cos\theta}$, then the steering vector can be rewritten as,

$$\mathbf{a}(\theta) = [1, z, z^2, \ldots, z^{N-1}]^T. \tag{2.55}$$

As what we mentioned above, when there is a signal coming from θ, we have

$$1/P(\theta) = \mathbf{a}^H(\theta)(\mathbf{E}_n\mathbf{E}_n^H)\mathbf{a}(\theta) = \mathbf{a}^H(\theta)\mathbf{C}\mathbf{a}(\theta) = 0, \tag{2.56}$$

where $\mathbf{C} = \mathbf{E}_n\mathbf{E}_n^H$. Substituting Eq. (2.55) into Eq. (2.56) yields,

$$1/P(\theta) = \sum_{m=0}^{N-1}\sum_{n=0}^{N-1} z^n C_{mn} z^{-m} = \sum_{m=0}^{N-1}\sum_{n=0}^{N-1} C_{mn} z^{n-m} = 0. \tag{2.57}$$

The final double summation can be simplified by rewriting it as a single sum by setting $l = n - m$, i.e.,

$$1/P(\theta) = \sum_{l=-(N-1)}^{N-1} C_l z^l = 0, \tag{2.58}$$

where $C_l = \sum_{n-m=l} C_{mn}$. Equation (2.58) defines a polynomial of degree $(2N - 2)$ with $(N - 1)$ pairs of zeros, where one is within the unit circle and the other outside. The phases of these zeros provide the DOAs of estimated sources.

2.3.5 ESPRIT

In this subsection, we introduce another super-resolution DOA estimation strategy, namely estimation of signal parameters via rotational invariance techniques, abbreviated as ESPRIT. This algorithm was first proposed by Roy [91, 92] and further discussed in some other publications [93–95]. The algorithm requires so-called invariance in the geometric structure of the array when estimating signal parameters. This invariance can be obtained by two methods: one is that the array itself has two or more identical subarrays; The second is to obtain two or more identical submatrixes through certain transformations. Due to its outstanding performance in terms of effectiveness and stability, this algorithm has been recognized as a classic algorithm for spatial spectrum estimation.

The basic idea of the ESPRIT algorithm is to select two subarrays with identical configurations. The corresponding offset distance of the two selected subarrays are equal, that is, the array elements are divided into a set of pairs, and each pair has the same translation distance between them.

Considering an N-element ULA, as shown in Fig. 2.3a. The first step of the ESPRIT algorithm is to select two subarrays of the same length. Suppose we select N_s antennas for each subarray and N_s need to satisfy $N_s > K$, where K is the number of signals to be estimated. Figure 2.3 depicts the two forms of subarray division. Figure 2.3b shows the non-overlapping scenario where the two subarrays do not share antennas. Figure 2.3c shows the overlapping scenario where the two subarrays share

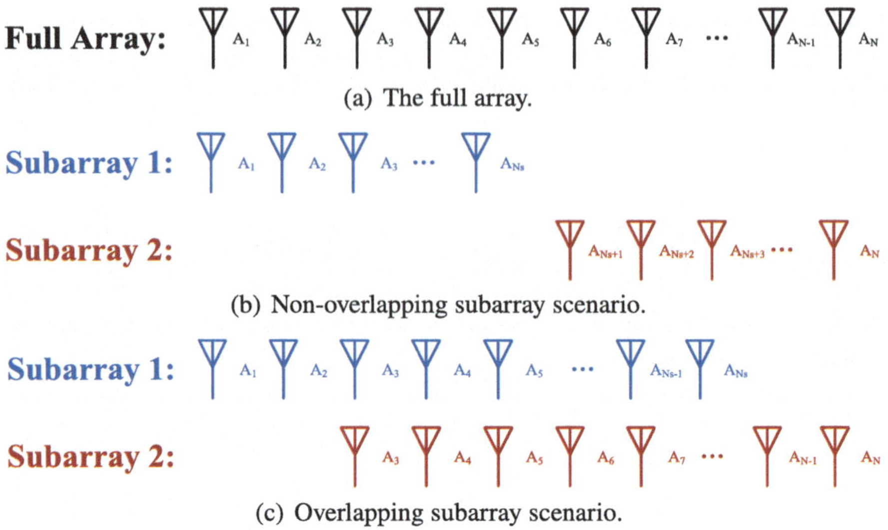

Fig. 2.3 Subarrays of ESPRIT algorithm

some common elements. It is not difficult to see that the distance offset between the pair of corresponding antennas in the two subarrays is the same under both scenarios.

As for the full array, the received data under multi-source scenario can be represented as,

$$\mathbf{x}(t) = \mathbf{A}\mathbf{s}(t) + \mathbf{n}(t), \tag{2.59}$$

where $\mathbf{A} = [\mathbf{a}(\theta_1), ..., \mathbf{a}(\theta_K)]$ is the matrix composed of the steering vectors pointing to the K sources and $\mathbf{s}(t)$ is the signal vector.

The received signal of the two subarrays can be expressed as,

$$\mathbf{x}_1(t) = \mathbf{Z}_1\mathbf{x}(t) = \mathbf{A}_1\mathbf{s}(t) + \mathbf{Z}_1\mathbf{n}(t), \tag{2.60}$$

and

$$\mathbf{x}_2(t) = \mathbf{Z}_2\mathbf{x}(t) = \mathbf{A}_2\mathbf{s}(t) + \mathbf{Z}_2\mathbf{n}(t), \tag{2.61}$$

respectively, where $\mathbf{Z}_1 \in \mathbb{R}^{N_s \times N}$ and $\mathbf{Z}_2 \in \mathbb{R}^{N_s \times N}$ are the selection matrices of the two subarrays with each row to be the selection vector of the corresponding antenna. $\mathbf{A}_1$ and $\mathbf{A}_2$ are the steering matrices of the two subarrays which are defined as $\mathbf{A}_1 = \mathbf{Z}_1\mathbf{A}$ and $\mathbf{A}_2 = \mathbf{Z}_2\mathbf{A}$, respectively.

Since the steering vector of the array has the property of motion invariant, thus we have,

$$\mathbf{A}_2 = \mathbf{A}_1\Phi, \tag{2.62}$$

where,

$$\Phi = \text{diag}[e^{j\phi_1}, e^{j\phi_2}, ..., e^{j\phi_K}], \tag{2.63}$$

with ϕ_k containing the DOA information of the kth signal source which is defined as,

$$\phi_k = \frac{2\pi}{\lambda}\Delta d \sin\theta_k, \tag{2.64}$$

where Δd is the offset distance between two subarrays.

Under the premise of ensuring that there is no fully correlated signal source, matrix $\mathbf{A}$ is a column full rank matrix. The signal subspace spanned by the column vectors of matrix $\mathbf{A}$ can be expressed by,

$$\mathbf{V} = \mathbf{AT}, \tag{2.65}$$

where $\mathbf{T} \in \mathbb{C}^{K\times K}$ is a non-singular matrix. Then the signal subspace of the two subarrays can be obtained utilizing the selection matrices, that is,

$$\mathbf{V}_1 = \mathbf{Z}_1\mathbf{V} = \mathbf{A}_1\mathbf{T}, \tag{2.66}$$

and

$$\mathbf{V}_2 = \mathbf{Z}_2\mathbf{V} = \mathbf{A}_2\mathbf{T}. \tag{2.67}$$

According to Eq. (2.66), we have,

$$\mathbf{A}_1 = \mathbf{V}_1\mathbf{T}^{-1}. \tag{2.68}$$

Substituting Eqs. (2.62), (2.68) and (2.67) is transformed into,

$$\mathbf{V}_2 = \mathbf{V}_1\mathbf{T}^{-1}\Phi\mathbf{T}. \tag{2.69}$$

Define $\Psi = \mathbf{T}^{-1}\Phi\mathbf{T}$, since Ψ and Φ are similar matrices, so they have the same eigenvalues. Combining Φ is a diagonal matrix, the eigenvalues of Ψ are exactly the elements of Φ. If we can obtain the matrix Ψ and calculate its eigenvalues, then we can obtain the estimated value of ϕ_k, $k = 1, ..., K$.

According to Eq. (2.52), we can estimate the signal subspace $\tilde{\mathbf{V}}$ by performing eigenvalue decomposition on the received data covariance matrix $\tilde{\mathbf{R}}_x$ and taking the eigenvectors corresponding to the largest K eigenvalues.

Utilizing Eqs. (2.66), (2.67) and (2.69), we further have,

$$\begin{aligned}\tilde{\mathbf{V}}_1 &= \mathbf{Z}_1\tilde{\mathbf{V}},\\ \tilde{\mathbf{V}}_2 &= \mathbf{Z}_2\tilde{\mathbf{V}},\\ \tilde{\mathbf{V}}_2 &= \tilde{\mathbf{V}}_1\Psi.\end{aligned} \tag{2.70}$$

Ψ can be estimated utilizing the least squares (LS) method as follows,

$$\begin{aligned}\tilde{\Psi}_{\mathrm{LS}} &= \arg\min_{\Psi}\{||\tilde{\mathbf{V}}_2 - \tilde{\mathbf{V}}_1\Psi||_F\}, \qquad (2.71)\\ &= \arg\min_{\Psi}\{\mathrm{tr}\{[\tilde{\mathbf{V}}_2 - \tilde{\mathbf{V}}_1\Psi]^H[\tilde{\mathbf{V}}_2 - \tilde{\mathbf{V}}_1\Psi]\}\}.\end{aligned}$$

The estimated result of the LS is given by,

$$\tilde{\Psi}_{\mathrm{LS}} = [\tilde{\mathbf{V}}_1^H\tilde{\mathbf{V}}_1]^{-1}\tilde{\mathbf{V}}_1^H\tilde{\mathbf{V}}_2 \tag{2.72}$$

2.4 Target Detection

Radar is an electromagnetic system for the detection and location of reflecting objects such as aircraft, ships, spacecraft, vehicles, people and the natural environment [96]. The term *radar* is a contraction of of the words *ra*dio *d*etection *a*nd *r*anging. It operates by radiating energy into space and detecting the echo signal reflected from an object or target. The range, or distance, from the radar to a target is found by measuring the time it takes for the radar signal to travel to target and return back to the radar. The target's location in angle can be found from the direction the narrow-beamwidth radar antenna points when the received echo signal is of maximum amplitude. If the target is in motion with respect to the radar, there is a shift in the frequency of the echo signal due to the Doppler effect. This frequency shift is proportional to the velocity of the target relative to the radar (also called the radial velocity). Radar can also provide information about the nature of the target being observed [64].

A pulse radar that employs the Doppler shift for detecting moving targets is either a moving target indication (MTI) radar or a pulse Doppler radar. In the real world, radars have to deal with more than receiver noise when detecting targets since they also receive echoes from the natural environment such as land, sea and weather. These echoes are called clutter since they can "clutter" the radar display. Clutter echoes can be many orders of magnitude larger than aircraft echoes. The need for joint space and time processing in MTI radar arises from the inherent two-dimensional nature of ground clutter as shown in Fig. 2.4. We could see that a jammer is only spread in Doppler frequency, whereas clutter is correlated in both spatial and Doppler domains. One could also recognize that the clutter spectrum focuses along a ridge on the Angle-Doppler plane (the ridge becomes a straight line for ULAs), whose shape depends on the array configuration and amplitude is modulated by the transmit and receive

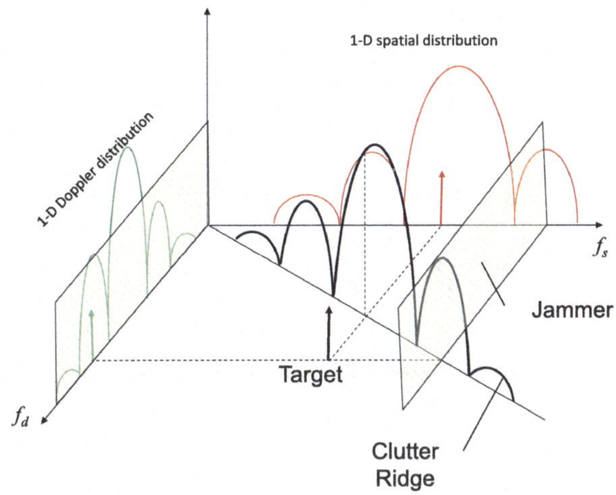

Fig. 2.4 Angle-Doppler structure of airborne clutter, target and interference

directivity pattern. It is clear from Fig. 2.4 that the straightforward 1-D processing strategy, either optimum temporal or spatial filter as shown in Eq. (2.1), is insufficient to detect a slow target from strong clutter due to the broad stopband of temporal or spatial filter. However, a space-time filter operates in the whole Angle-Doppler plane and it forms a very narrow ditch along the trajectory of the clutter spectrum so that even slow targets may fall into the passband and become detectable.

The term space-time adaptive processing (STAP) was first applied to multidimensional adaptive filtering of clutter and jamming in airborne MTI radars. A STAP beamformer has N antenna elements and a coherent processing interval (CPI) consisting of M pulses with a fixed pulse repetition interval (PRI) as shown in Fig. 2.5. Similar to the adaptive 1-D spatial filter in Eq. (2.1), the basic underlying optimum STAP beamformer is given by,

$$\mathbf{w}_{\mathrm{STAP}} = \gamma \mathbf{R}_n^{-1} \mathbf{a}, \tag{2.73}$$

where $\mathbf{R}_n$ is the positive-definite $MN \times MN$ dimensional covariance matrix associated with the total interference (clutter plus jammer plus receiver noise) and $\mathbf{a}$ is the MN-dimensional space-time steering vector of the target, which is a Kronecker product of the spatial and temporal steering vectors. Figure 2.6 shows the structure of a space-time data cube with L range bins and the lth one is the cell under test (CUT). The interference covariance matrix $\mathbf{R}_n$ is usually estimated using range cells that are adjacent to the CUT (usually centered around it), avoiding guard cells to account for target leakage also shown in Fig. 2.6. Similar to Eq. (2.7), the ML estimate of $\mathbf{R}_n$ is $\tilde{\mathbf{R}}_n = \frac{1}{L}\sum_{l=1}^{L} \mathbf{x}(l)\mathbf{x}^H(l)$ given L IID space-time training data.

The detection of a radar signal is based on establishing a threshold at the output of the receiver. If the threshold level is set too low, noise might exceed it and be mistaken for a target. This is called a false alarm. If the threshold is set to high, noise might not be large enough to cause false alarms, but weak target echoes might not exceed

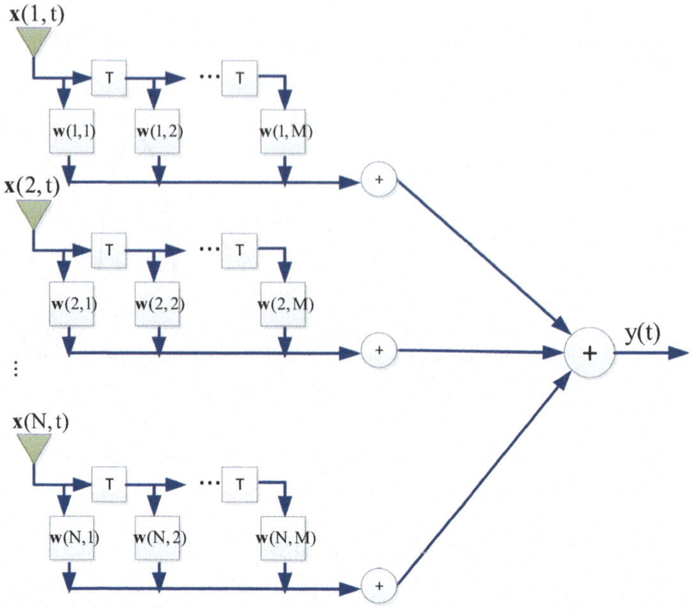

Fig. 2.5 The structure of a STAP beamformer

Fig. 2.6 The structure of a space-time data cube

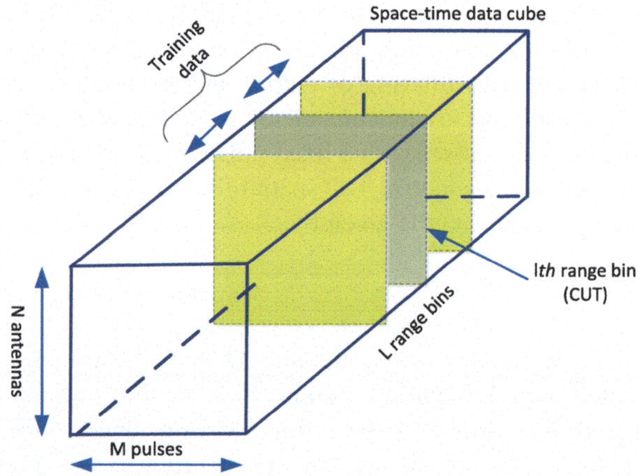

the threshold and would not be detected. When this occurs, it is called a missed detection. The detection problem is usually treated as a hypothesis test to determine the presence of a signal with a known steering vector. The signal model we adopt here is identical to that shown in Eq. (2.34). The null and alternative hypotheses are given by,

$$H_0 : \mathbf{x} = \mathbf{n}; \tag{2.74}$$

and

$$H_1 : \mathbf{x} = s\mathbf{a} + \mathbf{n}, \tag{2.75}$$

where s is the target signal. According to the Neyman-Pearson criterion [97], the detector decides the alternative hypothesis H_1 for a given false alarm rate (FAR), $P_{\text{fa}} = \alpha$, if

$$L(\mathbf{x}) = \frac{P(\mathbf{x};\, H_1)}{P(\mathbf{x};\, H_0)} > \mathcal{T}, \tag{2.76}$$

where the threshold $\mathcal{T}$ can be found from

$$P_{\text{fa}} = \int_{\mathbf{x}:L(\mathbf{x})>\mathcal{T}} P(\mathbf{x};\, H_0) = \alpha. \tag{2.77}$$

From the signal model, we know that $P(\mathbf{x};\, H_1) = \mathcal{N}(s\mathbf{a}, \mathbf{R}_n)$ and $P(\mathbf{x};\, H_0) = \mathcal{N}(\mathbf{0}, \mathbf{R}_n)$. Taking the logarithm of Eq. (2.76) yields,

$$(\mathbf{x} - s\mathbf{a})^H \mathbf{R}_n^{-1} (\mathbf{x} - s\mathbf{a}) - \mathbf{x}^H \mathbf{R}_n^{-1} \mathbf{x} > \mathcal{T}'. \tag{2.78}$$

Substituting the ML estimate of s in Eq. (2.50) and $\tilde{\mathbf{R}}_n$ into Eq. (2.78), we have that

$$L(\mathbf{x}) = \frac{|\mathbf{a}^H \tilde{\mathbf{R}}_n^{-1} \mathbf{x}|^2}{\mathbf{a}^H \tilde{\mathbf{R}}_n^{-1} \mathbf{a}} > \mathcal{T}', \tag{2.79}$$

which is the adaptive matched filter (AMF) detector [98]. The other kind of detector, named the generalized likelihood ratio test (GLRT) [99], also possesses the constant false alarm rate (CFRA) property. The GLRT detection statistic is given by,

$$L(\mathbf{x}) = \frac{|\mathbf{a}\tilde{\mathbf{R}}_n^{-1} \mathbf{x}|^2}{\mathbf{a}^H \tilde{\mathbf{R}}_n^{-1} \mathbf{a}(1 + \frac{1}{L}\mathbf{x}^H \tilde{\mathbf{R}}_n^{-1} \mathbf{x})}, \tag{2.80}$$

where L is the number of training snapshots. The difference between the two detectors, AMF and GLRT, is that the GLRT takes the statistics of the covariance matrix estimate $\tilde{\mathbf{R}}_n$ into account (and hence achieves better performance). The difference between the two detectors decreases as L increases as the estimate of the covariance matrix becomes better. It is clear that the GLRT reduces to the AMF for sufficiently large L. The theoretical false alarm rate P_{fa} of the AMF detector with L training data, is [98–100]

$$P_{\text{fa}} = \int_0^1 \left(1 + \frac{\gamma\eta}{L}\right)^{-K} f_\beta(\eta;\, K+1, N-1) d\eta, \tag{2.81}$$

where $K = L - N + 1$ and N is the dimension of the detector. η is a type I beta distributed variable with parameters $K + 1$ and $N - 1$. The detection probability $P_{\rm D}$, on the other hand, is

$$P_{\rm D} = 1 - \int_0^1 h(\eta) f_\beta(\eta; K + 1, N - 1) d\eta, \tag{2.82}$$

where

$$h(\eta) = \left(1 + \frac{\gamma\eta}{L}\right)^{-K} \sum_{l=1}^{K} \binom{K}{k} \left(\frac{\gamma\eta}{L}\right)^k G_k\left(\frac{\eta\rho}{1 + \frac{\gamma\eta}{L}}\right), \tag{2.83}$$

and

$$G_k(y) = e^{-y} \sum_{n=0}^{k} \frac{y^n}{n!}, \tag{2.84}$$

where ρ is the output SCNR and $\binom{K}{k} = \frac{K!}{k!(K-k)!}$. We can see from Eq. (2.83) that the relationship between the detection probability $P_{\rm D}$ and the output SCNR is nonlinear.

2.5 Summary

In this chapter, we briefly reviewed some important array applications, those are adaptive beamforming, paramater estimation and target detection. Adaptive beamforming is an effective method of closed-loop spatial filtering to eliminate unwanted interferences and improve the signal to interference-plus-noise ratio (SINR). However, the computational complexity of a fully adaptive beamformer is prohibitively high, and thus a practical technique, referred to as adaptive sidelobe canceller, was introduced subsequently. DOA estimation and target detection are two main applications of antenna arrays, where the former is used to characterize the propagation direction of the waveform and the latter is to determine whether a signal is present in a measurement. In the first application of DOA estimation, we first introduced the commonly-used theoretical bound, Cramer-Rao Bound, which can help to quantitatively evaluate the potential accuracy of any estimation algorithm, and then some famous DOA estimation algorithms were reviewed briefly. In the second application of target detection, we took the airborne MTI radar as an example and introduced the multi-dimensional space-time adaptive processing for ground clutter suppression. In the following chapters, we are going to examine optimal sparse array design in different array applications.

Part II
Sparse Sensing via Antenna Selection

Chapter 3
Sparse Sensing for Determininic Beamforming

Antenna arrays are capable of performing beamforming, which makes them an effective tool for combating interference while providing certain gains towards desired sources [4, 6, 101, 102]. The beamforming performance is not only dependent on the exciation weights but also on the array configuration [103–105]. For the same number of antennas, different array configurations yield various beampattern shapes, including mainlobe ripples, sidelobe level and null depth. The opposite is also true, that is, for the same array configuration, beamformers can assume different characteristics and performances. As such, sparse array design should utilize both the array structure and excitation weights towards achieving the optimum performance. Beamforming can be broadly classified into deterministic and adaptive [1]. The former focuses on synthesizing beampatterns with prescribed mainlobe width and reduced sidelobe levels [106, 107]. Adaptive beamforming, on the other hand, is influenced by the environment and incorporates, in addition to noise characteristics, the source and interference signals or data statistics which are present in the array field of view (FOV). In this chapter, we focus on the deterministic beamforming with a reduced number of antennas, that is thinned array beampattern synthesis, via iterative soft-thresholding-based optimization algorithms in Sect. 3.2 and the sparse quiescent beamformer design combining adaptive and deterministic constraints in Sect. 3.3.

3.1 Introduction

The last 50 years have seen a growing demand for large aperture arrays exhibiting increased capabilities in terms of flexibility and reconfigurability, yet simultaneously offering reduced hardware cost and computational complexity [108]. Thinned arrays are ideal for satisfying these requirements, as they can maintain the same mainlobe width and peak sidelobe level (PSL) with a significant reduction in cost, weight, and complexity. Moreover, thinned arrays are flexible and reconfigurable due to the periodic quantization of the element positions.

X. Wang et al., *Sparse Sensing and Sparsity Sensed in Multi-sensor Array Applications*, https://doi.org/10.1007/978-981-99-9558-5_3

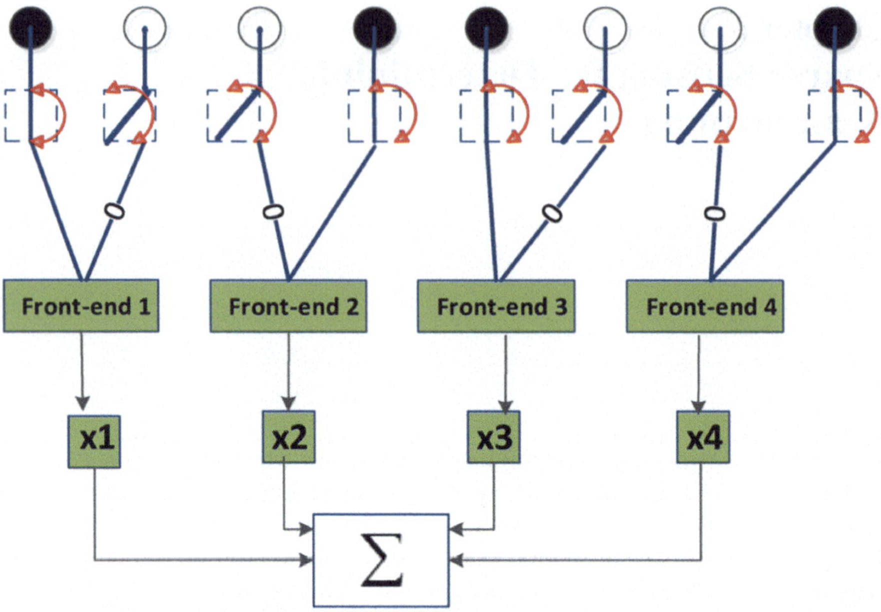

Fig. 3.1 Diagram of the reconfigurable array thinning strategy

Beamforming can be broadly classified into open-loop and closed-loop techniques [109, 110]. The former is also called deterministic beampattern synthesis and the latter is called adaptive beamforming. In this chapter, we consider array reconfigurability for open-loop null-steering algorithms and we will discuss sparse arrays for adaptive beamforming in the next chapter. The high cost of an entire front-end per antenna makes large array interference nulling quite expensive. In order to reduce the hardware cost, a smaller number of front-ends are installed in the receiver. Thinning strategies then adaptively select an optimum subset of antennas over a full array layout to connect to the following beam forming network (BFN), whereas those belonging to the complementary subset are connected to matched loads or removed. The on–off status of the array elements is controlled by acting on a set of radio frequency (RF) switches [111–113], giving an easily-reconfigurable and low-complexity antenna architecture. Therefore, we propose an adaptive thinned array framework for open-loop interference suppression, as shown in Fig. 3.1.

There are many effective methods for synthesizing thinned arrays existing in the literature. These include techniques such as heuristic method [114], convex optimization [115], Bayesian compressive sensing [116], matrix pencil [117] to list a few. These methods, however, are prohibitively computationally expensive or can even fail when the scale of the problem becomes large. Yang [118] combined the iterative fast Fourier transform (FFT) algorithm [119, 120] and array interpolation together to synthesize large antenna arrays, but the convergence rate of his method is highly dependent on the choice of parameter values and initial search points. It is desirable to adaptively achieve arbitrary sidelobe levels and the required deep nulls for combating interferences and noise according to continuously changing environmental scenarios [121]. Thus, developing fast array thinning algorithms becomes necessary for adaptive interference nulling [122].

In Sect. 3.2, we apply an iterative soft-thresholding based optimization method for array thinning, which has been shown to effectively address large-scale optimization problems in the areas of compressive sensing and image processing due to its simple structure. Each iteration of this algorithm comprises a multiplication by a matrix and its transpose, along with a scalar shrinkage step on the obtained result. However, when the problem scale is small or moderate with many constraints, the iterative soft-thresholding algorithms do not exhibit obvious advantages.

In Sect. 3.3, we consider two beamformers with controlled sidelobe levels for maximizing the output SNR. The first is fully adaptive and provides the maximum SNR (the MSNR beamformer) without concerning with homogeneity of array responses towards different sources. The second beamformer is semi-adaptive and strives to reach desired response values while utilizing the source spatial correlations adaptively to minimize output SNR degradation. Two convex relaxation approaches and an iterative linear fractional programming (ILFP) algorithm are proposed to solve underlying sparse array beamformer problems.

3.2 Sparse Array Beampattern Synthesis

In this section, we investigate array thinning for open-loop beampattern synthesis.

3.2.1 *Beampattern Synthesis Algorithms*

Consider an N-antenna array in the x-y plane of a Cartesian coordinate system and let the vector $\mathbf{P} \in \mathbb{R}^{N\times 2}$ contain the positions of the array elements. A plane wave with wavelength λ is impinging on the array from a direction of arrival (DOA) that is specified by elevation angle, θ, and azimuth angle, ϕ. The steering vector of the array in the look direction (θ, ϕ) is

$$\mathbf{a}(\theta, \phi) = e^{j\frac{2\pi}{\lambda}\mathbf{P}[u_x, u_y]^T}, \tag{3.1}$$

where $u_x = \cos\theta\cos\phi$ and $u_y = \cos\theta\sin\phi$. We are interested in synthesizing a desired beampattern for this array using a minimum number of antennas. This is formulated in the next subsection.

3.2.1.1 Problem Formulation

Let us assume that we have a desired reference pattern, $\mathbf{f}_\mathrm{d}$, over some $(K \times L)$ sampling of the elevation and azimuth angular ranges, where $KL > N$. That is $\mathbf{f}_\mathrm{d}$ is available for the KL positions $(\theta_i, \phi_j), i = 1, \ldots, K; j = 1, \ldots, L$. Let $\mathbf{x} \in \mathbb{C}^N$ denote the complex excitation vector. Then the array response can be expressed as,

$$\mathbf{f}_\mathrm{d} = \mathbf{A}\mathbf{x}, \tag{3.2}$$

where $\mathbf{A} \in \mathbb{C}^{KL \times N}$ and

$$\mathbf{A} = [\mathbf{a}(\theta_1, \phi_1), ..., \mathbf{a}(\theta_1, \phi_L), ..., \mathbf{a}(\theta_K, \phi_1), ..., \mathbf{a}(\theta_K, \phi_L)]^T.$$

Now we write the desired reference pattern comprised of two parts, the amplitude and phase, as follows:

$$\mathbf{f}_\mathrm{d} = \mathbf{f}_\mathrm{dM} \odot \mathbf{f}_\mathrm{dP} = \mathbf{F}_\mathrm{dM}\mathbf{f}_\mathrm{dP}, \tag{3.3}$$

where $\odot$ means Hadamard product, and $\mathbf{F}_\mathrm{dM} = \mathrm{diag}(\mathbf{f}_\mathrm{dM})$. The thinned array beam pattern synthesis problem can be formulated as composing a subarray with as few antennas as possible that gives the desired beampattern $\mathbf{f}_\mathrm{d}$, i.e.,

$$\begin{aligned} \min_{\mathbf{x}} \ & \|\mathbf{x}\|_0, \\ \text{s.t.} \ & \mathbf{A}\mathbf{x} = \mathbf{f}_\mathrm{d}. \end{aligned} \tag{3.4}$$

The l_0 norm, $\|\mathbf{x}\|_0$, is defined as the number of nonzero entries of the excitation vector $\mathbf{x}$. So if $x_i = 0$, the ith antenna is discarded; otherwise the nonzero value is taken as the corresponding complex excitation for it. The l_0-norm minimization problem is non-convex and generally very hard to solve, as its solution usually requires an intractable combinatorial search [123]. Therefore, we relax Eq. (3.4) by replacing the l_0 norm with the well-known l_1 norm [123], and replace the equality constraint by an inequality to give the new problem

$$\begin{aligned} \min_{\mathbf{x}} \ & \|\mathbf{x}\|_1, \\ \text{s.t.} \ & \|\mathbf{A}\mathbf{x} - \mathbf{f}_\mathrm{d}\|_2 \le \epsilon, \end{aligned} \tag{3.5}$$

for some fidelity parameter $\epsilon \ge 0$. In order to control the trade-off among the errors in the mainlobe ripple, sidelobe level and null depth, we introduce a weight vector $\mathbf{w}$ (the corresponding weight matrix is the diagonal matrix $\mathbf{W} = \mathrm{diag}(\mathbf{w})$) to reweight the l_2-norm of the error. Then, if we want to maintain a high fidelity (i.e. smaller error) in the synthesised beampattern in some looking directions, we make the entries of $\mathbf{w}$ that correspond to these look directions larger. Equation (3.5) can be re-cast into the following unconstrained weighted least squared form,

$$\min_{\mathbf{x}} \|\mathbf{x}\|_1 + \frac{\beta_2}{2}\|\mathbf{A}\mathbf{x} - \mathbf{f}_{\mathrm{d}}\|_{\mathbf{W}}^2, \tag{3.6}$$

where the matrix norm is defined as $\|\mathbf{A}\mathbf{x} - \mathbf{f}_{\mathrm{d}}\|_{\mathbf{W}}^2 = (\mathbf{A}\mathbf{x} - \mathbf{f}_{\mathrm{d}})^H \mathbf{W}(\mathbf{A}\mathbf{x} - \mathbf{f}_{\mathrm{d}})$. $\beta_2 > 0$ is a regularization parameter that balances between the solution sparsity and the minimization of the synthesized beampattern error. Generally, a smaller β_2 yields a sparser minimizer $\mathbf{x}$, but also a larger error between the synthesised and desired beampatterns.

3.2.1.2 Transformation from the Complex to the Real Domain

The problem presented above is complex. Therefore, we transform it to the real domain in order to implement the optimization algorithms directly. The transformation is defined as follows:

$$\tilde{\mathbf{A}} = \begin{bmatrix} \mathcal{R}(\mathbf{A}) & -\mathcal{I}(\mathbf{A}) \\ \mathcal{I}(\mathbf{A}) & \mathcal{R}(\mathbf{A}) \end{bmatrix}, \tag{3.7}$$

$$\tilde{\mathbf{x}} = [\mathcal{R}(\mathbf{x})^T, \mathcal{I}(\mathbf{x})^T]^T, \quad \tilde{\mathbf{f}_{\mathrm{d}}} = [\mathcal{R}(\mathbf{f}_{\mathrm{d}})^T, \mathcal{I}(\mathbf{f}_{\mathrm{d}})^T]^T, \tag{3.8}$$

$$\tilde{\mathbf{w}} = [\mathbf{w}^T, \mathbf{w}^T]^T, \quad \tilde{\mathbf{W}} = \mathrm{diag}(\tilde{\mathbf{w}}), \tag{3.9}$$

where $\mathcal{R}(\cdot)$ means the real part of $\cdot$ and $\mathcal{I}(\cdot)$ its imaginary part. Using Eqs. (3.7)–(3.9), the optimization problem in Eq. (3.6) becomes

$$\min_{\tilde{\mathbf{x}}} \ \|\tilde{\mathbf{x}}\|_{2,1} + \frac{\beta_2}{2}\|\tilde{\mathbf{A}}\tilde{\mathbf{x}} - \tilde{\mathbf{f}}_{\mathrm{d}}\|_{\tilde{\mathbf{W}}}^2. \tag{3.10}$$

In this formulation the $l_{2,1}$ norm is defined as

$$\|\tilde{\mathbf{x}}\|_{2,1} = \|\mathbf{x}\|_1 = \sum_{i=1}^{N} \sqrt{\mathcal{R}^2(\mathbf{x}(i)) + \mathcal{I}^2(\mathbf{x}(i))} = \sum_{i=1}^{N} \|\tilde{\mathbf{x}}_{g_i}\|_2, \tag{3.11}$$

where

$$\tilde{\mathbf{x}}_{g_i} = [\tilde{\mathbf{x}}(i), \tilde{\mathbf{x}}(i+N)]^T, i = 1, ..., N. \tag{3.12}$$

The weighted l_2 norm, on the other hand, remains the same:

$$\|\tilde{\mathbf{A}}\tilde{\mathbf{x}} - \tilde{\mathbf{f}}_{\mathrm{d}}\|_{\tilde{\mathbf{W}}}^2 = \|\mathbf{A}\mathbf{x} - \mathbf{f}_{\mathrm{d}}\|_{\mathbf{W}}^2. \tag{3.13}$$

3.2.1.3 Setting the Desired Beampattern Phase

In most applications of antenna arrays such as communications and radar [124], we do not have a phase requirement on the directional response of the array. Therefore, we can use the phase of the desired beampattern, $\mathbf{f}_{\mathrm{dP}}$, as an extra free parameter in the optimisation in order to make the synthesized beampattern approach the desired one. Substituting Eq. (3.3) into the objective function in Eq. (3.6) yields,

$$\|\mathbf{x}\|_1 + \frac{\beta_2}{2}\|\mathbf{A}\mathbf{x} - \mathbf{f}_{\mathrm{d}}\|_{\mathbf{W}}^2 = \|\mathbf{x}\|_1 + \frac{\beta_2}{2}\|\mathbf{A}\mathbf{x} - \mathbf{F}_{\mathrm{dM}}\mathbf{f}_{\mathrm{dP}}\|_{\mathbf{W}}^2. \tag{3.14}$$

It is clear that Eq. (3.14) is non-convex with respect to both variables $\mathbf{x}$ and $\mathbf{f}_{\mathrm{dP}}$. We utilise an alternating descent method that iteratively shifts between the two variables $\mathbf{x}$ and $\mathbf{f}_{\mathrm{dP}}$. The l_1 norm in the above equation does not depend on $\mathbf{f}_{\mathrm{dP}}$. Thus, we choose $\mathbf{f}_{\mathrm{dP}}$ that minimises the re-weighted l_2 norm, which reduces the error between the desired and synthesised beampattern. Given the excitation vector $\mathbf{x}$, we take the derivative of Eq. (3.14) with respect to $\mathbf{f}_{\mathrm{dP}}$ and set it to zero to give

$$\mathbf{f}_{\mathrm{dP}}^{(u)} = \mathbf{F}_{\mathrm{dM}}^{-1}\mathbf{A}\mathbf{x}. \tag{3.15}$$

Since the entries of the phase vector $\mathbf{f}_{\mathrm{dP}}$ have unit magnitude, and noting that the ith entry of $\mathbf{A}\mathbf{x}$ is $\mathbf{a}_{r,i}^T\mathbf{x}$, we normalise each entry of $\mathbf{f}_{\mathrm{dP}}^{(u)}$ as follows:

$$\mathbf{f}_{\mathrm{dP}}(i) = \frac{\mathbf{f}_{\mathrm{dP}}^{(u)}(i)}{|\mathbf{f}_{\mathrm{dP}}^{(u)}(i)|} = \frac{\mathbf{a}_{r,i}^T\mathbf{x}}{|\mathbf{a}_{r,i}^T\mathbf{x}|}. \tag{3.16}$$

Eq. (3.16) reveals an interesting insight of the complex beampattern synthesis problem, which should be formulated as the following power pattern synthesis,

$$\min_{\mathbf{x}} \ \|\mathbf{x}\|_1 + \frac{\beta_2}{2}\||\mathbf{A}\mathbf{x}| - \mathbf{f}_{\mathrm{dM}}\|_{\mathbf{W}}^2. \tag{3.17}$$

However, since the function $|\mathbf{A}\mathbf{x}|$ is non-differentiable with respect to the complex vector variable $\mathbf{x}$, it is difficult to minimize Eq. (3.17) directly. Therefore we retain the formulation in Eq. (3.14) and update the desired beampattern phase $\mathbf{f}_{\mathrm{dP}}$ in each iteration. Now let us denote $\mathbf{x}^{(k)}$ and $\mathbf{x}^{(k+1)}$ as solutions of the kth and $(k+1)$th iterations respectively. Then combining Eqs. (3.14) and (3.16) in the $(k+1)$th iteration yields,

$$\|\mathbf{x}^{(k+1)}\|_1 + \frac{\beta_2}{2}\|\mathbf{A}\mathbf{x}^{(k+1)} - \mathbf{F}_{\mathrm{dM}}(\mathbf{A}\mathbf{x}^{(k)} \oslash |\mathbf{A}\mathbf{x}^{(k)}|)\|_{\mathbf{W}}^2. \tag{3.18}$$

Here $\oslash$ denotes the element-wise division as defined in Eq. (3.16). Since the distance between two successive solutions $\mathbf{x}^{(k+1)}$ and $\mathbf{x}^{(k)}$ becomes small with the convergence of the algorithm, we have that $\mathbf{x}_o = \mathbf{x}^{(k)} \simeq \mathbf{x}^{(k+1)}$. Then proceeding from Eq. (3.18) yields

$$\begin{aligned}&\|\mathbf{x}_o\|_1 + \frac{\beta_2}{2}\|\mathbf{A}\mathbf{x}_o - \mathbf{F}_{\mathrm{dM}}(\mathbf{A}\mathbf{x}_o \oslash |\mathbf{A}\mathbf{x}_o|)\|_{\mathbf{W}}^2 \\ &= \|\mathbf{x}_o\|_1 + \frac{\beta_2}{2}\||\mathbf{A}\mathbf{x}_o| - \mathbf{f}_{\mathrm{dM}}\|_{\mathbf{W}}^2,\end{aligned} \tag{3.19}$$

where $\mathbf{A}\mathbf{x}_o = |\mathbf{A}\mathbf{x}_o| \odot (\mathbf{A}\mathbf{x}_o \oslash |\mathbf{A}\mathbf{x}_o|)$. We can see that Eq. (3.19) converges to the power pattern synthesis Eq. (3.17) finally.

3.2.1.4 Setting the Beampattern Weight

The challenge for the least-square optimization is the control of sidelobe levels. Unlike the mainlobe, where the goal is to match the desired reference pattern, the requirement for the sidelobes is to achieve an attenuation greater than or equal to the specification. As shown in [125], the shaped beampattern synthesis with PSL constraints is expressed as:

$$\begin{cases} -\delta \le |\mathbf{A}\mathbf{x}| - \mathbf{f}_{\mathrm{dM}} \le \delta, & \text{for mainlobe region,} \\ |\mathbf{A}\mathbf{x}| - \mathbf{f}_{\mathrm{dM}} \le 0, & \text{for sidelobe and null region,} \end{cases} \tag{3.20}$$

where δ is a fidelity factor of the beampattern error in the mainlobe region. From Eq. (3.20), we can observe that the desired beampattern amplitude $\mathbf{f}_{\mathrm{dM}}$ is actually the "mask" in the sidelobe region, as $\mathbf{f}_{\mathrm{dM}}$ sets the upper bound on the sidelobe level and the lower bound is $-\infty$. However, for the mainlobe region, $\mathbf{f}_{\mathrm{dM}}$ is not a "mask" but the desired reference pattern that should be achieved with a tolerance error δ. Thus the upper and lower bounds for the mainlobe region are $\mathbf{f}_{\mathrm{dM}} + \delta$ and $\mathbf{f}_{\mathrm{dM}} - \delta$ respectively. Therefore, large negative differences between the sidelobes of the synthesised and desired beampatterns satisfy the requirements. The least-square optimization, however, treats both the negative and positive errors in the same way, therefore trying to match the desired sidelobe level rather than bettering it. In order to circumvent this problem, we update the weight vector $\mathbf{w}$ iteratively (or $\tilde{\mathbf{w}}$) using a similar method to [22]. We adjust the weight vector value element-wise in the $(k+1)$th iteration as follows:

Let us denote the ith entry of the error $|\mathbf{A}\mathbf{x}^{(k)}| - \mathbf{f}_{\mathrm{dM}}$ by $\mathbf{e}^{(k)}(i)$. That is $\mathbf{e}^{(k)}(i) = |\mathbf{a}_{r,i}^T\mathbf{x}^{(k)}| - \mathbf{f}_{\mathrm{dM}}(i)$. Then for the mainlobe region,

$$\mathbf{w}^{(k+1)}(i) = \begin{cases} \mathbf{w}^{(k)}(i), & |\mathbf{e}^{(k)}(i)| \le \delta \\ \mathbf{w}^{(k)}(i) + k_m|\mathbf{e}^{(k)}(i)|, & \text{otherwise.} \end{cases} \tag{3.21}$$

For the sidelobe region,

$$\mathbf{w}^{(k+1)}(i) = \max\{k_s\mathbf{e}^{(k)}(i), 0\}, \tag{3.22}$$

and for the null region,

$$\mathbf{w}^{(k+1)}(i) = \max\{k_n \mathbf{e}^{(k)}(i), 0\}, \tag{3.23}$$

where k_m, k_s and k_n are positive gain factors corresponding to the mainlobe, sidelobe and null respectively. Observe that for the mainlobe region, $\mathbf{w}$ never decreases from its initial value. Whereas for both the sidelobe and null, the weight $\mathbf{w}$ is zero if the synthesised beampattern is lower than the desired level.

3.2.1.5 Group Sparse Optimization

In this section, we will apply the Group Sparse Optimization (GSO) method to deal with the mixed $l_{2,1}$-norm regularisation in Eq. (3.10). The GSO approach introduced in [126, 127] is based on a variable splitting strategy and the classic alternating direction method (ADM). The convergence rate of the GSO method is guaranteed by the existing ADM theory.

Now we introduce an auxiliary variable $\mathbf{z}$ to Eq. (3.10), giving

$$\begin{aligned} &\min_{\tilde{\mathbf{x}},\mathbf{z}} \ \|\mathbf{z}\|_{2,1} + \frac{\beta_2}{2} \| \tilde{\mathbf{A}}\tilde{\mathbf{x}} - \tilde{\mathbf{f}}_\mathrm{d} \|^2_{\tilde{\mathbf{W}}}, \\ &\text{s.t.} \ \ \mathbf{z} = \tilde{\mathbf{x}}. \end{aligned} \tag{3.24}$$

The augmented Lagrangian problem is of the form,

$$\min_{\tilde{\mathbf{x}},\mathbf{z}} \|\mathbf{z}\|_{2,1} - \boldsymbol{\lambda}^T(\mathbf{z} - \tilde{\mathbf{x}}) + \frac{\beta_1}{2}\|\mathbf{z} - \tilde{\mathbf{x}}\|_2^2 + \frac{\beta_2}{2}\|\tilde{\mathbf{A}}\tilde{\mathbf{x}} - \tilde{\mathbf{f}}_\mathrm{d}\|^2_{\tilde{\mathbf{W}}}, \tag{3.25}$$

where $\boldsymbol{\lambda}$ is a multiplier and β_1 is a penalty parameter. Using the ADM approach, we minimize the augmented Lagrangian in Eq.(3.25) with respect to $\tilde{\mathbf{x}}$ and $\mathbf{z}$ alternatively. Minimizing with respect to $\tilde{\mathbf{x}}$ yields,

$$\tilde{\mathbf{x}} = (\beta_1 \mathbf{I} + \beta_2 \tilde{\mathbf{A}}^T \tilde{\mathbf{W}} \tilde{\mathbf{A}})^{-1} (\beta_1 \mathbf{z} - \boldsymbol{\lambda} + \beta_2 \tilde{\mathbf{A}}^T \tilde{\mathbf{W}} \tilde{\mathbf{f}}_\mathrm{d}). \tag{3.26}$$

When the matrix $\tilde{\mathbf{A}}$ has more columns than rows, we can reduce the computational complexity of the matrix inversion by using the Sherman-Morrison-Woodbury formula:

$$(\beta_1 \mathbf{I} + \beta_2 \tilde{\mathbf{A}}^T \tilde{\mathbf{W}} \tilde{\mathbf{A}})^{-1} = \frac{1}{\beta_1}\mathbf{I} - \frac{\beta_2}{\beta_1}\tilde{\mathbf{A}}^T(\beta_1 \tilde{\mathbf{W}}^{-1} + \beta_2 \tilde{\mathbf{A}}\tilde{\mathbf{A}}^T)^{-1}\tilde{\mathbf{A}}. \tag{3.27}$$

Now minimizing over $\mathbf{z}$ yields the following problem

$$\min_{\mathbf{z}} \|\mathbf{z}\|_{2,1} - \boldsymbol{\lambda}^T \mathbf{z} + \frac{\beta_1}{2}\|\mathbf{z} - \tilde{\mathbf{x}}\|_2^2. \tag{3.28}$$

By some simple manipulations, the problem in Eq. (3.28) is equivalent to

$$\min_{\mathbf{z}} \sum_{i=1}^{N} \left[\|\mathbf{z}_{g_i}\|_2 + \frac{\beta_1}{2} \|\mathbf{z}_{g_i} - \tilde{\mathbf{x}}_{g_i} - \frac{1}{\beta_1} \boldsymbol{\lambda}_{g_i}\|_2^2 \right], \tag{3.29}$$

where similar to the definition of $\tilde{\mathbf{x}}_{g_i}$ in Eq. (3.12), we have that

$$\boldsymbol{\lambda}_{g_i} = [\boldsymbol{\lambda}(i), \boldsymbol{\lambda}(i+N)]^T, i = 1, ..., N, \tag{3.30}$$

and

$$\mathbf{z}_{g_i} = [\mathbf{z}(i), \mathbf{z}(i+N)]^T, i = 1, ..., N. \tag{3.31}$$

Equation (3.29) has a closed form solution using the vector-form soft-thresholding formula:

$$\mathbf{z}_{g_i} = \max\{\|\mathbf{r}_i\|_2 - \frac{1}{\beta_1}, 0\} \frac{\mathbf{r}_i}{\|\mathbf{r}_i\|_2}, \tag{3.32}$$

where

$$\mathbf{r}_i = \tilde{\mathbf{x}}_{g_i} + \frac{1}{\beta_1} \boldsymbol{\lambda}_{g_i}. \tag{3.33}$$

For brevity of notation, we refer to the above group-wise soft-thresholding operator by $\mathbf{z} = \text{Shrink}(\tilde{\mathbf{x}} + \frac{1}{\beta_1}\boldsymbol{\lambda}, \frac{1}{\beta_1})$. Finally, the multiplier $\boldsymbol{\lambda}$ is updated in the standard way:

$$\boldsymbol{\lambda} = \boldsymbol{\lambda} - \gamma\beta_1(\mathbf{z} - \tilde{\mathbf{x}}), \tag{3.34}$$

where γ is the step length. In this method, the preference of some specific antennas or the preservation of maximum aperture length can be realised by changing $\|\mathbf{z}\|_{2,1}$ to $\|\mathbf{z}\|_{\mathbf{t},2,1}$ in Eq. (3.24), where $\mathbf{t}(i)$ is the weight imposed on the ith antenna and $\|\mathbf{z}\|_{\mathbf{t},2,1} = \sum_{i=1}^{N} \mathbf{t}(i)\|\mathbf{z}_{g_i}\|_2$. Then the group-wise soft-thresholding operator becomes $\mathbf{z} = \text{Shrink}(\tilde{\mathbf{x}} + \frac{1}{\beta_1}\boldsymbol{\lambda}, \frac{\mathbf{t}}{\beta_1})$.

Now we summarise the implementation procedure of the GSO algorithm as follows:

- **Initialisation**: Set the initial desired beampattern phase to zero, i.e. $\mathbf{f}_{\text{dP}} = \mathbf{1}$; Initialise $\mathbf{z}^{(0)} = \tilde{\mathbf{x}}^{(0)} = \mathbf{0}, \boldsymbol{\lambda}^{(0)} = \mathbf{0}, \gamma = \left(\sqrt{5}+1\right)/2, \beta_1 = 10/\text{mean}(\mathbf{f}_{\text{dM}})$ and $\beta_2 = 1/\text{mean}(\mathbf{f}_{\text{dM}})$; Initialise k_m, k_s, k_n - for example set $k_m = 40, k_s = 4000, k_n = 10^5$;
- **Outer Iteration**: *while* the desired beampattern has not been achieved *do*

  1. **Inner Iteration:** *while* $\|\tilde{\mathbf{x}}^{(k+1)} - \tilde{\mathbf{x}}^{(k)}\|_2 > \nu$ *do*

     a. Transform the problem from the complex to the real domain according to Eqs. (3.7)–(3.9);
     b. $\mathbf{z}^{(k+1)} = \text{Shrink}(\tilde{\mathbf{x}}^{(k)} + \frac{1}{\beta_1}\boldsymbol{\lambda}^{(k)}, \frac{1}{\beta_1})$;
     c. $\tilde{\mathbf{x}}^{(k+1)} = (\beta_1\mathbf{I} + \beta_2\tilde{\mathbf{A}}^T\tilde{\mathbf{W}}\tilde{\mathbf{A}})^{-1}(\beta_1\mathbf{z}^{(k+1)} - \boldsymbol{\lambda}^{(k)} + \beta_2\tilde{\mathbf{A}}^T\tilde{\mathbf{W}}\tilde{\mathbf{f}}_{\text{d}})$;

d. $\boldsymbol{\lambda}^{(k+1)} = \boldsymbol{\lambda}^{(k)} - \gamma_1 \beta_1 (\mathbf{z}^{(k+1)} - \tilde{\mathbf{x}}^{(k+1)})$;

e. Transform the real solution $\tilde{\mathbf{x}}^{(k+1)}$ back to a complex solution $\mathbf{x}^{(k+1)}$ by $\mathbf{x}^{(k+1)}(i) = \tilde{\mathbf{x}}^{(k+1)}(i) + j\tilde{\mathbf{x}}^{(k+1)}(i+N)$. Update the desired beampattern phase $\mathbf{f}_{\mathrm{dP}}$ using Eq. (3.16). Go back to the start of the Inner Iteration.
End Inner Iteration.

2. Update the weight vector **w** according to Eqs. (3.21)–(3.23); Go back to Outer Iteration.
End Outer Iteration.

In the above procedure, the inner iterations are stopped when two consective solutions are close enough to each other. This is controlled by setting ν to the desired tolerance.

3.2.1.6 Fast Iterative Soft-Thresholding Algorithm

In this method, we relax the regularizer $\|\tilde{\mathbf{x}}\|_{2,1}$ to $\|\tilde{\mathbf{x}}\|_1$. Intuitively, this relaxation decouples the real and imaginary parts from each other and one would expect it not to be able to synthesize very sparse arrays with respect to the coupled case. In the extreme, the array weights can be either purely real or purely imaginary, which leads to small $\|\tilde{\mathbf{x}}\|_0$ norm (half the original value) but an entirely non-sparse array [128]. However, as we will show mathematically, the soft-thresholding operator prevents such problems from occurring. We will in fact elucidate the relationship between the two.

Let us rewrite the relaxed version of the optimisation problem in Eq. (3.10) as follows,

$$\min_{\tilde{\mathbf{x}}} \ \|\tilde{\mathbf{x}}\|_1 + \frac{\beta_2}{2} \parallel \tilde{\mathbf{A}}\tilde{\mathbf{x}} - \tilde{\mathbf{f}}_{\mathrm{d}} \parallel_{\tilde{\mathbf{W}}}^2 . \tag{3.35}$$

Using the function $G(\tilde{\mathbf{x}})$ to represent the second part of Eq. (3.35), the gradient of $G(\tilde{\mathbf{x}})$ denoted by $\nabla G(\tilde{\mathbf{x}})$ is

$$\nabla G(\tilde{\mathbf{x}}) = \tilde{\mathbf{A}}^H \tilde{\mathbf{W}}(\tilde{\mathbf{A}}\tilde{\mathbf{x}} - \tilde{\mathbf{f}}_{\mathrm{d}}). \tag{3.36}$$

It is well known that the minimizer of Eq. (3.35) can be obtained by an iterative soft-thresholding algorithm [129, 130]. That is

$$\tilde{\mathbf{x}}^{(k+1)} = \operatorname*{argmin}_{\tilde{\mathbf{x}}} \|\tilde{\mathbf{x}}\|_1 + \frac{\tau\beta_2}{2} \|\tilde{\mathbf{x}} - (\tilde{\mathbf{x}}^{(k)} - \frac{1}{\tau}\nabla G(\tilde{\mathbf{x}}^{(k)}))\|_2^2, \tag{3.37}$$

with $\tau \geq \lambda_{\max}(\tilde{\mathbf{A}}^H \tilde{\mathbf{W}} \tilde{\mathbf{A}})$. Then the unique minimizer of Eq. (3.37) is

$$\tilde{\mathbf{x}}^{(k+1)} = \mathrm{Shrink}\left(\tilde{\mathbf{x}}^{(k)} - \frac{1}{\tau}\nabla G(\tilde{\mathbf{x}}^{(k)}), \frac{1}{\tau\beta_2}\right). \tag{3.38}$$

Similarly to the vector-form soft-thresholding operator in Eq. (3.32), the scalar soft-thresholding operator is defined as,

$$\text{Shrink}\left(\cdot, \frac{1}{\tau\beta_2}\right) = \text{sign}(\cdot)\max\{|\cdot| - \frac{1}{\tau\beta_2}, 0\}. \tag{3.39}$$

It is well-known that the array aperture length is a crucial factor for the mainlobe width. In some applications, there are stringent requirements on the array resolution and mainlobe width. Here we propose a modification that allows us to preserve the maximum array aperture length, namely we replace the constant scalar β_2, by a constant vector $\boldsymbol{\beta}_2 \in \mathbb{R}^N$. Thus a larger entry of $\boldsymbol{\beta}_2$ implies lower threshold, and a higher likelihood of selecting the corresponding antenna. So for example, if we hope to preserve the full array aperture length, we can assign large values to the entries of $\boldsymbol{\beta}_2$ corresponding to the boundary antennas.

It has been shown in [130] that the convergence rate of the traditional soft-thresholding algorithm of Eq. (3.38) is linear, which is not fast enough for reconfigurable array applications. In order to accelerate the convergence speed, a fast iterative soft-thresholding algorithm (FISTA) that uses a linear combination of the previous two points was proposed in [129]. This method can achieve a quadratic convergence rate. Starting with $\mathbf{y}^{(1)} = \tilde{\mathbf{x}}^{(0)}$ and $t_1 = 1$, we make at the kth iteration the following calculations

$$\tilde{\mathbf{x}}^{(k)} = \text{Shrink}\left(\mathbf{y}^{(k)} - \frac{1}{\tau}\nabla G(\mathbf{y}^{(k)}), \frac{1}{\tau\beta_2}\right), \tag{3.40}$$

$$t^{(k+1)} = \frac{1 + \sqrt{1 + 4(t^{(k)})^2}}{2}, \tag{3.41}$$

$$\mathbf{y}^{(k+1)} = \tilde{\mathbf{x}}^{(k)} + \left(\frac{t^{(k)} - 1}{t^{(k+1)}}\right)\left(\tilde{\mathbf{x}}^{(k)} - \tilde{\mathbf{x}}^{(k-1)}\right). \tag{3.42}$$

It is desirable to design thinned arrays having no more elements than necessary. For this reason, the continuation strategy proposed in [130] is adopted here to adjust the value of β_2 for automatic determination of the required number of antennas during the design process. Now we give the detailed implementation procedure of the FISTA as follows:

- **Initialisation:** Initialise the desired beampattern phase to zero, i.e. $\mathbf{f}_{\text{dP}} = \mathbf{1}$. Set $\tilde{\mathbf{x}}^{(0)} = \mathbf{0}$. Initialise the trade-off parameter β_2 according to the sparsity requirement, and set the values for k_m, k_s, k_n.
- **Outer Iteration:** *while* the desired beampattern has not been achieved *do*

  1. **Inner Iteration:** *while* $\|\tilde{\mathbf{x}}^{(k+1)} - \tilde{\mathbf{x}}^{(k)}\|_2 > \nu$ *do*

     a. Transform the problem from the complex to the real domain according to Eqs. (3.7)–(3.9);
     b. Obtain $\tilde{\mathbf{x}}^{(k)}$ using Eqs. (3.40)–(3.42);

c. Transform the real solution $\tilde{\mathbf{x}}^{(k+1)}$ back to a complex solution $\mathbf{x}^{(k+1)}$ by $\mathbf{x}^{(k+1)}(i) = \tilde{\mathbf{x}}^{(k+1)}(i) + j\tilde{\mathbf{x}}^{(k+1)}(i+N)$. Update the desired beampattern phase $\mathbf{f}_{\mathrm{dP}}$ using Eq. (3.16). Go back to the start of the Inner Iteration;
End Inner Iteration.

2. Update the weight vector $\mathbf{w}$ according to Eqs. (3.21)–(3.23);
3. Adjust the value of β_2 utilising the continuation strategy outlined below; Go back to the Outer Iteration.
End Outer Iteration.

Continuation Strategy:

Let $N_a^{(i)}$, $\beta_2^{(i)}$ and $\Delta\beta_2^{(i)}$ denote respectively the number of selected antennas, the value of β_2 and the step size at the ith iteration. While the desired beampattern has not been reached, we adjust the value of β_2 in every iteration as follows:

1. If $N_a^{(i-1)} < K$ then $\beta_2^{(i)} = \beta_2^{(i-1)} + \Delta\beta_2^{(i)}$, where

   a. if $\beta_2^{(i-1)} > \beta_2^{(i-2)}$ (that is β_2 was increased in the previous iteration), then $\Delta\beta_2^{(i)} = \Delta\beta_2^{(i-1)}$;
   b. if $\beta_2^{(i-1)} < \beta_2^{(i-2)}$ (that is β_2 was decreased in the previous iteration), then $\Delta\beta_2^{(i)} = 0.5\Delta\beta_2^{(i-1)}$;

2. If $N_a^{(i-1)} > K$ then $\beta_2^{(i)} = \beta_2^{(i-1)} - \Delta\beta_2^{(i)}$ where

   a. if $\beta_2^{(i-1)} > \beta_2^{(i-2)}$ (that is β_2 was increased in the previous iteration), then $\Delta\beta_2^{(i)} = 0.5\Delta\beta_2^{(i-1)}$;
   b. if $\beta_2^{(i-1)} < \beta_2^{(i-2)}$ (that is β_2 was decreased in the previous iteration), then $\Delta\beta_2^{(i)} = \Delta\beta_2^{(i-1)}$.

Here K is the number of front-ends installed in the receiver. In our work, we have found that setting the initial step size, $\Delta\beta_2$ equal to the initial value of β_2, works quite well.

3.2.1.7 Sparsity Performance of the FISTA

As mentioned above, the decoupling of the real and imaginary parts from each other might be expected to produce a less sparse solution by suppressing only either the real or imaginary parts to zero. In this section, we give a simple proof of the optimality of Eq. (3.38) that gives insight into the comparable sparsity that is achieved by the FISTA and GSO approaches. Let us denote the optimum solutions obtained from the FISTA and the GSO to be $\tilde{\mathbf{x}}_{oF}$ and $\tilde{\mathbf{x}}_{oG}$ respectively. We define a mapping $\mathcal{M} : \mathbb{R}^{2N} \rightarrow \mathbb{R}_+^N$ as

$$\mathcal{M}(\tilde{\mathbf{x}}_{oT})(i) = \max\{|\tilde{\mathbf{x}}_{oT}(i)|, |\tilde{\mathbf{x}}_{oT}(i+N)|\}, \ T = F, G. \tag{3.43}$$

That is the mapping which selects the maximum absolute value between the real and imaginary components of the corresponding excitation vector element for each antenna. Then the number of selected antennas for each algorithm can be expressed as the l_0-norm of $\mathcal{M}(\tilde{\mathbf{x}}_{oT})$.

Theorem: The number of selected antennae of the FISTA is always less than or equal to that of the GSO for the same threshold value, i.e.

$$\|\mathcal{M}(\tilde{\mathbf{x}}_{oF})\|_0 \leq \|\mathcal{M}(\tilde{\mathbf{x}}_{oG})\|_0. \tag{3.44}$$

Proof: For simplicity, we rewrite the objective function of Eq. (3.37) as the following form,

$$\|\tilde{\mathbf{x}}\|_1 + \frac{\tau\beta_2}{2}\|\tilde{\mathbf{x}} - \mathbf{y}\|_2^2, \tag{3.45}$$

with $\mathbf{y} = \tilde{\mathbf{x}}^{(k)} - \frac{1}{\tau}\nabla G(\tilde{\mathbf{x}}^{(k)})$. It is well known that the condition for $\tilde{\mathbf{x}}_{oT}$ being one of the optimal solutions of Eq. (3.45) is

$$\mathbf{0} \in \mathrm{SIGN}(\tilde{\mathbf{x}}_{oT}) + \tau\beta_2(\tilde{\mathbf{x}}_{oT} - \mathbf{y}), \tag{3.46}$$

where $\mathbf{0}$ is the zero vector in $\mathbb{R}^{2N}$. The function $\mathrm{SIGN}(\tilde{\mathbf{x}}_{oT})$ is the subgradient function of the l_1-norm $\|\tilde{\mathbf{x}}_{oT}\|_1$, where $\tilde{\mathbf{x}}_{oT} \in \mathbb{R}^{2N}$. It is defined elementwise with respect to each entry of the vector $\tilde{\mathbf{x}}_{oT}$:

$$\mathrm{SIGN}(\tilde{\mathbf{x}}_{oT}(i)) \begin{cases} = 1 & \tilde{\mathbf{x}}_{oT}(i) > 0, \\ \in [-1, 1] & \tilde{\mathbf{x}}_{oT}(i) = 0, \\ = -1 & \tilde{\mathbf{x}}_{oT}(i) < 0. \end{cases} \tag{3.47}$$

Substituting Eq. (3.47) into Eq. (3.46) yields

$$\tau\beta_2(\tilde{\mathbf{x}}_{oT}(i) - \mathbf{y}(i)) \begin{cases} = -1 & \tilde{\mathbf{x}}_{oT}(i) > 0, \\ \in [-1, 1] & \tilde{\mathbf{x}}_{oT}(i) = 0, \\ = 1 & \tilde{\mathbf{x}}_{oT}(i) < 0. \end{cases} \tag{3.48}$$

Therefore we have that

$$\tilde{\mathbf{x}}_{oT}(i) = \begin{cases} \mathbf{y}(i) - \frac{1}{\tau\beta_2} & \mathbf{y}(i) > \frac{1}{\tau\beta_2}, \\ 0 & \mathbf{y}(i) \in [-\frac{1}{\tau\beta_2}, \frac{1}{\tau\beta_2}], \\ \mathbf{y}(i) + \frac{1}{\tau\beta_2} & \mathbf{y}(i) < -\frac{1}{\tau\beta_2}. \end{cases} \tag{3.49}$$

which is the same as Eq. (3.38). We can observe that the soft-thresholding operator works by suppressing the entries within the region bounded by the positive and negative thresholds, $[-\frac{1}{\tau\beta_2}, \frac{1}{\tau\beta_2}]$, to zero and using a linear mapping for the other

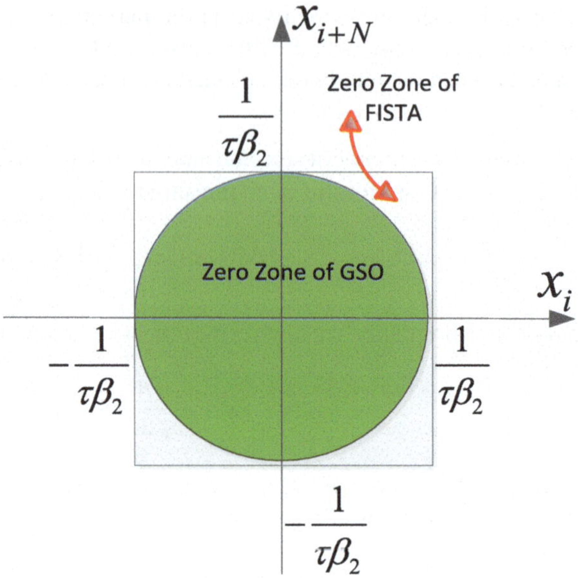

Fig. 3.2 The suppression to zero (nulling) regions for the FISTA and the GSO under the same threshold. The FISTA has a square nulling region whereas that of the GSO is a circle internally tangent to the square

areas, i.e. the magnitude of $\mathbf{y}(i)$ is reduced by an amount equal to $\frac{1}{\tau\beta_2}$. Thus, as shown in Fig. 3.2, the two-dimensional nulling region for the FISTA is a square with sides of length $\frac{2}{\tau\beta_2}$. The GSO algorithm, on the other hand, compares each pair of $\mathbf{y}$, i.e. $\sqrt{\mathbf{y}(i)^2 + \mathbf{y}(i+N)^2}$ to the threshold. Consequently, its two-dimensional nulling region is a circle with radius $\frac{1}{\tau\beta_2}$. Clearly, $\sqrt{\mathbf{y}(i)^2 + \mathbf{y}(i+N)^2} \in [0, \frac{1}{\tau\beta_2}]$ implies $|\mathbf{y}(i)| \in [0, \frac{1}{\tau\beta_2}]$ and $|\mathbf{y}(i+N)| \in [0, \frac{1}{\tau\beta_2}]$, whereas the converse does not hold (put another way, the nulling region of the GSO is circumscribed by that of the FISTA). This means that an antenna that is nulled by the GSO will also be discarded by FISTA, but an antenna discarded by FISTA may be kept by the GSO algorithm. Therefore, for the same threshold value $\frac{1}{\tau\beta_2}$, $\|\mathcal{M}(\tilde{\mathbf{x}}_{oF})\|_0 \leq \|\mathcal{M}(\tilde{\mathbf{x}}_{oG})\|_0$. This shows that the FISTA can generate a sparser solution than the GSO and prevent the purely real or imaginary solution from happening, in contrast to the Baysian Inference Solver of [128] that needs to utilise the multi-task strategy to address this problem. □

3.2.1.8 Amplitude Sparse Optimization

In this section, we propose another method, called Amplitude Sparse Optimization (ASO), which imposes the sparsity constraint on the amplitude of the excitation vector $\mathbf{x}$. We decompose the complex excitation $\mathbf{x}$ into two parts: the amplitude and the phase, i.e. $\mathbf{x} = \mathbf{X}_a\mathbf{x}_p = \mathbf{X}_p\mathbf{x}_a$, where $\mathbf{X}_a = \mathrm{diag}(\mathbf{x}_a)$ and $\mathbf{X}_p = \mathrm{diag}(\mathbf{x}_p)$. Now the problem is formulated as

$$\min_{\mathbf{x}_a, \mathbf{x}_p} \mathbf{1}^T \mathbf{x}_a + \frac{\beta_2}{2} \|\mathbf{A}\mathbf{X}_a \mathbf{x}_p - \mathbf{f}_d\|_{\mathbf{W}}^2, \tag{3.50}$$

where $\mathbf{1}$ is the vector with all entries 1. In this method, we assume the desired beampattern $\mathbf{f}_d$ is real. Firstly, we minimize the objective function in Eq. (3.50) with respect to $\mathbf{x}_p$. Setting the first derivative with respect to $\mathbf{x}_p^H$ to zero yields

$$\mathbf{x}_p = \mathbf{X}_a^{-1} (\mathbf{A}^H \mathbf{W} \mathbf{A})^{-1} (\mathbf{A}^H \mathbf{W}) \mathbf{f}_d. \tag{3.51}$$

After normalisation, we obtain

$$\mathbf{x}_p = [(\mathbf{A}^H \mathbf{W} \mathbf{A})^{-1} (\mathbf{A}^H \mathbf{W}) \mathbf{f}_d] \oslash |(\mathbf{A}^H \mathbf{W} \mathbf{A})^{-1} (\mathbf{A}^H \mathbf{W}) \mathbf{f}_d|. \tag{3.52}$$

Next we minimize the objective function in Eq. (3.50) with respect to $\mathbf{x}_a$, which can be rewritten as,

$$\begin{aligned} &\mathbf{1}^T \mathbf{x}_a + \frac{\beta_2}{2} \|\mathbf{A}\mathbf{X}_p \mathbf{x}_a - \mathbf{f}_d\|_{\mathbf{W}}^2 \\ &= \mathbf{1}^T \mathbf{x}_a + \frac{\beta_2}{2} (\mathbf{A}\mathbf{X}_p \mathbf{x}_a - \mathbf{f}_d)^H \mathbf{W} (\mathbf{A}\mathbf{X}_p \mathbf{x}_a - \mathbf{f}_d) \\ &= \mathbf{1}^T \mathbf{x}_a + \frac{\beta_2}{2} \mathbf{x}_a^T \mathcal{R}\{\mathbf{X}_p^H \mathbf{A}^H \mathbf{W} \mathbf{A} \mathbf{X}_p\} \mathbf{x}_a - \beta_2 \mathbf{x}_a^T \mathcal{R}\{\mathbf{X}_p^H \mathbf{A}^H \mathbf{W} \mathbf{f}_d\} + \frac{\beta_2}{2} \mathbf{f}_d^T \mathbf{W} \mathbf{f}_d. \end{aligned} \tag{3.53}$$

Setting the derivative with respect to $\mathbf{x}_a$ to zero and noting that $\mathbf{x}_a \geq 0$, we find that

$$\mathbf{x}_a = \max \left\{ \mathcal{R}\{\mathbf{X}_p^H \mathbf{A}^H \mathbf{W} \mathbf{A} \mathbf{X}_p\}^{-1} \left[\mathcal{R}\{\mathbf{X}_p^H \mathbf{A}^H \mathbf{W} \mathbf{f}_d\} - \frac{1}{\beta_2} \mathbf{1} \right], \mathbf{0} \right\}. \tag{3.54}$$

Now we summarise the implementation procedure of the proposed ASO method as follows:

- **Initialisation:** Initialise $\mathbf{w}(i) = 1/\mathbf{f}_d(i)$, the trade-off parameter β_2 as well as k_m, k_s, k_n.
- **Loop:** *while* the desired beampattern has not been achieved *do*

  1. Calculate the phase of the excitation vector $\mathbf{x}$ according to Eq. (3.52);
  2. Calculate the amplitude of the excitation vector $\mathbf{x}$ according to Eq. (3.54);
  3. Obtain the complex excitation vector by $\mathbf{x} = \mathbf{X}_a \mathbf{x}_p$;
  4. Update the weight $\mathbf{w}$ according to Eqs. (3.21)–(3.23);
  5. According to the sparsity of selected subarray, adjust β_2 utilising the continuation strategy;
  6. go back to Step 1) if stopping criterion is not met, otherwise terminate.

  End Loop.

In this method, the preservation of the maximum aperture length can be realised by changing $\mathbf{1}^T \mathbf{x}_a$ to $\mathbf{t}^T \mathbf{x}_a$, where smaller entries of $\mathbf{t}$ imply that the corresponding antennas have a higher likelihood of being selected. It should be noted that the assumption of real beampattern in the ASO algorithm results in a slightly worse performance compared to the other two methods, a fact that is borne by the simulation results.

Now we will summarise the merits and disadvantages of the proposed three methods for beampattern synthesis: (1) It is hard for the ASO algorithm to synthesize large antenna arrays due to the matrix inversion involved in the formulas. Moreover, since $\mathbf{A}^H\mathbf{W}\mathbf{A}$ may become singular due to the high coherency of the steering matrix $\mathbf{A}$, the ASO algorithm may not work under some scenarios. However, the ASO algorithm converges very fast when synthesizing small antenna arrays due to its simple iteration structure. (2) Because of the transformation from the group sparsity promoting variable $\mathbf{z}$ to $\tilde{\mathbf{x}}$ in Eq. (3.26), the GSO algorithm focuses on the least squared error more than the solution sparsity. Thus it is good at controlling the mainlobe ripple but produces less sparse arrays especially when the least squared solution does not have an approximate sparse structure. (3) The FISTA method can be used to synthesize large arrays with sparse solutions, but with more computational time compared to the GSO and ASO methods.

3.2.2 Simulations

In this section, we present extensive numerical results to validate the effectiveness and reliability of the proposed methods. We give a number of examples based on representative reference patterns, including both focused and shaped beampatterns, as well as arbitrary initial array layouts. We also compare their performance to the reweighted l_1-norm method described in [35] and the iterative FFT algorithm in [29]. The computational time, number of selected antennas, Dynamic Range Ratio (DRR) and performance of these examples are summarised in Table 3.2. Since we are aiming at synthesizing receiver array beampattern, we do not place any explicit constraints on the excitation weights. The computer used for simulation has an Intel-i5 CPU and 8GB RAM.

3.2.2.1 Focused Beampattern Synthesis

(a) Example 1

In the first example, we synthesize a nonsymmetric focused beampattern as shown in [125]. The mainlobe width is $|u| = 0.12$ with the left and right sidelobe level being -20dB and -30dB respectively. The synthesized beampattern is shown in Fig. 3.3. The 22-antenna positions are the same as those in [125]. The synthesis speeds of both the FISTA and the GSO algorithms are an order of 10 faster than the l_1-norm method, while the ASO is an order of 100 faster. Both the FISTA and the GSO algorithms can

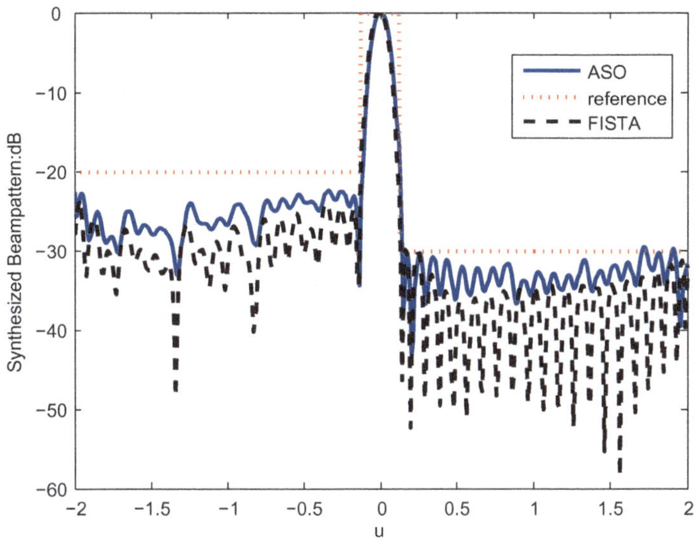

Fig. 3.3 Focused beampattern of Example 1 with mainlobe width $|u| = 0.12$, sidelobel level -20dB and -30dB. The beampattern of the GSO is similar to the FISTA and is omitted here for clarity

maintain a -24dB left sidelobe level with preserved array directivity. The complex antenna excitations for the FISTA are shown in Table 3.1.

(b) Example 2

In the second example, We compare our method with iterative FFT based algorithm [29] for large array synthesis problems that CVX cannot handle. The array aperture length is 180λ and [29] has 326 randomly placed antennas with inter-element spacing greater than $\lambda/4$. Since arbitrarily placed antennas are not practical in real applications, we apply array thinning to a 721-antenna uniform linear array with inter-element spacing $\lambda/4$. The 3dB mainlobe width and sidelobe level given in [29] are $[-0.36°, 0.36°]$ and -24.6dB respectively. Since there is a stringent requirement on the mainlobe width, the maximum aperture length should be used. Therefore, instead of using the scalar trade-off parameter β_2 in Eq. (3.6), we employ a vector $\boldsymbol{\beta}_2$ with the first and last entries set much larger than the others (with values 100 versus 0.05). The beampattern that is obtained from the proposed FISTA method is shown in Fig. 3.4. Neither the ASO nor the GSO method work well in this example, where the least squared solution does not have an approximate sparse structure. We can see that the sidelobe level of the FISTA method is -26dB and the mainlobe width $[-0.34°, 0.34°]$. Both of these are better than the results obtained by the FFT method. The cost of this, however, is that the computational time of our method is 16.18 seconds, which is slower than the 6 seconds reported in [29] with an Intel i5-Core and 4GB RAM. The computational time of the FFT method may be less than 6 seconds if run on our computer, but it is important to point out that their method is

Table 3.1 Complex Excitation Weights Obtained by the FISTA method for example 1

Element No.	Position (λ)	Amplitude	Phase (rad)
1	0	0.3479	0.6421
2	0.34	0.4089	1.4605
3	0.80	0.4486	1.4876
4	1.27	0.5137	1.454
5	1.74	0.6071	1.4385
6	2.20	0.7003	1.4961
7	2.67	0.7817	1.4876
8	3.14	0.8717	1.4774
9	3.60	0.9415	1.5365
10	4.07	0.9814	1.5291
11	4.53	1	1.5824
12	5.00	0.9796	1.5777
13	5.47	0.9617	1.5708
14	5.93	0.9126	1.6332
15	6.40	0.8382	1.6277
16	6.86	0.7461	1.6925
17	7.33	0.6376	1.6884
18	7.80	0.5429	1.6918
19	8.26	0.4457	1.7492
20	8.73	0.35	1.729
21	9.20	0.264	1.6907
22	9.66	0.2204	1.9198

fast only when best values of the parameters are chosen. The search for right parameters requires many trials which will dramatically increase the computational load. The number of selected antennas is 325 for the FISTA, one antenna fewer than the arbitrarily placed antennas in [29].

3.2.2.2 Shaped Beampattern Synthesis

(a) Example 3

In the third example, we synthesise a non-uniform 41-antenna linear array as was done in [35]. This is a symmetrical array with an aperture length of 20 wavelengths. The antenna positions are shown in [25]. The synthesized beampatterns are shown in Fig. 3.5. The mainlobe width and ripple are 40° and −0.4455dB respectively and the sidelobe level is −30dB in [35] with 31 selected antennas. The ASO gives the worst performance with −0.5dB mainlobe ripple and 35 selected antennas. The GSO

Table 3.2 The computational time, sparsity, DRR and performance of the FISTA, the GSO, the ASO and the reweighted l_1-norm method

	Focused beampattern						
Exam no.	Method name	Time (sec)	Antenna number	Main (u/deg)	Side (dB)	Null (dB)	DRR 1
1	l_1-norm	3.05	22	±0.12 [2]	−30, −20	–	no [1]
	FISTA	0.57	22	±0.12 [2]	−30, −24	–	4.54
	ASO	0.013	22	±0.12 [2]	−30, −22.5	–	8.64
	GSO	0.17	22	±0.12 [2]	−30, −24	–	4.63
2	l_1-norm	–	–	–	–	–	–
	FFT [3]	6.1	326	±0.36°	−24.6	-	no [1]
	FISTA	16.18	325	±0.34°	-26	-	5.43
	ASO	–	–	–	–	–	–
	GSO	–	–	–	–	–	–
	Shaped Beampattern						
Exam No.	Method Name	Time (sec)	Antenna Number	Main (dB)	Side (dB)	Null (dB)	DRR 1
3	l_1-norm	2.08	31	−0.4455	−30	–	61.35
	FISTA	0.89	24	−0.3845	−30	–	39.9
	ASO	0.023	35	−0.5	−30	–	119.05
	GSO	1.32	26	−0.4	−31.23	–	81.18
4	l_1-norm	2.05	37	−0.91	−40	−60	140.85
	FISTA	1.1	37	−0.309	−40.8	−63.32	157.14
	ASO	0.01	37	−0.3448	−42.61	−58.12	184.62
	GSO	0.043	37	−0.145	−42.4	−61.67	157.14
5	l_1-norm	2.28	39	−0.8965	−30	–	68.03
	FISTA	1.33	35	-0.8938	−30	–	35.46
	ASO	0.01	37	−1.972	−30	–	50.76
	GSO	1.43	35	−0.597	−30	–	34.36
6	l_1-norm	146.72	76	−1.69	−25.85	–	34.22
	FISTA	119.05	76	−1.72	−26.68	–	11.45
	ASO	–	–	–	–	-	–
	GSO	112.48	165	−0.7751	−23.32	–	19.92

[1] Papers [29, 125] do not report the DRR value;
[2] Paper [125] and its reference papers define the mainlobe width in terms of $u = \sin\theta$, i.e. $|u| = 0.12$;
[3] Simulation results for the iterative FFT algorithm are only given for example 2

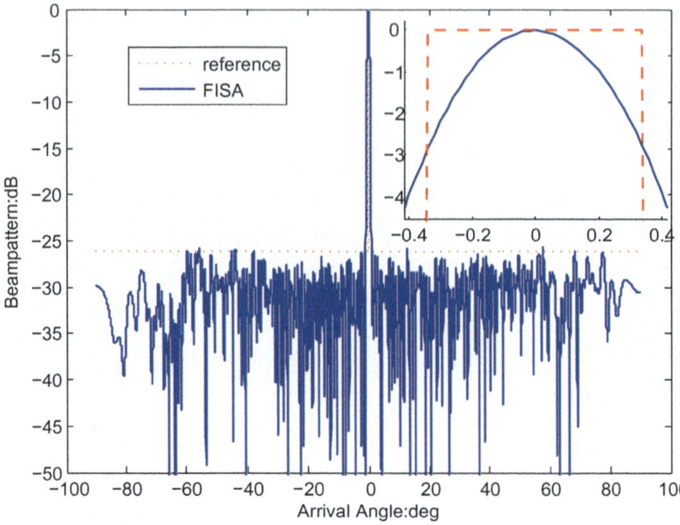

Fig. 3.4 Focused Beampattern of Example 2: the sidelobe level is −26dB, the mainlobe width is [−0.34, 0.34] degrees

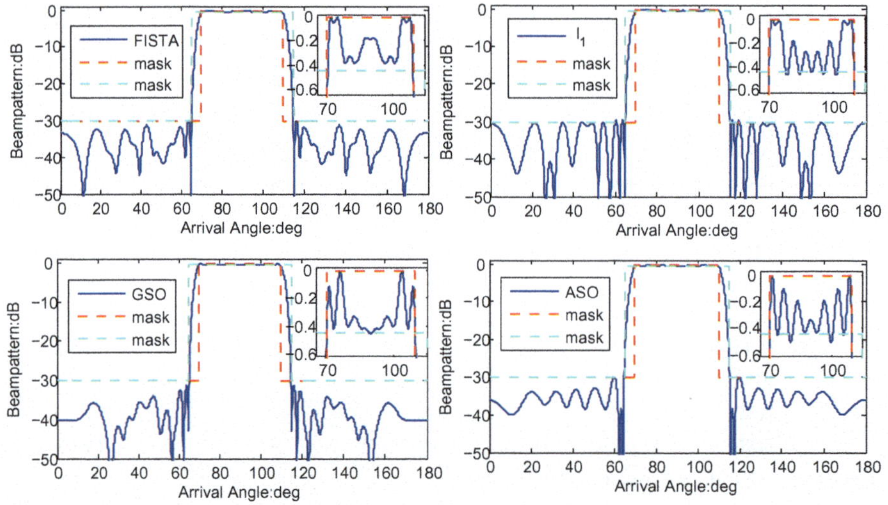

Fig. 3.5 Flap-top beampattern of Example 3

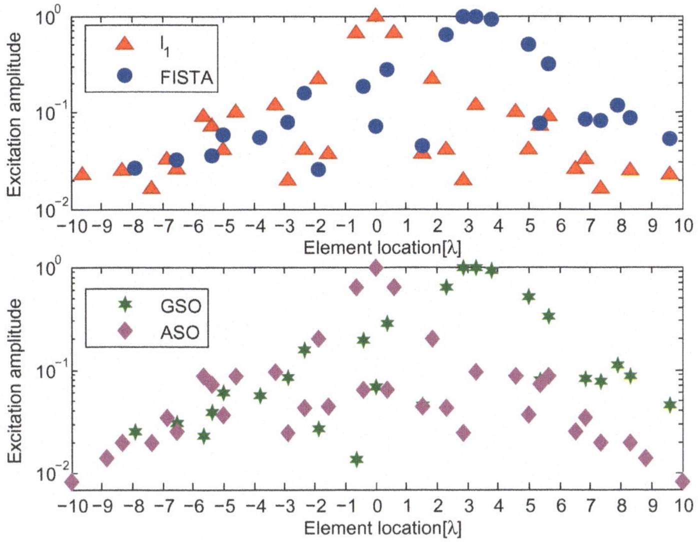

Fig. 3.6 Antenna positions and corresponding normalized excitation amplitude of Example 3

presents the lowest sidelobe level −31.23dB and −0.4dB mainlobe ripple with 30 selected antennas. The proposed FISTA method has the smallest mainlobe ripple −0.3845dB with only 24 selected antennas. Thus in this example, the FISTA produces a much sparser array than other methods while still fulfilling the requirements. For the computational time, the ASO method is the fastest. The FISTA is slower than the ASO and a little faster than the GSO, but much faster than the reweighted l_1-norm. The selected antenna positions and corresponding normalized excitation amplitude are shown in Fig. 3.6.

(b) Example 4

In order to show the proposed method can deal with arbitrary shaped beampatterns, we synthesise in the fourth example a flat-top pattern with a notch, again using the same non-uniform 41-antenna linear array. The synthesised beampattern is shown in Fig. 3.7. The sidelobe level is −40dB and the notch depth is −60dB. We can see that all four methods used select 37 antennas. Note that the GSO gives the smallest mainlobe ripple of −0.145dB and has very fast convergence rate, while the FISTA method has the deepest null depth of −63.32dB with −0.309dB mainlobe ripple. The computational time of the FISTA is still half of the reweighted l_1-norm. The ASO method is again the fastest, but with unsatisfactory null depth. The reweighted l_1-norm exhibits the highest mainlobe ripple and the slowest convergence rate.

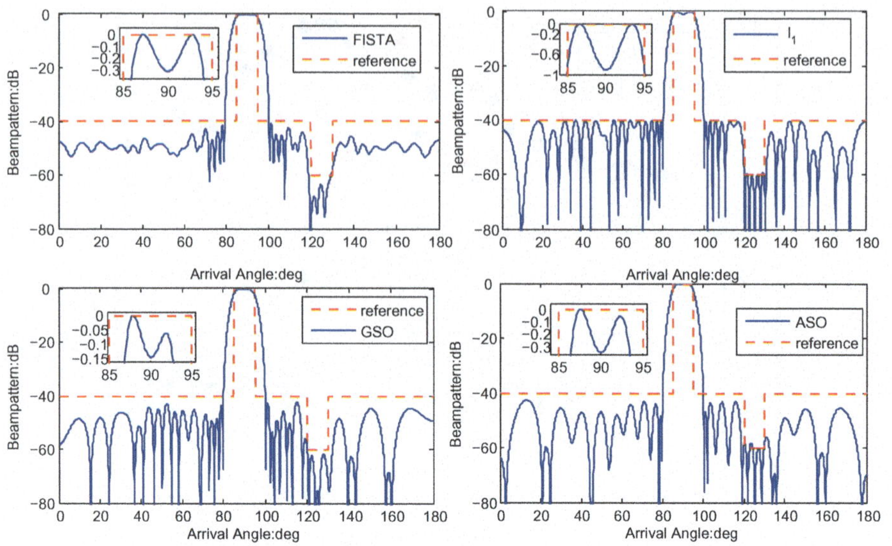

Fig. 3.7 Flap-top beampattern with notch of Example 4

(c) Example 5

In the fifth example, we again use the same non-uniform 41-antenna linear array. The desired beampattern has two flat-top mainlobes around 90° and 125° respectively with a sidelobe level of −30dB as shown in Fig. 3.8. We can see that the beampattern of the ASO method has the largest mainlobe ripple, around −1.972dB. The GSO method shows the best performance, with only −0.597dB mainlobe ripple and 35 selected antennas. The proposed FISTA method also selects 35 antennas, but with −0.8938dB mainlobe ripple similar to the reweighted l_1-norm. Looking at the computational time, the ASO is always the fastest in the above three examples, while the reweighted l_1-norm method is the slowest. The FISTA is slightly faster than the GSO and saves approximately half the computational time compared with the reweighted l_1-norm method.

(d) Example 6

Finally, a quarter wavelength spaced rectangular planar array is considered. We employ the same 21 × 21 antenna array that was used in [35]. The required mainlobe region is $(u_x^2 + u_y^2) \leq 0.2^2$, and the sidelobe region is $u_x^2 + u_y^2 \geq 0.4^2$. The synthesised beampattern reported in [35] has a mainlobe ripple of −1.69dB and sidelobe level of −25.85dB as shown in Fig. 8 of that reference. The resulting beampattern using the proposed FISTA method is shown in Fig. 3.9 and the contour plot of the synthesized beampattern is shown in Fig. 3.10. Although the mainlobe ripple is slightly higher with −1.72dB, the sidelobe level is maintained by FISTA under −26.68dB. The selected subarray is shown in Fig. 3.11 which has similar configuration to the selected subarray in [35]. Both comprise of an inner circle and an outer region that

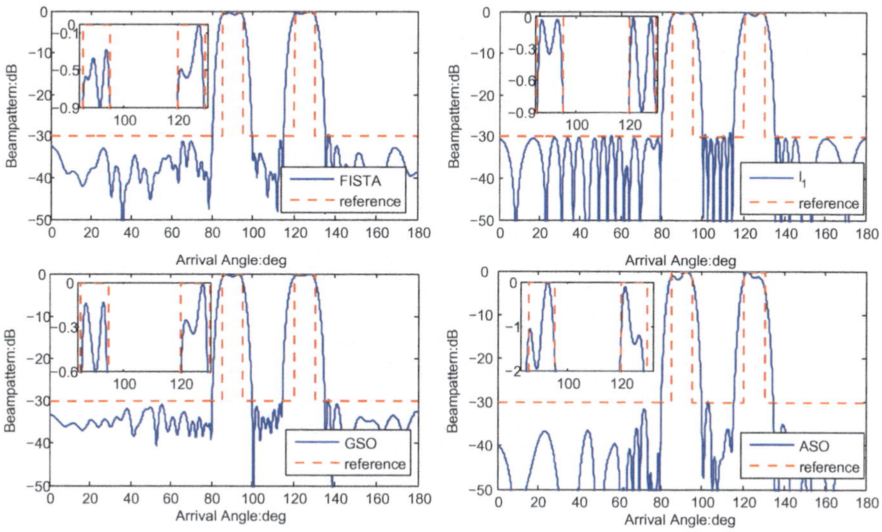

Fig. 3.8 Two flap-top beams of Example 5

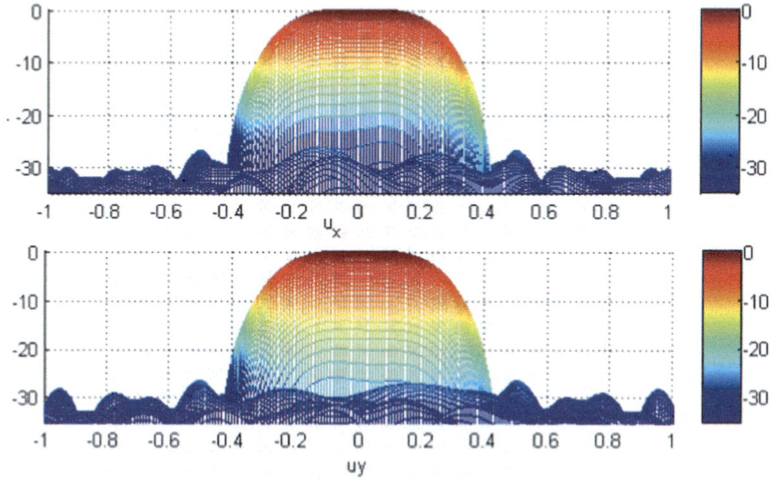

Fig. 3.9 Synthesized beampattern of a 21 × 21 planar array, the mainlobe region is $u_x^2 + u_y^s \leq 0.04$, the sidelobe level is −26.68 dB

is a square here and circle in [35]. Although the GSO method can also be utilised here and its mainlobe ripple is the smallest with −0.7751dB, its sidelobe level is only −23.32dB with 165 selected antennas. The computational time in this example is nearly same for both the FISTA and the GSO, and both are much faster than the reweighted l_1-norm method.

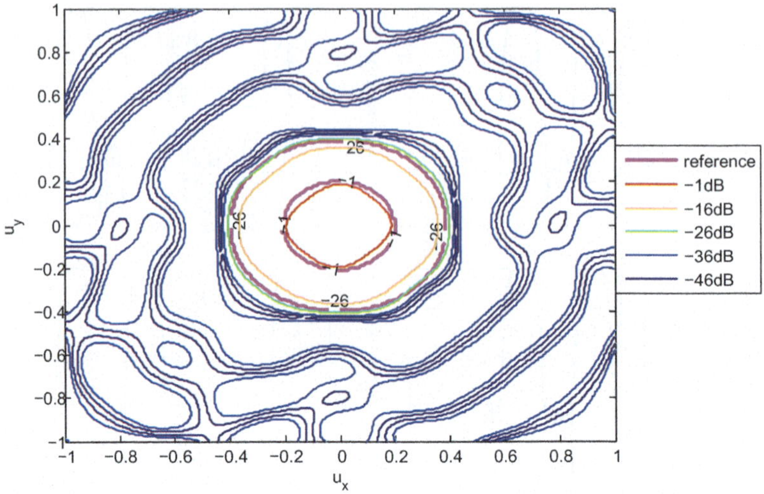

Fig. 3.10 Contour plot of the planar array beampattern

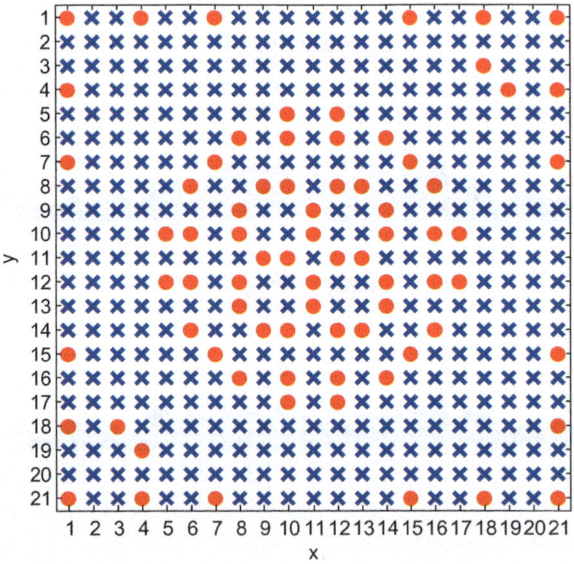

Fig. 3.11 76-antenna subarray selected in a 21 × 21 planar array, with circle being selected and cross being abandoned

3.3 Sparse Quiescent Beamformer Design with MaxSNR

In this section, we examine sparse array quiescent beamforming for multiple sources in interference-free environment.

3.3.1 Problem Formulation

Consider a linear array of N isotropic antennas with positions specified by multiple integer of unit inter-element spacing $x_n d$, $x_n \in \mathbb{N}$, $n = 1, \ldots, N$. Note that the linear array configuration is adopted for simplicity, and all proposed algorithms in this paper are applicable to two dimensional arrays. Suppose that p source signals are impinging on the array from directions $\{\theta_1, \ldots, \theta_p\}$ with spatial steering vectors specified by,

$$\mathbf{u}_k = [e^{jk_0 x_1 d \cos\theta_k}, \ldots, e^{jk_0 x_N d \cos\theta_k}]^T, k = 1, \ldots, p. \tag{3.55}$$

The wavenumber is defined as $k_0 = 2\pi/\lambda$ with λ being the wavelength and ^{T} denotes transpose operation. The received signal at time instant t is given by,

$$\mathbf{x}(t) = \mathbf{U}\mathbf{s}(t) + \mathbf{n}(t), \tag{3.56}$$

where $\mathbf{U} = [\mathbf{u}_1, \ldots, \mathbf{u}_p] \in \mathbb{C}^{N \times p}$ is the source array manifold matrix. In the above equation, $\mathbf{s}(t) \in \mathbb{C}^p$ denotes the source vector at time instant t and $\mathbf{n}(t) \in \mathbb{C}^N$ the received noise vector. The output of the N-antenna beamformer is given by,

$$y(t) = \mathbf{w}^H \mathbf{x}(t), \tag{3.57}$$

where $\mathbf{w} \in \mathbb{C}^N$ is the complex vector of beamformer weights and ^{H} stands for Hermitian operation. With additive Gaussian noise, i.e., $\mathbf{n}(t) \sim \mathcal{CN}(\mathbf{0}, \sigma_n^2 \mathbf{I})$, where σ_n^2 denotes the noise power level, and in the absence of interfering sources, the optimal weight vector for maximizing the output SNR is given by [1],

$$\mathbf{w} = \mathcal{P}\{\mathbf{R}_s\} = \mathcal{P}\{\mathbf{U}\mathbf{C}_s\mathbf{U}^H\}. \tag{3.58}$$

where $\mathcal{P}\{\cdot\}$ denotes the principal eigenvector of the matrix, $\mathbf{R}_s$ is defined as $\mathbf{R}_s = \mathbf{U}\mathbf{C}_s\mathbf{U}^H$ with $\mathbf{C}_s = E\{\mathbf{s}(t)\mathbf{s}^H(t)\}$ being the source auto-correlation matrix. The output SNR is,

$$\text{SNR} = \frac{\mathbf{w}^H \mathbf{R}_s \mathbf{w}}{\mathbf{w}^H \mathbf{R}_n \mathbf{w}} = \frac{\lambda_{\max}\{\mathbf{R}_s\}}{\sigma_n^2} = \frac{\|\mathbf{R}_s\|_2}{\sigma_n^2}. \tag{3.59}$$

Clearly, array configuration affects the output SNR of the MSNR beamformer through the term of $\|\mathbf{R}_s\|_2$. Note that the optimal weight vector in Eq. (3.58) is a linear combination of the source steering vectors.

Intuitively, the MSNR beamformer favours the strong and closely-spaced sources for maximizing the output SNR. Therefore, one disadvantage of the MSNR beamformer is that it cannot guarantee equal sensitivities towards all sources, which may not be desirable in practice, specifically when all sources should be equally served by the receiver. Although the deterministic beamformer with predefined mainlobes and controlled sidelobe level can provide exact and equal sensitivities, it does not strive to maximize the output SNR. To illustrate, suppose the array response towards the p desired sources is determined by the vector $\mathbf{r} \in \mathbb{C}^p$, i.e.,

$$\mathbf{U}^H \mathbf{w} = \mathbf{r} \text{ and } \mathbf{w}^H \mathbf{w} = 1. \tag{3.60}$$

Then, the output SNR of the deterministic beamformer can be calculated as,

$$\text{SNR} = \frac{\mathbf{w}^H \mathbf{R}_s \mathbf{w}}{\mathbf{w}^H \mathbf{R}_n \mathbf{w}} = \frac{\mathbf{r}^H \mathbf{C}_s \mathbf{r}}{\sigma_n^2}, \tag{3.61}$$

which is a fixed value, independent of the array configuration.

In order to define a role for SNR in deterministic sparse array beamformer design, and strike a compromise between preset beamformer characteristics and SNR performance, we first relax the condition of exact array response and attempt to only offer approximate equal gains towards all sources of interest. This is achieved by first expressing the weight vector as a function of sources steering vectors, i.e.,

$$\mathbf{w} = \sum_{k=1}^{p} \beta_k \mathbf{u}_k, \tag{3.62}$$

with specified coefficients β_k, $k = 1, \ldots, p$. Accordingly, the array response towards the ith source is given by,

$$r_i = \mathbf{u}_i^H \mathbf{w} = N\beta_i + \sum_{k=1,k\neq i}^{p} \beta_k \rho_{ik}, \tag{3.63}$$

where $\rho_{ik} = \mathbf{u}_i^H \mathbf{u}_k$ denotes the spatial correlation between the ith and kth sources. Notably, the array response cannot be exactly described by the coefficient vector $\boldsymbol{\beta} = [\beta_1, \ldots, \beta_p]^T$ due to the non-zero correlations, and as such, the case of equal coefficients will not necessarily lead to equal gain, unless the sources are spatially orthogonal. Using Eq. (3.62), the output SNR of the beamformer becomes,

$$\text{SNR} = \frac{\mathbf{w}^H \mathbf{R}_s \mathbf{w}}{\mathbf{w}^H \mathbf{R}_n \mathbf{w}} = \frac{\boldsymbol{\beta}^H \mathbf{U}^H \mathbf{U} \mathbf{C}_s \mathbf{U}^H \mathbf{U} \boldsymbol{\beta}}{\sigma_n^2 \boldsymbol{\beta}^H \mathbf{U}^H \mathbf{U} \boldsymbol{\beta}}, \tag{3.64}$$

which is influenced by the array configuration through the spatial correlations among sources in the matrix $\mathbf{U}^H \mathbf{U} = [\rho_{ik}, i, k = 1, \ldots, p]$. In essence, spatial correlations cause the array response to deviate from the specified coefficient vector $\boldsymbol{\beta}$ and, thus, leverage the array configuration to further improve output SNR compared to deterministic beamformers.

3.3.2 Unconstrained Sparse MSNR Beamformer Design

Given the number of antennas and a specific beamformer, array configurations remain a source of flexibility and can allocate DoFs, i.e. antenna positions, towards achieving the optimum design of maximizing the output SNR of quiescent beamformers. Suppose that there are N grid points out of which K antennas can be placed. Denote an antenna selection vector $\mathbf{z} = [z_i, i = 1, \dots, N] \in \{0, 1\}^N$ with "zero" entry for a discarded antenna and "one" entry for a selected antenna. As steering vectors are directional, the steering vector of the selected sparse array can be expressed as $\mathbf{z} \odot \mathbf{u}_k, k = 1, \dots, p$, with $\odot$ denoting element-wise product. In addition, we define a selection matrix as $\mathbf{Z} = \{0, 1\}^{K \times N}$ for a simple mathematical derivation in the sequel, with a "one" entry in the ith row and the jth column, where $i = 1, \dots, K$ and $z_j = 1, j \in \{1, \dots, N\}$. Thus, the selection vector $\mathbf{z}$ and the selection matrix $\mathbf{Z}$ are inner-connected and their relationship can be expressed as $\mathbf{Z}\mathbf{Z}^T = \mathbf{I}$ and $\mathbf{Z}^T\mathbf{Z} = \text{diag}(\mathbf{z})$. The implementation of antenna selection can also be expressed in terms of the selection matrix $\mathbf{Z}$ as $\mathbf{u}_k \odot \mathbf{z} = \mathbf{Z}\mathbf{u}_k, k = 1, \dots, p$ and, accordingly, $\mathbf{U}(\mathbf{z}) = [\mathbf{u}_k \odot \mathbf{z}, k = 1, \dots, p] = \mathbf{Z}\mathbf{U}$. Ideally and from Eq. (3.59), the optimum sparse array MSNR beamformer should be reconfigured through antenna selection such that the spectral norm of reduced-dimensional source covariance matrix $\|\mathbf{R}_s(\mathbf{z})\|_2$ is maximized. That is,

$$\begin{aligned} \max_{\mathbf{z}} \quad & \left\{ \|\mathbf{R}_s(\mathbf{z})\|_2 = \|\mathbf{U}(\mathbf{z})\mathbf{C}_s\mathbf{U}^H(\mathbf{z})\|_2 \right\}, \\ \text{s.t.} \quad & \mathbf{z} \in \{0, 1\}^N, \ \mathbf{1}^T\mathbf{z} = K, \end{aligned} \tag{3.65}$$

where $\mathbf{1}$ is a column vector of all ones.

The method of implementing eigenvalue decomposition for each subset of K sensor locations is computationally prohibitive, even with a small number of sensors. Towards solving Eq. (3.65), we first relax the binary constraints $\mathbf{z} \in \{0, 1\}^N$ to a box constraint $0 \le \mathbf{z} \le 1$, as the spectral norm of a matrix is convex and the global maximizer of a convex function locates at the extreme points of the polyhedron $0 \le \mathbf{z} \le 1, \mathbf{1}^T\mathbf{z} = K$ [131]. This satisfies the boolean property of the selection variable automatically and eliminates the binary constraints in the problem formulation. Furthermore, maximizing a convex function directly as per Eq. (3.65) is a non-convex optimization problem, thus we resort to the following two methods for convex relaxation.

3.3.2.1 Iterative Affine Approximation

Define the column vectors of $\mathbf{U}^H$ as $\bar{\mathbf{u}}_i, i = 1, \ldots, N$, and $\tilde{\mathbf{u}}_i = \mathbf{C}_s^{1/2}\bar{\mathbf{u}}_i$. According to the definition of matrix spectral norm [132], the objective function in Eq. (3.65) is equivalent to,

$$f(\mathbf{z}) = \| \sum_{i=1}^{N} z_i \tilde{\mathbf{u}}_i \tilde{\mathbf{u}}_i^H \|_2, \tag{3.66}$$
$$= \max_{\|\mathbf{b}\|_2=1} \sum_{i=1}^{N} z_i \mathbf{b}^H \tilde{\mathbf{u}}_i \tilde{\mathbf{u}}_i^H \mathbf{b},$$
$$= \sum_{i=1}^{N} z_i \tilde{\mathbf{b}}^H \tilde{\mathbf{u}}_i \tilde{\mathbf{u}}_i^H \tilde{\mathbf{b}},$$

where $\tilde{\mathbf{b}}$ is clearly the principal eigenvector of the matrix $\sum_{i=1}^{N} z_i \tilde{\mathbf{u}}_i \tilde{\mathbf{u}}_i^H$. Since the convex objective function can be approximated iteratively by its affine global underestimator, $f(\mathbf{z})$ in the $(k+1)$th iteration can be approximated based on the previous solution $\mathbf{z}^{(k)}$ by,

$$f(\mathbf{z}) \approx f(\mathbf{z}^{(k)}) + g(\mathbf{z}^{(k)})^T (\mathbf{z} - \mathbf{z}^{(k)}), \tag{3.67}$$

where $g(\mathbf{z}^{(k)})$ is the gradient of $f(\mathbf{z})$ evaluated at the point $\mathbf{z}^{(k)}$. The ith entry of $g(\mathbf{z}^{(k)})$ is obtained from Eq. (3.66),

$$g(z_i^{(k)}) = \frac{\partial f}{\partial z_i} |_{\mathbf{z}^{(k)}} = \tilde{\mathbf{b}}^{(k)H} \tilde{\mathbf{u}}_i \tilde{\mathbf{u}}_i^H \tilde{\mathbf{b}}^{(k)}. \tag{3.68}$$

Here, $\tilde{\mathbf{b}}^{(k)}$ denotes the principal eigenvector of the matrix $\sum_{i=1}^{N} z_i^{(k)} \tilde{\mathbf{u}}_i \tilde{\mathbf{u}}_i^H$. Thus, the non-convex selection problem can be relaxed iteratively as,

$$\max_{\mathbf{z}} \; f(\mathbf{z}^{(k)}) + g(\mathbf{z}^{(k)})^T (\mathbf{z} - \mathbf{z}^{(k)}), \tag{3.69}$$
$$\text{s.t.} \; 0 \le \mathbf{z} \le 1, \; \mathbf{1}^T \mathbf{z} = K.$$

This transforms the non-convex MSNR beamformer design in Eq. (3.65) into a linear programming (LP) problem as Eq. (3.69), and, in turn, significantly alleviates the computational complexity. The iterative approximations are referred to as sequential convex progamming (SCP) [133]. Note that SCP is a local heuristic and its performance depends on the initial point $\mathbf{z}^{(0)}$. It is, therefore, typical to initialize the algorithm with several feasible points $\mathbf{z}^{(0)}$ and the final choice is the one with the maximum objective value over the different runs.

3.3.2.2 Frame Based Approximation

Frames are signal representation tools that are redundant, which means that the number of frames is more than the dimension of the denoted signal space, and thus no longer linearly independent as bases. They are often used when there is a need for more flexibility in choosing a representation[134]. A family $\hat{\mathbf{E}} = \{\hat{\mathbf{e}}_i\}_{i \in \mathcal{I}}$ with the index set $\mathcal{I}$ in a Hilbert space $\mathcal{H}$ is called a frame, if there exists two constants $0 < \alpha \leq \beta < \infty$, such that for all vectors $\mathbf{h}$ in $\mathcal{H}$, we have

$$a\|\mathbf{h}\|^2 \leq \sum_{i \in \mathcal{I}} |\langle \hat{\mathbf{e}}_i, \mathbf{h} \rangle|^2 \leq b\|\mathbf{h}\|^2, \tag{3.70}$$

where $\langle \hat{\mathbf{e}}_i, \mathbf{h} \rangle = \hat{\mathbf{e}}_i^H \mathbf{h}$ and a, b are named frame bounds. Tight frames (TFs) are frames with equal frame bounds, that is, $a = b$. Parseval TFs (PTFs) are TFs with frame bound $a = b = 1$. Denote the frame operator as $\mathbf{\Phi} = \hat{\mathbf{E}}\hat{\mathbf{E}}^H$, then the definition of frame implies that,

$$a\mathbf{I} \leq \mathbf{\Phi} \leq b\mathbf{I}. \tag{3.71}$$

Accordingly, the frame operator $\mathbf{\Phi} = \hat{\mathbf{E}}\hat{\mathbf{E}}^H = \mathbf{I}$ for PTF. Note that the Grammian $\mathbf{G} = \hat{\mathbf{E}}^H \hat{\mathbf{E}} \neq \mathbf{I}$.

Naimark Theorem: ([135] A set $\hat{\mathbf{E}} = \{\hat{\mathbf{e}}_i\}_{i \in \mathcal{I}}$ in a Hilbert space $\mathcal{H}$ is a PTF for $\mathcal{H}$ if and only if there is a larger Hilbert space $\mathcal{K}$, $\mathcal{H} \subset \mathcal{K}$, and a set of orthonormal bases $\{\mathbf{e}_i\}_{i \in \mathcal{I}}$ for $\mathcal{K}$ so that the orthogonal projection $\mathbf{P}$ of $\mathcal{K}$ onto $\mathcal{H}$ satisfies: $\mathbf{P}\mathbf{e}_i = \hat{\mathbf{e}}_i$, for all $i \in \mathcal{I}$.

The Naimark Theorem implies every PTF can be obtained by projecting a set of orthonormal bases from a larger space to a lower space. This process is called seeding. Recalling the definition of the selection matrix $\mathbf{Z}$, the columns of $\mathbf{Z}$ are obtained by projecting the N-dimensional identity matrix $\mathbf{I}$ to the K-dimensional lower space, and they constitute a PTF. Assume that the source auto-correlation matrix $\mathbf{C}_s$ is known a priori or previously estimated. Then, the principal eigenvector of the source covariance matrix associated with the array of fully populated grid of N sensors can be provided through eigenvalue decomposition of $\mathbf{R}_s$, i.e.,

$$\mathbf{R}_s = \mathbf{U}\mathbf{C}_s\mathbf{U}^H = \mathbf{E}\mathbf{\Lambda}\mathbf{E}^H, \tag{3.72}$$

where $\mathbf{\Lambda} = \text{diag}(\lambda_1, \ldots, \lambda_p, 0, \ldots, 0)$ with the p eigenvalues along its diagonal in a descending order and $\mathbf{e}_i$, $i = 1, \ldots, N$ are corresponding eigenvectors. According to the Naimark Theorem, a PTF can be obtained by seeding from a set of eigenbasis $\mathbf{E}$ by deleting a subset of rows of $\mathbf{E}$. We denote the result as,

$$\hat{\mathbf{E}} = [\hat{\mathbf{e}}_1, \ldots, \hat{\mathbf{e}}_N] = [\mathbf{E}^T(\mathcal{J})]^T, \tag{3.73}$$

where $\mathcal{J} \subset \{1, \ldots, N\}$ is the index set of remained columns of the matrix $\mathbf{E}^T$. Clearly, $\|\hat{\mathbf{e}}_i\|_2 \neq 1, i = 1, \ldots, N$, and thus $\mathbf{G} = \hat{\mathbf{E}}^H \hat{\mathbf{E}} \neq \mathbf{I}$. However, we still have $\mathbf{\Phi} = \hat{\mathbf{E}}\hat{\mathbf{E}}^H = \mathbf{I}$. Therefore, the column vectors of $\hat{\mathbf{E}}$ constitute a PTF and the source covariance matrix after antenna selection can be represented in terms of the PTF as,

$$\hat{\mathbf{R}}_s = \mathbf{ZUC}_s\mathbf{U}^H\mathbf{Z}^T = (\mathbf{ZE})\mathbf{\Lambda}(\mathbf{ZE})^H = \hat{\mathbf{E}}\mathbf{\Lambda}\hat{\mathbf{E}}^H. \tag{3.74}$$

Furthermore, the frame $\hat{\mathbf{e}}_1 = \mathbf{Z}\mathbf{e}_1$ still possesses the largest coefficient λ_1. For this reason and based on the PFT argument, we consider $\hat{\mathbf{e}}_1/\|\hat{\mathbf{e}}_1\|_2$ to closely mimic the principal eigenvector of the selected sparse array. Accordingly, the optimum weight vector for maximizing output SNR of the sparse array can be approximated by $\mathbf{w} \approx \hat{\mathbf{e}}_1/\|\hat{\mathbf{e}}_1\|_2$. Combining this result with the definition of the spectral norm of the source covariance matrix, we obtain the lower bound, i.e.,

$$\begin{aligned}\|\mathbf{U}(\mathbf{z})\mathbf{C}_s\mathbf{U}^H(\mathbf{z})\|_2 &= \max\nolimits_{\|\mathbf{b}\|_2=1} \mathbf{b}^H\mathbf{U}(\mathbf{z})\mathbf{C}_s\mathbf{U}^H(\mathbf{z})\mathbf{b}, \\ &\geq \frac{\hat{\mathbf{e}}_1^H\mathbf{U}(\mathbf{z})\mathbf{C}_s\mathbf{U}^H(\mathbf{z})\hat{\mathbf{e}}_1}{\|\hat{\mathbf{e}}_1\|_2^2}, \\ &= \frac{\mathbf{e}_1^H\text{diag}(\mathbf{z})\mathbf{UC}_s\mathbf{U}^H\text{diag}(\mathbf{z})\mathbf{e}_1}{\mathbf{e}_1^H\text{diag}(\mathbf{z})\mathbf{e}_1}.\end{aligned} \tag{3.75}$$

Therefore, the sparse array MSNR beamformer design can be formulated as,

$$\begin{aligned}&\max_{\mathbf{z}} \ \frac{\mathbf{e}_1^H\text{diag}(\mathbf{z})\mathbf{UC}_s\mathbf{U}^H\text{diag}(\mathbf{z})\mathbf{e}_1}{\mathbf{e}_1^H\text{diag}(\mathbf{z})\mathbf{e}_1}, \\ &\text{s.t.} \ \mathbf{z} \in \{0, 1\}^N, \ \mathbf{1}^T\mathbf{z} = K.\end{aligned} \tag{3.76}$$

The objective function in Eq. (3.76) can be manipulated into a quasi-convex quadratic fractional problem [133], solved by the algorithm proposed in Sect. 3.3.4.

3.3.3 *Unconstrained Sparse Quiescent Beamformer Design*

We still consider the p discrete desired sources. Denote the sidelobe region as Ω, and sample Ω with a set of predefined discrete angles $\{\theta_{s1}, \ldots, \theta_{sL}\}$. Their respective steering vectors of the fully populated array are denoted as $\mathbf{a}_i, i = 1, \ldots, L$ and the corresponding steering matrix is $\mathbf{A} = [\mathbf{a}_1, \ldots, \mathbf{a}_L]$. The controlled sidelobes can be formulated as $|\mathbf{w}^H\mathbf{A}| \leq \epsilon$, where ϵ denotes the desired sidelobe level. Clearly, the sample number L should be sufficiently large for a better-shaped quiescent pattern, which inevitably increases the number of constraints and reducing the DoF to shape the pattern over other angular regions. Based on the frame theory explained in Sect. 3.3.2.2, we implement eigenvalue decomposition $\mathbf{AA}^H = \mathbf{V\Gamma V}^H$ and utilize the set of dominant frames $\mathbf{V}(\mathcal{J}) = \{\mathbf{v}_i, i \in \mathcal{J}\}$ to represent sidelobe region. The sidelobe constraints can then be expressed as $\mathbf{V}^H(\mathcal{J})\mathbf{w} = 0$ [1]. The deterministic array design with exactly K selected antennas is formulated as,

$$\begin{aligned} \text{Find} \;\; & \mathbf{w}, \\ \text{s.t.} \;\; & \mathbf{w}^H \mathbf{u}_i \mathbf{u}_i^H \mathbf{w} \leq 1 + \delta, i = 1, \ldots, p, \\ & \mathbf{w}^H \mathbf{u}_i \mathbf{u}_i^H \mathbf{w} \geq 1 - \delta, i = 1, \ldots, p, \\ & \mathbf{V}^H(\mathcal{J})\mathbf{w} = 0, \\ & \text{or } \mathbf{w}^H \mathbf{a}_i \mathbf{a}_i^H \mathbf{w} \leq \epsilon, i = 1, \ldots, L, \\ & \|\mathbf{w}\|_0 = K. \end{aligned} \tag{3.77}$$

where δ represents the acceptable mainlobe ripple. Different from the adaptive beamformer design with a binary selection variable $\mathbf{z}$, the variable $\mathbf{w}$ in the deterministic design in Eq. (3.77) is only required to be sparse with cardinality K. We split the $N \times 1$ weight variable into real and imaginary parts and stack them as a $2N \times 1$ vector, i.e., $\tilde{\mathbf{w}} = [\mathbb{R}(\mathbf{w})^T, \mathbb{I}(\mathbf{w})^T]^T$. Then, the problem of deterministic array design can be transformed from the complex domain to the real domain, i.e.,

$$\text{Find} \;\; \tilde{\mathbf{w}}, \tag{3.78a}$$
$$\text{s.t.} \;\; \tilde{\mathbf{w}}^H \tilde{\mathbf{U}}_i \tilde{\mathbf{w}} \leq 1 + \delta, i = 1, \ldots, p, \tag{3.78b}$$
$$\tilde{\mathbf{w}}^H \tilde{\mathbf{U}}_i \tilde{\mathbf{w}} \geq 1 - \delta, i = 1, \ldots, p, \tag{3.78c}$$
$$\tilde{\mathbf{V}}^H(\mathcal{J})\tilde{\mathbf{w}} = 0, \tag{3.78d}$$
$$\text{or } \tilde{\mathbf{w}}^H \tilde{\mathbf{A}}_i \tilde{\mathbf{w}} \leq \epsilon, i = 1, \ldots, L \tag{3.78e}$$
$$\|\mathbf{P}_u(\tilde{\mathbf{w}} \odot \tilde{\mathbf{w}}) + \mathbf{P}_l(\tilde{\mathbf{w}} \odot \tilde{\mathbf{w}})\|_0 = K. \tag{3.78f}$$

where the matrices $\tilde{\mathbf{U}}_i$, $\tilde{\mathbf{A}}_i$ and $\tilde{\mathbf{V}}(\mathcal{J})$ are defined as,

$$\tilde{\mathbf{U}}_i = \begin{bmatrix} \mathbb{R}(\mathbf{u}_i\mathbf{u}_i^H) & -\mathbb{I}(\mathbf{u}_i\mathbf{u}_i^H) \\ \mathbb{I}(\mathbf{u}_i\mathbf{u}_i^H) & \mathbb{R}(\mathbf{u}_i\mathbf{u}_i^H) \end{bmatrix}, \; i = 1, \ldots, p \tag{3.79}$$

$$\tilde{\mathbf{A}}_i = \begin{bmatrix} \mathbb{R}(\mathbf{a}_i\mathbf{a}_i^H) & -\mathbb{I}(\mathbf{a}_i\mathbf{a}_i^H) \\ \mathbb{I}(\mathbf{a}_i\mathbf{a}_i^H) & \mathbb{R}(\mathbf{a}_i\mathbf{a}_i^H) \end{bmatrix}, \; i = 1, \ldots, L \tag{3.80}$$

and

$$\tilde{\mathbf{V}}(\mathcal{J}) = \begin{bmatrix} \mathbb{R}(\mathbf{V}(\mathcal{J})) & -\mathbb{I}(\mathbf{V}(\mathcal{J})) \\ \mathbb{I}(\mathbf{V}(\mathcal{J})) & \mathbb{R}(\mathbf{V}(\mathcal{J})) \end{bmatrix}, \tag{3.81}$$

respectively. The selection matrices $\mathbf{P}_u = [\mathbf{I}_{N\times N}, \mathbf{0}_{N\times N}]$ and $\mathbf{P}_l = [\mathbf{0}_{N\times N}, \mathbf{I}_{N\times N}]$ in Eq. (3.78f). As the formulation Eq. (3.77) is homogeneous in terms of $\mathbf{w}$, we assume that the complex weights are included in the unit Euclidean ball. We define a slack variable $\mathbf{t}$ to relax the non-convex cardinality constraint in Eq. (3.78f) and transform the l_0 norm to the second-order cone programming (SOCP) as shown in Eq. (3.82e).

$$\max_{\tilde{\mathbf{w}},\mathbf{t}} \ \mathbf{t}^T(\mathbf{t}-\mathbf{1}), \tag{3.82a}$$

$$\text{s.t.} \ \tilde{\mathbf{w}}^H\tilde{\mathbf{U}}_i\tilde{\mathbf{w}} \leq 1+\delta, i=1,\ldots,p, \tag{3.82b}$$

$$\tilde{\mathbf{w}}^H\tilde{\mathbf{U}}_i\tilde{\mathbf{w}} \geq 1-\delta, i=1,\ldots,p, \tag{3.82c}$$

$$\tilde{\mathbf{V}}^H(\mathcal{J})\tilde{\mathbf{w}} = 0, \tag{3.82d}$$

$$\text{or } \tilde{\mathbf{w}}^H\tilde{\mathbf{A}}_i\tilde{\mathbf{w}} \leq \epsilon, i=1,\ldots,L,$$

$$\tilde{\mathbf{w}}_i^2 + \tilde{\mathbf{w}}_{i+N}^2 \leq t_i, i=1,\ldots,N, \tag{3.82e}$$

$$0 \leq \mathbf{t} \leq 1, \tag{3.82f}$$

$$\mathbf{1}^T\mathbf{t} = K. \tag{3.82g}$$

Note that both the objective function in Eq. (3.82a) and the lower bound constraint imposed on the mainlobe in Eq. (3.82c) are non-convex. Similar to Eq. (3.69), we utilize the SCP as a local heuristic that leverages the ability to efficiently solve convex optimization problems. The $(k+1)$th iteration of the deterministic design based on the kth solution $\tilde{\mathbf{w}}^{(k)}$ and $\mathbf{t}^{(k)}$ can be written as,

$$\begin{aligned}
\max_{\tilde{\mathbf{w}},\mathbf{t}} \ & (2\mathbf{t}^{(k)}-\mathbf{1})^T\mathbf{t} - \mathbf{t}^{(k)T}\mathbf{t}^{(k)}, \\
\text{s.t.} \ & 2\tilde{\mathbf{w}}^{(k)H}\tilde{\mathbf{U}}_i\tilde{\mathbf{w}} - \tilde{\mathbf{w}}^{(k)H}\tilde{\mathbf{U}}_i\tilde{\mathbf{w}}^{(k)} \leq 1+\delta, i=1,\ldots,p, \\
& 2\tilde{\mathbf{w}}^{(k)H}\tilde{\mathbf{U}}_i\tilde{\mathbf{w}} - \tilde{\mathbf{w}}^{(k)H}\tilde{\mathbf{U}}_i\tilde{\mathbf{w}}^{(k)} \geq 1-\delta, i=1,\ldots,p, \\
& \tilde{\mathbf{V}}^H(\mathcal{J})\tilde{\mathbf{w}} = 0, \\
& \text{or } \tilde{\mathbf{w}}^H\tilde{\mathbf{A}}_i\tilde{\mathbf{w}} \leq \epsilon, i=1,\ldots,L, \\
& \tilde{\mathbf{w}}_i^2 + \tilde{\mathbf{w}}_{i+N}^2 \leq t_i, i=1,\ldots,N \\
& 0 \leq \mathbf{t} \leq 1, \\
& \mathbf{1}^T\mathbf{t} = K.
\end{aligned} \tag{3.83}$$

Several runs with different feasible points should be carried out and the best K-antenna sparse array with the weight $\mathbf{w}$ is chosen as the final choice. A phase-only reconfigurable array can be achieved by restricting the modulus of weight coefficients to a constant and synthesizing the desired pattern by changing the phase only [136].

3.3.4 *Sparse Quiescent Beamformer Design with MSNR and Controlled Sidelobes*

Although superior in maximizing the output SNR, the MSNR beamformer ignores the shape of the quiescent pattern, such as sidelobe levels. Below, we combine the adaptive and deterministic approaches, in Sects. 3.3.2 and 3.3.3 respectively, to offer a generalized metric for sparse array beamformer design. The associated optimization problem becomes more involved due to the additional sidelobe controlling constraints. We take the form of the sidelobe constraints in Eq. (3.78d) as an example in the following derivation.

3.3.4.1 Sparse MSNR Beamformer Design with Controlled Quiescent Pattern

Since it is not straightforward to adapt the iterative affine approximation to be compatible with the deterministic design, we utilize the frame based approximation to design the MSNR beamformer with controlled quiescent pattern. Thus, the formulation is similar to Eq. (3.76) with an additional sidelobe level constraint. That is,

$$\begin{aligned} \max_{\mathbf{z}} \quad & \frac{\mathbf{e}_1^H \text{diag}(\mathbf{z}) \mathbf{U} \mathbf{C}_s \mathbf{U}^H \text{diag}(\mathbf{z}) \mathbf{e}_1}{\mathbf{e}_1^H \text{diag}(\mathbf{z}) \mathbf{e}_1}, \\ \text{s.t.} \quad & \mathbf{e}_1^H \text{diag}(\mathbf{z}) \mathbf{V}(\mathcal{J}) = \mathbf{0}, \\ & \mathbf{z} \in \{0, 1\}^N, \ \mathbf{1}^T \mathbf{z} = K. \end{aligned} \tag{3.84}$$

Define the vector $\bar{\mathbf{e}}_1 = \mathbf{e}_1^* \odot \mathbf{e}_1$, the matrices $\bar{\mathbf{V}} = [\mathbf{e}_1^* \odot \mathbf{v}_1, \ldots, \mathbf{e}_1^* \odot \mathbf{v}_{|\mathcal{J}|}]$ and $\bar{\mathbf{U}} = [\mathbf{e}_1^* \odot \mathbf{u}_1, \ldots, \mathbf{e}_1^* \odot \mathbf{u}_p]$ with * denoting conjugate operation. The problem in Eq. (3.84) can be rewritten in the form of quadratic fractional,

$$\begin{aligned} \max_{\mathbf{z}} \quad & \frac{\mathbf{z}^H \bar{\mathbf{U}} \mathbf{C}_s \bar{\mathbf{U}}^H \mathbf{z}}{\mathbf{z}^H \bar{\mathbf{e}}_1}, \\ \text{s.t.} \quad & \mathbf{z}^H \bar{\mathbf{V}} = 0, \\ & 0 \leq \mathbf{z} \leq 1, \ \mathbf{1}^T \mathbf{z} = K. \end{aligned} \tag{3.85}$$

We relax the binary constraints $\mathbf{z} \in \{0, 1\}^N$ to the box constraint $0 \leq \mathbf{z} \leq 1$, as the global maximizer of a quasi-convex function locates at the extreme points of the polyhedron [131, 137]. We propose an iterative linear fractional programming (ILFP) algorithm to solve the quadratic fractional. The problem in the $(k+1)$th iteration based on the kth solution $\mathbf{z}^{(k)}$ is written as,

$$\begin{aligned} \max_{\mathbf{z}} \quad & \frac{2\mathbf{z}^{(k)H} \bar{\mathbf{U}} \mathbf{C}_s \bar{\mathbf{U}}^H \mathbf{z} - \mathbf{z}^{(k)H} \bar{\mathbf{U}} \mathbf{C}_s \bar{\mathbf{U}}^H \mathbf{z}^{(k)}}{\mathbf{z}^H \bar{\mathbf{e}}_1}, \\ \text{s.t.} \quad & \mathbf{z}^H \bar{\mathbf{V}} = 0, \\ & 0 \leq \mathbf{z} \leq 1, \ \mathbf{1}^T \mathbf{z} = K. \end{aligned} \tag{3.86}$$

The linear fractional programming in Eq. (3.86) can be further transformed into a LP utilizing the Charnes-Cooper transformation [138] as follows,

$$\max_{\mathbf{y},\alpha} \; 2\mathbf{z}^{(k)H}\bar{\mathbf{U}}\mathbf{C}_s\bar{\mathbf{U}}^H\mathbf{y} - \mathbf{z}^{(k)H}\bar{\mathbf{U}}\mathbf{C}_s\bar{\mathbf{U}}^H\mathbf{z}^{(k)}\alpha, \tag{3.87}$$
$$\text{s.t.} \; \mathbf{y}^H\bar{\mathbf{V}} = 0,$$
$$\mathbf{1}^T\mathbf{y} = K\alpha, \; 0 \le \mathbf{y} \le \alpha,$$
$$\alpha > 0, \; \bar{\mathbf{e}}_1^H\mathbf{y} = 1.$$

The optimum selection vector is finally obtained by $\mathbf{z} = \mathbf{y}/\alpha$. The highest computational complexity of the ILFP is of order kn^2m using an interior-point based method, with n, m and k denoting the respective numbers of variables, constraints and iterations [133]. Empirical results show that no more than ten iterations, that is $k \le 10$, are required for the ILFP algorithm to converge. Thus, the computational complexity of the ILFP is comparable to that of the LP, which can be accelerated by diverse solvers in the literature [139, 140].

3.3.4.2 Sparse Array Semi-Adaptive Beamformer Design Using ILFP

Let $\boldsymbol{\beta}$ in Eq. (3.62) be $\boldsymbol{\beta} = \text{diag}(\boldsymbol{\gamma})\boldsymbol{\eta} = [\gamma_1 e^{j\phi_1}, \ldots, \gamma_p e^{j\phi_p}]$ with $\boldsymbol{\gamma} = [\gamma_1, \ldots, \gamma_p]$ denoting amplitude vector and $\boldsymbol{\eta} = [e^{j\phi_i}, \ldots, e^{j\phi_p}]$ denoting phase vector. For approximately equal response beamforming, $\gamma_i = 1, i = 1, \ldots, p$. The desired beamformer can be expressed as,

$$\mathbf{w} = \mathbf{U}\text{diag}(\boldsymbol{\gamma})\boldsymbol{\eta} = \sum_{i=1}^{p} \gamma_i e^{j\phi_i}\mathbf{u}_i. \tag{3.88}$$

Note that the coefficient amplitudes $\gamma_i, i = 1, \ldots, k$ are user-specified to approximate a desired pattern response. The coefficient phases, on the other hand, can be utilized as a free parameter for enhancing the semi-adaptive beamformer's performance. The optimum coefficient phase can be obtained by maximizing the output SNR in Eq. (3.64), i.e.,

$$\hat{\boldsymbol{\eta}} = \text{argmax}\left\{\frac{\boldsymbol{\eta}^H\text{diag}(\boldsymbol{\gamma})\mathbf{U}^H\mathbf{R}_s\mathbf{U}\text{diag}(\boldsymbol{\gamma})\boldsymbol{\eta}}{\boldsymbol{\eta}^H\text{diag}(\boldsymbol{\gamma})\mathbf{U}^H\mathbf{U}\text{diag}(\boldsymbol{\gamma})\boldsymbol{\eta}}\right\}. \tag{3.89}$$

Clearly, $\hat{\boldsymbol{\eta}}$ is the principal eigenvector of the following matrix,

$$\hat{\boldsymbol{\eta}} = \mathcal{P}\{(\mathbf{U}^H\mathbf{U})^{-1}(\mathbf{U}^H\mathbf{R}_s\mathbf{U})\} = \mathcal{P}\{\mathbf{C}_s\mathbf{U}^H\mathbf{U}\}. \tag{3.90}$$

Subsequently, element-wise normalization $\oslash$ is required,

$$\boldsymbol{\eta} = \hat{\boldsymbol{\eta}} \oslash |\hat{\boldsymbol{\eta}}| = \left[\hat{\boldsymbol{\eta}}_1/|\hat{\boldsymbol{\eta}}_1|, \ldots, \hat{\boldsymbol{\eta}}_p/|\hat{\boldsymbol{\eta}}_p|\right]. \tag{3.91}$$

Note that the coefficient phase $\boldsymbol{\eta}$ is array-dependent. The optimum semi-adaptive beamformer should be configured in a way such that the output SNR in Eq. (3.64) is maximized, i.e.,

$$\begin{aligned} \max_{\mathbf{z},\eta} \quad & \frac{\boldsymbol{\eta}^H \hat{\mathbf{U}}^H \text{diag}(\mathbf{z}) \mathbf{R}_s \text{diag}(\mathbf{z}) \hat{\mathbf{U}} \boldsymbol{\eta}}{\boldsymbol{\eta}^H \hat{\mathbf{U}}^H \text{diag}(\mathbf{z}) \hat{\mathbf{U}} \boldsymbol{\eta}}, \\ \text{s.t.} \quad & \mathbf{z} \in \{0, 1\}^N, \ \mathbf{1}^T \mathbf{z} = K, \end{aligned} \tag{3.92}$$

where $\hat{\mathbf{U}} = \mathbf{U}\text{diag}(\boldsymbol{\gamma})$. The design of semi-adaptive beamformer with controlled sidelobe level is formulated as,

$$\begin{aligned} \max_{\mathbf{z},\eta} \quad & \frac{\boldsymbol{\eta}^H \hat{\mathbf{U}}^H \text{diag}(\mathbf{z}) \mathbf{R}_s \text{diag}(\mathbf{z}) \hat{\mathbf{U}} \boldsymbol{\eta}}{\boldsymbol{\eta}^H \hat{\mathbf{U}}^H \text{diag}(\mathbf{z}) \hat{\mathbf{U}} \boldsymbol{\eta}}, \\ \text{s.t.} \quad & \mathbf{V}^H(\mathcal{J}) \text{diag}(\mathbf{z}) \hat{\mathbf{U}} \boldsymbol{\eta} = \mathbf{0}; \\ & \mathbf{z} \in \{0, 1\}^N, \ \mathbf{1}^T \mathbf{z} = K. \end{aligned} \tag{3.93}$$

Clearly, the sparse array design problem with respect to the two variables, $\mathbf{z}$ and $\boldsymbol{\eta}$, is highly non-convex. We adopt an alternating optimization method, which iteratively shifts between $\mathbf{z}$ and $\boldsymbol{\eta}$, to solve Eq. (3.93). First, assuming the fixed coefficient phase η and utilizing the following property of Khatri-Rao product $\circ$,

$$\mathbf{A}\text{diag}(\mathbf{x})\mathbf{b} = (\mathbf{b}^T \circ \mathbf{A})\mathbf{x}, \tag{3.94}$$

we obtain

$$\mathbf{U}^H \text{diag}(\mathbf{z}) \hat{\mathbf{U}} \boldsymbol{\eta} = [(\hat{\mathbf{U}} \boldsymbol{\eta})^T \circ \mathbf{U}^H] \mathbf{z}, \tag{3.95}$$

$$\boldsymbol{\eta}^H \hat{\mathbf{U}}^H \text{diag}(\mathbf{z}) \hat{\mathbf{U}} \boldsymbol{\eta} = \mathbf{z}^T [(\hat{\mathbf{U}} \boldsymbol{\eta}) \circ (\hat{\mathbf{U}} \boldsymbol{\eta})^*]. \tag{3.96}$$

and

$$\mathbf{V}^H(\mathcal{J}) \text{diag}(\mathbf{z}) \hat{\mathbf{U}} \boldsymbol{\eta} = [(\hat{\mathbf{U}} \boldsymbol{\eta})^T \circ \mathbf{V}^H(\mathcal{J})] \mathbf{z}. \tag{3.97}$$

Define the vector $\tau = (\hat{\mathbf{U}}\boldsymbol{\eta}) \circ (\hat{\mathbf{U}}\boldsymbol{\eta})^*$, the matrices $\underline{\mathbf{U}} = (\hat{\mathbf{U}}\boldsymbol{\eta})^T \circ \mathbf{U}^H$ and $\underline{\mathbf{V}} = (\hat{\mathbf{U}}\boldsymbol{\eta})^T \circ \mathbf{V}^H(\mathcal{J})$. The problem in Eq. (3.93) can be rewritten in the form of quadratic fractional,

$$\begin{aligned} \max_{\mathbf{z}} \quad & \frac{\mathbf{z}^T \underline{\mathbf{U}}^H \mathbf{C}_s \underline{\mathbf{U}} \mathbf{z}}{\mathbf{z}^T \tau} \\ \text{s.t.} \quad & \underline{\mathbf{V}} \mathbf{z} = \mathbf{0}, \\ & \mathbf{0} \le \mathbf{z} \le 1, \ \mathbf{1}^T \mathbf{z} = K. \end{aligned} \tag{3.98}$$

Similar to the problem in Eq. (3.85), the ILFP algorithm can then be utilized to obtain the optimal semi-adaptive beamformer with controlled sidelobe level. Given

Table 3.3 The design procedure of semi-adaptive beamformer

Step 1	Set iteration number $k = 0$; Generate set $\mathcal{J}_K \subset \{1, \ldots, N\}$
	Set $\boldsymbol{\eta} = \mathbf{1}$, $\mathbf{z}^{(0)} = \mathbf{0}$ and $\mathbf{z}^{(0)}(\mathcal{J}_K) = 1$
Step 2	Employ ILFP to solve the selection problem in Eq. (3.98)
Step 3	Compute $\hat{\boldsymbol{\eta}} = \mathcal{P}\{\mathbf{C}_s\mathbf{U}^H \text{diag}(\mathbf{z}^{(k)})\mathbf{U}\}$ and normalize $\boldsymbol{\eta} = \hat{\boldsymbol{\eta}} \oslash \lvert\hat{\boldsymbol{\eta}}\rvert$
Step 4	If $\lVert\mathbf{z}^{(k)} - \mathbf{z}^{(k-1)}\rVert_2 \geq \delta$, set $k = k + 1$ and go to Step 2
	Otherwise, terminate

the selected sparse array, the optimum coefficient phase can be calculated from Eqs. (3.90) and (3.91). The semi-adaptive beamformer design procedure is summarized in Table 3.3. Note that the construction of sparse array beamformers comprises two intertwined steps, optimum sparse array design through antenna selection as elucidated above, and weight calculations either by Eqs. (3.58) or (3.88) depending on the implemented beamformers.

3.3.5 Simulations

In this subsection, simulation results are presented to validate the proposed sparse array quiescent beamformer design.

3.3.5.1 Example 1

Consider $K = 8$ available antennas and $N = 16$ uniformly spaced positions with an inter-element spacing of $d = \lambda/2$. Assume that three uncorrelated source signals are impinging on the array from directions $\theta_1 = 65^\circ, \theta_2 = 75^\circ, \theta_3 = 125^\circ$ with the respective SNR being 6dB, 3dB and 0dB. Assuming that all the information of the sources is known to the receiver, thus the MSNR beamformer weights can be calculated as the principal eigenvector of the source correlation matrix as stated by Eq. (3.58). We enumerate all the 12870 different configurations based on the MSNR beamformer and the results are plotted in Fig. 3.12 in an ascending order. Moreover, a semi-adaptive beamformer $\mathbf{w} = \mathbf{U}\boldsymbol{\beta}$ with $|\gamma_i| = 1, i = 1, \ldots, 3$ is implemented for each sparse array and the corresponding output SNRs are also depicted in Fig. 3.12 for comparison. Note that the array coefficient phases $\eta_i, i = 1, \ldots, 3$ are calculated according to Eq. (3.90) for each sparse array.

The following remarks are in order: (1) The optimum sparse array implementing the MSNR beamformer can attain 16.5dB output SNR, which is 1.45dB higher than the worst array configuration. This underscores the importance of array configurations in determining the output SNR in interference-free and quiescent scenarios. (2) The semi-adaptive beamformer performs worse than the MSNR beamformer in

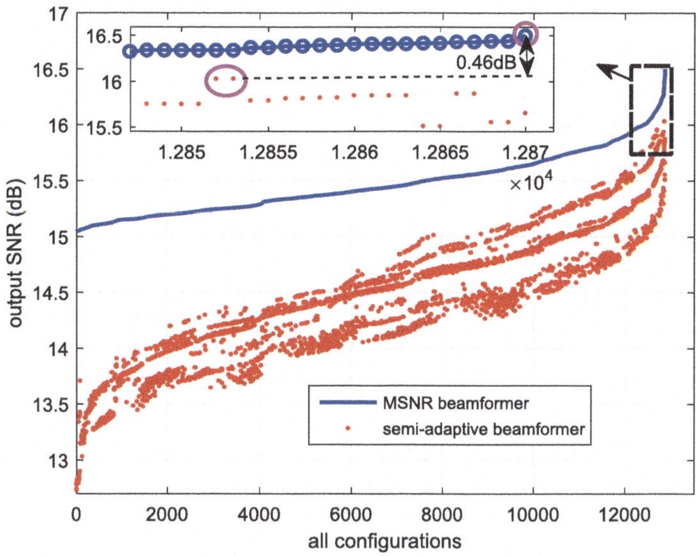

Fig. 3.12 Output SNRs of the MSNR and semi-adaptive beamfomers for all sparse arrays. The inner plot shows the output SNR difference

terms of output SNR. The maximum output SNR of the semi-adaptive beamformer is 16.04dB, which is the best sparse array that can offer approximately equal gain towards each source. (3) Array configurations also affect the performance of the semi-adaptive beamformer significantly, where the SNR difference between the best and worst sparse arrays is 3.31dB. (4) The output SNR of semi-adaptive beamformers demonstrates an increasing trend with regard to array configurations sorted ascendingly according to that of the MSNR beamformer. The optimum sparse arrays for the two beamformers are shown in Fig. 3.13, with filled dots denoting selected positions for locating antennas and cross for discarded. The beampatterns of the MSNR and semi-adaptive beamformers based on the respective optimum sparse arrays are plotted in Fig. 3.14. Clearly, the MSNR beamformer ignores the far and weak source in order to maximize the output SNR. This disadvantage is overcome by the semi-adaptive beamformer with a 0.46dB performance degradation. Note that both sparse arrays exhibit high sidelobe levels especially in the angular region around the three sources.

3.3.5.2 Example 2

In this example, we verify the effectiveness of proposed antenna selection methods. There are two methods for constructing sparse MSNR beamformer, namely iterative affine approximation and frame based approximation solved by the ILFP algorithm. The ILFP algorithm is also used to reconfigure the semi-adaptive beamformer. We

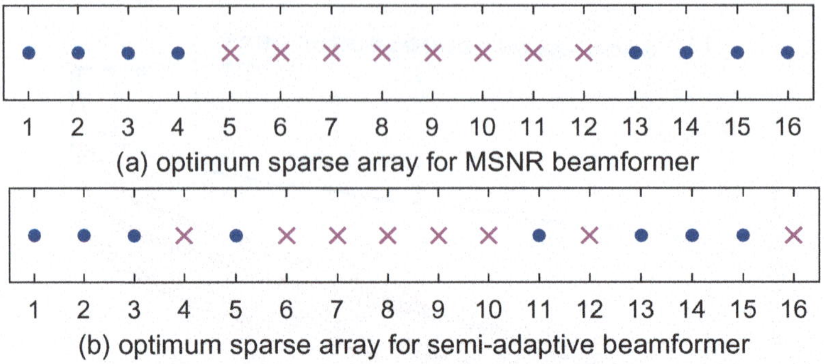

Fig. 3.13 Optimum sparse arrays for the MSNR and semi-adaptive beamformers

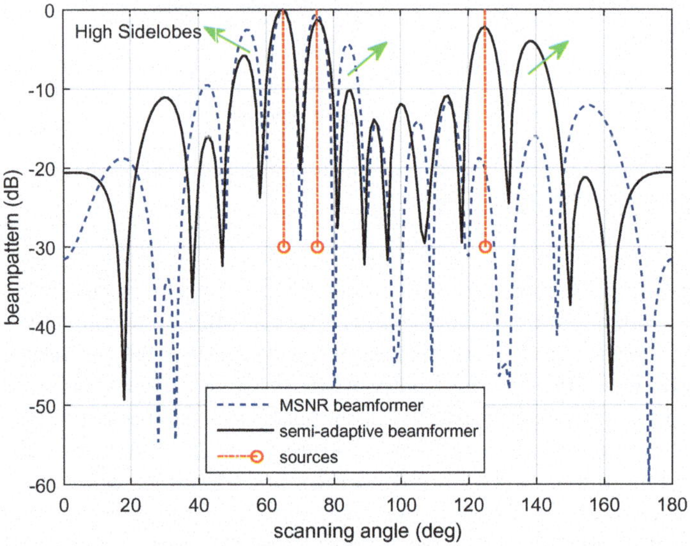

Fig. 3.14 Beampatterns of the MSNR and semi-adaptive beamfomers based on respective optimum sparse arrays. The green arrows indicate the unwanted high sidelobes

run 200 Monte-Carlo trials and three random integers uniformly generated within the range [0, 180] are used for the sources' arriving angles in each trial. Optimum 8-antenna sparse arrays are selected based on the MSNR and semi-adaptive beamformers utilizing the two relaxation methods and the ILFP algorithm in each scenario. The output SNRs corresponding to the selected sparse arrays are calculated and those of true optimum arrays obtained through enumeration is used for the benchmark. The error distances between the two are plotted in the form of histogram in Fig. 3.15a–c. Furthermore, a summary of the simulation results is depicted in Fig. 3.15d, where the percentage of trials with the error distance smaller than a set of predefined threshold values is calculated. We can observe that the iterative affine approximation exhibits

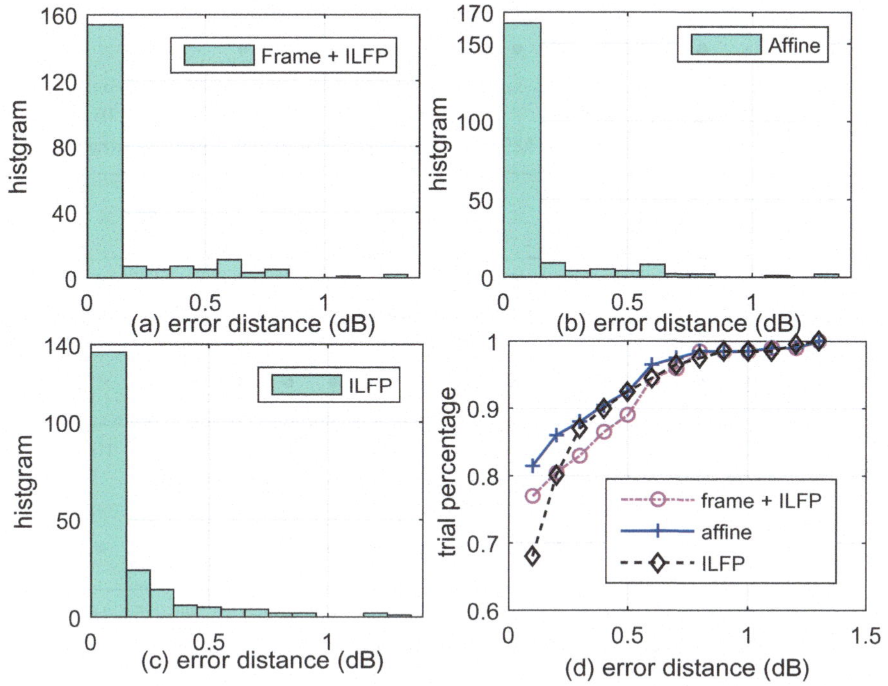

Fig. 3.15 **a** and **b** The error distance between output SNRs of true optimum arrays by enumeration and those of selected sparse array MSNR beamformer through frame based approximation and iterative affine approximation, respectively; **c** The error distance between output SNRs of true optimum arrays by enumeration and those of selected sparse array semi-adaptive beamformer through the ILFP algorithm; **d** The summary of simulation results in (**a**), (**b**), and (**c**)

a slightly higher accuracy than the fame based approximation for the MSNR beamformer design, while 94.5% of the trials demonstrates less than 0.6dB error using both methods. For the semi-adaptive beamformer design, 90% of the trials achieve less than 0.4dB error using the ILFP algorithm. Nevertheless, the two relaxation methods and the ILFP algorithm perform equally well apart from a negligible and acceptable deviation from the true global optimum solutions. This again validates the effectiveness of proposed convex relation and iterative optimization methods.

3.3.5.3 Example 3

As already demonstrated in the Example 1, the beampatterns of selected optimum sparse arrays (a) and (b) exhibit high sidelobes for both beamformers since sidelobe constraints were not considered in the design procedure. We then calculate the optimum 8-antenna sparse MSNR and semi-adaptive beamformers with controlled sidelobes utilizing Eq. (3.87) and the method summarized in Table 3.3. The side-

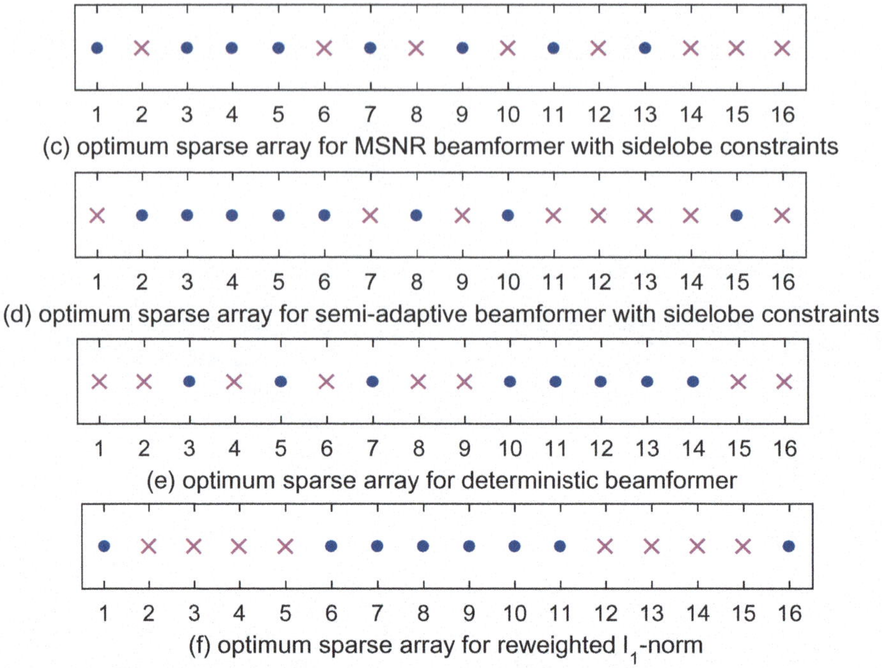

Fig. 3.16 Optimum sparse arrays for the MSNR, semi-adaptive, deterministic beamformers and reweighted l_1-norm method with sidelobe constraints

lobes are required not to exceed -10dB with respect to the main peak. Smaller values of sidelobe levels reduce the output SNR and yield highly distorted beampattern at unconstrained angles. The two selected optimum sparse arrays are shown in Fig. 3.16c, d. Their respective beampatterns are depicted in Fig. 3.17. Clearly, both arrays circumvent the high sidelobes in the angular region around the three sources. The weight calculation methods for both MSNR and semi-adaptive beamformers do not change, the sidelobe level are well maintained only through the proper choice of array reconfiguration. Once again, the array configuration is key to determine the beamforming performance.

In order to compare the (semi-)adaptive beamformer design with deterministic array synthesis, we implement the array thinning according to Eq. (3.83). The thinned array is shown in Fig. 3.16e and the synthesized beampattern is plotted in pink dash-dot curve in Fig. 3.17, with the weights obtained from Eq. (3.83). For a comprehensive comparison, we also implement the deterministic sparse array synthesis based on the well-known reweighted l_1-norm method [35, 141], where the largest $K = 8$ coefficients are remained with others setting to zero. The corresponding sparse array is depicted in Fig. 3.16f and its beampattern is plotted in green dot curve in Fig. 3.17. Clearly, all the three proposed beamformers successfully suppress the unwanted sidelobes by choosing respective suitable configurations. The determin-

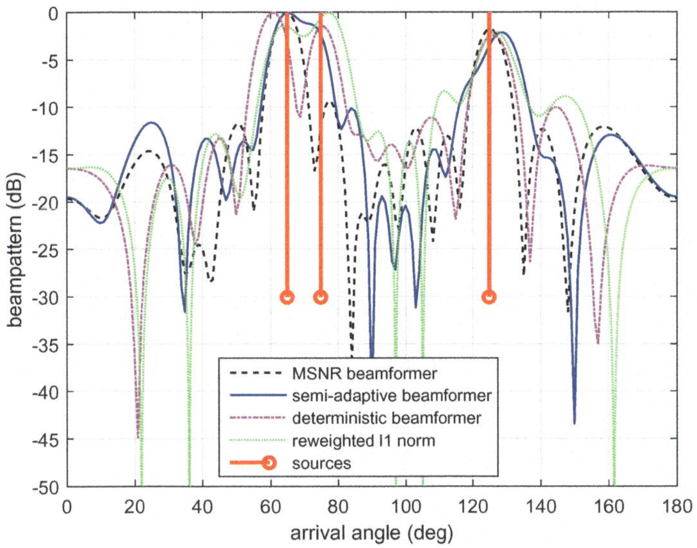

Fig. 3.17 Beampatterns of the MSNR, semi-adaptive, deterministic beamfomers and reweighted l_1-norm method based on respective optimum sparse arrays with sidelobe constraints

istic beamformer exhibits a best-shaped response pattern, whereas the MSNR beamformer completely ignores the source arriving from 75° due to the additional sidelobe constraints. The semi-adaptive beamformer demonstrates a satisfactory compromise between the deterministic and the MSNR beamformers, although it fails to distinguish the first two closely-spaced sources. The sparse array configured through the reweighted l_1-norm method exhibits as high as -8.3dB sidelobes around the third source and broadened mainlobes around the first two closely-spaced sources.

3.3.5.4 Example 4

We examine the beamformer design in the case of correlated sources. Consider a distributed source arriving within the angular region [75°, 105°]. The source correlation is generated randomly with its absolute value bounded between zero and one. Suppose there are 12 antennas that can be placed in 20 uniformly-spaced positions. We implement antenna selection for the MSNR, semi-adaptive and deterministic beamformers with sidelobe constraints, and compare with the reweighted l_1-norm method. The four selected optimum sparse arrays are plotted in Fig. 3.18g, h, k and l. Their respective beampatterns are depicted in Fig. 3.19. We can observe the notable trade-off between mainlobe shape, such as mainlobe width and transition bandwidth, and sidelobe level. Again, the deterministic beamformer exhibits the best-shaped quiescent pattern, and slightly better than that of the sparse array configured through

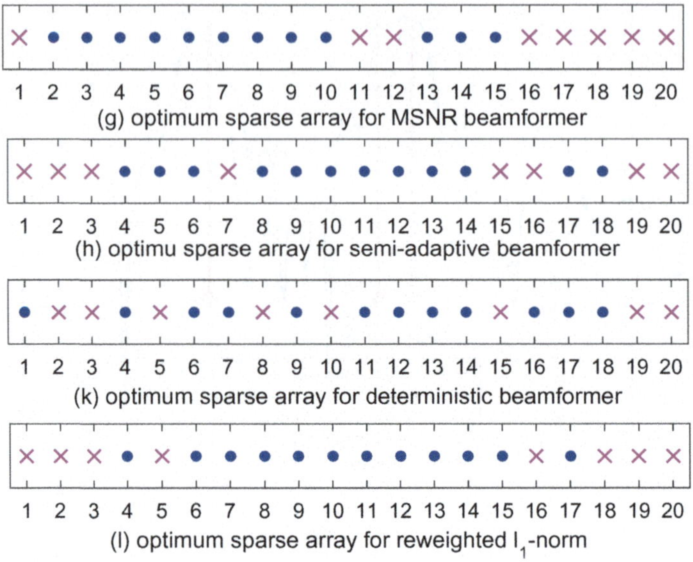

Fig. 3.18 Optimum sparse arrays for the MSNR, semi-adaptive, deterministic beamformers and reweighted l_1-norm with sidelobe constraints

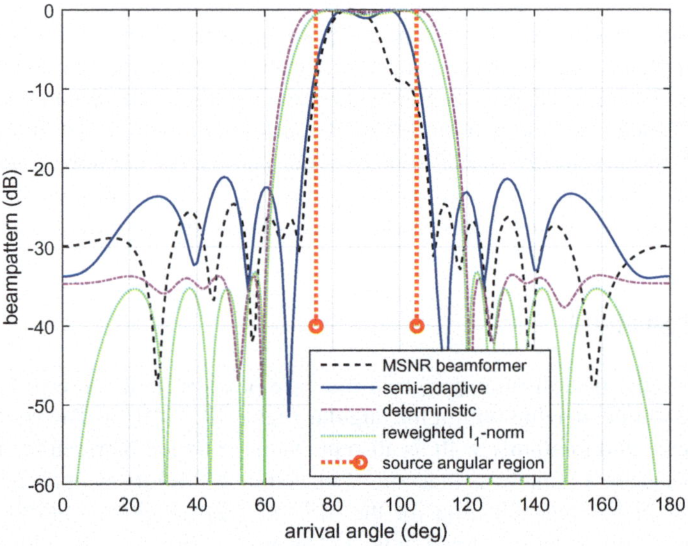

Fig. 3.19 Beampatterns of the MSNR, semi-adaptive, deterministic beamfomers and reweighted l_1-norm based on respective optimum sparse arrays with sidelobe constraints

the reweighted l_1-norm. The semi-adaptive beamformer demonstrates high sidelobe levels and the MSNR beamformer presents a poor mainlobe shape.

3.3.5.5 Example 5

For a further demonstration of the proposed sparse array quiescent beamformer design, we select 51 antennas out of 11×11 square array in terms of the MSNR beamformer with controlled sidelobe level. There are four discrete uncorrelated sources with 0dB SNR, arriving from $v_x = [-0.2, -0.3, 0.8, 0.65]$ and $v_y = [0.4, -0.6, 0.6, -0.4]$ in two-dimensional electronic angles, where $v_x = \cos\theta \cos\psi$ and $v_y = \cos\theta \sin\psi$ with θ and ψ denoting elevation and azimuth angles, respectively. The normalized beampattern of the selected 51-antenna optimum sparse array is plotted in Fig. 3.20. The sparse array is shown on the left of Fig. 3.21 and the contour of the beampattern is on the right. The output SNR of the sparse array MSNR beamformer is 20.72dB with -14dB peak sidelobe level (PSL) as indicated in the last row of the Table 3.4.

Finally, we compare the output SNR and PSL of the eleven selected sparse arrays (a)–(n) in Table 3.4. No doubt that the MSNR beamformer achieves the maximum output SNR, however it ignores some weak sources and exhibits poor quiescent pattern. The semi-adaptive beamformer overcomes the first disadvantage of the MSNR beamformer with an acceptable performance loss. The quiescent beampattern of both beamformers can be regularized by adding sidelobe constraints. The deterministic

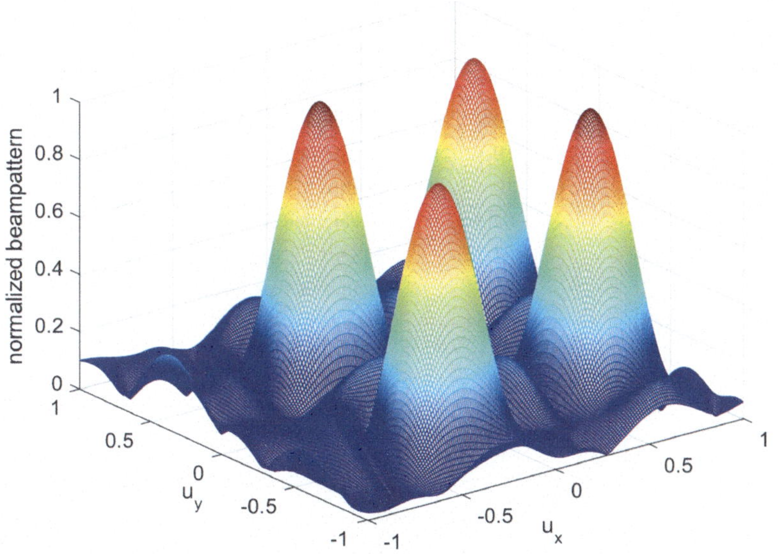

Fig. 3.20 Normalized beampattern of the MSNR beamfomer based on the selected 51-antenna sparse array with sidelobe constraints

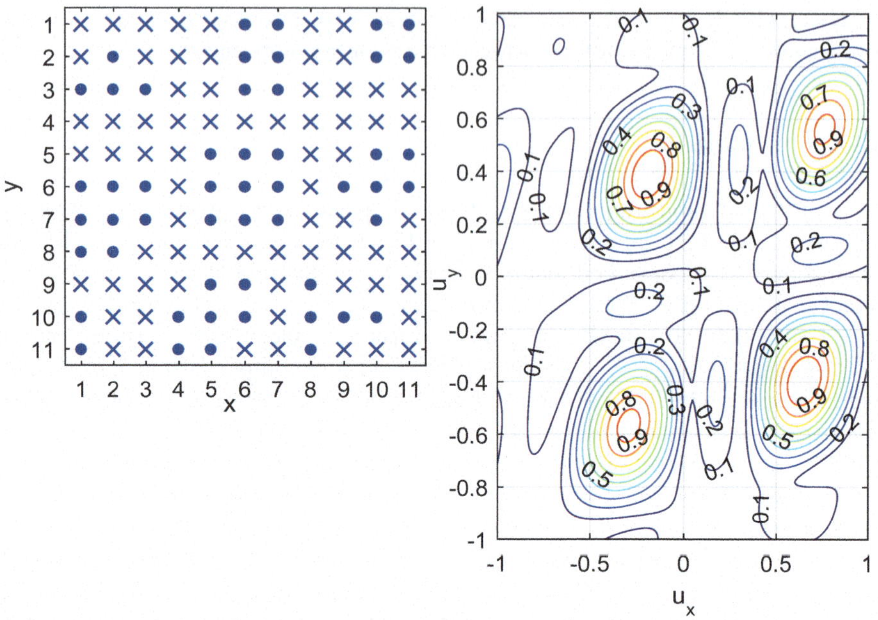

Fig. 3.21 The left: the selected 51-antenna sparse array (n); The right: the contour of the beampattern

Table 3.4 The output SNR, PSL and Computational time of each array

Array Name	SNR (dB)	PSL (dB)	Beamformer	Sidelobe Constraints	Time Sec
(a)	16.5	−2.54	MSNR	No	0.52
(b)	16.04	−4.05	semi-adaptive	No	0.65
(c)	16.17	−9.89	MSNR	Yes	1.19
(d)	14.84	−10.12	semi-adaptive	Yes	1.20
(e)	13.55	−10	deterministic	Yes	1.80
(f)	13.66	−8.3	reweighted l_1	Yes	1.67
(g)	15.56	−24.72	MSNR	Yes	1.86
(h)	15.31	−21.34	semi-adaptive	Yes	1.95
(k)	10.91	−33.54	deterministic	Yes	2.37
(l)	11.18	−33.18	reweighted l_1	Yes	1.6
(n)	20.72	−14	MSNR	Yes	3.13

beamformer demonstrates the worst output SNR, whereas its advantage is manifested by the well-controlled sidelobes. The beamformer design combining both adaptive and deterministic constraints offers a compromise between the output SNR and quiescent beampatterns. The reweighted l_1-norm performs worse especially when dealing with unsymmetric beampattern as it assumes symmetric array configuration and conjugate weight coefficients. The optimum sparse array is calculated by the CVX software [142] embedded in MATLAB using a desktop with 3.4GHz Intel i7-CPU and 16GB RAM. The computational times are listed in the last column of Table 3.4 and demonstrate comparable values to the benchmark of the well-known reweighted l_1-norm algorithm.

3.4 Summary

In this chapter, we considered the sparse sensing for determininic beamforming via antenna selection. In Sect. 3.2, three soft-thresholding based optimization methods were proposed to synthesize thinned arrays with arbitrary shaped beampatterns including multi-beam forming and several angular regions of suppression. The proposed methods were highly flexible, easily reconfigurable and can handle arbitrarily shaped arrays. They exhibited significant improvement with respect to convex optimization in regard to the mainlobe ripple control, sidelobe level reduction and also the computational time. Moreover, the FISTA could be utilized to synthesize large scaled antenna array that the CVX cannot handle. Therefore, the proposed FISTA based array thinning method could be utilized in applications requiring a fast method to adaptively reconfigure large antenna arrays with specified beampattern requirements. In Sect. 3.3, we examined the design of optimum sparse array beamformer in the presence of multiple desired sources in interference-free environment. Two beamformers, namely the MSNR and semi-adaptive beamformers, were considered and compared with the deterministic design. Unlike other work in the literature, we utilized the array configuration as an additional DoF to improve the beamforming performance without concerning with complicated beamformer design. We proposed a general sparse array design metric combining adaptive and deterministic approaches, where the reconfigured beamformers exhibited well-controlled sidelobes with an acceptable performance loss.

Chapter 4
Sparse Sensing for Adaptive Beamforming

In this chapter, we consider sparse array design for adaptive beamforming, which is a more effective method for interference suppression compared to deterministic beamforming in Chap. 3. A new parameter, referred to as spatial correlation coefficient (SCC), is proposed to characterize the spatial separation between the desired signal and the interference subspace in both single-source and multi-source scenarios. The squared absolute value of the SCC is directly related to the output SINR, and thus maximizing the output SINR is equivalent to minimizing the SCC. In Sect. 4.2, we first consider the sparse adaptive beamforming in the single-source case. In Sect. 4.3, we consider the general case of multiple sources and multiple interferences in the field of view, and extend the parameter SCC into the multi-source scenario. We formulate three angles, namely, eigenspace angle, conventional angle, and minimum canonical angle for characterizing spatial separation between the source and interference subspaces. Three sparse array design methods, incorporating these angles, are proposed in terms of minimizing the SCC to increase the separation between these two subspaces. The problems are transformed into the convex form utilizing the properties of Schur complement and/or convex relaxation.

4.1 Introduction

Adaptive array processing is an effective tool for counteracting interferences while providing the necessary degrees of freedom (DoFs) towards the directions of desired sources. It finds applications in radar, sonar, wireless communications, radio astronomy, and satellite navigation, to list a few [143–147]. Although the nominal array configuration is uniform, sparse arrays have recently emerged to play a fundamental role in various sensing systems involving multi-antenna transmitters and receivers [148–150]. Diverse metrics have been proposed to design optimum sparse arrays with different objectives, including direction finding, beampattern synthesis, target detection, and spatial filtering. For example, nested and coprime arrays have been

X. Wang et al., *Sparse Sensing and Sparsity Sensed in Multi-sensor Array Applications*, https://doi.org/10.1007/978-981-99-9558-5_4

proposed in radar and sonar as structured sparse arrays where sensor locations follow prescribed spatial arrangements. These arrays offer high resolution direction finding and can deal with more sources than physical sensors [151, 152]. Cramer-Rao Bound is another metric utilized to guide determinations of antenna positions in seeking high-accuracy direction-of-arrival (DOA) estimation [153–155]. Sparse arrays also permit high flexibility in beampattern mainlobe and sidelobe design and specifications [35, 156, 157]. In all the above systems, the main task is, in essence, to decide on where to place a given number of sensors to deliver optimum performance, with different optimization metrics leading to differed sparse array configurations [158–162]. For example, an optimum array configuration for high spatial resolution does not lend itself to high accuracy DOA estimation, as demonstrated in [153]. From signal enhancement and interference suppression perspective, an optimum array configuration is the one that yields high signal-to-interference-plus-noise ratio (SINR). The analysis and design of such arrays are the focus of this chapter.

Whether the problem is cast as optimum placement of a given number of antennas or equivalently selecting a subset of antennas from a large set to connect with front-end receivers, the optimum array, in this work, is sparse and environment-dependent. The antenna selection perspective of the problem relies on low-complexity Radio Frequency (RF) switches [104, 163–165], with the fundamental goal of reducing hardware cost associated with expensive RF chains. Furthermore, optimum sparse array corresponding to the changing environment can be adaptively reconfigured through switching on/off antennas.

The sparse array design for MaxSINR assumes priori knowledge of source and interference DOAs. This knowledge can be provided by first estimating the DOAs of all incoming signals, then reconfigure the array accordingly. In this respect, we adopt, in essence, the two-step open-loop adaptive filtering approach that was devised for linear uniform arrays [13, 84]. The antennas corresponding to the full array or those selected for optimum DOA estimation [153–155] are first switched on. Then, multiplexing to a different set of sensors for optimum beamforming based on the sensed field of view (FOV) is followed.

The effect of array configurations in interference mitigation in the case of a single source was investigated in Sect. 4.2, where a parameter, spatial correlation coefficient (SCC), was proposed to characterize the spatial angle between the source and interference subspaces and it was shown to be directly related to the output SINR. In this respect, sensor placement which maximizes the spatial separation leads to optimum sparse arrays. In Sect. 4.3, we consider the general case in which the dimension of the desired signal subspace is arbitrary and not necessarily confined to a unit value. This is the case when multiple communication emitters or broadcasters are present in the FOV. It can also occur in lock-in beams in radar tracking of multiple targets, or when operating in environments with rich scattering and multipath signals [166–169]. In these situations, the techniques devised for a single-dimensional signal subspace are no longer suitable for problem formulation, performance analysis, or array design. Accordingly, new metrics and approaches, accounting for higher dimensional signal subspace, must be introduced. Note that the generalization of single-source case to the multi-source multi-interference scenario is a standalone problem and is not a straightforward or ready extension of the techniques in Sect. 4.2.

With multiple sources and interfering signals impinging on the receiver, the optimum weight vector, which provides the maximum output SINR (MaxSINR), is the principal eigenvector of the product of the inverse noise covariance and source correlation matrices [1]. In order to quantitatively analyze the effect of array configurations on adaptive beamforming performance, we derive the lower and upper bounds of the MaxSINR. These bounds reveal that the array configuration influences the functionality of adaptive beamformers in two aspects. It decides the maximum output SNR (MaxSNR) in interference-free environments and defines the performance loss inevitably incurred by interference nulling in interference-active environments. The MaxSNR coincides with the upper bound of MaxSINR, while the performance loss depends on the spatial separation between source and interference subspaces. The latter is fully characterized by three formulated spatial angles.

In this chapter, we deal with the sparse sensing for adaptive beamforming under two scenarios. In Sect. 4.2, we start from the simple case of single source and seek for the effective antenna selection strategy. In Sect. 4.3, we extend the problem to a more complex multi-source case.

4.2 Reconfigurable Sparse Arrays in Single-Source Case

In this section, we investigate the reconfigurable sparse array design by acting on a sequence of RF switches in the single-source case. We start with the multi-interference case to introduce the important parameter, spatial separation coefficient (SCC), and then degeneralize the concept of SCC to the single-interference case to unveil its prominent role in manipultating the output SINR.

4.2.1 Spatial Correlation Coefficient

The impact of the interference on the desired signal can be quantitatively characterized by the output SINR [170, 171]. In addition to the Doppler separation, the spatial dimension also plays a pivotal part in the output SINR in the antenna array case [172, 173]. The array configuration, in particular, is fundamental to the overall performance of the receiver. Its effect on the performance is embodied in the spatial separation coefficient (SCC) [103, 174], which expresses the spatial separation between the desired signal and interferences with respect to the array. In this subsection, we start with the case of multiple interferences to formulate the SCC.

Let the direction of arrival (DOA) be specified by (θ, ϕ) with θ and ϕ being the elevation and azimuth angles, respectively. Then the electronic u-space DOA parameter corresponding to (θ, ϕ) is defined as $\mathbf{u} = [\cos\theta\cos\phi \ \cos\theta\sin\phi]^T$ where ^{T} is the transpose operation. Assume the number of antennas is N and that the matrix $\mathbf{P} = [\mathbf{p}_1, \mathbf{p}_2, \ldots, \mathbf{p}_N]^T \in \mathbb{R}^{N\times 2}$ contains the coordinates $\mathbf{p}_n = [x_n, y_n]^T$ of the antenna elements for $n = 1, \ldots, N$. Suppose the desired signal has a DOA $\mathbf{u}_s$ and that a total of L interferences have DOAs $\mathbf{u}_j$, $j = 1, \ldots, L$ respectively. Then the spatial steering vectors are,

$$\mathbf{s} = e^{jk_0\mathbf{P}\mathbf{u}_s}, \quad \mathbf{v}_j = e^{jk\cdot\mathbf{P}\mathbf{u}_j}, j = 1, \ldots, L, \tag{4.1}$$

where $k_0 = 2\pi/\lambda$ is the wavenumber and λ is the wavelength. Under the assumption that the interferences are uncorrelated with one another and with the noise, the covariance matrix of the interference plus noise becomes,

$$\mathbf{R}_n = \sigma^2\mathbf{I} + \sum_{j=1}^{L} \sigma_j^2 \mathrm{v}_j \mathrm{v}_j^H, \tag{4.2}$$

where $\mathbf{I} \in \mathbb{R}^{N\times N}$ is the identity matrix and σ^2 the thermal noise power and σ_j^2, $j = 1, \ldots, L$ the power of the jth inter-ference. Now we arrange the desired signal and L interference steering vectors into the matrices,

$$\mathbf{V}_I = [\mathbf{v}_1, \ \mathbf{v}_2, \ \ldots, \ \mathbf{v}_L], \ \mathbf{V}_s = [\mathbf{s}, \ \mathbf{v}_1, \ \mathbf{v}_2, \ \ldots, \ \mathbf{v}_L]. \tag{4.3}$$

Putting $\Sigma_I = \mathrm{diag}[\sigma_1^2, \sigma_2^2, \ldots, \sigma_L^2]$, we can write the interference plus noise covariance matrix into a more concise form,

$$\mathbf{R}_n = \sigma^2\mathbf{I} + \mathbf{V}_I \Sigma_I \mathbf{V}_I^H. \tag{4.4}$$

Applying the Woodbury matrix identity to the inverse noise covariance matrix, $\mathbf{R}_n$, yields,

$$\mathbf{R}_n^{-1} = \frac{1}{\sigma^2}(\mathbf{I} - \frac{1}{\sigma^2}\mathbf{V}_I(\Sigma_I^{-1} + \frac{1}{\sigma^2}\mathbf{V}_I^H\mathbf{V}_I)^{-1}\mathbf{V}_I^H). \tag{4.5}$$

Now let us assume that the interferences are much stronger than white noise, i.e. $\sigma_j^2 \gg \sigma^2, \forall j$, then Eq. (4.5) can be further simplified as

$$\mathbf{R}_n^{-1} \approx \frac{1}{\sigma^2}(\mathbf{I} - \mathbf{V}_I(\mathbf{V}_I^H\mathbf{V}_I)^{-1}\mathbf{V}_I^H). \tag{4.6}$$

We see that, when the interference to noise ratio (INR) is high, $\mathbf{R}_n^{-1}$ approximates the interference nullspace. Accordingly, the optimum minimum variance distortionless response (MVDR) adaptive beamforming filter,

$$\mathbf{w}_{\mathrm{opt}} = \gamma\mathbf{R}_n^{-1}\mathbf{s} = \frac{\gamma}{\sigma^2}(\mathbf{I} - \mathbf{V}_I(\mathbf{V}_I^H\mathbf{V}_I)^{-1}\mathbf{V}_I^H)\mathbf{s}, \tag{4.7}$$

resides in the interference nullspace and has become the interference eigencanceler. Here γ is a constant that does not affect the $\mathrm{SINR}_{\mathrm{out}}$. The relationship between the

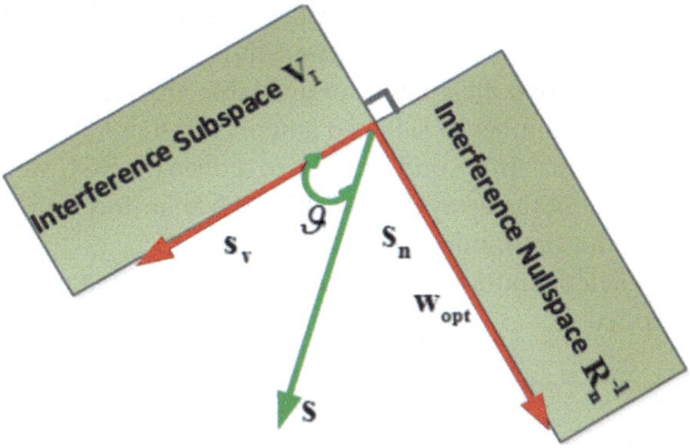

Fig. 4.1 The relationship between the optimum beamforming filter $\mathbf{w}_{\text{opt}}$, the interference subspace $\mathbf{V}_I$ and the nullspace $\mathbf{R}_n^{-1}$

optimum beamforming filter $\mathbf{w}_{\text{opt}}$, the interference subspace $\mathbf{V}_I$ and the interference nullspace $\mathbf{R}_n^{-1}$ is shown in Fig. 4.1. The steering vector of the desired signal $\mathbf{s}$ can be decomposed into two orthogonal components: the interference subspace component $\mathbf{s}_v$ and the nullspace component $\mathbf{s}_n$, i.e. $\mathbf{s} = \mathbf{s}_v + \mathbf{s}_n$ with $\mathbf{s}_v = (\mathbf{V}_I(\mathbf{V}_I^H\mathbf{V}_I)^{-1}\mathbf{V}_I^H)\mathbf{s}$ and $\mathbf{s}_n = (\mathbf{I} - \mathbf{V}_I^H\mathbf{V}_I)^{-1}\mathbf{V}_I^H)\mathbf{s}$ respectively. The optimum beamforming filter, i.e. eigencancelor weight vector $\mathbf{w}_{\text{opt}}$, is along the interference nullspace direction $\mathbf{s}_n$. We define the SCC as the absolute value of the cosine of the angle between the desired signal $\mathbf{s}$ and interference subspace component $\mathbf{s}_v$,

$$|\alpha| = |\cos\theta| = \left|\frac{\mathbf{s}^H\mathbf{s}_v}{||\mathbf{s}||_2||\mathbf{s}_v||_2}\right|. \tag{4.8}$$

Here the length of $\mathbf{s}$ is $||\mathbf{s}||_2 = \sqrt{N}$ under the assumption of isotropic antennas. Since the SINR_{out} is directly related to the squared value of the SCC, substituting the expression of $\mathbf{s}_v$ into Eq. (4.8) and taking the squared value yields

$$\begin{aligned}|\alpha|^2 &= \left|\frac{\mathbf{s}^H\mathbf{s}_v}{\sqrt{N}||\mathbf{s}_v||_2}\right|^2 = \frac{|\mathbf{s}^H\mathbf{V}_I(\mathbf{V}_I^H\mathbf{V}_I)^{-1}\mathbf{V}_I^H\mathbf{s}|^2}{N||\mathbf{V}_I(\mathbf{V}_I^H\mathbf{V}_I)^{-1}\mathbf{V}_1^H\mathbf{s}||_2^2}, \\ &= \frac{1}{N}\mathbf{s}^H\mathbf{V}_I(\mathbf{V}_I^H\mathbf{V}_I)^{-1}\mathbf{V}_I^H\mathbf{s}.\end{aligned} \tag{4.9}$$

It is clear that $|\alpha|^2 \leq 1$ and the SCC can be interpreted as the cosine of the angle between the desired signal and the interference subspace as shown in Fig. 4.1. Small values of absolute SCC indicate spatial dissimilarity between the desired signal and interference, and orthogonality is achieved when SCC is zero.

Finally, the SINR_{out} becomes

$$\text{SINR}_{\text{out}} = \sigma_s^2\mathbf{s}^H\mathbf{R}_n^{-1}\mathbf{s} = \text{SNR}\cdot N(1 - |\alpha|^2), \tag{4.10}$$

where σ_s^2 denotes the power of the desired signal and SNR is the signal-to-noise ratio. Equation (4.10) shows that the SINR_{out} of the interference eigencancelor depends on two factors: the number of available antennas N and the squared SCC value $|\alpha|^2$. When the number of selected antennas is fixed, the performance can be improved by changing the array configuration to reduce the SCC value. Thus the SCC characterizes the effect of the array geometry on the beamforming performance and is an effective metric for optimum subarray selection.

When the number of interferences is reduced to one, the SCC in Eq. (4.8) can be simplied as,

$$|\alpha| = \frac{|\mathbf{s}^H \mathbf{v}|}{N}. \tag{4.11}$$

The inverse covariance in Eq. (4.5) is simplified as,

$$\mathbf{R}_n^{-1} = \sigma^{-2}\left(\mathbf{I} - \frac{\mathbf{v}^H \mathbf{v}}{\sigma^2 \sigma_j^2 + N}\right). \tag{4.12}$$

Accordingly, the output SINR in Eq. (4.10) can be rewritten as,

$$\text{SINR}_{\text{out}} = N\text{SNR}(1 - |\alpha|^2 \rho), \tag{4.13}$$

where $\text{SNR} = \sigma_s^2/\sigma_j^2$ and ρ is an interference to noise figure given in terms of the interference to noise ratio $\text{INR} = \sigma_j^2/\sigma^2$ as,

$$\rho = \frac{N\text{INR}}{1 + N\text{INR}}. \tag{4.14}$$

The interference to noise figure ρ characterizes the relative effects of the white noise and the interference on the array performance. It satisfies $0 \leq \rho < 1$. A value that is close to 0 indicates that the white noise is the dominant source of performance degradation, and a value close to 1 implies that the interference dominates the noise. We now make the following observations:

1. When the noise is dominant, ρ is close to 0 and the effect of the SCC is suppressed. In this case, the array gain is linear with the number of array elements . Therefore, designing a sparse array with a smaller number of antennas can incur a significant performance loss.
2. If ρ is large (close to 1), the interference is dominant and the effect of the SCC is pronounced. Then the optimum weight vector $\mathbf{w}_{\text{opt}}$ lies in the interference nullspace. As a result, the spatial filter will also suppress the interference directional component of the desired signal. Therefore, designing a sparse array with a smaller number of antennas that minimises the SCC, i.e. makes the interference and the desired signal identically or approximately orthogonal with respect to the array configuration, may only incur a small performance loss as compared with the full array.

Since the focus of adaptive beamforming is interference mitigation, we assume the interference is much stronger than the noise in the rest part of the chapter, which is also the typical scenario in practice.

4.2.2 Antenna Selection for Adaptive Beamforming

In this section, we first incorporate antenna selection into the SCC expression and then utilize the properties of Schur complement and iterative reweighted algorithm to solve the antenna selection problem.

We implement antenna selection on the derived SCC parameter. Define the binary selection vector $\mathbf{x} \in \{0, 1\}^N$ with "one" meaning the corresponding antenna is selected and "zero" meaning discarded. Then, the SCC of the selected subarray can be expressed as

$$|\alpha(\mathbf{x})|^2 = \frac{1}{K}\mathbf{s}^H \text{diag}(\mathbf{x})\mathbf{V}_I(\mathbf{V}_I^H \text{diag}(\mathbf{x})\mathbf{V}_I)^{-1}\mathbf{V}_I^H \text{diag}(\mathbf{x})\mathbf{s}, \tag{4.15}$$

where K is the number of selected antennas. Thus the antenna selection problem in terms of minimizing the SCC is

$$\begin{aligned} \min_{\mathbf{x}} \quad & |\alpha(\mathbf{x})|^2, \\ \text{s.t.} \quad & \mathbf{x} \in \{0, 1\}^N, \quad \mathbf{1}^T\mathbf{x} = K. \end{aligned} \tag{4.16}$$

Utilizing the Schur complement of a block matrix, we can transform the minimum value of SCC into the following equivalent linear matrix inequality (LMI) constraint optimization problem,

$$\begin{aligned} \min_{\mathbf{x}} \quad & \delta, \\ \text{s.t.} \quad & \begin{bmatrix} \mathbf{V}_I^H \text{diag}(\mathbf{x})\mathbf{V}_I & \mathbf{V}_I^H \text{diag}(\mathbf{x})\mathbf{s} \\ \mathbf{s}^H \text{diag}(\mathbf{x})\mathbf{V}_I & K\delta \end{bmatrix} \succeq 0, \end{aligned} \tag{4.17}$$

Let us define the feasible set $\mathcal{S} = \{\mathbf{x} \in \{0, 1\}^N : \mathbf{1}^T\mathbf{x} = K\}$, which comprises the extreme points of the polytope $\mathcal{D} = \{\mathbf{x} : 0 \preceq \mathbf{x} \preceq 1,\ \mathbf{1}^T\mathbf{x} = K\}$. Substituting Eq. (4.17) into Eq. (4.16) and relaxing the binary constraints by replacing the feasible set $\mathcal{S}$ by the polytope $\mathcal{D}$, we can obtain,

$$\begin{aligned} \min_{\mathbf{x}} \quad & \delta, \\ \text{s.t.} \quad & \begin{bmatrix} \mathbf{V}_I^H \text{diag}(\mathbf{x})\mathbf{V}_I & \mathbf{V}_I^H \text{diag}(\mathbf{x})\mathbf{s} \\ \mathbf{s}^H \text{diag}(\mathbf{x})\mathbf{V}_I & K\delta \end{bmatrix} \succeq 0, \\ & \mathbf{x} \in \mathcal{D}. \end{aligned} \tag{4.18}$$

As for the above problem, we can solve it in a convex form utilizing a modified iterative reweighted l_1-norm to promote the binary sparsity via a set of judiciously designed reweights. In the jth iteration, Eq. (4.18) can be rewritten into,

$$\begin{aligned} \min_{\mathbf{x}} \quad & \delta + \iota_0 \mathbf{p}^{(j)T} \mathbf{x}, \\ \text{s.t.} \quad & \begin{bmatrix} \mathbf{V}_I^H \text{diag}(\mathbf{x}) \mathbf{V}_I & \mathbf{V}_I^H \text{diag}(\mathbf{x}) \mathbf{s} \\ \mathbf{s}^H \text{diag}(\mathbf{x}) \mathbf{V}_I & K\delta \end{bmatrix} \succeq 0, \\ & 0 \leq \mathbf{x} \leq 1, \\ & \mathbf{1}^T \mathbf{x} = K, \end{aligned} \tag{4.19}$$

where ι_0 is an artificially-set trade-off parameters between the SCC and the boolean sparsity and $\mathbf{p}^{(j)}$ is modified reweighted coefficient vectors. For the j-th iteration, the reweighted coefficient of the i-th antenna, that is the i-th element of $\mathbf{p}^{(j)}$, is updated by,

$$[\mathbf{p}^{(j)}]_i = \frac{1 - [\mathbf{x}^{(j-1)}]_i}{1 - e^{-\beta_1 [\mathbf{x}^{(j-1)}]_i} + \epsilon_0} - (\frac{1}{\epsilon_0})([\mathbf{x}^{(j-1)}]_i)^{\beta_2}, \tag{4.20}$$

where $\mathbf{x}^{(j-1)}$ is the optimal solution of the $(j-1)$-th iteration, β_1 and β_2 are two artificially-set parameters and ϵ_0 is preset to avoid the value of the denominator from being "0". According to the characteristics of the reweighted iterative strategy, Eq. (4.19) can converge to a local optimal solution of Eq. (4.18).

4.2.3 Simulations

In the following, we present simulation results to show the advantage of our approach. The desired signal is fixed at $\theta_s = 0.1\pi, \phi_s = 0.2\pi$ radian. We consider two scenarios: In the first, the desired signal is close to the interference subspace, whereas in the second, the two are well separated. In the first scenario, the first and second interferences are arriving from $\theta_1 = 0.15\pi, \phi_1 = 0.25\pi$ radian and $\theta_2 = 0.2\pi, \phi_2 = 0.3\pi$ radian. In the second scenario, the two interferences arrive from $\theta_1 = 0.3\pi, \phi_1 = 0.4\pi$ radian and $\theta_2 = 0.35\pi, \phi_2 = 0.3\pi$ radian. We adopt a 4×4 square array as the full layout and present the trade-off curve between the output SINR and cost in Fig. 4.2 to illustrate the determination of the number of selected antennas. The number K is changing from 3 to 16 in steps of 1. We calculate the normalised output SINR and computational cost by taking the entire full array as a reference, where the computational cost is of order K^3. Observe that using an 8-antenna subarray saves 87.5% of computational cost with only 1 dB performance degradation in the close scenario and 2.8 dB SINR loss in the far scenario. This is a significant saving in computational load for a modest performance loss. Note that

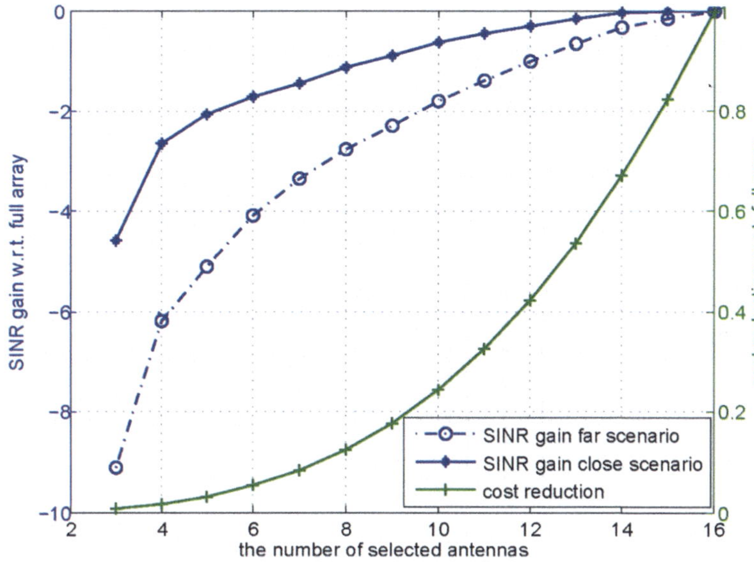

Fig. 4.2 The trade-off curve between the performance and the computational cost for both far and close scenarios

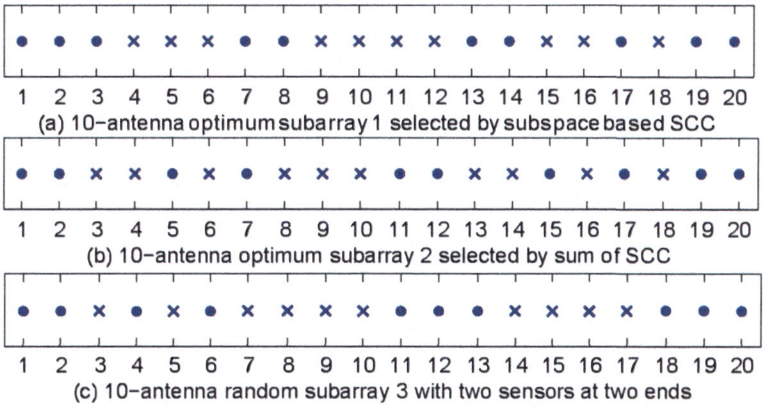

Fig. 4.3 The structure of three 10-antenna subarrays

this does not take into account the additional hardware saving due to the reduction in the number of front ends, which is equal to the reduction in the number of antennas.

Next, we select 10 antennas from a 20-antenna uniform linear array for enhanced interference nulling. The desired signal arrives from 60° in elevation with SNR being –20 dB and four interferences coming from 45°, 55°, 65°, 70° respectively with INR all being 30 dB. We select two optimum subarrays through minimizing the proposed subspace based SCC ignoring the fact that the interferences are not orthogonal with each other as shown in Fig. 4.3. We also select a third subarray with two antennas fixed

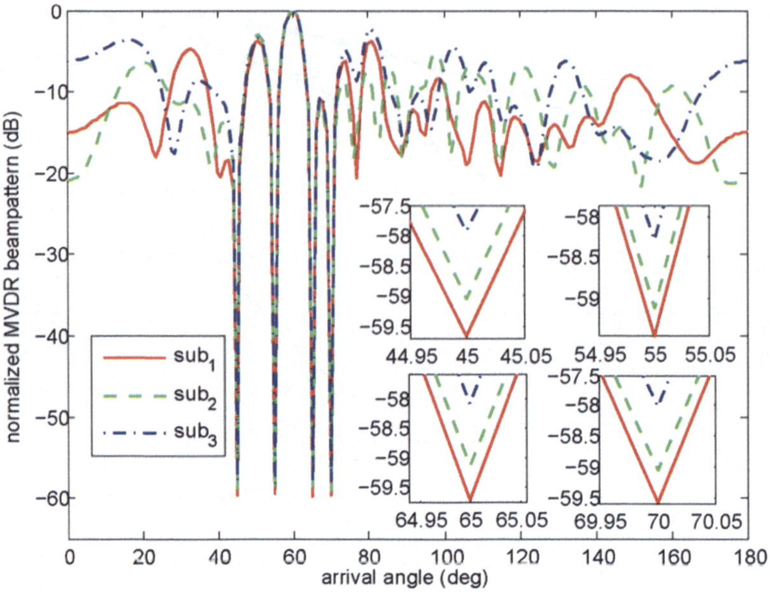

Fig. 4.4 MVDR beampatterns of three lO-antenna subarrays as shown in Fig. 4.3. The number of time snapshots is 100 and 1000 Monte Carlo runs are used

at two ends and other eight randomly spaced in between for comparison. The MVDR beampatterns of the three arrays are shown in Fig. 4.4 by averaging 1000 Monte-Carlo simulations. We can see that the first subarray, denoted as "sub1", produces deeper nulls than the other two subarrays, but exhibits nearly same mainlobe width and peak sidelobe level.

4.3 Reconfigurable Sparse Arrays in Multi-source Case

In this section, we analyze the effect of non-uniform array configurations on adaptive beamforming for enhanced SINR in the case of multi-beam forming. The array is configured using a given number of antennas or via a selection of subset of antennas from a larger available set, leading to sparse arrays in both cases. The bounds on the highest achievable SINR for a given number of antennas are formulated and used to offer new insights into adaptive beamforming. The upper and lower bounds underscore the role of array configurations in optimizing performance for interference-free and interference-active environments, respectively. This section considers the general case of multiple sources and interferences in the FoV. We formulate three angles, namely, eigenspace angle, conventional angle, and minimum canonical angle for characterizing spatial separation between the source and interference subspaces, and as such, represents performance loss incurred by interference nulling. Three sparse array design methods, incorporating these angles, are proposed. Simulation examples confirm the role of subspace angles in optimum beamforming and validate the utility of antenna selection algorithms for sparse array design.

4.3.1 Problem Formulation

Consider a non-uniform linear array of K antennas with positions specified by multiple integer of unit inter-element spacing $d_n d$, $d_n \in \mathbb{N}, n = 1, \ldots, K$. The symbol $\mathbb{N}$ denotes the set of integer numbers. Suppose that p target sources are impinging on the array from directions $\{\psi_1, \ldots, \psi_p\}$ and q strong interfering signals are arriving from directions $\{\phi_1, \ldots, \phi_q\}$. The corresponding steering vectors are,

$$\begin{aligned} \mathbf{s}_k &= [e^{jk_0 d_1 d \cos\psi_k}, \ldots, e^{jk_0 d_N d \cos\psi_k}]^T, k = 1, \ldots, p, \\ \mathbf{v}_l &= [e^{jk_0 d_1 d \cos\phi_l}, \ldots, e^{jk_0 d_N d \cos\phi_l}]^T, l = 1, \ldots, q, \end{aligned} \tag{4.21}$$

where the wavenumber is defined as $k_0 = 2\pi/\lambda$ with λ being the wavelength and ^{T} denotes the transpose operation. The received signal at time instant t is given by,

$$\mathbf{x}(t) = \mathbf{S}\mathbf{u}_s(t) + \mathbf{V}\mathbf{u}_i(t) + \mathbf{n}(t), \tag{4.22}$$

where $\mathbf{S} = [\mathbf{s}_1, \ldots, \mathbf{s}_p]$, $\mathbf{V} = [\mathbf{v}_1, \ldots, \mathbf{v}_q]$ are the source and interference array manifold matrices with the full column rank. In the above equation, $\mathbf{u}_s(t) \in \mathbb{C}^p$, $\mathbf{u}_i(t) \in \mathbb{C}^q$ are, respectively, the statistically independent source and interfering signals, $\mathbf{n}(t) \in \mathbb{C}^K$ denotes the received Gaussian noise vector. The symbol $\mathbb{C}$ denotes the set of complex numbers. The output of the K-antenna beamformer is given by,

$$y(t) = \mathbf{w}^H \mathbf{x}(t); \tag{4.23}$$

where $\mathbf{w} \in \mathbb{C}^K$ is the complex vector of beamformer weights and ^{H} stands for the Hermitian operation. The optimum weight vector for minimizing the output variance while keeping the desired signal distortionless can be expressed as the following optimization problem [1],

$$\begin{aligned} \min_{\mathbf{w}} \quad & \mathbf{w}^H \mathbf{R}_n \mathbf{w}, \\ \text{s.t.} \quad & \mathbf{w}^H \mathbf{R}_s \mathbf{w} = 1. \end{aligned} \tag{4.24}$$

The abbreviation "s.t." stands for "subject to", $\mathbf{R}_s$ and $\mathbf{R}_n$ are defined as $\mathbf{R}_s = \mathbf{S}\mathbf{C}_s\mathbf{S}^H$ and $\mathbf{R}_n = \mathbf{V}\mathbf{C}_v\mathbf{V}^H + \sigma_n^2\mathbf{I}$. Here, $\mathbf{C}_s = E\{\mathbf{u}_s(t)\mathbf{u}_s^H(t)\}$ and $\mathbf{C}_v = E\{\mathbf{u}_i(t)\mathbf{u}_i^H(t)\}$ denote the source and interference correlation matrices, respectively, and σ_n^2 is the noise power level. Without loss of generality, we assume unit noise power $\sigma_n^2 = 1$.

Utilizing the Lagrange multiplier, the minimum variance distortionless response (MVDR) beamformer can be found as,

$$\mathbf{w}_{\text{opt}} = \mathcal{P}\{\mathbf{R}_n^{-1}\mathbf{R}_s\}, \tag{4.25}$$

where $\mathcal{P}\{\cdot\}$ denotes the principal eigenvector of the matrix. Under the assumption of Gaussian noise, the MVDR beamformer also exhibits MaxSINR, which can be expressed as,

$$\text{SINR}_{\text{opt}} = \frac{\mathbf{w}_{\text{opt}}^H \mathbf{R}_s \mathbf{w}_{\text{opt}}}{\mathbf{w}_{\text{opt}}^H \mathbf{R}_n \mathbf{w}_{\text{opt}}} = \lambda_{\max}\{\mathbf{R}_n^{-1}\mathbf{R}_s\}, \tag{4.26}$$

The above SINR is the maximum eigenvalue of matrix product $\mathbf{R}_n^{-1}\mathbf{R}_s$. With no interference present, the optimum quiescent adaptive beamformer in Eq. (4.25) becomes,

$$\mathbf{w}_{\text{q}} = \mathbf{e}_0 = \mathcal{P}\{\mathbf{R}_s\}. \tag{4.27}$$

The associated MaxSNR is,

$$\text{SNR}_{\text{q}} = \frac{\lambda_{\max}\{\mathbf{R}_s\}}{\sigma_n^2} = \lambda_0, \tag{4.28}$$

where λ_0 denotes the maximum eigenvalue of $\mathbf{R}_s$ and $\mathbf{e}_0$ is the corresponding eigenvector. Clearly, from Eq. (4.28), array configurations affect the output SNR in the interference-free case through the term $\lambda_0 = \|\mathbf{R}_s\|_2$.

4.3.2 *Performance Analysis of MVDR Beamformer*

Utilizing the matrix inversion lemma, the interference plus noise inverse covariance matrix can be written as,

$$\begin{aligned} \mathbf{R}_n^{-1} &= \frac{1}{\sigma_n^2}\left[\mathbf{I} - \mathbf{V}(\mathbf{C}_j + \mathbf{V}^H\mathbf{V})^{-1}\mathbf{V}^H\right], \\ &= \frac{1}{\sigma_n^2}\left[\mathbf{I} - \hat{\mathbf{P}}_{\mathbf{V}}\right], \end{aligned} \tag{4.29}$$

where $\mathbf{C}_j = \sigma_n^2\mathbf{C}_v^{-1}$. Note that $\hat{\mathbf{P}}_{\mathbf{V}}$ approximates the projection matrix, $\mathbf{P}_{\mathbf{V}} = \mathbf{V}(\mathbf{V}^H\mathbf{V})^{-1}\mathbf{V}^H$, into the interference subspace. When the interferences are much stronger than noise, the matrix $\mathbf{C}_j \approx 0$, and in turn, $\hat{\mathbf{P}}_{\mathbf{V}} = \mathbf{P}_{\mathbf{V}}$ [175, 176]. In this paper, we consider the case of open-loop adaptive beamforming for sparse arrays in strong interference environments, and derive the bounds on the respective SINR. Substituting Eq. (4.29) into Eq. (4.26), we formulate the upper bound on the output SINR,

$$
\begin{aligned}
\text{SINR}_{\text{opt}} &= \lambda_{\max}\{\mathbf{R}_n^{-1}\mathbf{R}_s\}, \\
&= \frac{1}{\sigma_n^2}\lambda_{\max}\{\mathbf{R}_s - \mathbf{P}_{\mathbf{V}}\mathbf{R}_s\}, \\
&\leq \frac{1}{\sigma_n^2}\lambda_{\max}\{\mathbf{I} - \mathbf{P}_{\mathbf{V}}\}\lambda_{\max}\{\mathbf{R}_s\}, \\
&= \lambda_{\max}\{\mathbf{R}_s\} = \lambda_0.
\end{aligned} \tag{4.30}
$$

Clearly, the upper bound of MaxSINR in interference-active environments is essentially the MaxSNR when no interference is present. Nulling or significantly reducing interference power results in some degradation of the desired sources, unless the source and interference subspaces are mutually orthogonal,

$$
\mathbf{P}_{\mathbf{V}}\mathbf{S} = \mathbf{0} \rightarrow \mathbf{P}_{\mathbf{V}}\mathbf{R}_s = \mathbf{0}. \tag{4.31}
$$

From Eq. (4.30), the MaxSINR is essentially the maximum eigenvalue of the component of the matrix $\mathbf{R}_s$ projected onto the interference null subspace. It is also viewed as the spectral norm of the difference between two positive definite matrices, $\mathbf{R}_s$ and $\mathbf{R}_s\mathbf{P}_{\mathbf{V}}$. To maximize the spectral norm, we need to set the latter to zero. Accordingly, array reconfiguration for performance enhancement mandates two tasks to be carried out simultaneously. The first task is to maximize the upper bound defined by $\lambda_0 = \lambda_{\max}\{\mathbf{R}_s\}$. The second task is attempting to reach orthogonality between the interference subspace $\tilde{\mathbf{V}} = \text{span}(\mathbf{V})$ and the source subspace $\tilde{\mathbf{S}} = \text{span}(\mathbf{S})$. The first task was the subject of our investigation in Sect. 3.2 of Chap. 3. In this section, we focus on the intricate problem of considering the two tasks combined with emphasis on minimizing the loss of beamforming sensitivity to the desired signals as a consequence of interference nulling.

4.3.3 Spatial Separation Between Two Subspaces

As stated in Sect. 4.3.2, the performance loss incurred from the interference-free to interference-active environments is caused by the non-orthogonality between the source and interference subspaces. The separation between two vector subspaces are typically characterized by conventional angle and canonical angle [177–180]. In this section, we employ these two angles to describe the spatial separation between the source and interference subspaces. Further, we introduce the eigenspace angle to transform the multi-source multi-interference problem into a single component case that enables simplified performance analysis. The relationship between the above three angles and the beamformer output SINR are elucidated as well.

4.3.3.1 Eigenspace Angle

The definition of the eigenspace angle is tantamount to the case of single source and multiple interferences in Sect. 4.2. The principal eigenvector of the source covariance matrix $\mathbf{e}_0$ can be decomposed into two orthogonal components, as shown in Fig. 4.5: the interference subspace component $\mathbf{e}_v$ and the interference nullspace component $\mathbf{e}_n$, i.e., $\mathbf{e}_0 = \mathbf{e}_v + \mathbf{e}_n$ with $\mathbf{e}_v = \mathbf{P_V}\mathbf{e}_0$ and $\mathbf{e}_n = (\mathbf{I} - \mathbf{P_V})\mathbf{e}_0$, respectively. The squared cosine of the eigenspace angle is defined as follows,

$$\cos^2\vartheta = \frac{|\mathbf{e}_0^H\mathbf{e}_v|^2}{\|\mathbf{e}_0\|_2^2\|\mathbf{e}_v\|_2^2} = \mathbf{e}_0^H\mathbf{P_V}\mathbf{e}_0, \tag{4.32}$$

where we used the fact that $\|\mathbf{e}_0\|_2 = 1$. In the following, we show that the output SINR can be characterized by its lower bound, formulated in terms of the eigenspace angle, i.e.,

$$\begin{aligned} \mathrm{SINR_{opt}} &\approx \frac{1}{\sigma_n^2}\lambda_{\max}\{(\mathbf{I}-\mathbf{P_V})\mathbf{R}_s\} \\ &= \max\nolimits_{\|\mathbf{x}\|_2=1}\mathbf{x}^H(\mathbf{I}-\mathbf{P_V})\mathbf{R}_s\mathbf{x} \\ &= \max\nolimits_{\|\mathbf{x}\|_2=1}\mathbf{x}^H\mathbf{R}_s\mathbf{x} - \mathbf{x}^H\mathbf{P_V}\mathbf{R}_s\mathbf{x} \\ &\geq \lambda_0 - \mathbf{e}_0^H\mathbf{P_V}\mathbf{R}_s\mathbf{e}_0, \\ &= \lambda_0(1-\mathbf{e}_0^H\mathbf{P_V}\mathbf{e}_0), \\ &= \lambda_0(1-\cos^2\vartheta). \end{aligned} \tag{4.33}$$

Combining Eqs. (4.30) and (4.33) gives upper and lower bounds of the output SINR, i.e.,

$$\mathcal{B} = \lambda_0(1-\cos^2\vartheta) \leq \mathrm{SINR_{opt}} \leq \lambda_0, \tag{4.34}$$

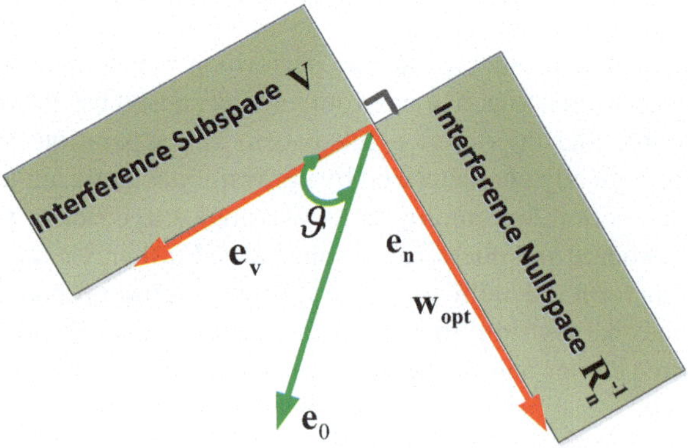

Fig. 4.5 The illustration of the eigenspace angle in multi-source cases

The above bounds are applicable for any array configuration. To achieve optimum sparse array configuration employing a limited number of antennas, we must consider all available antenna positions. This is equivalent to having access to a fully populated array, followed by antenna selections. However, it must be pointed out that, unlike steering vectors, eigenvectors are not directional, where the corresponding entries can be easily removed with regard to discarded antennas. Consequently, in the process of array thinning, the eigenvectors of selected sparse arrays are not the selected sparse eigenvectors of fully populated array and, as such, must be recomputed through eigenvalue decomposition. It is then desirable that the eigenspace angle be defined in terms of the source and interference array manifold matrices. This can be accomplished by representing the eigenvector $\mathbf{e}_0$ as a linear combination of the source steering vectors, i.e., $\mathbf{e}_0 = \mathbf{SN}\boldsymbol{\nu}$ with $\boldsymbol{\nu}$ denoting the coefficient vector and $\mathbf{N} = (\mathbf{S}^H\mathbf{S})^{-1/2}$ being the normalization matrix, such that $\|\boldsymbol{\nu}\|_2 = 1$. Then, the squared cosine value of the eigenspace angle can be rewritten in a more useful form, as follows,

$$\cos^2\vartheta = \boldsymbol{\nu}^H\mathbf{N}\mathbf{S}^H\mathbf{P}_\mathbf{V}\mathbf{S}\mathbf{N}\boldsymbol{\nu}. \tag{4.35}$$

4.3.3.2 Conventional Angle

Define the p-dimensional source and the q-dimensional interference subspaces as $\tilde{\mathbf{S}} = \text{span}\{\mathbf{S}\}$ and $\tilde{\mathbf{V}} = \text{span}\{\mathbf{V}\}$, respectively. The commonly used formula for conventional angle is expressed in terms of the generalized inner product and the generalized norm, as stated in Proposition 4.1 [178].

Proposition 4.1 *An explicit formula for the squared cosine of the conventional angle θ between the two subspaces $\tilde{\mathbf{S}}$ and $\tilde{\mathbf{V}}$ for arbitrary $p \le q$ is,*

$$\cos^2\theta = \frac{|\tilde{\boldsymbol{W}}\boldsymbol{W}^H|}{\|\boldsymbol{s}_1, \ldots, \boldsymbol{s}_p\|^2 \cdot \|\boldsymbol{v}_1, \ldots, \boldsymbol{v}_q\|^{2p}}, \tag{4.36}$$

where $\boldsymbol{W}$ is defined in terms of Euclidean inner product, i.e.,

$$\boldsymbol{W} = [\langle \boldsymbol{s}_i, \boldsymbol{v}_k\rangle, i = 1, \ldots, p, k = 1, \ldots, q] = \boldsymbol{S}^H\boldsymbol{V}$$

and

$$\tilde{\boldsymbol{W}} = \left[\langle \boldsymbol{s}_i, \boldsymbol{v}_k | \boldsymbol{v}_{i_2(k)}, \ldots, \boldsymbol{v}_{i_q(k)}\rangle, i = 1, \ldots, p, k = 1, \ldots, q, \right]$$

is defined in terms of generalized inner product, which can be calculated from the matrix determinant,

$$\langle \boldsymbol{s}_i, \boldsymbol{v}_k | \boldsymbol{v}_{i_2(k)}, \ldots, \boldsymbol{v}_{i_q(k)}\rangle = |\boldsymbol{V}_{(k)}^H\boldsymbol{V}|. \tag{4.37}$$

Here $\{i_2(k), \ldots, i_q(k)\} = \{1, \ldots, q\} \setminus \{k\}$ and $\boldsymbol{V}_{(k)}$ is the matrix with the kth column replaced by the vector $\boldsymbol{s}_i$. The denominator in Eq. (4.36) comprises two generalized norms, defined as $\|\boldsymbol{s}_1, \ldots, \boldsymbol{s}_p\| = \langle \boldsymbol{s}_1, \boldsymbol{s}_1 | \boldsymbol{s}_2, \ldots, \boldsymbol{s}_p\rangle^{1/2} = |\boldsymbol{S}^H\boldsymbol{S}|^{1/2}$ and

$\|v_1, \ldots, v_q\| = \langle v_1, v_1 | v_2, \ldots, v_q \rangle^{1/2} = |\mathbf{V}^H \mathbf{V}|^{1/2}$. *For the physical meaning of generalized inner product and proof of Proposition 4.1, we refer the reader to [178] and references therein.*

In the following, we present a simplified form of Eq. (4.36) which is both expressible in terms of source and interference manifold matrices and is more relevant to the problem at hand. The cosine value of the conventional angle θ between the source and interference subspaces is, in essence, the ratio between the volume of the p-dimensional parallelepiped spanned by the projection of $\mathbf{S}$ on the interference subspace $\tilde{\mathbf{V}}$ and the p-dimensional parallelepiped spanned by $\mathbf{S}$ [181], i.e.,

$$\cos^2\theta := |\hat{\mathbf{S}}^H \hat{\mathbf{S}}| / |\mathbf{S}^H \mathbf{S}|, \tag{4.38}$$

where $\hat{\mathbf{s}}_i$'s denote the projection vector of $\mathbf{s}_i$'s on $\tilde{\mathbf{V}}$, i.e., $\hat{\mathbf{s}}_i = \mathbf{P}_\mathbf{V}\mathbf{s}_i, i = 1, \ldots, p$ and $\hat{\mathbf{S}} = [\hat{\mathbf{s}}_1, \ldots, \hat{\mathbf{s}}_p]$. Note that the projection of $\mathbf{s}_i$'s on $\tilde{\mathbf{V}}$ is independent of the choice of basis of $\tilde{\mathbf{V}}$. Further, since projections are linear transformations, the above ratio is also invariant with the basis of $\tilde{\mathbf{S}}$. Proceeding from Eq. (4.38) and substituting $\hat{\mathbf{S}}$ with $\mathbf{P}_\mathbf{V}\mathbf{S}$, we obtain,

$$\cos^2\theta = \frac{|\mathbf{S}^H \mathbf{P}_\mathbf{V} \mathbf{S}|}{|\mathbf{S}^H \mathbf{S}|} = \frac{|\mathbf{S}^H \mathbf{V}(\mathbf{V}^H \mathbf{V})^{-1} \mathbf{V}^H \mathbf{S}|}{|\mathbf{S}^H \mathbf{S}|}. \tag{4.39}$$

Since $\mathbf{S}^H\mathbf{S} = \hat{\mathbf{S}}^H\hat{\mathbf{S}} + \mathbf{S}^H(\mathbf{I} - \mathbf{P}_\mathbf{V})\mathbf{S}$, then $\mathbf{S}^H\mathbf{S} - \hat{\mathbf{S}}^H\hat{\mathbf{S}} \geq 0$. This implies that every eigenvalue of the matrix $\hat{\mathbf{S}}^H\hat{\mathbf{S}}$ is no higher than the corresponding eigenvalue of the matrix $\mathbf{S}^H\mathbf{S}$, i.e., $\lambda_i \geq \hat{\lambda}_i \geq 0, i = 1, \ldots, p$ [181]. Utilizing this fact in Eq. (4.39) yields,

$$\cos^2\theta = \frac{\prod_{i=1}^{p} \hat{\lambda}_i}{\prod_{i=1}^{p} \lambda_i} \leq 1. \tag{4.40}$$

Thus, the ratio in Eq. (4.38) is a fraction number in [0, 1] and is independent of the choice of basis of $\tilde{\mathbf{S}}$ and $\tilde{\mathbf{V}}$.

Note that the expression of the conventional angle in Eq. (4.39) involves the matrix inversion term $(\mathbf{V}^H\mathbf{V})^{-1}$. The matrix $\mathbf{V}^H\mathbf{V}$ may become singular in the search for optimum sparse arrays. This property is a byproduct of the identification issue of nonuniform arrays. Accordingly, a more reliable and explicit formula of the conventional angle between two subspaces becomes essential. Define the matrix in the numerator of Eq. (4.39) as $\mathbf{Q} = \mathbf{S}^H\mathbf{P}_\mathbf{V}\mathbf{S}$,

$$\mathbf{Q} = \begin{bmatrix} \mathbf{s}_1^H \mathbf{P}_\mathbf{V} \mathbf{s}_1 & \cdots & \mathbf{s}_1^H \mathbf{P}_\mathbf{V} \mathbf{s}_p \\ \vdots & \ddots & \vdots \\ \mathbf{s}_p^H \mathbf{P}_\mathbf{V} \mathbf{s}_1 & \cdots & \mathbf{s}_p^H \mathbf{P}_\mathbf{V} \mathbf{s}_p \end{bmatrix}. \tag{4.41}$$

Each entry of $\mathbf{Q}$ is basically the Euclidean inner product of two projections of $\mathbf{s}_i$ and $\mathbf{s}_j$ on the interference subspace $\tilde{\mathbf{V}}$, i.e., $\hat{\mathbf{s}}_i^H \hat{\mathbf{s}}_j = \mathbf{s}_i^H \mathbf{P}_\mathbf{V} \mathbf{s}_j, i, j = 1, \ldots, p$.

Consider the determinant of the matrix $|\mathbf{V}_{(\mathbf{s}_i)}^H \mathbf{V}_{(\mathbf{s}_j)}|$, $i, j = 1, \ldots, p$ with $\mathbf{V}_{(\mathbf{s}_i)} = [\mathbf{s}_i, \mathbf{V}]$,

$$
\begin{aligned}
|\mathbf{V}_{(\mathbf{s}_i)}^H \mathbf{V}_{(\mathbf{s}_j)}| &= \begin{vmatrix} \mathbf{s}_i^H \mathbf{s}_j & \mathbf{s}_i^H \mathbf{V} \\ \mathbf{V}^H \mathbf{s}_j & \mathbf{V}^H \mathbf{V} \end{vmatrix} \\
&= |\mathbf{V}^H \mathbf{V}|(\mathbf{s}_i^H \mathbf{s}_j - \mathbf{s}_i^H \mathbf{P}_{\mathbf{V}} \mathbf{s}_j).
\end{aligned} \tag{4.42}
$$

Thus, each entry of the matrix $\mathbf{Q}$ can be obtained by a simple manipulation of Eq. (4.42),

$$
\mathbf{s}_i^H \mathbf{P}_{\mathbf{V}} \mathbf{s}_j = \mathbf{s}_i^H \mathbf{s}_j - \frac{|\mathbf{V}_{(\mathbf{s}_i)}^H \mathbf{V}_{(\mathbf{s}_j)}|}{|\mathbf{V}^H \mathbf{V}|}. \tag{4.43}
$$

Define the matrix $\hat{\mathbf{Q}} = \left[|\mathbf{V}_{(\mathbf{s}_i)}^H \mathbf{V}_{(\mathbf{s}_j)}|, i, j = 1, \ldots, p\right]$. Replacing each entry of Eq. (4.41) with (4.43), we obtain

$$
\mathbf{Q} = \mathbf{S}^H \mathbf{S} - \frac{1}{|\mathbf{V}^H \mathbf{V}|} \hat{\mathbf{Q}}. \tag{4.44}
$$

Substituting Eq. (4.44) into Eq. (4.39) yields the expression of conventional angle between source and interference subspaces,

$$
\cos^2 \theta = \frac{||\mathbf{V}^H \mathbf{V}|(\mathbf{S}^H \mathbf{S}) - \hat{\mathbf{Q}}|}{|\mathbf{S}^H \mathbf{S}||\mathbf{V}^H \mathbf{V}|^p}. \tag{4.45}
$$

Utilizing the definition of generalized norms, the denominator of Eq. (4.45) is the same as that of Eq. (4.36). Thus, the numerator involved with the generalized inner-product in Eq. (4.36) can be replaced by its counterpart in Eq. (4.45), that is $|\mathbf{V}^H \mathbf{V}|(\mathbf{S}^H \mathbf{S}) - \hat{\mathbf{Q}} = \tilde{\mathbf{W}} \mathbf{W}^H$.

For the case where $\tilde{\mathbf{S}} = \text{span}(\mathbf{s})$, then the desired signal subspace is of rank one. In this case, Eq. (4.39) simplifies to

$$
\cos^2 \theta = \frac{1}{K} \mathbf{s}^H \mathbf{P}_{\mathbf{V}} \mathbf{s}, \tag{4.46}
$$

which agrees with Eq. (21) of the single-source case discussed in our previous work [176]. Here, we utilize the fact that $\mathbf{s}^H \mathbf{s} = K$. For Eq. (4.45), the matrix $\hat{\mathbf{Q}}$ becomes a scalar, that is $\hat{Q} = |\mathbf{V}_{\mathbf{s}}^H \mathbf{V}_{\mathbf{s}}|$. Substituting $\hat{Q}$ into Eq. (4.45) yields,

$$
\cos^2 \theta = \frac{K|\mathbf{V}^H \mathbf{V}| - |\mathbf{V}_{\mathbf{s}}^H \mathbf{V}_{\mathbf{s}}|}{K|\mathbf{V}^H \mathbf{V}|} = 1 - \frac{|\mathbf{V}_{\mathbf{s}}^H \mathbf{V}_{\mathbf{s}}|}{K|\mathbf{V}^H \mathbf{V}|}, \tag{4.47}
$$

which coincides with Eq. (27) in [176]. Clearly, conventional angle includes the single source scenario as a special case.

Next, we formulate the relationship between the conventional angle θ and the eigenspace angle ϑ. Assume the p eigenvalues of the matrix $\mathbf{NS}^H\mathbf{P}_\mathbf{V}\mathbf{SN}$ in Eq. (4.35) to be $1 \geq \zeta_1 \geq \zeta_2 \geq \cdots \geq \zeta_p$. Then,

$$\cos^2\theta = \frac{|\mathbf{S}^H\mathbf{P}_\mathbf{V}\mathbf{S}|}{|\mathbf{S}^H\mathbf{S}|} = |\mathbf{NS}^H\mathbf{P}_\mathbf{V}\mathbf{SN}| = \Pi_{i=1}^{p}\zeta_i \leq \zeta_p. \tag{4.48}$$

Moreover, for the eigenspace angle ϑ, we apply the definition of eigenvalue to Eq. (4.35) [132] to obtain that

$$\zeta_p \leq \cos^2\vartheta = \boldsymbol{\nu}^H\mathbf{NS}^H\mathbf{P}_\mathbf{V}\mathbf{SN}\boldsymbol{\nu} \leq \zeta_1. \tag{4.49}$$

Combining Eqs. (4.48) and (4.49), we can state that

$$\cos^2\theta \leq \cos^2\vartheta \Rightarrow 90^\circ \geq \theta \geq \vartheta \geq 0^\circ. \tag{4.50}$$

4.3.3.3 Canonical Angles

Canonical angles have been also utilized to describe the separation between two subspaces [180]. Let $\theta_1 \leq \theta_2 \leq \cdots \leq \theta_p$ be the canonical angles between $\tilde{\mathbf{S}}$ and $\tilde{\mathbf{V}}$, which are defined recursively by,

$$\begin{aligned} \cos\theta_1 &:= \max_{\mathbf{s}\in\tilde{\mathbf{S}}, \|\mathbf{s}\|_2=1} \max_{\mathbf{v}\in\tilde{\mathbf{V}}, \|\mathbf{v}\|_2=1} |\langle \mathbf{s}, \mathbf{v}\rangle|, \\ \cos\theta_{i+1} &:= \max_{\mathbf{s}\in\tilde{\mathbf{S}}_i, \|\mathbf{s}\|_2=1} \max_{\mathbf{v}\in\tilde{\mathbf{V}}_i, \|\mathbf{v}\|_2=1} |\langle \mathbf{s}, \mathbf{v}\rangle|, \end{aligned} \tag{4.51}$$

where $i = 1, \ldots, p-1$ and $\tilde{\mathbf{S}}_i$ is the orthogonal complement of $\mathbf{s}_i$ relative to $\tilde{\mathbf{S}}_{i-1}$ and $\tilde{\mathbf{V}}_i$ is the orthogonal complement of $\mathbf{v}_i$ relative to $\tilde{\mathbf{V}}_{i-1}$ (with $\tilde{\mathbf{S}}_0 = \tilde{\mathbf{S}}$ and $\tilde{\mathbf{V}}_0 = \tilde{\mathbf{V}}$). We utilize the minimum canonical angle $\tilde{\theta} = \theta_1$ to characterize the spatial separation between two subspaces. The relationship between the conventional angle and canonical angles is elucidated in Proposition 4.2 [180], which also provides a method of calculating the minimum canonical angle.

Proposition 4.2 *Assume that the two sets of basis vectors of* $\boldsymbol{S} = [\boldsymbol{s}_1, \ldots, \boldsymbol{s}_p]$ *and* $\boldsymbol{V} = [\boldsymbol{v}_1, \ldots, \boldsymbol{v}_q]$ *are orthonormal. Denote the eigenvalue decomposition of* $\boldsymbol{WW}^H = \boldsymbol{S}^H\boldsymbol{VV}^H\boldsymbol{S}$ *as* $\boldsymbol{A\Gamma A}^H$, *where each column* $\boldsymbol{a}_i, i = 1, \ldots, p$ *of* $\boldsymbol{A}$ *is an eigenvector, and* $\boldsymbol{\Gamma}$ *is a* $p \times p$ *diagonal matrix with real eigenvalues* $\gamma_1, \ldots, \gamma_p$ *in a descending order along the diagonal. Then,*

1. The cosine values of canonical angles equal to the square roots of eigenvalues, i.e., $\cos\theta_i = \sqrt{\gamma_i}, i = 1, \ldots, p$.
2. The squared cosine value of the conventional angle $\cos^2\theta$ is equal to the product of all squared cosine of canonical angles, i.e., $\cos^2\theta = \Pi_{i=1}^{p}\cos^2\theta_i$.

Proof Canonical angles can be expressed as the maximum values of $f(\mathbf{s}) := \max_{\mathbf{v}\in\tilde{\mathbf{V}}, \|\mathbf{v}\|_2=1} |\langle \mathbf{s}, \mathbf{v}\rangle|$, $\mathbf{s} \in \tilde{\mathbf{S}}$, $\|\mathbf{s}\|_2 = 1$. For each $\mathbf{s}$, the absolute value of the Euclidean inner product $|\langle \mathbf{s}, \mathbf{v}\rangle|$ is maximized on $\{\mathbf{v} \in \tilde{\mathbf{V}} : \|\mathbf{v}\|_2 = 1\}$ at

$$\mathbf{v} = \frac{\hat{\mathbf{s}}}{\|\hat{\mathbf{s}}\|_2} = \frac{\mathbf{V}\mathbf{V}^H\mathbf{s}}{\|\mathbf{V}\mathbf{V}^H\mathbf{s}\|_2}, \tag{4.52}$$

and the maximum value is,

$$f(\mathbf{s}) = \max_{\mathbf{v}\in\tilde{\mathbf{V}}, \|\mathbf{v}\|_2=1} |\langle \mathbf{s}, \mathbf{v}\rangle| = \|\hat{\mathbf{s}}\|_2. \tag{4.53}$$

Let us consider the function $g(\mathbf{s}) := f(\mathbf{s})^2$ first, given by,

$$g(\mathbf{s}) = \|\hat{\mathbf{s}}\|_2^2 = \|\mathbf{V}\mathbf{V}^H\mathbf{s}\|_2^2,\ \mathbf{s} \in \tilde{\mathbf{S}},\ \|\mathbf{s}\|_2 = 1. \tag{4.54}$$

Writing $\mathbf{s} = \sum_{i=1}^{p} a_i \mathbf{s}_i$, we have

$$\hat{\mathbf{s}} = (\mathbf{V}\mathbf{V}^H)\mathbf{s} = \sum_{i=1}^{p} a_i (\mathbf{V}\mathbf{V}^H)\mathbf{s}_i. \tag{4.55}$$

Then the function $g(\mathbf{s})$ can be rewritten as $g(\mathbf{a})$ with $\mathbf{a} = [a_1, \ldots, a_p]^T \in \mathbb{C}^p$, given by,

$$\begin{aligned} g(\mathbf{a}) &= \|\sum_{i=1}^{p} a_i (\mathbf{V}\mathbf{V}^H)\mathbf{s}_i\|_2^2 \\ &= \sum_{k=1}^{p}\sum_{i=1}^{p} a_i^* a_k \mathbf{s}_i^H (\mathbf{V}\mathbf{V}^H)\mathbf{s}_k. \end{aligned} \tag{4.56}$$

The function $g(\mathbf{a})$ can be further simplified in terms of the matrix $\mathbf{W} = \mathbf{S}^H\mathbf{V}$ as,

$$g(\mathbf{a}) = \mathbf{a}^H\mathbf{W}\mathbf{W}^H\mathbf{a} = \|\mathbf{W}^H\mathbf{a}\|_2^2,\ \|\mathbf{a}\|_2 = 1. \tag{4.57}$$

To find the maximum values of $g(\mathbf{a})$, we use the Lagrange method as follows,

$$h_\gamma(\mathbf{a}) := \|\mathbf{W}^H\mathbf{a}\|_2^2 + \gamma(1 - \|\mathbf{a}\|_2^2). \tag{4.58}$$

Setting the gradient of $h_\gamma(\mathbf{a})$ with respect to $\mathbf{a}$ to zero yields,

$$\nabla h_\gamma(\bar{\mathbf{a}}) = 2\mathbf{W}\mathbf{W}^H\bar{\mathbf{a}} - 2\bar{\gamma}\bar{\mathbf{a}} = 0. \tag{4.59}$$

Here $\bar{\mathbf{a}}$ and $\bar{\gamma}$ denote the optimum primal and dual variables respectively. Hence we have that,

$$\mathbf{W}\mathbf{W}^H\bar{\mathbf{a}} = \bar{\gamma}\bar{\mathbf{a}}. \tag{4.60}$$

Multiplying both sides by $\bar{\mathbf{a}}^H$ in Eq. (4.60), we have that

$$\bar{\gamma} = \bar{\mathbf{a}}^H\mathbf{W}\mathbf{W}^H\bar{\mathbf{a}} = g(\bar{\mathbf{a}}), \tag{4.61}$$

due to $\|\bar{\mathbf{a}}\|_2 = 1$. Note that $g(\bar{\mathbf{a}})$ corresponds to the squared value of the maximum of $f(\mathbf{s})$, i.e. $\cos^2\theta_i, i = 1, \ldots, p$. Hence the optimum Lagrange multipliers $\bar{\gamma}$'s are essentially squared cosine values of canonical angles. Moreover, Eq. (4.60) also shows that the optimum Lagrange multipliers $\bar{\gamma}$'s are eigenvalues of the matrix $\mathbf{W}\mathbf{W}^H$ and $\bar{\mathbf{a}}$'s are corresponding eigenvectors. Thus, the cosine canonical angles, $\cos\theta_i, i = 1, \ldots, p$ are singular values of the matrix $\mathbf{W}$ and the principle vectors are corresponding eigenvectors. Furthermore,

$$|\mathbf{W}\mathbf{W}^H| = \Pi_{i=1}^{p}\bar{\gamma}_i = \Pi_{i=1}^{p}\cos^2\theta_i. \tag{4.62}$$

Utilizing the orthonormal property of two basis vectors $\mathbf{S}$ and $\mathbf{V}$ in Eq. (4.39), we have that

$$\cos^2\theta = |\mathbf{S}^H\mathbf{V}\mathbf{V}^H\mathbf{S}| = |\mathbf{W}\mathbf{W}^H|. \tag{4.63}$$

Combining Eqs. (4.62) and (4.63) yields,

$$\cos^2\theta = \Pi_{i=1}^{p}\cos^2\theta_i. \;\square \tag{4.64}$$

Proposition 4.2 **is proved.**

Proposition 4.2 provides a method of calculating the minimum canonical angle $\tilde{\theta}$ between two subspaces, that is, calculating the 2-norm of the product matrix of two sets of orthonormal basis $\|\mathbf{W}\|_2 = \|\mathbf{S}^H\mathbf{V}\|_2$. However, often neither the source steering vectors $\{\mathbf{s}_1, \ldots, \mathbf{s}_p\}$ nor the interference steering vectors $\{\mathbf{v}_1, \ldots, \mathbf{v}_q\}$ are orthonormal. The following proposition provides a simple approach to calculate the minimal canonical angle without the assumption of orthonormality.

Proposition 4.3 *If $\boldsymbol{P}_S$ and $\boldsymbol{P}_V$ are the orthogonal projectors onto $\tilde{\mathbf{S}}$ and $\tilde{\mathbf{V}}$, respectively, then*

$$\cos^2\tilde{\theta} = \|\boldsymbol{P}_S\boldsymbol{P}_V\|_2 = \|\boldsymbol{P}_V\boldsymbol{P}_S\|_2. \tag{4.65}$$

Proof According to the definition of canonical angles in Eq. (4.51), for each $\mathbf{s} \in \tilde{\mathbf{S}}$, the absolute value of the Euclidean inner product $|\langle\mathbf{s}, \mathbf{v}\rangle|$ is maximized on $\{\mathbf{v} \in \tilde{\mathbf{V}} : \|\mathbf{v}\|_2 = 1\}$ at

$$\hat{\mathbf{v}} = \frac{\hat{\mathbf{s}}}{\|\hat{\mathbf{s}}\|_2} = \frac{\mathbf{P}_V\mathbf{s}}{\|\mathbf{P}_V\mathbf{s}\|_2}, \tag{4.66}$$

and the maximum value is,

$$f(\mathbf{s}) = \max_{\mathbf{v}\in\tilde{\mathbf{V}}, \|\mathbf{v}\|_2=1} |\langle \mathbf{s}, \mathbf{v}\rangle| = \|\mathbf{P}_{\mathbf{V}}\mathbf{s}\|_2. \tag{4.67}$$

Now, let us consider the function

$$g(\mathbf{s}) = f^2(\mathbf{s}) = \|\mathbf{P}_{\mathbf{V}}\mathbf{s}\|_2^2 = \mathbf{s}^H\mathbf{P}_{\mathbf{V}}\mathbf{s}. \tag{4.68}$$

We set $\mathbf{s} = \mathbf{Sb}/\|\mathbf{Sb}\|_2$, and substituting it into Eq. (4.68) yields,

$$g(\mathbf{b}) = \frac{\mathbf{b}^H\mathbf{S}^H\mathbf{P}_{\mathbf{V}}\mathbf{Sb}}{\mathbf{b}^H\mathbf{S}^H\mathbf{Sb}}. \tag{4.69}$$

Clearly, $g(\mathbf{b})$ represents the Rayleigh quotient of two symmetric matrices $\mathbf{S}^H\mathbf{P}_{\mathbf{V}}\mathbf{S}$ and $\mathbf{S}^H\mathbf{S}$. To find the optimum vector $\mathbf{b}$ corresponding to the extreme value of $g(\mathbf{b})$, we calculate its derivative with respect to $\mathbf{b}$,

$$\frac{\partial g(\mathbf{b})}{\partial \mathbf{b}} = \frac{2\mathbf{S}^H\mathbf{P}_{\mathbf{V}}\mathbf{Sb}(\mathbf{b}^H\mathbf{S}^H\mathbf{Sb}) - 2\mathbf{S}^H\mathbf{Sb}(\mathbf{b}^H\mathbf{S}^H\mathbf{P}_{\mathbf{V}}\mathbf{Sb})}{(\mathbf{b}^H\mathbf{S}^H\mathbf{Sb})^2}. \tag{4.70}$$

Setting it to zero yields,

$$\mathbf{S}^H\mathbf{P}_{\mathbf{V}}\mathbf{Sb} = \frac{\mathbf{b}^H\mathbf{S}^H\mathbf{P}_{\mathbf{V}}\mathbf{Sb}}{\mathbf{b}^H\mathbf{S}^H\mathbf{Sb}}\mathbf{S}^H\mathbf{Sb} = g(\mathbf{b})\mathbf{S}^H\mathbf{Sb}. \tag{4.71}$$

Multiplying both sides of Eq. (4.71) by $(\mathbf{S}^H\mathbf{S})^{-1}$, we obtain

$$(\mathbf{S}^H\mathbf{S})^{-1}\mathbf{S}^H\mathbf{P}_{\mathbf{V}}\mathbf{Sb} = g(\mathbf{b})\mathbf{b}. \tag{4.72}$$

Therefore, the maximum value of $g(\mathbf{b})$ is the maximum eigenvalue of the matrix $(\mathbf{S}^H\mathbf{S})^{-1}\mathbf{S}^H\mathbf{P}_{\mathbf{V}}\mathbf{S}$, which is equivalent to $\|\mathbf{P}_{\mathbf{V}}\mathbf{P}_{\mathbf{S}}\|_2$, according to the definition of matrix spectral norm. This completes the proof of Proposition 4.3. The equality $\|\mathbf{P}_{\mathbf{S}}\mathbf{P}_{\mathbf{V}}\|_2 = \|\mathbf{P}_{\mathbf{V}}\mathbf{P}_{\mathbf{S}}\|_2$ is a result of the symmetry of orthogonal projectors. Note that for two arbitrary sets of basis vectors $\mathbf{S} = [\mathbf{s}_1, \ldots, \mathbf{s}_p]$ and $\mathbf{V} = [\mathbf{v}_1, \ldots, \mathbf{v}_q]$, we have $\mathbf{P}_{\mathbf{S}} = \mathbf{S}(\mathbf{S}^H\mathbf{S})^{-1}\mathbf{S}^H$ and $\mathbf{P}_{\mathbf{V}} = \mathbf{V}(\mathbf{V}^H\mathbf{V})^{-1}\mathbf{V}^H$.

It is clear that $\cos\theta \leq \cos\tilde{\theta}$ from Proposition 4.2, which implies $\theta \geq \tilde{\theta}$. Moreover,

$$\cos^2\tilde{\theta} = \|\mathbf{P}_{\mathbf{V}}\mathbf{P}_{\mathbf{S}}\|_2 = \|\mathbf{N}\mathbf{S}^H\mathbf{P}_{\mathbf{V}}\mathbf{S}\mathbf{N}\|_2 = \zeta_1. \tag{4.73}$$

Comparing with Eq. (4.49) produces,

$$\cos^2\tilde{\theta} \geq \cos^2\vartheta \Rightarrow \tilde{\theta} \leq \vartheta. \tag{4.74}$$

The relationship among the three angles can then be described by the following inequality,

$$\cos^2\theta \leq \cos^2\vartheta \leq \cos^2\tilde{\theta}. \tag{4.75}$$

That is, the eigenspace angle is upper bounded by the conventional angle θ, while lower bounded by the minimum canonical angle $\tilde{\theta}$, i.e., $\tilde{\theta} \leq \vartheta \leq \theta$. When the source subspace is of single dimension, i.e., $\tilde{\mathbf{S}} = \text{span}(\mathbf{s})$, the three angles coincide and its squared cosine value equals to the SCC in single-source cases.

The three angles considered provide a full-view characterization of the spatial separation between the source and interference subspaces. The conventional angle defines the maximum angle between two arbitrary components from two subspaces. The minimum canonical angle provides an estimate of the worst performance loss. The eigenspace angle describes the separation between the source dominant component and the interference subspace and offers a reasonable assessment of performance loss.

4.3.4 Sparse Array Design by Antenna Selection

Three sparse array design approaches by antenna selection are proposed based on the spatial angles formulated in Sect. 4.3.3. Given the number of antennas and the employed beamformer, array configurations remain a source of flexibility and can allocate DoFs, i.e. antenna positions, towards achieving the optimum design of MaxSINR of adaptive MVDR beamformers. Suppose that there are N grid points for possible sensor positions and K available antennas. Denote a selection vector $\mathbf{z} = [z_i, i = 1, \ldots, N] \in \{0, 1\}^N$ with "zero" entry for a discarded position and "one" entry for a selected position. For each selection vector $\mathbf{z}$, there is a corresponding selection matrix $\mathbf{Z} = \{0, 1\}^{K \times N}$ with a "one" entry in the ith row and the jth column, where $i = 1, \ldots, K$ and $z_j = 1, j \in \{1, \ldots, N\}$. The selection vector $\mathbf{z}$ and selection matrix $\mathbf{Z}$ are inner-connected by $\mathbf{Z}^T\mathbf{Z} = \text{diag}(\mathbf{z})$. The diagonal matrix $\text{diag}(\mathbf{z})$ is the antenna selection operator with the vector $\mathbf{z}$ populating along the diagonal. Since all the N grid locations are known, the full array manifold matrices corresponding to sources and interferences, termed as $\bar{\mathbf{S}}$ and $\bar{\mathbf{V}}$, can be calculated in advance. As steering vectors are directional, the array manifold matrices of a sparse array with K antennas in relation to the full array of N antennas, can be expressed as $\mathbf{S} = \mathbf{Z}\bar{\mathbf{S}}$ and $\mathbf{V} = \mathbf{Z}\bar{\mathbf{V}}$ for sources and interferences, respectively. We formulate the antenna selection as a convex optimization problem in the sequel and solve it through MATLAB embedded software CVX [133, 142].

4.3.4.1 The First Approach Utilizing Conventional Angle

The first approach is aiming at minimizing the squared cosine of conventional angle. Since the logarithm is a monotone increasing function, we take the logarithm of Eq. (4.39) as the objective function for maximizing the spatial separation between the source and interference subspaces. The problem is formulated as follows,

$$\min_{\mathbf{z},\mathbf{W}} \ \log|\mathbf{W}| - \log|\bar{\mathbf{S}}^H \text{diag}(\mathbf{z})\bar{\mathbf{S}}|, \tag{4.76}$$
$$\text{s.t.} \ \begin{bmatrix} \bar{\mathbf{V}}^H \text{diag}(\mathbf{z})\bar{\mathbf{V}} & \bar{\mathbf{V}}^H \text{diag}(\mathbf{z})\bar{\mathbf{S}} \\ \bar{\mathbf{S}}^H \text{diag}(\mathbf{z})\bar{\mathbf{V}} & \mathbf{W} \end{bmatrix} \succeq \mathbf{0},$$
$$\mathbf{1}^T\mathbf{z} = K,$$
$$0 \le \mathbf{z} \le 1.$$

In the first constraint, we utilize Schur complement condition on matrix positive semi-definiteness to express the matrix inversion term $(\bar{\mathbf{V}}^H \text{diag}(\mathbf{z})\bar{\mathbf{V}})^{-1}$ in the numerator of Eq. (4.39) as linear matrix inequality, which provides the upper bound, $\log|\bar{\mathbf{S}}^H \text{diag}(\mathbf{z})\bar{\mathbf{V}}(\bar{\mathbf{V}}^H \text{diag}(\mathbf{z})\bar{\mathbf{V}})^{-1}\bar{\mathbf{V}}^H \text{diag}(\mathbf{z})\bar{\mathbf{S}}| \le \log|\mathbf{W}|$. The second and third constraints restrict only K grid locations selected for antenna placement. The concave part of the objective function, $\log|\mathbf{W}|$, requires to be iteratively linearized. Let $\mathbf{W}^{(k)}$ denote the kth iteration of the optimization variable $\mathbf{W}$. The first-order Taylor series expansion of $\log|\mathbf{W}|$ around $\mathbf{W}^{(k)}$ is given by,

$$\log|\mathbf{W}| \approx C + \text{tr}\{(\mathbf{W}^{(k)} + \epsilon\mathbf{I})^{-1}\mathbf{W}\}, \tag{4.77}$$

where C is a constant that does not affect the minimization and ϵ can be interpreted as a small regularization constant to provide robustness against singularity of the matrix $\mathbf{W}$ [182]. Utilizing Eq. (4.77) in Eq. (4.76) leads to,

$$\min_{\mathbf{z},\mathbf{W}} \ \text{tr}\{(\mathbf{W}^{(k)} + \epsilon\mathbf{I})^{-1}\mathbf{W}\} - \log|\bar{\mathbf{S}}^H \text{diag}(\mathbf{z})\bar{\mathbf{S}}|, \tag{4.78}$$
$$\text{s.t.} \ \begin{bmatrix} \bar{\mathbf{V}}^H \text{diag}(\mathbf{z})\bar{\mathbf{V}} & \bar{\mathbf{V}}^H \text{diag}(\mathbf{z})\bar{\mathbf{S}} \\ \bar{\mathbf{S}}^H \text{diag}(\mathbf{z})\bar{\mathbf{V}} & \mathbf{W} \end{bmatrix} \succeq \mathbf{0},$$
$$\mathbf{1}^T\mathbf{z} = K,$$
$$0 \le \mathbf{z} \le 1.$$

We solve a weighted trace minimization problem at each iteration. The iterative approximations are referred to as sequential convex programming (SCP) [133], and SCP is a local heuristic with its performance depending on the initial point. It is, therefore, common to initialize the algorithm with several feasible points and the final solution $\mathbf{z}^*$ is the one found with the minimum objective value over the different runs. Since the function $\log|\mathbf{W}|$ is concave in $\mathbf{W}$, its value decreases by an amount more than the decrease in the linearized objective function at each iteration. Based on this observation, it can be shown that the iterative procedure in Eq. (4.78) converges to a local minimum of the problem in Eq. (4.76) [182, 183].

However, as the boolean constraints $\mathbf{z} \in \{0, 1\}^N$ are relaxed to box constraints $0 \le \mathbf{z} \le 1$ in Eqs. (4.76) and (4.78), the returned selection variable $\mathbf{z}$ does not necessarily satisfy the boolean property. We adopt a randomization technique [184] to obtain an approximate selection vector based on the solution $\mathbf{z}^*$ of the relaxed problem in Eq.

Table 4.1 The procedure of randomization technique

Step 0:	Set $\text{SINR}_\text{m} = 0$, $k = 0$, maximum iteration number k_m
Step 1:	Generate an N-dimensional random vector $\mathbf{w}$ from uniform distribution in the interval (0, 1)
Step 2:	Implement element-wise logical comparison, $\mathbf{t} := (\mathbf{w} < \mathbf{z}^*)$
Step 3:	If $\text{card}(\mathbf{t}) = K$, calculate its corresponding output SINR
Step 4:	If $\text{SINR} > \text{SINR}_\text{m}$, set $\text{SINR}_\text{m} := \text{SINR}$ and $\mathbf{z}_\text{opt} := \mathbf{t}$ Set $k = k + 1$. If $k < k_\text{m}$, go to Step 1; otherwise, terminate

(4.78). The randomization technique is summarized in Table 4.1. Empirical results suggest that no more than ten randomizations be required to obtain a stable boolean antenna selection vector.

4.3.4.2 The Second Approach Utilizing Minimum Canonical Angle

Clearly, the approach in Sect. 4.3.4.1 only considers the second task, that is maximizing the spatial separation between the source and interference subspaces, while it ignores the first task. The optimum sparse array design, however, mandates the incorporation of the two tasks simultaneously. Proceeding from Eq. (4.26), MaxSINR can be rewritten in terms of matrix spectral norm as,

$$\begin{aligned} \text{SINR}_\text{opt} &= \lambda_\text{max}\{\mathbf{R}_n^{-1}\mathbf{R}_s\}, \\ &= \|(\mathbf{I} - \mathbf{P}_\mathbf{V})\mathbf{S}\mathbf{C}_s\mathbf{S}^H\|_2, \\ &= \|\mathbf{S}_g^H(\mathbf{I} - \mathbf{P}_\mathbf{V})\mathbf{S}_g\|_2, \\ &= \|\mathbf{S}_g^H\mathbf{S}_g - \mathbf{S}_g^H\mathbf{P}_\mathbf{V}\mathbf{S}_g\|_2, \\ &\geq \|\mathbf{S}_g^H\mathbf{S}_g\|_2 - \|\mathbf{S}_g^H\mathbf{P}_\mathbf{V}\mathbf{S}_g\|_2, \end{aligned} \tag{4.79}$$

where $\mathbf{S}_g = \mathbf{S}\mathbf{C}_s^{1/2}$. Clearly, Eq. (4.79) provides a lower bound on output SINR. The antenna selection can be formulated as the maximization of the lower bound, i.e.,

$$\begin{aligned} \max_{\mathbf{z},\mathbf{W}} \quad & \|\bar{\mathbf{S}}_g^H\text{diag}(\mathbf{z})\bar{\mathbf{S}}_g\|_2 - \|\mathbf{W}\|_2, \\ \text{s.t.} \quad & \begin{bmatrix} \bar{\mathbf{V}}^H\text{diag}(\mathbf{z})\bar{\mathbf{V}} & \bar{\mathbf{V}}^H\text{diag}(\mathbf{z})\bar{\mathbf{S}}_g \\ \bar{\mathbf{S}}_g^H\text{diag}(\mathbf{z})\bar{\mathbf{V}} & \mathbf{W} \end{bmatrix} \succeq \mathbf{0}, \\ & \mathbf{1}^T\mathbf{z} = K, \\ & 0 \leq \mathbf{z} \leq 1, \end{aligned} \tag{4.80}$$

where $\bar{\mathbf{S}}_g = \bar{\mathbf{S}}\mathbf{C}_s^{1/2}$. Set the convex part of the objective in Eq. (4.80) as $f(\mathbf{z}) = \|\bar{\mathbf{S}}_g^H\text{diag}(\mathbf{z})\bar{\mathbf{S}}_g\|_2$, which is obviously the maximum eigenvalue of $\mathbf{R}_s$, i.e. $f(\mathbf{z}) = \lambda_0$.

According to the definition of matrix spectrum norm, we have

$$\lambda_0 = f(\mathbf{z}) = \max_{\|\mathbf{b}\|_2=1} \mathbf{b}^H \bar{\mathbf{S}}_g^H \text{diag}(\mathbf{z}) \bar{\mathbf{S}}_g \mathbf{b}, \tag{4.81}$$
$$= \tilde{\mathbf{b}}^H \bar{\mathbf{S}}_g^H \text{diag}(\mathbf{z}) \bar{\mathbf{S}}_g \tilde{\mathbf{b}},$$

where $\tilde{\mathbf{b}}$ denotes principal eigenvector of matrix $\bar{\mathbf{S}}_g^H \text{diag}(\mathbf{z}) \bar{\mathbf{S}}_g$. As it is difficult to maximize a convex function, we approximate it iteratively by its affine surrogate. Let $\mathbf{z}^{(k)}$ denote the kth iterate of the optimization variable $\mathbf{z}$. The first-order Taylor series expansion of $f(\mathbf{z})$ around $\mathbf{z}^{(k)}$ is given by,

$$f(\mathbf{z}) = f(\mathbf{z}^{(k)}) + \nabla f(\mathbf{z}^{(k)})^T (\mathbf{z} - \mathbf{z}^{(k)}), \tag{4.82}$$

where $\nabla f(\mathbf{z}^{(k)})$ denotes the gradient of $f(\mathbf{z})$ evaluated at the point $\mathbf{z}^{(k)}$ and $\nabla f(\mathbf{z}^{(k)}) = [\nabla f(\mathbf{z}_1^{(k)}), \ldots, \nabla f(\mathbf{z}_N^{(k)})]^T$ with

$$\nabla f(\mathbf{z}_i^{(k)}) = \tilde{\mathbf{b}}^{(k)H} \bar{\mathbf{s}}_{g,i} \bar{\mathbf{s}}_{g,i}^H \tilde{\mathbf{b}}^{(k)}, \tag{4.83}$$

where $\bar{\mathbf{s}}_{g,i}$ denotes the ith column of matrix $\bar{\mathbf{S}}_g^H$ and $\tilde{\mathbf{b}}^{(k)}$ is the principal eigenvector of matrix $\bar{\mathbf{S}}_g^H \text{diag}(\mathbf{z}^{(k)}) \bar{\mathbf{S}}_g$. Based on Eqs. (4.82)–(4.83), the antenna selection problem can be formulated iteratively as,

$$\begin{aligned} \max_{\mathbf{z},\mathbf{W}} \quad & f(\mathbf{z}^{(k)}) + \nabla f(\mathbf{z}^{(k)})^T (\mathbf{z} - \mathbf{z}^{(k)}) - \|\mathbf{W}\|_2, \\ \text{s.t.} \quad & \begin{bmatrix} \bar{\mathbf{V}}^H \text{diag}(\mathbf{z}) \bar{\mathbf{V}} & \bar{\mathbf{V}}^H \text{diag}(\mathbf{z}) \bar{\mathbf{S}}_g \\ \bar{\mathbf{S}}_g^H \text{diag}(\mathbf{z}) \bar{\mathbf{V}} & \mathbf{W} \end{bmatrix} \succeq \mathbf{0}, \\ & \mathbf{1}^T \mathbf{z} = K, \\ & 0 \leq \mathbf{z} \leq 1. \end{aligned} \tag{4.84}$$

The randomization procedure introduced in Table 4.1 is then utilized to obtain an appropriate boolean solution. The algorithm can be run for several times with different initializing points $\mathbf{z}^{(0)}$ and the final choice is the one found with the maximum objective value over the different runs.

4.3.4.3 The Third Approach Utilizing Eigenspace Angle

Consider the lower bound of output SINR in Eq. (4.34) in terms of eigenspace angle. Substituting $\lambda_0 = \lambda_{\max}\{\mathbf{R}_s\}$ by the function $f(\mathbf{z})$ in Eq. (4.81), the lower bound can be rewritten as,

$$\mathcal{B} = (\tilde{\mathbf{b}}^H \bar{\mathbf{S}}_g^H \text{diag}(\mathbf{z}) \bar{\mathbf{S}}_g \tilde{\mathbf{b}}) - \lambda_0 \mathbf{e}_0^H \mathbf{P}_{\mathbf{V}} \mathbf{e}_0. \tag{4.85}$$

As the maximum eigenvalue of a matrix is a non-linear function of the selection vector $\mathbf{z}$, we resort to its linear approximation based on the previous iterative solution. Denote $\lambda_0^{(k)} = \lambda_{\max}(\bar{\mathbf{S}}_g^H \text{diag}(\mathbf{z}^{(k)})\bar{\mathbf{S}}_g)$ as the maximum eigenvalue and $\tilde{\mathbf{b}}^{(k)}$ as the principal eigenvector in the previous iteration, the lower bound can be approximated by,

$$\mathcal{B} \approx \tilde{\mathbf{b}}^{(k)H} \bar{\mathbf{S}}_g^H \text{diag}(\mathbf{z})\bar{\mathbf{S}}_g \tilde{\mathbf{b}}^{(k)} - \lambda_0^{(k)}(\mathbf{e}_0^H \mathbf{P}_{\bar{\mathbf{V}}} \mathbf{e}_0). \tag{4.86}$$

The antenna selection can be formulated as a maximization problem based on the objective in Eq. (4.86),

$$\begin{aligned}
\max_{\mathbf{z},\gamma} \quad & \tilde{\mathbf{b}}^{(k)H} \bar{\mathbf{S}}_g^H \text{diag}(\mathbf{z})\bar{\mathbf{S}}_g \tilde{\mathbf{b}}^{(k)} - \lambda_0^{(k)}\gamma, \\
\text{s.t.} \quad & \begin{bmatrix} \bar{\mathbf{V}}^H \text{diag}(\mathbf{z})\bar{\mathbf{V}} & \bar{\mathbf{V}}^H \text{diag}(\mathbf{z})\bar{\mathbf{e}}_0 \\ \bar{\mathbf{e}}_0^H \text{diag}(\mathbf{z})\bar{\mathbf{V}} & \gamma \end{bmatrix} \succeq \mathbf{0}, \\
& \mathbf{1}^T\mathbf{z} = K, \\
& 0 \le \mathbf{z} \le 1,
\end{aligned} \tag{4.87}$$

where $\bar{\mathbf{e}}_0$ is the principal eigenvector of the source covariance matrix $\bar{\mathbf{S}}\mathbf{C}_s\bar{\mathbf{S}}^H$ for the full array, $\mathbf{P}_{\bar{\mathbf{V}}} = \bar{\mathbf{V}}(\bar{\mathbf{V}}^H\bar{\mathbf{V}})^{-1}\bar{\mathbf{V}}^H$ and $\mathbf{e}_0^H \mathbf{P}_{\mathbf{V}} \mathbf{e}_0 = \bar{\mathbf{e}}_0^H \text{diag}(\mathbf{z})\bar{\mathbf{V}}(\bar{\mathbf{V}}^H \text{diag}(\mathbf{z})\bar{\mathbf{V}})^{-1}\bar{\mathbf{V}}^H$ $\text{diag}(\mathbf{z})\bar{\mathbf{e}}_0$. Note that the objective function in Eq. (4.87) is the lower bound of Eq. (4.86) from Schur complement condition, i.e. $\gamma \ge \mathbf{e}_0^H \mathbf{P}_{\mathbf{V}} \mathbf{e}_0$. This formulation cannot guarantee the boolean property of the selection variable either. The randomization procedure in Table 4.1 is then utilized to obtain the feasible solution.

4.3.5 Simulations

In this section, simulation results are presented to validate the effectiveness of proposed metrics.

4.3.5.1 Interference-Free Beamforming

Consider $K = 8$ available antennas and $N = 14$ uniformly spaced positions with half-wavelength inter-element spacing, i.e., $d = \lambda/2$. There are three equal powered sources arriving from directions $\psi_1 = 40°$, $\psi_2 = 65°$, $\psi_3 = 125°$ with 0 dB SNR. There are no interferences present in the considered example. Clearly, the adaptive beamformer puts more emphasis on the first two closely-spaced sources compared to the third one so as to obtain MaxSNR. We enumerate all the 3003 different configurations to calculate output SNRs according to Eq. (4.28), which are plotted in Fig. 4.6 in pink dots. We can observe that the worst array attains 9.77 dB output SNR, compared to the highest 11.58 dB SNR achieved by the optimum array depicted in the

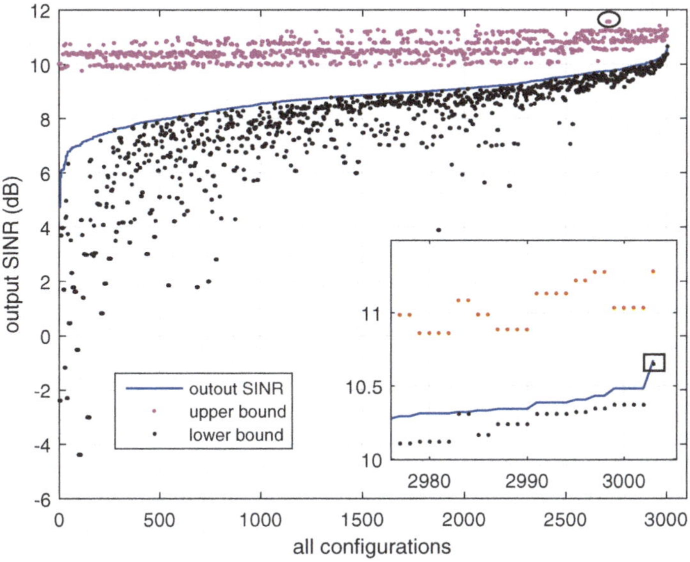

Fig. 4.6 The output SINR of all enumerated configurations and their corresponding upper and lower bounds. The circle in the figure indicates the array **a** with MaxSNR, and the square indicates the array **b** with MaxSINR. The sources are arriving from $\psi_1 = 40°$, $\psi_2 = 65°$, $\psi_3 = 125°$ and there are no interferences

upper plot of Fig. 4.7. We refer to this configuration as array (a). The dB difference between the highest and lowest achievable SNRs verifies the important role of array configuration for determining the performance in interference-free beamforming.

4.3.5.2 Interference-Active Beamforming

(a) Bounds of MaxSINR

We add to the above example four strong interfering signals impinging on the array from directions $\phi_1 = 50°$, $\phi_2 = 62°$, $\phi_3 = 122°$, $\phi_4 = 150°$ with 30 dB interference-to-noise ratio (INR). We enumerate all the 3003 different possible configurations and calculate both the corresponding output SINR, as given in Eq. (4.26) and the lower bound of Eq. (4.34). Note that the maximum eigenvalue λ_0 and principal eigenvector $\mathbf{e}_0$ are calculated for each enumerated sparse array. The results are plotted in Fig. 4.6 in an ascending order of the output SINR. It is worth noting that the upper bound is the output SNR of the above interference-free case. The optimum 8-antenna array presents 10.68 dB output SINR compared to 4.75 dB for the worst array. This difference again underscores the dependency of the adaptive beamforming performance on array configurations. We can also observe from Fig. 4.6 that the two SINR bounds are reasonably tight for each configuration, especially for those with large SINR. For the optimum 8-antenna array, the lower bound is only 0.03 dB

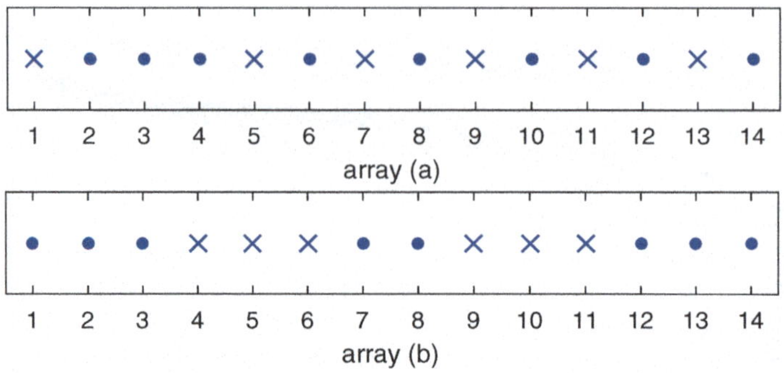

Fig. 4.7 The array **a** is the 8-antenna array with MaxSNR of interference-free beamforming; The array **b** is the optimum 8-antenna array with MaxSINR of interference-active beamforming. Dots denote selected and cross for discarded

below the optimum SINR value. These close values demonstrate that the bounds are effective measures for analyzing and predicting the highest attainable SINR for any given array configuration. Most important, the lower bound is an effective metric for configuring the array, achieving the maximum attainable output SINR. The optimum 8-antenna array with the MaxSINR, referred to as "array (b)" is depicted in the lower plot of Fig. 4.7. The beampatterns of the two arrays using the MVDR weights in Eq. (4.25) are plotted in Fig. 4.8. Clearly, array (a) demonstrates larger gains towards the sources than array (b), whereas array (b) places deeper nulls against the interferences compared to array (a). The reason is that the design metric of array (a) is blind to the presence of interferences.

(a) Upper Bound and Spatial Separation

We first calculate, for each configuration, the three spatial angles, namely, the eigenspace angle in Eq. (4.32), the conventional angle in Eq. (4.38), and the minimum canonical angle in Eq. (4.65). The results are plotted in Fig. 4.9 in a descending order of the eigenspace angle. It is evident that the eigenspace angle is bounded between the conventional angle and the minimum canonical angle for each array.

In order to investigate the interplay between the upper bound and the spatial separation in determining MaxSINR, we plot in Fig. 4.10 the output SINR, its upper bound and the squared cosine of eigenspace angle in an ascending order of the output SINR. The output SINR and its upper bound are normalized by the maximum value of upper bound. Observe that the array with the maximum upper bound does not generate the MaxSINR due to its large performance loss caused by interference nulling. The array with the largest eigenspace angle is the optimum configuration, as it minimizes the performance loss while offering a modest upper bound. The optimization problem works as such that when the sources and interferences have close angular separation, as in the example considered, the objective of maximizing the spatial separation takes priority over maximizing the upper bound.

To compare with the case where the sources and interferences are well separated, we change the sources' arriving angle to $\psi_1 = 70°$, $\psi_2 = 85°$, $\psi_3 = 100°$. We, again,

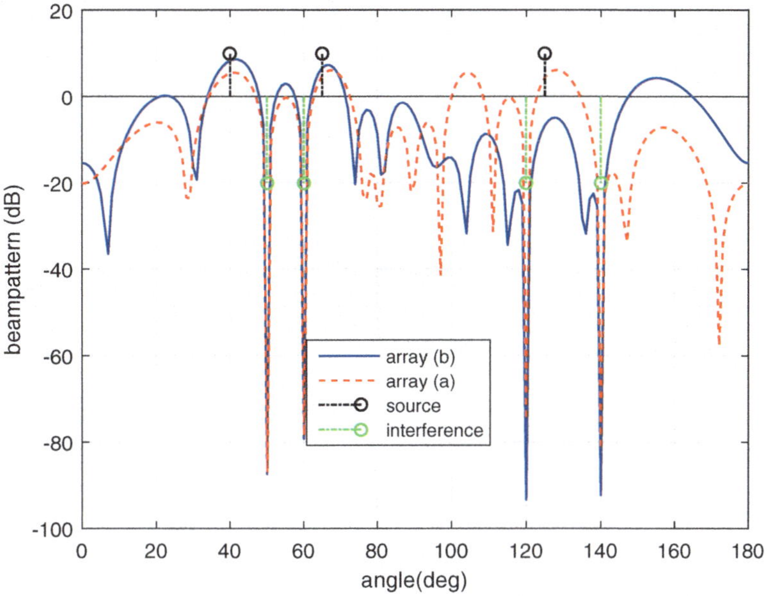

Fig. 4.8 The beampatterns of the two arrays (**a**) and (**b**) in Fig. 4.7

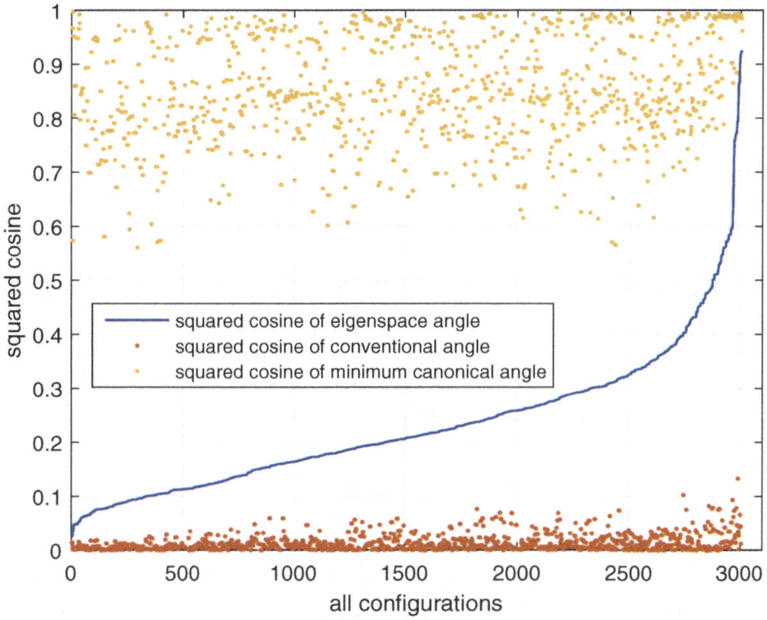

Fig. 4.9 The squared cosine of three angles corresponding to each configuration: eigenspace angle, conventional angle and minimum canonical angle. The sources are arriving from $\psi_1 = 40°$, $\psi_2 = 65°$, $\psi_3 = 125°$ and the interferences are coming from $\phi_1 = 50°$, $\phi_2 = 62°$, $\phi_3 = 122°$, $\phi_4 = 150°$

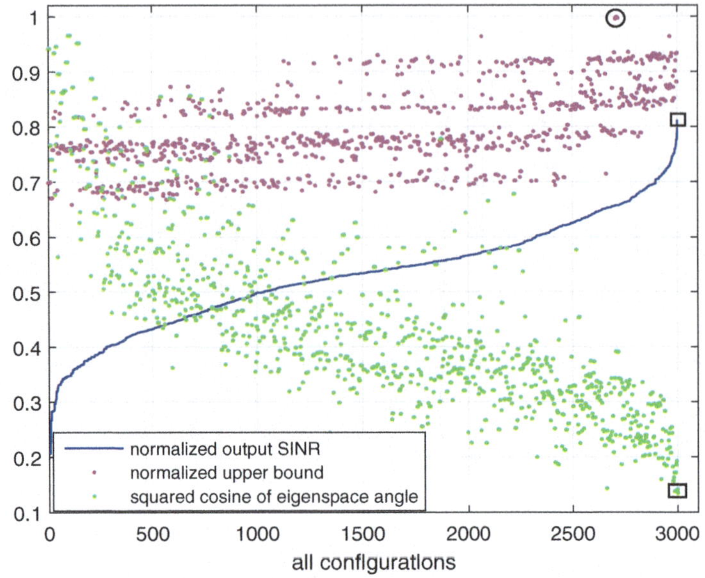

Fig. 4.10 The output SINR, the squared cosine of eigenspace angle and the upper bound of all enumerated configurations. The circle in the figure indicates the array **a** with the maximum upper bound, MaxSNR, and the square indicates the array **b** with the largest spatial separation and MaxSINR. The sources are arriving from $\psi_1 = 40°$, $\psi_2 = 65°$, $\psi_3 = 125°$ and the interferences are coming from $\phi_1 = 50°$, $\phi_2 = 62°$, $\phi_3 = 122°$, $\phi_4 = 150°$

plot the output SINR, the upper bound, and the squared cosine of eigenspace angle in Fig. 4.11 in an ascending order of the output SINR. Clearly, more than 90% of the configurations exhibit an eigenspace angle no smaller than 60°. In this scenario, the array with the maximum upper bound performs the best and manifests the optimum configuration. This array is termed as array (c) and shown in the upper plot of Fig. 4.12, whereas the array with the minimum squared cosine of eigenspace angle is named as array (d) and shown in the lower plot of Fig. 4.12.

4.3.5.3 Three Antenna Selection Approaches

In this section, we demonstrate the merits and effectiveness of three proposed antenna selection approaches. We fix the first source at $\psi_1 = 42°$ and change the incident angle of the second source ψ_2 from 0° to 180° in steps of 4°. The four strong interferences are impinging on the array from directions $\phi_1 = 50°$, $\phi_2 = 62°$, $\phi_3 = 122°$, $\phi_4 = 150°$ with an INR of 30 dB. We select 8 out of 16 uniformly spaced antennas utilizing the three proposed approaches in each scenario and calculate the corresponding output SINR. The results are depicted in Fig. 4.13, and compared with that of the true optimum sparse array obtained through enumeration. It is evident that the sparse array selected based on the MaxSNR metric performs the worst, especially

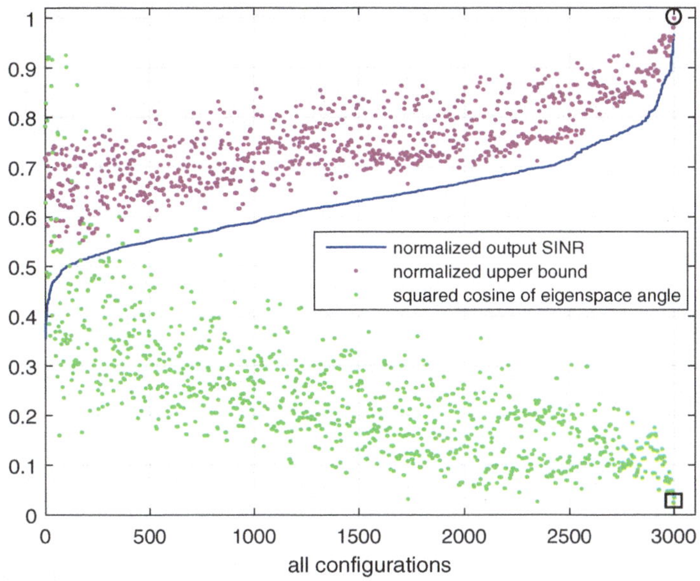

Fig. 4.11 The output SINR, the squared cosine of eigenspace angle, and the upper bound of all enumerated configurations. The circle in the figure indicates the array **c** with the maximum upper bound, MaxSNR, and the square indicates the array **d** with the largest spatial separation. The sources are arriving from $\psi_1 = 70°$, $\psi_2 = 85°$, $\psi_3 = 100°$ and the interferences are coming from $\phi_1 = 50°$, $\phi_2 = 62°$, $\phi_3 = 122°$, $\phi_4 = 150°$

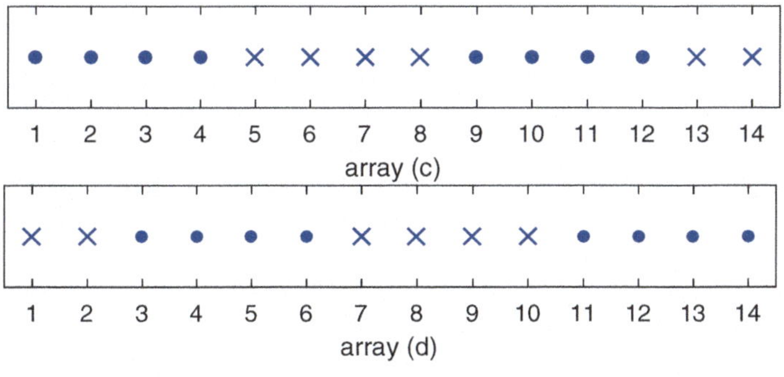

Fig. 4.12 The array **c** is the 8-antenna array with MaxSNR, also with MaxSINR; The array **d** is the optimum 8-antenna array with the largest spatial separation. Dots denote selected and cross for discarded

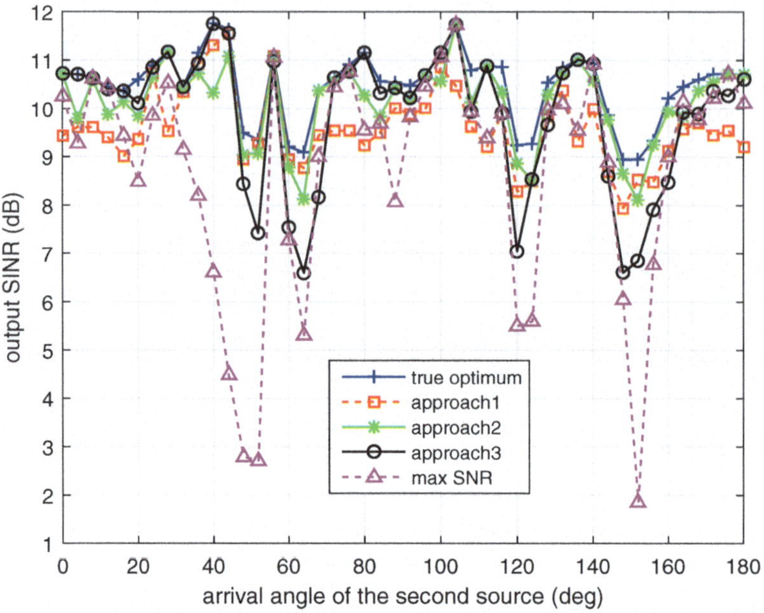

Fig. 4.13 The output SINR of selected arrays using different antenna selection methods. The interferences are coming from $\phi_1 = 50°$, $\phi_2 = 62°$, $\phi_3 = 122°$, $\phi_4 = 150°$. There are two sources, the first source is fixed at $\psi_1 = 42°$ and the angle of the second one is changing from $0°$ to $180°$ in steps of $4°$

when the second source approaches the interference subspace, as this metric is blind to interference presence. On the contrary, the first approach behaves better in the angular region around interferences than in other regions, as it focuses on alleviating the performance loss associated with interference nulling. The second approach, which deals directly with SINR expression, behaves the best and is superior to the third approach in most cases, especially when the second source is close to the interference subspace. It is favourable that both approaches are implemented and choose the one with a larger output SINR as the final solution.

In order to fully demonstrate the effectiveness of proposed antenna selection approaches, we increase the scale of antenna array, that is selecting 8 out of 21 antenna candidates. The simulation scenario remains the same as mentioned above. Note that enumeration does not work here due to its prohibitive computational complexity. Since the underlying problem is about non-uniform sparse arrays, we employ the eight-antenna nested and coprime arrays shown in Fig. 4.14 for comparison with the understanding that only one set of weights is used for each structure according to the MVDR beamforming. We then implement antenna selection using the second approach for each arrival angle ψ_2 and depicts the output SINR of selected sparse arrays comparing with that of nested and coprime arrays in Fig. 4.15. It is clear that neither the nested nor the coprime structured sparse array is capable of

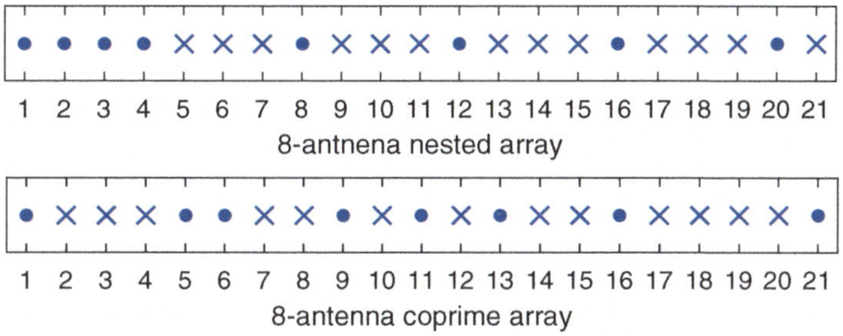

Fig. 4.14 The eight-antenna nested and coprime structured sparse arrays

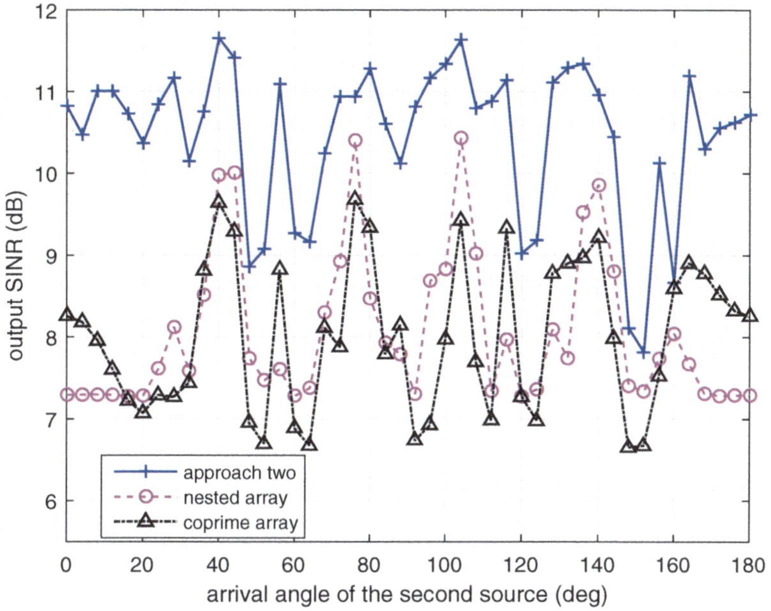

Fig. 4.15 The output SINR of selected sparse arrays using the second approach, the nested array and the coprime array. The interferences are coming from $\phi_1 = 50°$, $\phi_2 = 62°$, $\phi_3 = 122°$, $\phi_4 = 150°$. There are two sources, the first source is fixed at $\psi_1 = 42°$ and the angle of the second one is changing from 0° to 180° in steps of 4°

maximizing the beamformer's output SINR, although the two exhibit excellence in high-resolution direction finding. The instinctive explanation is that the optimum sparse array is designed based on the metric of maximum output SINR, whereas both the nested and the coprime arrays are constructed with the goal of achieving high spatial resolution and ability of dealing with more sources than physical sensors. There is not omnipotent sparse array structure that is optimum in every application. Different objectives require the proposition of different optimizing metrics.

4.4 Summary

In this chapter, we consider the sparse sensing design for adaptive beamforming. In Sect. 4.2, we studied reconfigurable adaptive antenna arrays by antenna selection in the single-source scenario. We introduced the parameter SCC and formulated the relationship between the output SINR and the SCC. The closed-form formula revealed that the output SINR is soly determined by the squared absolute SCC value for a fixed number of antennas. Thus, minimizing the SCC value is equivalent to maximizing the output SINR. In Sect. 4.3, we extended the problem into multi-source scenario. The effect of array configuration on adaptive beamforming was analytically characterized including general-rank signal subspace models. The two-folded role of antenna arrays in adaptive beamforming towards desired and undesired signals was clearly revealed from the formulated bounds of MaxSINR. It was shown that the MaxSINR is determined by the MaxSNR of interference-free beamforming and the spatial separation between source and interference subspaces. Three spatial angles between subspaces were considered to provide performance loss assessment caused by the process of interference nulling. Based on subspace angles, three sparse array design methods by antenna selection were proposed. Simulation results manifested the important role of array configurations in determining the output SINR of adaptive beamformers and the effectiveness of proposed antenna selection algorithms.

Chapter 5
Cognitive-Driven Optimization of Sparse Sensing

Beamformers employ an array of antenna elements to collect the electromagnetic wave in the spatial domain and filter the corrupted signal in either the element-space or the beam-space. The spatial filtering performance of both element-space and beam-space beamformers is jointly determined by two key factors, beamformer geometry and excitation weights. As mentioned in Chap. 4, the sparse array design for maximum output SINR assumes priori knowledge of source and interference DOAs. In this respect, a two-step procedure is employed in Chap. 4, where the antennas selected for optimum DOA estimation are first switched on and multiplexing to a different set of sensors for optimum beamforming based on the sensed FOV is followed. In this chapter, we introduce the cognitive-driven optimization of sparse sensing, which can fulfill the task of environmental sensing automatically and adaptively.

In Sect. 5.2, we propose the adaptive beamformer design by regularized complementary antenna switching (RCAS), which first collects the full array data via a set of deterministic complementary sparse arrays and then reconfigures the sparse array as the optimum configuration for adaptive beamforming. In Sect. 5.3, we propose the cognitive sparse beamformer design via regularized switching network. The proposed beamformer is capable of swiftly adapting the beamformer geometry via antenna removement and addition according to the environmental dynamics through a "perception-action" cycle. Finally, a summary is provided in Sect. 5.4.

5.1 Introduction

Antenna arrays use a set of antennas to sample signals in the spatial domain from the sources in their field of view (FoV). Different objectives can be achieved using antenna arrays, among them, beamforming has found extensive use for several decades in diverse applications, such as radar, sonar, telescope and wireless communications to list a few [185–187]. Beamforming can be performed in either the element-space or the beam-space. Element-space beamformers weight the signals

X. Wang et al., *Sparse Sensing and Sparsity Sensed in Multi-sensor Array Applications*, https://doi.org/10.1007/978-981-99-9558-5_5

arriving at each of the elements and sum the results to obtain an output [68, 188, 189]. However, full-dimensional element-space beamformers are expensive attributed to the front-end channel associated with each antenna and complicated digital signal processing. One potential solution is sparse array beamformers which optimally select a subset of "best" antennas from the larger set of available antennas [190–192]. However, antenna selection exhibits performance degradation compared to the full array system. The other method is sparse beam-space beamformers, which construct a set of sparse beams as a preliminary step and further processing is implemented on the beam outputs [193–195]. The spatial filtering performance of both element-space and beam-space beamformers is jointly determined by two factors, beamformer geometry and excitation weights. Driven by the importance of array configurations, research into sparse beamformer design techniques tailored for different applications continue unabated. In order to facilitate the continuity of beamformer geometry and practical implementation, we employ regularized switching network, which divides the full array of elements/beams into groups and each group connects with one front-end channel. Precisely one antenna/beam is switched on in each group to compose the sparse beamformer.

Beampattern and array gain (AG) are two key metrics for beamformer performance evaluation. Sparse arrays designed in terms of beampattern synthesis are termed as deterministic sparse arrays, the antenna positions of which are fixed once calculated off-line. Although deterministic sparse arrays are blind to the situation, they possess resilient though average interference suppression performance regardless of surrounding environment, and thus are good choices when no environmental information is available. Sparse arrays designed in terms of maximizing output signal-to-interference-plus-noise-ratio (MaxSINR) are referred to as adaptive sparse arrays, where a subset of different sensors are adaptively switched on in accordance with changing environmental situations. Therefore, adaptive sparse arrays are superior to deterministic arrays in the metric of interference suppression thanks to their desirable situational awareness and exclusive focus on strong interferences. Beamformers, which maximize array gain while maintaining well-controlled beampatterns, exhibit high robustness against interferences suddenly intruded on and thus are preferred in dynamic environment [196].

There are extensive works in the literature on sparse beamformer design, where some focus on synthesizing a desired-shaped beampattern using a smallest number of antennas [197–199], and others strive to select an optimum subset of antennas or beams from a large set for maximizing the AG [200, 201]. However, all these designs require some priori real-time knowledge of the operating environment, such as the number of interferences and their directions of arrival (DOAs). Most of the works [197, 198] adopted a two-step open-loop design, where the sparse configuration tailored for DOA estimation is first switched on and multiplexing to a different sparse beamformer based on the sensed situation is followed. The work in [202] configured a fully augmentable sparse array to estimate the received data correlation matrix corresponding to the full array aperture, and then reconfigured another sparse array along with weights simultaneously for beamforming. The subsequent work in [203] proposed a matrix completion method to interpolate all missing spatial lags, whereas the matrix completion accuracy heavily affects the sparse array design. Although, deep learning approaches for antenna selection have been studied recently for DOA estimation and beamforming for efficient communications [204, 205], it is difficult to secure sufficient training data mimicking different kinds of true environment. The problem becomes more pronounced in dynamic environments, where beamformers being cognisant of situational changes and making corresponding reconfigurations adaptively would be favoured.

To this end, we introduce the cognitive-driven optimization of sparse sensing in this chapter.

In Sect. 5.2, we propose the adaptive beamformer design by regularized complementary antenna switching (RCAS), to swiftly adapt both array configuration and excitation weights in accordance to the dynamic environment for enhancing interference suppression. In order to achieve an implementable design of array reconfiguration, the RCAS is conducted in the framework of regularized antenna switching, whereby the full array aperture is collectively divided into separate groups and only one antenna in each group is switched on to connect with the processing channel. A set of deterministic complementary sparse arrays with good quiescent beampatterns is first designed by RCAS and full array data is collected by switching among them while maintaining resilient interference suppression. Subsequently, adaptive sparse array tailored for the specific environment is calculated and reconfigured based on the information extracted from the full array data.

In Sect. 5.3, we provide the cognitive sparse beamformer design via regularized switching network. The proposed beamformer is capable of swiftly adapting entwined beamformer geometry and excitation weights according to the environmental dynamics through a "perception-action" cycle. In the "perception" step, situational information is extracted from the collected real-time data and the sparse beamformer is updated in the "action" step via a regularized switching network, which divides the large array into groups and one antenna or beam is replaced with other candidates in the same group in the metric of array gain during each iteration. Cognitive beamformers with situational awareness are becoming increasingly desirable, as they are cognisant of environmental changes and capable of swiftly adapting entwined beamformer geometry and excitation weights accordingly via a continuous "perception-action" cycle [206–208]. When no environmental information is available at the start, cognitive beamformers are initialized with a good quiescent beampattern, and then reconfigured to the optimal beamformers gradually with the accumulated learning of environmental information from the collected data.

5.2 Adaptive Beamformer Design by Regularized Complementary Antenna Switching

In this section, we investigate adaptive sparse array beamformer design and propose a novel regularized complementary antenna switching (RCAS) strategy. The difference between the proposed RCAS and the work in Chap. 4 is that (1) It does not require the assumption of known environmental information by alternate switching among complementary sparse arrays; (2) the filtering performance and normal functionality of the system are guaranteed even in the sensing stage; (3) regularized antenna switching is considered instead of unrestricted antenna selection to address practical circuitry issues and limitation on inter-element spacing. The hardware schematic of the RCAS is illustrated in Fig. 5.1. A large array of N uniformly spaced antennas is divided into L groups and each group contains M antennas. There is one front-end processing channel in each group, and the RF switch can connect any antenna in this group with the corresponding front-end within a sufficiently small time. The signal processing unit is responsible for spatial filtering and information extraction, and data storage will be started when the system is working in the sensing stage. Antenna selection unit calculates and determines the optimal array configuration in the specific environment and mandates the status change of the RF switches.

The proposed RCAS comprises two steps, deterministic complementary sparse array (DCSA) design, addressed in Sect. 5.2.2, and regularized adaptive sparse array (RASA) design, addressed in Sect. 5.2.3. In the first step, the full array is divided into complementary sparse arrays by regularized antenna switching, all with good quiescent patterns and collectively spanning the full aperture. In the second step, the optimum sparse array for adaptive beamforming is then calculated and reconfigured based on the information sensed by all antennas. The workflow of the RCAS strategy is depicted in Fig. 5.2. In the case of "cold start", the set of complementary deterministic sparse arrays are first sequentially switched on for environmental sensing. The adaptive sparse array that is optimum in the specific environment can then be calculated and reconfigured, and spatial filtering is conducted to suppress the interferences. When spatial filtering experiences dramatic performance degradation (for example output SINR has dropped by 3dB), the currently employed adaptive sparse array loses its optimality for the specific environment. Complementary sparse arrays can be reconfigured back again for sensing the environmental dynamics, and the corresponding optimum adaptive sparse array needs to be re-designed.

5.2.1 *Spatial Filtering Techniques*

In this section, a full array is assumed and three beamforming techniques are reviewed as follows.

5.2.1.1 Deterministic Beamforming

Assume a linear array with N antennas placed at the positions of $[x_1, \ldots, x_N]$ with an inter-element spacing of d. Deterministic beamforming essentially focuses on synthesizing a desired beampattern, which can be formulated as,

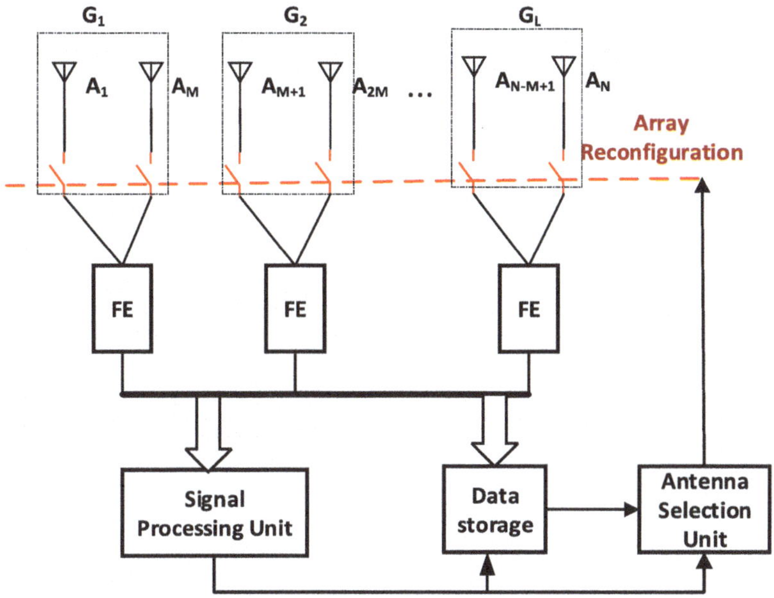

Fig. 5.1 The hardware schematic of RCAS strategy: N antennas are divided into L groups and each group comprises M antennas. Only one antenna in each group is switched on

$$\min_{\mathbf{w}} \|\mathbf{w}^H \mathbf{A}_s - \mathbf{f}\|_2^2, \quad \text{s.t. } \mathbf{w}^H \mathbf{a}(\theta_0) = 1, \tag{5.1}$$

where $\mathbf{w} \in \mathbb{C}^N$ is the N-dimensional weight coefficient vector with $\mathbb{C}$ denoting the set of complex number and $\mathbf{a}(\theta_0)$ is the steering vector towards the desired direction of θ_0,

$$\mathbf{a}(\theta_0) = [e^{jk_0 dx_1 \sin\theta_0}, \ldots, e^{jk_0 dx_N \sin\theta_0}]^T, \tag{5.2}$$

where wavenumber is defined as $k_0 = 2\pi/\lambda$ with λ denoting wavelength and the steering direction θ_0 is measured relative to the array broadside. Also, $\mathbf{A}_s \in \mathbb{C}^{N\times K}$ denotes the array manifold matrix of sidelobe region $\Omega_s = [\theta_1, \ldots, \theta_K]$, and $\mathbf{f} \in \mathbb{C}^{1\times K}$ represents the desired complex sidelobe beampattern, which can be expressed as the element-wise product between beampattern magnitude and phase. That is $\mathbf{f} = \mathbf{f}_d \odot \mathbf{f}_p$, where $\odot$ denotes element-wise product. The beampattern phase $\mathbf{f}_p$ can be fully employed as additional DoFs to enhance the beampattern synthesis and can be iteratively updated as,

$$\mathbf{f}_p^{(k+1)} = \mathbf{f}^{(k)} \oslash |\mathbf{f}^{(k)}|, \tag{5.3}$$

where the superscript $^{(k)}$ denotes the kth iteration and the operator $\oslash$ denotes element-wise division. The complex beampattern in the $(k+1)$th iteration is updated as $\mathbf{f}^{(k+1)} = \mathbf{f}_d \odot \mathbf{f}_p^{(k+1)}$. It was proved in [209] that this updating rule will converge to

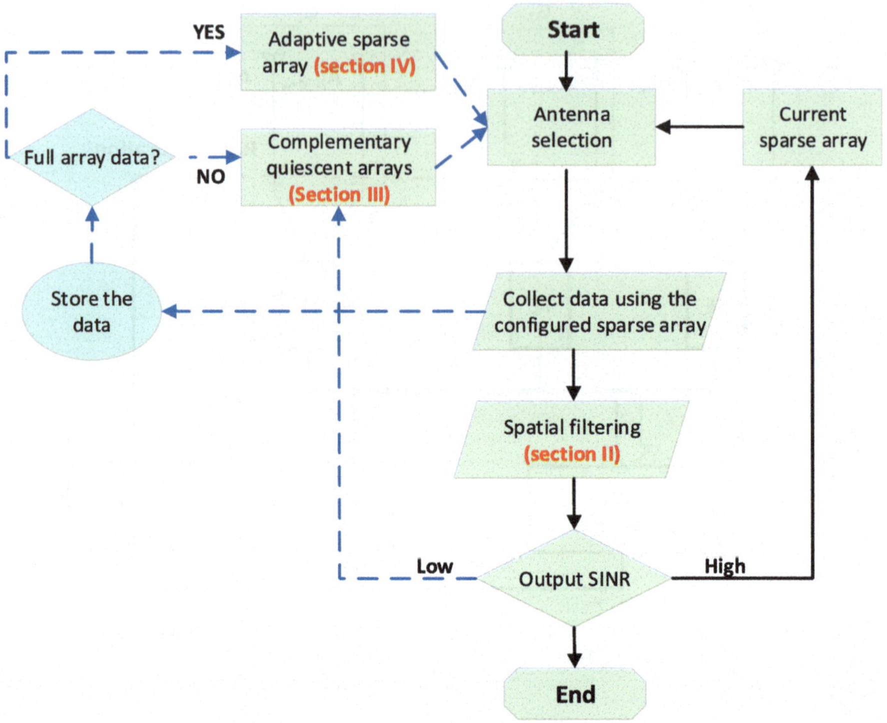

Fig. 5.2 The workflow of the RCAS strategy: Complementary sparse arrays are sequentially switched on for environmental sensing, followed by adaptive sparse array design

a non-differential power pattern synthesis, i.e., min $\||\mathbf{w}^H\mathbf{A}_s| - \mathbf{f}_d\|_2$. Note that the beampattern $\mathbf{f}$ is the only row vector and others are all column vectors in this paper.

Define the following matrix and vector,

$$\mathbf{B} = \begin{bmatrix} \mathbf{f}\mathbf{f}^H & -\mathbf{f}\mathbf{A}_s^H \\ -\mathbf{A}_s\mathbf{f}^H & \mathbf{A}_s\mathbf{A}_s^H \end{bmatrix}, \tag{5.4}$$

and $\tilde{\mathbf{w}} = [1, \mathbf{w}^T]^T$. Then Eq. (5.1) can be rewritten as,

$$\min_{\tilde{\mathbf{w}}} \tilde{\mathbf{w}}^H\mathbf{B}\tilde{\mathbf{w}}, \quad \text{s.t. } \tilde{\mathbf{w}}^H\mathbf{C} = \mathbf{g}, \tag{5.5}$$

where $\mathbf{C} = [\tilde{\mathbf{a}}(\theta_0), \mathbf{e}]$ with $\tilde{\mathbf{a}}(\theta_0) = [-1, \mathbf{a}^T(\theta_0)]^T$ is the extended steering vector, $\mathbf{e} \in \{0, 1\}^{(N+1)\times 1}$ has a single one at the first entry and other N entries being zero, and $\mathbf{g} = [0, 1]^T$. The optimal weight vector can be obtained from Lagrangian multiplier method, that is,

$$\tilde{\mathbf{w}}_o = \mathbf{B}^{-1}\mathbf{C}(\mathbf{C}^H\mathbf{B}^{-1}\mathbf{C})^{-1}\mathbf{g}. \tag{5.6}$$

Essentially, a weak interference is reckoned to come from each angle in the sidelobe region by deterministic beamformer, which attenuates all of them equally and simultaneously by a factor of $\mathbf{f}_d$, while maintaining a unit directional gain towards the desired signal [1, 210].

5.2.1.2 Adaptive Beamfoming

The well-known adaptive Capon beamformer aims to minimize the array output power while maintaining a unit gain towards the look direction [1],

$$\min_{\mathbf{w}} \mathbf{w}^H \mathbf{R} \mathbf{w}, \quad \text{s.t. } \mathbf{w}^H \mathbf{a}(\theta_0) = 1, \tag{5.7}$$

where $\mathbf{R} \in \mathbb{C}^{N \times N}$ is the covariance matrix of the received data by the employed antenna array and can be theoretically written as,

$$\mathbf{R} = \sigma_s^2 \mathbf{a}(\theta_0)\mathbf{a}^H(\theta_0) + \sum_{j=1}^{J} \sigma_j^2 \mathbf{a}(\theta_j)\mathbf{a}^H(\theta_j) + \sigma_n^2 \mathbf{I}, \tag{5.8}$$

where J is the number of interferences, and σ_s^2, σ_j^2 and σ_n^2 denote the power of source, the jth interference and white noise, respectively. The Capon beamformer weight vector is,

$$\mathbf{w}_o = \eta \mathbf{R}^{-1} \mathbf{a}(\theta_0), \tag{5.9}$$

where $\eta = 1/\mathbf{a}^H(\theta_0)\mathbf{R}^{-1}\mathbf{a}(\theta_0)$. Capon beamformer emphasizes on the strong interferences in order to minimize the total output power, which, in turn, unavoidably lifts the sidelobe level in other angular regions as a result of energy suppression towards the interferences' direction.

5.2.1.3 Combined Adaptive and Deterministic Beamforming

Although adaptive beamforming is superior to deterministic beamforming in terms of MaxSINR, high sidelobes of adaptive beamforming constitute a potential nuisance, especially, when directional interferers are suddenly turned on. As such, it is crucial to combine the merits of two types of sparse arrays, those are MaxSINR and well-controlled sidelobes. Inspired by our previous work in [197], a combined beamformer with a regularized adaptive and deterministic objectives is formulated as follows,

$$\min_{\mathbf{w}} \mathbf{w}^H \mathbf{R} \mathbf{w} + \beta \|\mathbf{w}^H \mathbf{A}_s - \mathbf{f}\|_2^2, \quad \text{s.t. } \mathbf{w}^H \mathbf{a}(\theta_0) = 1, \tag{5.10}$$

where the trade-off parameter β is adjusted to control the relative emphasis between quiescent pattern and output power. Combining Eqs. (5.5), (5.7) and (5.10) rewrites the formulation as,

$$\min_{\tilde{\mathbf{w}}} \tilde{\mathbf{w}}^H \mathbf{R}_b \tilde{\mathbf{w}}, \quad \text{s.t. } \tilde{\mathbf{w}}^H \mathbf{C} = \mathbf{g}, \tag{5.11}$$

where $\mathbf{R}_b = \tilde{\mathbf{R}} + \beta \mathbf{B}$ with

$$\tilde{\mathbf{R}} = \begin{bmatrix} 0 & \mathbf{0}_N^T \\ \mathbf{0}_N & \mathbf{R} \end{bmatrix}. \tag{5.12}$$

Here, $\mathbf{0}_N$ is a N-dimensional vector of all zeros. The optimal weight vector is therefore,

$$\tilde{\mathbf{w}}_o = \mathbf{R}_b^{-1}\mathbf{C}(\mathbf{C}^H \mathbf{R}_b^{-1}\mathbf{C})^{-1}\mathbf{g}. \tag{5.13}$$

By comparing Eqs. (5.13) with (5.6), we can observe that combined beamformer is capable of maintaining well-controlled beampattern while suppressing the interferences by adjusting the trade-off parameter β. When β is small, the combined beamformer prioritizes strong interferences over white noise for achieving MaxSINR. Otherwise, the minimization of white noise gain, that is sidelobe level, becomes a superior task.

As mentioned above, sparse array beamformer design techniques are generally divided into deterministic and adaptive techniques. Deterministic sparse array design aims to minimize the sidelobe level leveraging both array configuration and beamforming weights. No environmental information is required by the deterministic sparse array design, as it views all the sidelobe angular region equally important. Thus, deterministic sparse array is capable of suppressing the interferences from arbitrary directions, though the null depth may not be sufficient for MaxSINR. In a nutshell, deterministic sparse arrays possess resilient interference suppression performance regardless of surrounding environment. While adaptive sparse array is designed in terms of MaxSINR, which focuses on nulling strong interferences and thus mandates timely environmental information, such as the total number of interferences J and their respective arrival angles $\theta_j, j = 1, \ldots, J$. The environmental situation is usually unknown and needs to be estimated periodically, especially in dynamic environment. The lack of environmental information is regarded as an impediment to the optimum design of adaptive sparse arrays, which is solved by the proposed RCAS strategy in this section.

5.2.2 Deterministic Complementary Sparse Array Design

In the first step of the proposed RCAS strategy, the full array is divided into separate complementary sparse arrays all with good quiescent patterns, as a result, switching among them can collect the data of full array while maintaining a moderate interference suppression performance. Note that the set of complementary sparse arrays are fixed once designed, regardless of environmental dynamics. We delineate the design of deterministic complementary sparse arrays in this section.

The N-antenna full array is split into M sparse arrays by regularized antenna switching and each array consists of L antennas. Note that these M complementary sparse arrays do not have overlapping antennas and each array is capable of synthesizing a well-controlled quiescent pattern. The regularized array splitting can be formulated as,

$$\begin{aligned} \min_{\mathbf{w}_1,\ldots,\mathbf{w}_M} \quad & \sum_{m=1}^{M} \|\mathbf{w}_m^H \mathbf{A}_s - \mathbf{f}\|_2 \\ \text{s.t.} \quad & \mathbf{w}_m^H \mathbf{a}(\theta_0) = 1, m = 1, \ldots, M, \\ & \|\mathbf{P}_l \mathbf{w}_m\|_0 \le 1, l = 1, \ldots, L, m = 1, \ldots, M \\ & \sum_{m=1}^{M} \|\mathbf{q}_i^T \mathbf{w}_m\|_0 \le 1, i = 1, \ldots, N \end{aligned} \tag{5.14}$$

where $\mathbf{f} \in \mathbb{C}^{1\times K}$ is a row vector of the desired beampattern in the sidelobe angular region, and $\mathbf{w}_m \in \mathbb{C}^{N\times 1}, m = 1, \ldots, M$ is a sparse weight vector with only L non-zero coefficients corresponding to L selected antennas of the mth sparse array. The group selection matrix $\mathbf{P}_l \in \{0, 1\}^{M\times N}$ are all zeros except those entries of $\mathbf{P}_l(1, 1 + (l-1)M), \mathbf{P}_l(2, 2 + (l-1)M), \ldots, \mathbf{P}_l(M, lM)$. The vector $\mathbf{q}_i \in \{0, 1\}^{N\times 1}$ are all zeros except for the ith entry being one. The last two constraints of Eq. (5.14) are combined together to restrain that only one antenna in each group is selected for each sparse array and no overlapping antennas are selected for different sparse arrays. For the split M sparse arrays, the same quiescent beampattern is desired that the main beams be steered towards θ_0 with the prescribed sidelobes. For the sake of an easy exposition, we mostly omit the intended angle θ_0 in the rest of the paper, unless we specifically refer to a particular angle.

Put the M beamforming weight vectors into a matrix $\mathbf{W} = [\mathbf{w}_1, \ldots, \mathbf{w}_M] \in \mathbb{C}^{N\times M}$. Equivalently, problem in Eq. (5.14) can be rewritten as,

$$\begin{aligned} (P_0) \min_{\mathbf{W}} \quad & \|\mathbf{W}^H \mathbf{A}_s - \mathbf{F}\|_\mathrm{F} \\ \text{s.t.} \quad & \mathbf{W}^H \mathbf{a} = \mathbf{1}_M, \\ & \|\mathbf{P}_l \mathbf{W} \mathbf{c}_m\|_0 \le 1, l = 1, \ldots, L, m = 1, \ldots, M \\ & \|\mathbf{q}_i^T \mathbf{W}\|_0 \le 1, i = 1, \ldots, N \end{aligned} \tag{5.15}$$

where $\mathbf{F} = \mathbf{1}_M \mathbf{f}$ with $\mathbf{1}_M$ being a M-dimensional column vector of all ones, $\| \bullet \|_\mathrm{F}$ is the Frobenius norm of a matrix, and $\mathbf{c}_m \in \{0, 1\}^M$ is a column selection vector with all entries being zero except for the mth entry being one.

The notorious cardinality constraints in the third and fourth lines of Eq. (5.15) render the problem difficult to solve. In the following, we propose an algorithm to solve

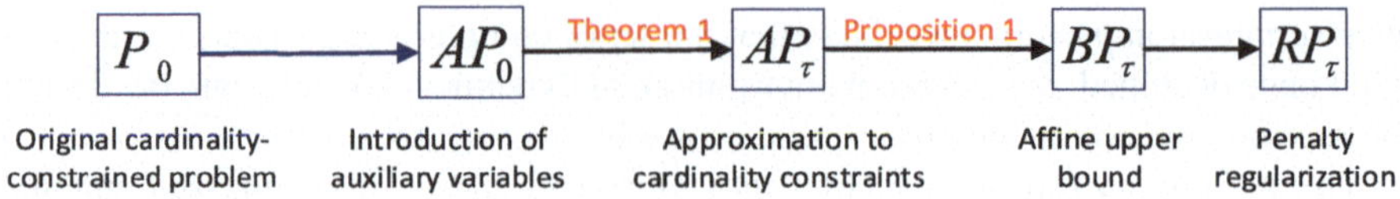

Fig. 5.3 The derivation procedure of deterministic complementary sparse array design

the cardinality-constrained optimization according to the derivation procedure illustrated in Fig. 5.3, where $P_0 \rightarrow AP_0$: an auxiliary variable is introduced to transform the complex-cardinality constraint to the real domain; $AP_0 \rightarrow AP_\tau$: a piece-wise linear function is utilized to approximate the l_0-norm; $AP_\tau \rightarrow BP_\tau$: an affine upper bound is derived to iteratively relax the non-convex constraints; $BP_\tau \rightarrow RP_\tau$: a regularized penalty is formulated to eliminate the requirement of feasible start search points.

5.2.2.1 Introduction of Auxiliary Variables

As the problem (P_0) is formulated in the complex domain, we first introduce an auxiliary variable $\mathbf{Z} \in \mathbb{R}_+^{N\times M}$ to transform the complex cardinality constraint to the real domain. Here, $\mathbb{R}_+$ denotes the set of non-negative numbers. That is,

$$(AP_0) \ \min_{\mathbf{W},\mathbf{Z}} \ \|\mathbf{W}^H\mathbf{A}_s - \mathbf{F}\|_\mathrm{F} \tag{5.16a}$$

$$\text{s.t.} \ \ \mathbf{W}^H\mathbf{a} = \mathbf{1}_M,$$

$$|\mathbf{W}| \le \mathbf{Z}, \tag{5.16b}$$

$$\|\mathbf{P}_l\mathbf{Z}\mathbf{c}_m\|_0 \le 1, \tag{5.16c}$$

$$l = 1, \ldots, L, m = 1, \ldots, M$$

$$\mathbf{1}_M^T\mathbf{P}_l\mathbf{Z}\mathbf{c}_m = 1, \tag{5.16d}$$

$$l = 1, \ldots, L, m = 1, \ldots, M$$

$$\mathbf{Z}\mathbf{1}_M = \mathbf{1}_N, \tag{5.16e}$$

where the inequality constraint in Eq. (5.16b) is element-wise, i.e., the absolute value of each element of $\mathbf{W}$ is bounded by the corresponding element of $\mathbf{Z}$. Constraints in Eqs. (5.16c)–(5.16e) combined to restrain that different antennas are selected by the M sparse arrays in each group. Furthermore, we can infer $\mathbf{Z} \in \{0, 1\}^{N\times M}$ from the constraints in Eqs. (5.16c) and (5.16d), which implies that $\mathbf{Z}$ is an $N \times M$-dimensional matrix with entries being either 0 or 1. To put it differently, the binary property of the auxiliary variable $\mathbf{Z}$ is implicitly imposed in the formulation (AP_0).

We can observe from problem (AP_0) that there is only one cardinality constraint left by introducing the auxiliary variable $\mathbf{Z}$.

Clearly, $\mathbf{W}$ solves (P_0) if and only if $(\mathbf{W}, \mathbf{Z})$ solves (AP_0) with $\mathbf{Z} = \text{sign}(|\mathbf{W}|)$, where $\text{sign}(\cdot) = 0$ only when $\cdot = 0$, otherwise $\text{sign}(\cdot) = 1$. The proof is as follows: (1) Suppose that $\mathbf{W}$ solves (P_0), then set $\mathbf{Z} = \text{sign}(|\mathbf{W}|)$ and $(\mathbf{W}, \mathbf{Z})$ solves (AP_0) as well. (2) Suppose that $(\mathbf{W}, \mathbf{Z})$ solves (AP_0). When $Z_{ij} = 1$, W_{ij} must be non-zero for minimizing the objective function, whereas when $Z_{ij} = 0$, we must have $W_{ij} = 0$ for all $i = 1, \ldots, N$, $j = 1, \ldots, M$. Therefore, $\mathbf{W}$ solves (P_0) as well. This proves the equivalence between (P_0) and (AP_0).

5.2.2.2 Approximation to Cardinality Constraints

As the l_0-norm is a notorious cardinality constraint and the problem involved is difficult to solve, we resort to a piece-wise linear function to approximate it. Assume that $\mathbf{b} \in \mathbb{R}_+^M$ is a M-dimensional non-negative real vector, and then the function is defined as,

$$\|\mathbf{b}\|_0 \approx \phi(\mathbf{b}, \tau) = \sum_{i=1}^{M} (1/\tau_i)(b_i - (b_i - \tau_i)^+). \tag{5.17}$$

where $\tau = [\tau_1, \ldots, \tau_M]^T > 0$ is a threshold vector, and $(b_i - \tau_i)^+ = \max\{b_i - \tau_i, 0\}$. Regarding to the properties of the approximation function $\phi(\mathbf{b}, \tau)$, we have the following lemma.

Lemma 5.1 *(1) For any $\tau > 0$, $\phi(\mathbf{b}, \tau)$ is a piecewise linear under-estimator of $\|\mathbf{b}\|_0$, i.e. $\phi(\mathbf{b}, \tau) \leq \|\mathbf{b}\|_0, \forall \mathbf{b} \in \mathbb{R}_+^M$, and $\phi(\mathbf{b}, \tau)$ is a non-increasing function of τ.*

(2) For any fixed $\mathbf{b} \in \mathbb{R}_+^M$, it holds that

$$\lim_{\tau \to 0^+} \phi(\mathbf{b}, \tau) = \|\mathbf{b}\|_0. \tag{5.18}$$

(3) The sub-gradient of the function $\phi(\mathbf{b}, \tau)$ is $\mathbf{g}(\mathbf{b}, \tau) = \frac{\partial \phi(\mathbf{b}, \tau)}{\partial \mathbf{b}} = [\frac{\partial \phi(\mathbf{b}, \tau)}{b_1}, \cdots, \frac{\partial \phi(\mathbf{b}, \tau)}{b_M}]^T$, where

$$\frac{\partial \phi(\mathbf{b}, \boldsymbol{\tau})}{b_i} = \begin{cases} 0, & \text{if } b_i > \tau_i \\ [0, 1/\tau_i], & \text{if } b_i = \tau_i \\ 1/\tau_i, & \text{if } b_i < \tau_i \end{cases} \tag{5.19}$$

and $[0, 1/\tau_i]$ denotes any number between 0 and $1/\tau_i$. The proof of Lemma 5.1 is provided as follows.

Proof of Lemma 5.1:

(1) The cardinality of a vector $\mathbf{b} \in \mathbb{R}_+^M$ is expressed as,

$$\|\mathbf{b}\|_0 = \mathbf{1}_M^T \text{sign}(\mathbf{b}), \tag{5.20}$$

where the sign function is defined as

$$\operatorname{sign}(\cdot) = \begin{cases} 1, & \text{if} \cdot > 0 \\ 0. & \text{if} \cdot = 0 \end{cases} \tag{5.21}$$

According to the definition of $\phi(\mathbf{b}, \tau)$, it is separable and can be rewritten as

$$\phi(\mathbf{b}, \tau) = \sum_{i=1}^{M} \phi(b_i, \tau_i), \tag{5.22}$$

where $\phi(b_i, \tau_i) = (1/\tau_i)(b_i - (b_i - \tau_i)^+)$ and thus

$$\phi(b_i, \tau_i) = \begin{cases} 1 & \text{if } b_i > \tau_i, \\ b_i/\tau_i & \text{if } b_i \leq \tau_i, \end{cases} \tag{5.23}$$

Here, $(b_i - \tau_i)^+ = \max\{b_i - \tau_i, 0\}$. Comparing the sign function in Eq. (5.21) with the function $\phi(b_i, \tau_i)$ in Eq. (5.23), we obtain that, $\forall\ b_i \geq 0$

$$0 \leq \phi(b_i, \tau_i) \leq \operatorname{sign}(b_i) \leq 1. \tag{5.24}$$

Combining Eqs. (5.20), (5.22) and (5.24) yields $\phi(\mathbf{b}, \tau) \leq \|\mathbf{b}\|_0$, that is $\phi(\mathbf{b}, \tau)$ is a piecewise linear under-estimator of $\|\mathbf{b}\|_0$. Moreover, function $\phi(b, \tau)$ is obviously a non-increasing function of τ, that is,

$$\phi(b, \tau_1) \geq \phi(b, \tau_2), \text{ if } \tau_1 \leq \tau_2 \tag{5.25}$$

Therefore, for any given vector $\mathbf{b}$, we have that $\phi(\mathbf{b}, \tau_2) \leq \phi(\mathbf{b}, \tau_1)$ from Eq. (5.22), given that $\tau_1 \leq \tau_2$. Here, $\tau_1 \leq \tau_2$ implies that $\tau_{1,i} \leq \tau_{2,i}, \forall i = 1, \ldots, M$. Thus, $\phi(\mathbf{b}, \tau)$ is a non-increasing function of τ.

(2) Comparing Eqs. (5.21) with (5.23), we also have that

$$\lim_{\tau \to 0^+} \phi(\cdot, \tau) = \operatorname{sign}(\cdot), \tag{5.26}$$

Thus, utilizing Eqs. (5.20) and (5.22), we have that

$$\lim_{\tau \to 0^+} \phi(\mathbf{b}, \tau) = \|\mathbf{b}\|_0. \tag{5.27}$$

(3) For arbitrary two points $\mathbf{b}_1$ and $\mathbf{b}_2$. We have that $\forall\ 0 \leq \alpha \leq 1$,

$$\alpha\phi(\mathbf{b}_1, \tau) + (1 - \alpha)\phi(\mathbf{b}_2, \tau) \leq \phi[\alpha\mathbf{b}_1 + (1 - \alpha)\mathbf{b}_2, \tau]. \tag{5.28}$$

This proves that the piece-wise linear function $\phi(\mathbf{b}, \tau)$ is a concave function with respect to the variable $\mathbf{b}$. The definition of sub-gradient is that for any $\mathbf{g}(\mathbf{b}_0, \tau) = \frac{\partial \phi(\mathbf{b}, \tau)}{\partial \mathbf{b}} |_{\mathbf{b}_0}$ such that $\forall$ $\mathbf{b}$, we have the following,

$$\phi(\mathbf{b}, \tau) \leq \phi(\mathbf{b}_0, \tau) + \mathbf{g}(\mathbf{b}_0, \tau)^T (\mathbf{b} - \mathbf{b}_0). \tag{5.29}$$

When $b_i > \tau_i$, the function $\phi(b_i, \tau_i)$ is differentiable and the gradient is zero. When $0 \leq b_i < \tau_i$, the function $\phi(b_i, \tau_i)$ is also differentiable and the gradient is $1/\tau_i$. When $b_i = \tau_i$, the function $\phi(b_i, \tau_i)$ is non-differentiable, and the sub-gradient can be any number between $[0, 1/\tau_i]$ according to Eq. (5.29).
The proof of Lemma 5.1 is complete.

Utilizing the approximation function of the l_0-norm in Lemma 5.1, problem (AP_0) in Eq. (5.16a) can be expressed as,

$$(AP_\tau) \min_{\mathbf{W},\mathbf{Z}} \ \|\mathbf{W}^H \mathbf{A}_s - \mathbf{F}\|_\text{F} \tag{5.30a}$$

$$\text{s.t.} \ \ \mathbf{W}^H \mathbf{a} = \mathbf{1}_M,$$

$$|\mathbf{W}| \leq \mathbf{Z}, \tag{5.30b}$$

$$\phi(\mathbf{P}_l \mathbf{Z} \mathbf{c}_m, \mathbf{P}_l \Pi \mathbf{c}_m) \leq 1, \tag{5.30c}$$

$$l = 1, \ldots, L, m = 1, \ldots, M$$

$$\mathbf{1}_M^T \mathbf{P}_l \mathbf{Z} \mathbf{c}_m = 1, \tag{5.30d}$$

$$l = 1, \ldots, L, m = 1, \ldots, M$$

$$\mathbf{Z}\mathbf{1}_M = \mathbf{1}_N. \tag{5.30e}$$

Here, $\Pi \in \mathbb{R}_+^{N \times M}$ is the threshold matrix. The relationship between (AP_0) and (AP_τ) is given in Theorem 5.1.

Theorem 5.1 *When the threshold matrix* $\Pi < 1$ *(that is,* $\Pi_{i,j} < 1, \forall i = 1, \ldots, N$, $j = 1, \ldots, M$*), the formulation* (AP_τ) *is equivalent to the formulation* (AP_0).

The proof of Theorem 5.1 is provided as follows.

Proof of Theorem 5.1:

By utilizing the implicit binary constraint of the auxiliary variable $\mathbf{Z} \in \{0, 1\}^{N \times M}$, the complementary sparse array design in Eq. (5.16a) can be rewritten as follows,

$$\begin{aligned}(AP_b)\ \min_{\mathbf{W},\mathbf{Z}}\ & \|\mathbf{W}^H\mathbf{A}_s - \mathbf{F}\|_F \\ \text{s.t.}\ & \mathbf{W}^H\mathbf{a} = \mathbf{1}_M, \\ & |\mathbf{W}| \leq \mathbf{Z}, \\ & \mathbf{Z} \in \{0, 1\}^{N\times M}, \\ & \mathbf{1}_M^T\mathbf{P}_l\mathbf{Z}\mathbf{c}_m = 1, l = 1, \ldots, L, m = 1, \ldots, M \\ & \mathbf{Z}\mathbf{1}_M = \mathbf{1}_N.\end{aligned} \tag{5.31}$$

By comparing the two problems in Eqs. (5.30a) and (5.31), we can observe that though the third constraint exhibits difference, the two formulations are essentially the same.

Let us define a set $\underline{\Omega} = \{\mathbf{Z} \geq 0 : \mathbf{Z}\mathbf{1}_M = \mathbf{1}_N, \mathbf{1}_M^T\mathbf{P}_l\mathbf{Z}\mathbf{c}_m = 1, l = 1, \ldots, L, m = 1, \ldots, M\}$, then the domain of problem (AP_τ) can be expressed as $\hat{\Omega}_\tau = \{\mathbf{Z} : \mathbf{Z} \in \underline{\Omega}, \phi(\mathbf{P}_l\mathbf{Z}\mathbf{c}_m, \mathbf{P}_l\Pi\mathbf{c}_m) \leq 1, l = 1, \ldots, L, m = 1, \ldots, M\}$. Similarly, the domain of the problem (AP_b) is same as that of (AP_0), that is $\hat{\Omega}_0 = \{\mathbf{Z} : \mathbf{Z} \in \underline{\Omega}, \mathbf{Z} \in \{0, 1\}^{N\times M}\}$. Then, we are going to prove that $\hat{\Omega}_\tau = \hat{\Omega}_0$.

Suppose that there exists some $\mathbf{Z}$ such that $\mathbf{Z} \in \hat{\Omega}_\tau$ and $\mathbf{Z} \notin \hat{\Omega}_0$. First, $\mathbf{Z} \in \hat{\Omega}_\tau$ implies that $\forall l \in \{1, \ldots, L\}$ and $\forall m \in \{1, \ldots, M\}$ such that,

$$\mathbf{1}_M^T\mathbf{P}_l\mathbf{Z}\mathbf{c}_m = 1, \tag{5.32}$$

and

$$\phi(\mathbf{P}_l\mathbf{Z}\mathbf{c}_m, \mathbf{P}_l\Pi\mathbf{c}_m) \leq 1. \tag{5.33}$$

Since $\mathbf{Z} \notin \hat{\Omega}_0$, it implies that $\exists l \in \{1, \ldots, L\}$ and $\exists m \in \{1, \ldots, M\}$, such that $\mathbf{P}_l\mathbf{Z}\mathbf{c}_m \notin \{0, 1\}^M$, i.e. it does not have a single entry equal to one and others being zero.

Set $\mathbf{b} = \mathbf{P}_l\mathbf{Z}\mathbf{c}_m$ and $\tau = \mathbf{P}_l\Pi\mathbf{c}_m$, we consider the following three cases:
(1) When there are more than one entries of $\mathbf{b}$ greater than τ, then $\phi(\mathbf{b}, \tau) > 1$, violating Eq. (5.33).
(2) When there are multiple entries of $\mathbf{b}$, all smaller than τ, according to the definition of the approximation function, we have that,

$$\phi(\mathbf{b}, \tau) = \sum_{i=1}^{M} b_i/\tau_i. \tag{5.34}$$

According to Eqs. (5.32) and (5.33), we have that

$$\phi(\mathbf{b}, \tau) \leq \mathbf{1}_M^T\mathbf{b}/\tau_{\min} = 1/\tau_{\min} \leq 1 \Rightarrow \tau_{\min} \geq 1. \tag{5.35}$$

where $\tau_{\min} = \min \tau_i, i = 1, \ldots, M$. Contradiction to $\tau < 1$.
(3) When there is only one entry of $\mathbf{b}$ greater than the corresponding entry of τ, we assume $b_k > \tau_k, k \in \{1, \ldots, M\}$. Utilizing Eq. (5.32), we have that

$$\sum_{i=1,i\neq k}^{M} b_i = 1 - b_k. \tag{5.36}$$

According to the definition of the approximation function and Eq. (5.33), we further have that

$$\phi(\mathbf{b}, \tau) = 1 + (1 - b_k)/\tau_k \leq 1 \Rightarrow b_k \geq 1. \tag{5.37}$$

Combining with Eq. (5.32), we further obtain that $b_k = 1$ and $b_i = 0, \forall i \neq k$. Contradiction to the assumption of $\mathbf{P}_l\mathbf{Z}\mathbf{c}_m \notin \{0, 1\}^M$, or equivalently the assumption of $\mathbf{Z} \notin \hat{\Omega}_0$. Thereby, $\mathbf{Z} \in \hat{\Omega}_\tau \Rightarrow \mathbf{Z} \in \hat{\Omega}_0$, i.e., $\hat{\Omega}_\tau \subseteq \hat{\Omega}_0$.

Assume that there exists some $\mathbf{Z}$ such that $\mathbf{Z} \in \hat{\Omega}_b$. That means there exists only one entry $p \in \{1 + (l - 1)M, \ldots, lM\}$ for each $l \in \{1, \ldots, L\}$ and $m \in \{1, \ldots, M\}$ such that $Z_{pm} = 1$ and $Z_{im} = 0, \forall i \in \{1 + (l - 1)M, \ldots, lM\}$ and $i \neq p$. According to the property of the approximation function in Lemma 5.1, we have that

$$\phi(Z_{im}) = \begin{cases} 1 & i = p, \\ 0 & i \neq p, \end{cases} \tag{5.38}$$

Thus, $\phi(\mathbf{P}_l\mathbf{Z}\mathbf{c}_m, \mathbf{P}_l\Pi\mathbf{c}_m) = 1, l = 1, \ldots, L, m = 1, \ldots, M$. That implies that $\mathbf{Z} \in \hat{\Omega}_0 \rightarrow \mathbf{Z} \in \hat{\Omega}_\tau$, i.e., $\hat{\Omega}_0 \subseteq \hat{\Omega}_\tau$. Thereby, we prove that $\hat{\Omega}_0 = \hat{\Omega}_\tau$. In a nutshell, the problem (AP_τ) is equivalent to the problem (AP_0).

The proof of Theorem 5.1 **is complete**.

5.2.2.3 Affine Upper Bound

As proved in Lemma 5.1, the approximation function $\phi(\mathbf{b}, \tau)$ is concave with respect to $\mathbf{b}$, thus the relaxed cardinality constraint in (AP_τ) is still nonconvex. We then resort to the following affine function to iteratively upper-bound $\phi(\mathbf{b}, \tau)$,

$$\phi(\mathbf{b}, \tau) \leq \bar{\phi}(\mathbf{b}; \mathbf{b}_0, \tau) = \phi(\mathbf{b}_0, \tau) + \mathbf{g}^T(\mathbf{b}_0, \tau)(\mathbf{b} - \mathbf{b}_0), \tag{5.39}$$

where $\mathbf{g}(\mathbf{b}_0, \tau) = [\frac{\partial\phi}{\partial b_1}, \ldots, \frac{\partial\phi}{\partial b_M}]^T\big|_{\mathbf{b}_0}$ denotes the sub-gradient of function $\phi(\mathbf{b}, \tau)$ evaluated at the point $\mathbf{b}_0$ and its definition is provided in the Lemma 5.1(3). Tracing back to problem (AP_τ), the subgradient of the function $\phi(\mathbf{P}_l\mathbf{Z}\mathbf{c}_m, \mathbf{P}_l\Pi\mathbf{c}_m)$ with respect to the variable $\mathbf{Z}$ is,

$$\frac{\partial\phi(\mathbf{P}_l\mathbf{Z}\mathbf{c}_m, \mathbf{P}_l\Pi\mathbf{c}_m)}{\partial\mathbf{Z}} = \mathbf{P}_l^T\tilde{\mathbf{r}}_{m,l}\mathbf{c}_m^T, \tag{5.40}$$

where $\tilde{\mathbf{r}}_{m,l} = \sum_{i=1}^{M} g(b_i, \tau_i)\mathbf{r}_i$ with $b_i = \mathbf{r}_i^T\mathbf{P}_l\mathbf{Z}\mathbf{c}_m$ and $\tau_i = \mathbf{r}_i^T\mathbf{P}_l\Pi\mathbf{c}_m$, and $\mathbf{r}_i \in \{0, 1\}^{M\times 1}$ is a row selection vector with all entries being zero except for the ith entry. The proof of Eq. (5.40) is provided as follows.

Proof of Eq. (5.40):

Define $\mathbf{b} = \mathbf{P}_l\mathbf{Z}\mathbf{c}_m$ and $\tau = \mathbf{P}_l\Pi\mathbf{c}_m$. Based on Eq. (5.17), $\phi(\mathbf{b}, \tau)$ is defined as,

$$\phi(\mathbf{b}, \tau) = \sum_{i=1}^{M} \phi(b_i, \tau) = \sum_{i=1}^{M} \frac{1}{\tau_i}[b_i - (b_i - \tau_i)^+],$$

where $b_i = \mathbf{r}_i^T\mathbf{P}_l\mathbf{Z}\mathbf{c}_m$ and $\tau_i = \mathbf{r}_i^T\mathbf{P}_l\Pi\mathbf{c}_m$. Proceeding from Eq. (5.41) and utilizing Lemma 5.1(3) and the chain rule, we have that,

$$\begin{aligned}
\frac{\partial\phi(\mathbf{b}, \tau)}{\partial\mathbf{Z}} &= \sum_{i=1}^{M} \frac{\partial\phi(b_i, \tau_i)}{\partial\mathbf{Z}}, \\
&= \sum_{i=1}^{M} \frac{\partial\phi(b_i, \tau_i)}{\partial b_i} \cdot \frac{\partial b_i}{\partial\mathbf{Z}}, \\
&= \sum_{i=1}^{M} g(b_i, \tau_i)(\mathbf{P}_l^T\mathbf{r}_i\mathbf{c}_m^T), \\
&= \mathbf{P}_l^T\tilde{\mathbf{r}}_{m,l}\mathbf{c}_m^T,
\end{aligned} \tag{5.41}$$

where $\tilde{\mathbf{r}}_{m,l} = \sum_{i=1}^{M} g(\mathbf{r}_i^T\mathbf{P}_l\mathbf{Z}\mathbf{c}_m, \mathbf{r}_i^T\mathbf{P}_l\Pi\mathbf{c}_m)\mathbf{r}_i$ □.

The proof of Eq. (5.40) **is complete.**

Substituting $\mathbf{b} = \mathbf{P}_l\mathbf{Z}\mathbf{c}_m$ and $\mathbf{b}_0 = \mathbf{P}_l\mathbf{Z}_0\mathbf{c}_m$ into Eq. (5.39) and utilizing the gradient in Eq. (5.40), we have that

$$\begin{aligned}
\phi(\mathbf{P}_l\mathbf{Z}\mathbf{c}_m, \mathbf{P}_l\Pi\mathbf{c}_m) &\le \bar{\phi}(\mathbf{P}_l\mathbf{Z}\mathbf{c}_m; \mathbf{P}_l\mathbf{Z}_0\mathbf{c}_m, \mathbf{P}_l\Pi\mathbf{c}_m), \\
&= \phi(\mathbf{P}_l\mathbf{Z}_0\mathbf{c}_m, \mathbf{P}_l\Pi\mathbf{c}_m) \\
&\quad + \mathrm{tr}\{(\mathbf{c}_m\tilde{\mathbf{r}}_{m,l}^T\mathbf{P}_l)(\mathbf{Z} - \mathbf{Z}_0)\}.
\end{aligned} \tag{5.42}$$

Here, the variable changes from the vector $\mathbf{b}$ to the matrix $\mathbf{Z}$, thus the trace operator $\mathrm{tr}(\cdot)$ is utilized instead of inner product.

Utilizing the upper bound in Eqs. (5.42)–(5.30c) yields,

$$\begin{aligned}
(BP_\tau)\ \min_{\mathbf{W},\mathbf{Z}}\ & \|\mathbf{W}^H\mathbf{A}_s - \mathbf{F}\|_F \\
\text{s.t.}\ & \mathbf{W}^H\mathbf{a} = \mathbf{1}_M, \\
& |\mathbf{W}| \le \mathbf{Z}, \\
& \phi(\mathbf{P}_l\mathbf{Z}_0\mathbf{c}_m, \mathbf{P}_l\Pi\mathbf{c}_m) \\
& +\mathrm{tr}\{(\mathbf{c}_m\tilde{\mathbf{r}}_{m,l}^T\mathbf{P}_l)(\mathbf{Z} - \mathbf{Z}_0)\} \le 1, \\
& l = 1, \ldots, L, m = 1, \ldots, M \\
& \mathbf{1}_M^T\mathbf{P}_l\mathbf{Z}\mathbf{c}_m = 1, l = 1, \ldots, L, m = 1, \ldots, M \\
& \mathbf{Z}\mathbf{1}_M = \mathbf{1}_N,
\end{aligned} \tag{5.43}$$

The following proposition proves that the problem (BP_τ) is equivalent to problem (AP_τ).

Proposition 5.1 *When the initial search point* $\mathbf{Z}^{(0)}$ *for solving* (BP_τ) *is feasible for* (AP_τ)*, the problem* (BP_τ) *will converge to the problem* (AP_τ) *iteratively.*

The proof of Proposition 5.1 is shown as follows.

Proof of Proposition 5.1:
Assume that $\mathbf{Z}^{(k)} \in \hat{\Omega}_\tau$ is a feasible point for (AP_τ), that is,

$$\begin{aligned} &\phi(\mathbf{P}_l\mathbf{Z}^{(k)}\mathbf{c}_m, \mathbf{P}_l\Pi\mathbf{c}_m) \le 1, l = 1, \ldots, L, m = 1, \ldots, M \\ &\mathbf{1}_M^T\mathbf{P}_l\mathbf{Z}^{(k)}\mathbf{c}_m = 1, l = 1, \ldots, L, m = 1, \ldots, M \\ &\mathbf{Z}^{(k)}\mathbf{1}_M = \mathbf{1}_N. \end{aligned} \tag{5.44}$$

Proceeding from the third constraint of (BP_τ), we can obtain that

$$\mathrm{tr}\{(\mathbf{c}_m\tilde{\mathbf{r}}_{m,l}^T\mathbf{P}_l)(\mathbf{Z} - \mathbf{Z}^{(k)})\} \le 0, l = 1, \ldots, L, m = 1, \ldots, M. \tag{5.45}$$

From Eq. (5.40), we can see that $\tilde{\mathbf{r}}_{m,l}$ is relevant to the gradient of the approximation function $\phi(\mathbf{P}_l\mathbf{Z}^{(k)}\mathbf{c}_m, \mathbf{P}_l\Pi\mathbf{c}_m)$. According to Lemma 5.1, the gradients corresponding to the "one" entries of $\mathbf{Z}^{(k)}$ are 0, while the gradients corresponding to the "zero" entries of $\mathbf{Z}^{(k)}$ are $1/\tau$. In order to satisfy Eq. (5.45), variable $\mathbf{Z}$ has to keep the "zero" entries of $\mathbf{Z}^{(k)}$. This implies that $\phi(\mathbf{P}_l\mathbf{Z}\mathbf{c}_m, \mathbf{P}_l\Pi\mathbf{c}_m) \le 1, l = 1, \ldots, L, m = 1, \ldots, M$ and the obtained solution $\mathbf{Z}^{(k+1)}$ is a feasible point for (AP_τ). Therefore, once the initial search point $\mathbf{Z}^{(0)}$ is chosen feasible, all iterates are feasible and the problem (BP_τ) will converge to the problem (AP_τ).
The proof of Proposition 5.1 **is complete.**

5.2.2.4 Penalty Regularization

The main drawback of the formulation (BP_τ) is that the initial search point is required to be feasible to (AP_τ). We next formulate an extension of (BP_τ), which is capable of removing the requirement of an feasible start point. We add a penalty regularization by removing the upper bound of the piece-wise linear approximation function in the fourth line of Eq. (5.43) to the objective function, that is,

$$
\begin{aligned}
(RP_\tau)\ \min_{\mathbf{W},\mathbf{Z}}\ & \|\mathbf{W}^H\mathbf{A}_s - \mathbf{F}\|_F + \rho \sum_{l=1}^{L}\sum_{m=1}^{M} \mathrm{tr}\{(\mathbf{c}_m \tilde{\mathbf{r}}_{m,l}^T \mathbf{P}_l)\mathbf{Z}\} \\
\text{s.t. }\ & \mathbf{W}^H \mathbf{a} = \mathbf{1}_M, \\
& |\mathbf{W}| \leq \mathbf{Z}, \\
& \mathbf{1}_M^T \mathbf{P}_l \mathbf{Z} \mathbf{c}_m = 1, l = 1, \ldots, L, m = 1, \ldots, M \\
& \mathbf{Z}\mathbf{1}_M = \mathbf{1}_N,
\end{aligned}
\tag{5.46}
$$

where ρ is a predetermined parameter. Note that the regularized penalty in the objective of (RP_τ) removes the constant terms $\phi(\mathbf{P}_l\mathbf{Z}_0\mathbf{c}_m, \tau) - \mathrm{tr}\{(\mathbf{c}_m\tilde{\mathbf{r}}_{m,l}^T\mathbf{P}_l)\mathbf{Z}_0\}$ in the third constraint of (BP_τ). Observe that for sufficiently large ρ, the second term in the objective function of (RP_τ) becomes hard constraints and the two problems (RP_τ) and (BP_τ) are equivalent [211]. Assume that $\mathbf{Z}^{(k)}$ is the returned solution to (RP_τ) in the kth iteration, and $\mathbf{Z}^{(k)}$ is infeasible to (AP_τ). That is,

$$
\bar{\phi}(\mathbf{Z}^{(k)}; \mathbf{Z}^{(k)}, \Pi) = \phi(\mathbf{Z}^{(k)}, \Pi) > 1. \tag{5.47}
$$

Since $\bar{\phi}(\mathbf{Z}; \mathbf{Z}^{(k)}, \Pi)$ is decreasing iteratively for sufficiently large ρ, i.e.,

$$
\bar{\phi}(\mathbf{Z}^{(k+1)}; \mathbf{Z}^{(k)}, \Pi) \leq \bar{\phi}(\mathbf{Z}^{(k)}; \mathbf{Z}^{(k)}, \Pi). \tag{5.48}
$$

The concavity of the function ϕ implies that

$$
\phi(\mathbf{Z}, \Pi) \leq \bar{\phi}(\mathbf{Z}; \mathbf{Z}^{(k)}, \Pi). \tag{5.49}
$$

Combining Eqs. (5.48) and (5.49), we obtain that,

$$
\phi(\mathbf{Z}^{(k+1)}, \Pi) \leq \phi(\mathbf{Z}^{(k)}, \Pi). \tag{5.50}
$$

We can argue from Eq. (5.50) that the approximated cardinality function is decreasing iteratively. When $\phi(\mathbf{Z}^{(k)}, \Pi)$ decreases to 1, $\mathbf{Z}^{(k)}$ becomes feasible to (AP_τ), the problem (RP_τ) will converge to the problem (AP_τ) according to Proposition 5.1.

We then propose deterministic complementary sparse array design in Algorithm 1, which successively solves the problem (BP_τ) and generates a sequence of $\{\mathbf{W}^{(k)}, \mathbf{Z}^{(k)}\}$. Proceeding from Eq. (5.3), the updating formula of desired beampattern $\mathbf{F}$ is $\mathbf{F}_p = (\mathbf{W}^{(k+1)H}\mathbf{A}_s) \oslash |\mathbf{W}^{(k+1)H}\mathbf{A}_s|$ and $\mathbf{F} = \mathbf{F}_d \odot \mathbf{F}_p$. To better approximate the l_0-norm, the threshold $\Pi \in \mathbb{R}_+^{N\times M}$ is updated iteratively according to $\mathbf{Z}$ in the previous iteration, per Eq. (5.51) in line 3. The upper bound of the approximated cardinality constraints in (BP_τ) is tight only in the neighbourhood of the initial point $\mathbf{Z}_0$, which renders the proposed algorithm a local heuristic [212], and thus, the final solution depends on the choice of the initial point. We can therefore initialize the algorithm with different initial points $\mathbf{Z}^{(0)}$ randomly from the range [0, 1] and take the one with the lowest objective value over different runs as the final solution.

Algorithm 1: Deterministic Complementary Sparse Array (DCSA) Design

Input : Parameter $\kappa = 0.5$, $\zeta = 0.001$, k=0, trade-off parameter ρ and desired beampattern $\mathbf{F} = \mathbf{F}_d \odot \mathbf{F}_p$ with $\mathbf{F}_p = \mathbf{1}$ and $\mathbf{F}_d$ specifies desired sidelobe level.

Output: A set of complementary sparse arrays denoted by $\mathbf{Z}$.

Initialize the selection matrix $\mathbf{Z}^{(0)} \in [0, 1]^{N \times M}$,

repeat

(1) Calculate the threshold matrix Π,

$$\Pi_{i,j} = \begin{cases} Z_{i,j}^{(k)} - \zeta, & \text{if } Z_{i,j}^{(k)} \geq \kappa, \\ Z_{i,j}^{(k)} + \zeta, & \text{if } Z_{i,j}^{(k)} < \kappa, \end{cases} \tag{5.51}$$

for $i = 1, \ldots, N$ and $j = 1, \ldots, M$.

(2) Calculate $\tilde{\mathbf{r}}_{m,l}$, $m = 1, \ldots, M, l = 1, \ldots, L$ according to the following formula,

$$\tilde{\mathbf{r}}_{m,l} = \sum_{i=1}^{M} g(z_i, \tau_i)\mathbf{r}_i,$$

where $z_i = \mathbf{r}_i^T \mathbf{P}_l \mathbf{Z}^{(k)} \mathbf{c}_m$, $\tau_i = \mathbf{r}_i^T \mathbf{P}_l \Pi \mathbf{c}_m$ and,

$$g(z_i, \tau_i) = \begin{cases} 0 & \text{if } z_i > \tau_i, \\ 1/\tau_i & \text{if } z_i \leq \tau_i. \end{cases}$$

(3) Solve problem (RP_τ) based on $\mathbf{Z}^{(k)}$ to get $(\mathbf{W}^{(k+1)}, \mathbf{Z}^{(k+1)})$, set iteration number k = k+1,

(4) Update $\mathbf{F}_p = (\mathbf{W}^{(k+1)H}\mathbf{A}_s) \oslash |\mathbf{W}^{(k+1)H}\mathbf{A}_s|$ and $\mathbf{F} = \mathbf{F}_d \odot \mathbf{F}_p$.

until $\|W^{(k+1)} - W^{(k)}\|_F$ *is sufficiently small*;

5.2.3 *Regularized Adaptive Sparse Array Design*

In the second step of the RCAS, the adaptive sparse array beamformer for the specific environment can then be calculated and reconfigured. The received data of the full array can be obtained by switching between the set of complementary sparse arrays designed in Sect. 5.2.2. Specifically, when each sparse array is switched on, a length of T samples are obtained and stored. After MT sampling intervals, the data of the full array is obtained and stacked into a matrix $\mathbf{Y} \in \mathbb{C}^{N \times T}$. The covariance of the full array is then $\mathbf{R}_f = (1/T)\mathbf{Y}\mathbf{Y}^H$. Based on the covariance, the optimum sparse array for combined beamforming as revisited in Sect. II-C can be obtained by,

$$\begin{aligned} \min_{\mathbf{w}} \ & \mathbf{w}^H \mathbf{R}_f \mathbf{w} + \beta \|\mathbf{w}^H \mathbf{A}_s - \mathbf{f}\|_2^2, \\ \text{s.t.} \ & \mathbf{w}^H \mathbf{a}(\theta_0) = 1, \ \|\mathbf{P}_l \mathbf{w}\|_0 \leq 1, l \leq 1, \ldots, L, \end{aligned} \tag{5.52}$$

where the second constraint is imposed to guarantee only one antenna is switched on in each group. Utilizing the same derivation procedure shown in Fig. 3, the problem is first transformed into the following formulation by introducing an auxiliary variable $\mathbf{z}$,

$$\begin{aligned} \min_{\mathbf{w},\mathbf{z}} \ & \mathbf{w}^H \mathbf{R}_f \mathbf{w} + \beta \|\mathbf{w}^H \mathbf{A}_s - \mathbf{f}\|_2^2, \\ \text{s.t.} \ & \mathbf{w}^H \mathbf{a}(\theta_0) = 1, \ |\mathbf{w}| \le \mathbf{z}, \\ & \mathbf{1}_M^T \mathbf{P}_l \mathbf{z} = 1, l = 1, \ldots, L, \\ & \|\mathbf{P}_l \mathbf{z}\|_0 \le 1, l = 1, \ldots, L, \end{aligned} \tag{5.53}$$

where $\mathbf{z} \in \{0, 1\}^N$ is the selection vector and bounds the absolute value of $\mathbf{w}$. Utilizing the piece-wise linear function to approximate the l_0-norm and upper-bounding the concave approximation function iteratively yield,

$$\begin{aligned} \min_{\mathbf{w},\mathbf{z}} \ & \mathbf{w}^H \mathbf{R}_f \mathbf{w} + \beta \|\mathbf{w}^H \mathbf{A}_s - \mathbf{f}\|_2^2, \\ \text{s.t.} \ & \mathbf{w}^H \mathbf{a}(\theta_0) = 1, \ |\mathbf{w}| \le \mathbf{z}, \\ & \mathbf{1}_M^T \mathbf{P}_l \mathbf{z} = 1, l = 1, \ldots, L, \\ & \phi(\mathbf{P}_l \mathbf{z}^{(k)}, \mathbf{P}_l \tau) + \mathbf{g}^T(\mathbf{P}_l \mathbf{z}^{(k)}, \mathbf{P}_l \tau) \mathbf{P}_l (\mathbf{z} - \mathbf{z}^{(k)}) \le 1, \\ & l = 1, \ldots, L \end{aligned} \tag{5.54}$$

where $\mathbf{g}(\mathbf{P}_l \mathbf{z}^{(k)}, \mathbf{P}_l \tau)$ is the gradient vector as defined in Lemma 5.1(3) and $\tau \in \mathbb{R}_+^N$ is an N-dimensional threshold vector. Similar to (RP_τ), we remove the third set of constraints as regularized penalties to the objective as follows,

$$\begin{aligned} \min_{\mathbf{w},\mathbf{z}} \ & \mathbf{w}^H \mathbf{R}_f \mathbf{w} + \beta \|\mathbf{w}^H \mathbf{A}_s - \mathbf{f}\|_2^2 + \rho \mathbf{g}^T(\mathbf{z}^{(k)}, \tau) \mathbf{z} \\ \text{s.t.} \ & \mathbf{w}^H \mathbf{a}(\theta_0) = 1, \ |\mathbf{w}| \le \mathbf{z}, \\ & \mathbf{1}_M^T \mathbf{P}_l \mathbf{z} = 1, l = 1, \ldots, L, \end{aligned} \tag{5.55}$$

where $\sum_{l=1}^{L} \mathbf{g}^T(\mathbf{P}_l \mathbf{z}^{(k)}, \mathbf{P}_l \tau) \mathbf{P}_l \mathbf{z} = \mathbf{g}^T(\mathbf{z}^{(k)}, \tau) \mathbf{z}$ utilizing the definition of the matrix $\mathbf{P}_l, l = 1, \ldots, L$. As complementary sparse arrays are deterministic design, it is acceptable to try different initial search points to secure the final optimum. For the design of adaptive sparse array, a good initial search point should be selected because of the requirement of real-time reconfiguration. The selection variable $\mathbf{z}$ can be initialized by a reweighted l_1-norm minimization [213]. The optimization in the kth iteration is formulated as,

$$\begin{aligned} \min_{\mathbf{w},\bar{\mathbf{z}}} \ & \mathbf{w}^H \mathbf{R}_f \mathbf{w} + \beta \|\mathbf{w}^H \mathbf{A}_s - \mathbf{f}\|_2^2 + \rho \mathbf{c}^T(\bar{\mathbf{z}}^{(k)}) \bar{\mathbf{z}} \\ \text{s.t.} \ & \mathbf{w}^H \mathbf{a}(\theta_0) = 1, \ |\mathbf{w}| \le \bar{\mathbf{z}}, \end{aligned} \tag{5.56}$$

where $\bar{\mathbf{z}} \in \{0, 1\}^N$ is the antenna selection vector and $\mathbf{c}(\bar{\mathbf{z}}^{(k)}) = \mathbf{1}_N \oslash (\bar{\mathbf{z}}^{(k)} + \gamma)$ with γ a small value for preventing explosion. Though formulation in Eq. (5.56) is robust against the choice of initial search point by initializing $\mathbf{c} = \mathbf{1}_N$, it fails to control the cardinality of the selection vector. The detailed implementation procedure of regularized adaptive sparse array (RASA) design is presented in Algorithm 3. When the iterative minimization of Eq. (5.56) converges, the obtained selection vector $\bar{\mathbf{z}}^{(k+1)}$ can be used as an initial point of Eq. (33). Though $\bar{\mathbf{z}}^{(k+1)}$ does not satisfy the regularized antenna positions, the iterative optimization of Eq. (5.55) will still converge as proved by Proposition 5.1.

Algorithm 2: Regularized Adaptive Sparse Array (RASA) Design

Input : A set of complementary sparse arrays $\mathbf{Z}$ designed by DCSA and parameter β.
Output: Optimum adaptive sparse array $\mathbf{z}$.

for $m = 1, \dots, M$ **do**
- Switch on the mth quiescent sparse array;
- Implement spatial filtering per Eq. (5.65);
- Collect T samples and store in the matrix $\mathbf{Y}$;

end
Calculate the covariance $\mathbf{R}_f = (1/T)\mathbf{Y}\mathbf{Y}^H$;
Set iteration number $k = 0$ and initialize $\bar{\mathbf{z}}^{(k)} = \mathbf{1}$;
repeat
  Calculate $\mathbf{c}(\bar{\mathbf{z}}^{(k)}) = \mathbf{1}_N \oslash (\bar{\mathbf{z}}^{(k)} + \gamma)$;
  Solve Eq. (5.56) to get $\bar{\mathbf{z}}^{(k+1)}$ and $k = k + 1$;
until $\|\bar{z}^{(k+1)} - \bar{z}^{(k)}\|_2$ *is sufficiently small*;
Set iteration number $k = 0$ and $\mathbf{z}^{(k)} = \bar{\mathbf{z}}^{(k+1)}$;
repeat
  Update τ according to

$$\tau_n = \begin{cases} z_n^{(k)} - \zeta, & \text{if } z_n^{(k)} \geq \kappa, \\ z_n^{(k)} + \zeta, & \text{if } z_n^{(k)} < \kappa, \end{cases} \tag{5.57}$$

  Calculate $\mathbf{g}(\mathbf{z}^{(k)}, \tau)$ per Eq. (19);
  Solve Eq. (5.55) to get $\mathbf{z}^{(k+1)}$ and $k = k + 1$;
until $\|z^{(k+1)} - z^{(k)}\|_2$ *is sufficiently small*;

5.2.4 Simulations

Extensive simulation results are presented in this section to validate the proposed strategy for adaptive sparse array beamformer design. We first demonstrate the effectiveness of proposed DCSA algorithm for solving the cardinality-constrained optimization problem, followed by RCAS strategy validation in both static and dynamic environment.

5.2.4.1 Algorithm Validation

First, we utilize a small array to validate the effectiveness of our proposed algorithm for solving the cardinality-constrained optimization problem. Assume a uniform linear array (ULA) comprising 16 antennas with an inter-element spacing of a quarter wavelength. This linear array is divided into 8 groups and each group contains two consecutive antennas. There are totally 8 front-ends installed and each front-end is responsible to connect with the two adjacent antennas in one group. We split this linear array into two sparse arrays through regularized antenna switching and each sparse array comprises 8 elements. The array is steered towards the broadside direction and the sidelobe angular region is defined as $[-90°, -12°] \cup [12°, 90°]$ and the desired sidelobe level is -15 dB. It is desirable that both split sparse arrays exhibit good quiescent beampatterns. As there are 128 different splitting (resulting in 128 pairs of complementary sparse arrays) in total, we enumerate the peak sidelobe levels (PSLs) of each pair among these 128 pairs for the lowest one. The optimum pair of sparse arrays, named as array 1 and 2, are shown in the upper plot of Fig. 5.4, followed by the worst pair of splitting. Clearly, arrays 1 and 2 are the same except of a reversal antenna

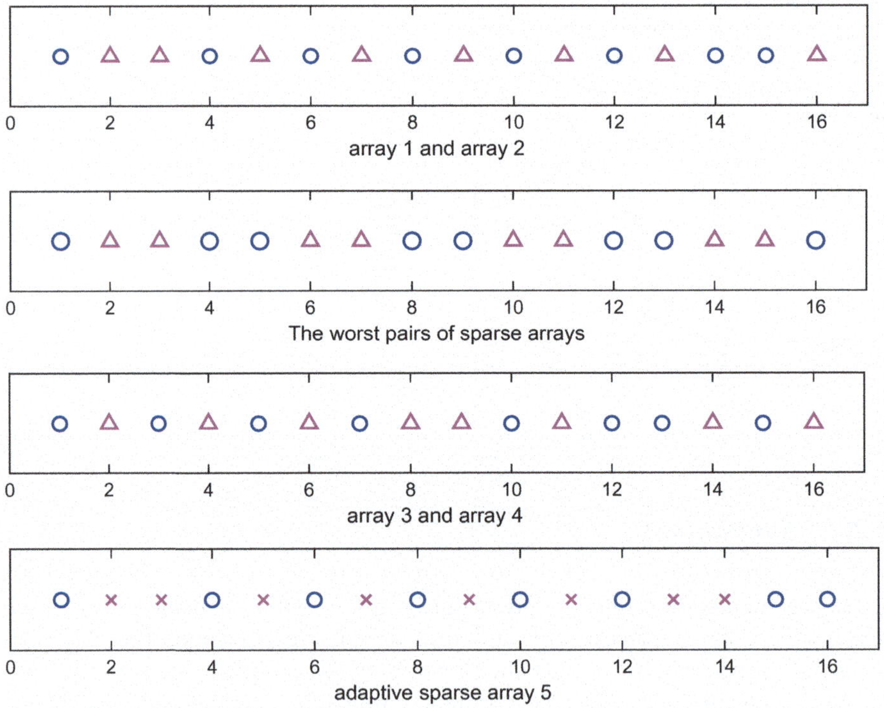

Fig. 5.4 The obtained optimum and worst array splitting by enumeration (upper plot) and Algorithm 1 (middle plot). The marker "circle" denotes one array and "triangle" denotes the other. The bottom plot is the adaptive sparse array, where "circle" denotes selected and "cross" denotes discarded

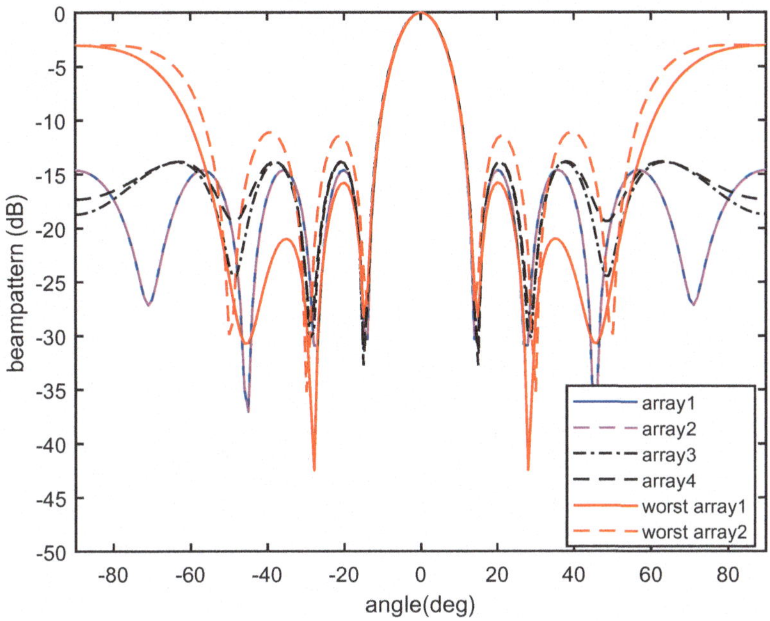

Fig. 5.5 The quiescent beampatterns of four sparse arrays 1–4 shown in Fig. 5.4 and the worst pair of sparse arrays

placement. We then run Algorithm 1, and the finally converged splitting is shown in the lower plot of Fig. 5.4. The quiescent beampatterns of these four sparse arrays are depicted in Fig. 5.5, where the worst pair of splitting is plotted as well. We can see that the PSL of the first pair returned by enumeration is -15dB, while that of the second pair returned by Algorithm 1 is slightly larger than –15 dB. Although inferior to the global optimum splitting, the proposed cardinality-constrained optimization algorithm could seek a sub-optimal solution with an acceptable performance.

We continue to investigate the convergence performance of the proposed algorithm. The trade-off parameter in (RP_τ) is first set as $\rho = 1$ in this considered example. We plot two parts of the objective function in (RP_τ), those are $\|\mathbf{W}^H\mathbf{A}_s - \mathbf{F}\|_\text{F}$ and $\|\mathbf{Z}^{(k+1)} - \mathbf{Z}^{(k)}\|_\text{F}$, versus the iteration number. The results are plotted in Fig. 5.6. We can see that the auxiliary variable $\mathbf{Z}$ requires maximally three iterations to converge, while the beamforming weight matrix $\mathbf{W}$ requires more iterations. The reason is explained as that the iterative updating of the beampattern phase $\mathbf{F}_p$ will further decrease the beampattern deviation after the sparse array is designed. We then change the value of the trade-off parameter ρ discretely to four different values of 0.1, 1, 2, 20. The curves of two parts of the objective function versus the iteration number are plotted in Fig. 5.6 as well. We can see that when $\rho \geq 1$, the effect of ρ on the convergence rate and synthesized beampattern shape is diminished and negligible.

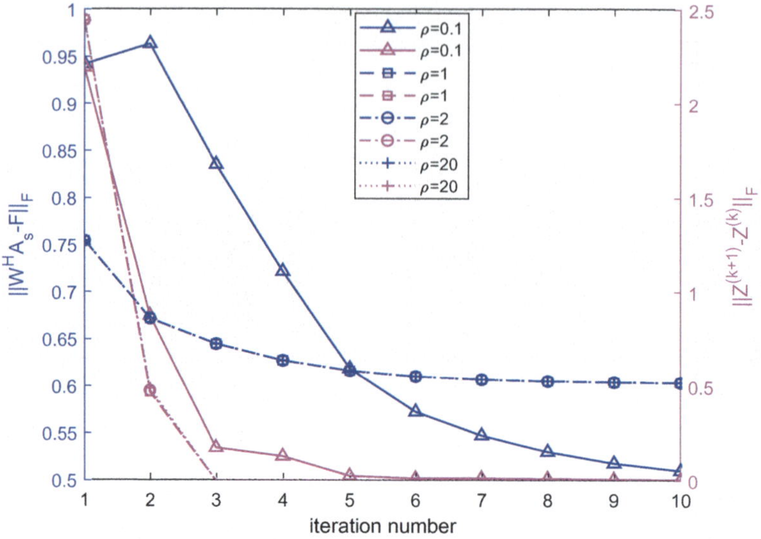

Fig. 5.6 Convergence performance: $\|\mathbf{W}^H\mathbf{A}_s - \mathbf{F}\|_\mathrm{F}$ and $\|\mathbf{Z}^{(k+1)} - \mathbf{Z}^{(k)}\|_\mathrm{F}$ versus iteration number in four cases of different ρ

5.2.4.2 Static Environment

As explained in our previous work [104], array configuration plays a very sensitive and viral role in the scenario of closely-spaced mainlobe interferences, whereby the importance of sparse arrays is amplified. Hence, we intentionally consider such cases in this work. Assume that four interferences are coming from the directions of $-28°, -12°, 10°, 25°$ with an interference-to-noise ratio (INR) of 20 dB. The proposed RCAS is conducted, where two quiescent sparse arrays 3 and 4 are first sequentially switched on to collect the data of the full array, and then the optimum adaptive sparse array is configured according to Algorithm 2 with the desired SLL setting as -5 dB. The structure of the adaptive sparse array is shown in the bottom plot of Fig. 5.4. The beampatterns of three sparse arrays (3)–(5) using both adaptive beamforming in Eq. (5.9) and combined beamforming in Eq. (5.13) are presented in Fig. 5.7. We can observe that (1) Adaptive beamforming, though excellent in suppressing interferences, exhibits catastrophic grating lobes except for the sparse array 5. This inadvertently manifests the superiority of the designed sparse array 5, irrespective of no explicit sidelobe constraints imposed in adaptive beamforming. (2) Combined beamforming is capable of controlling the sidelobe level with the sacrifice of shallow nulls towards the unwanted interferences; (3) Arrays 3 and 4 with combined beamforming cannot form nulls towards the two closely-spaced interferences from $-18°$ and $-12°$. The output SINR of five sparse arrays are listed in Table 5.1. We can obtain the corresponding results with Fig. 5.7, that is array 5 is supreme in

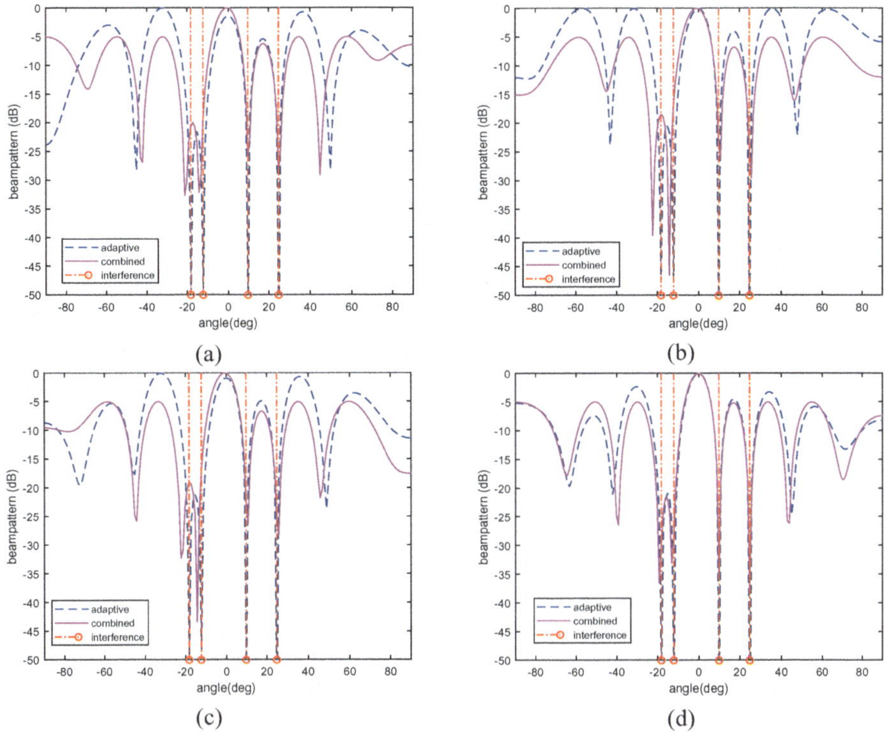

Fig. 5.7 The beampatterns of three sparse arrays (1)–(5) using both adaptive beamforming and combined beamforming: **a** Array 1 and 2 **b** Array 3 **c** Array 4 **d** Array 5

Table 5.1 Output SINR of five sparse arrays in Fig. 5.4

B/A	1 (dB)	2 (dB)	3 (dB)	4 (dB)	5 (dB)
Adaptive	4.28	4.28	3.84	4.28	6.52
Combined	−4.44	−4.44	−6.58	−6.2	3.92

"B" denotes beamforming method and "A" denotes array numbering. The unit is in dB

terms of the output SINR regardless of beamforming methods, while arrays 1(2) and 3(4) exhibit much worse performance especially for combined beamforming.

To thoroughly examine the performance of the proposed RCAS strategy, we proceed to enlarge the array size to a ULA with 32 antennas. This linear array is divided into 16 groups and each group contains two consecutive antennas. In the first step of RCAS, we design a set of complementary sparse arrays, which are obtained by the extended Algorithm 1 and shown in the upper plot of Fig. 5.8. The sidelobe angular region is defined as $[-90°, -7°] \cup [7°, 90°]$ and the desired sidelobe level is set as -5 dB for combined beamforming. Six interferences are impinging on the array from $-18°, -12°, -6°, 5°, 10°, 25°$ with an INR of 20 dB. The configuration of adaptive sparse array is shown in the lower plot of Fig. 5.8. The adaptive and combined beam-

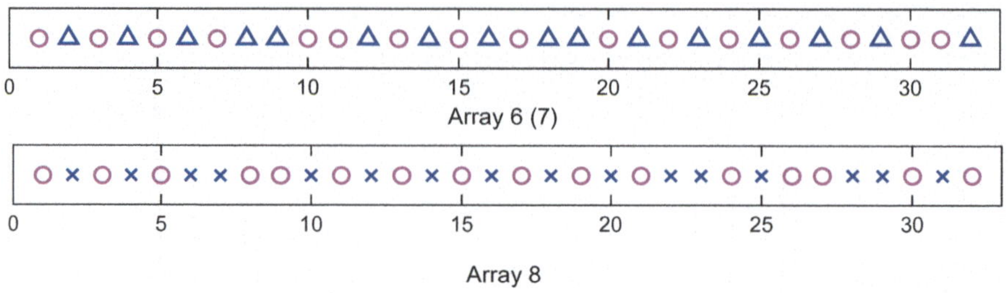

Fig. 5.8 Configurations of sparse arrays 6–8: arrays 6 and 7 are quiescent complementary and array 8 is adaptive

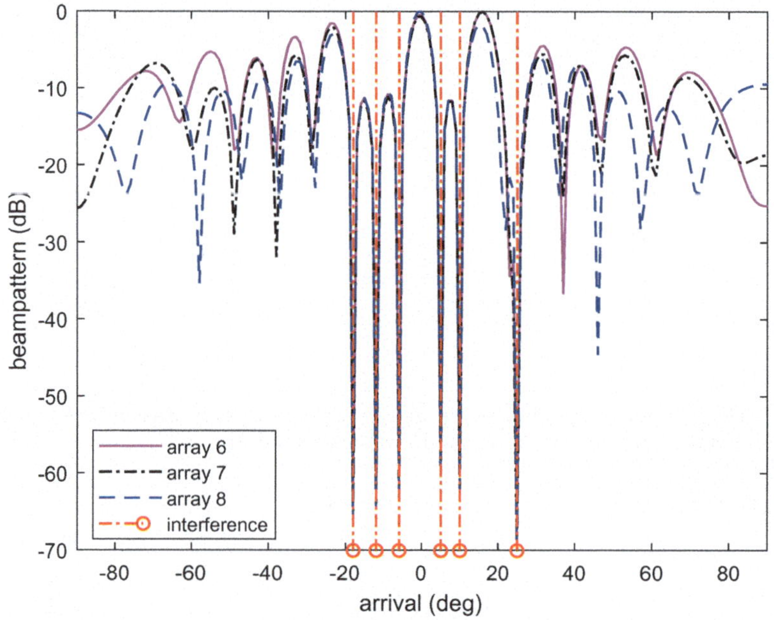

Fig. 5.9 Adaptive beamforming of sparse arrays 6, 7 and 8

patterns of three arrays 6–8 are compared in Figs. 5.9 and 5.10, respectively. Again, adaptive beamforming avoidably produces very high sidelobes (even grating lobes in a severe condition), which adversely affect the output performance in the dynamic environment where an unintentional interference is suddenly switched on. We can see that array 8 produces the deepest nulls towards the interferences, in turn yielding the highest SINR. Comparatively, the output SINR of sparse arrays 6–8 are listed in Table 5.2 for both adaptive and combined beamforming.

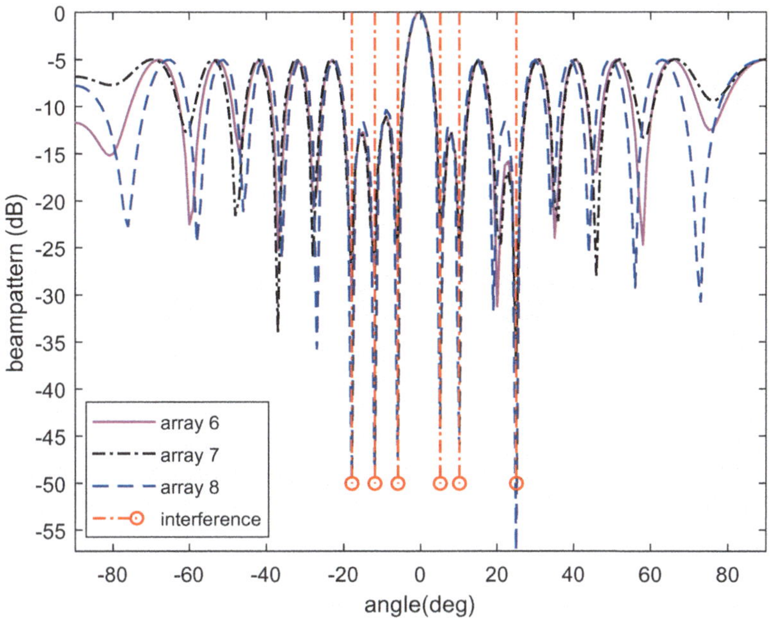

Fig. 5.10 Combined beamforming of sparse arrays 6, 7 and 8

Table 5.2 Output SINR of three sparse arrays in Fig. 5.8

Beamforming/Array	6 (dB)	7 (dB)	8 (dB)
Adaptive	6.8	6.99	8.44
Combined	−1.78	−2.95	7.44

5.2.4.3 Dynamic Environment

Different from the proposed RCAS strategy, the environmental information is obtained from the augmented covariance matrix of a virtual full co-array [43, 202, 203] and is referred to as coarray strategy. To give a brief explanation, the covariance matrix **R** of the physical array is first vectorized and then reordered according to the spatial lags. Both spatial smoothing and Toepliz matrix augmentation can then be utilized to obtain the full covariance matrix of the virtual co-array. It would be intriguing to compare the difference of array configuration and filtering performance based on the knowledge acquired from the two strategies. As it is impossible to construct a fully-augmentable sparse array under the constraint of regularized antenna switching, a nested array in Fig. 5.11 is deliberately employed for situational sensing and performance comparison.

We first examine the effect of snapshot number on both strategies. The simulation scenario remains the same as that of the above small array. We choose the sparse arrays

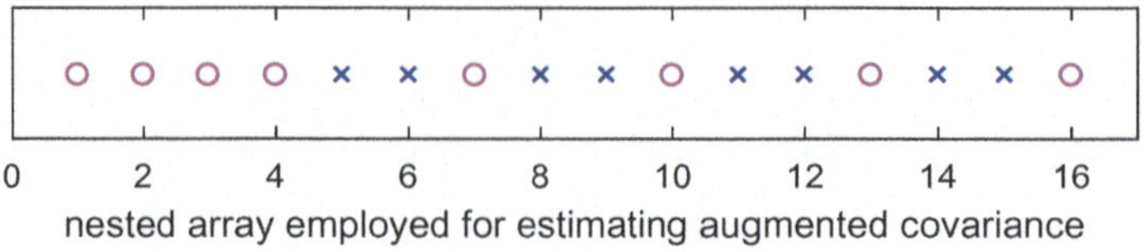

Fig. 5.11 The nested array deliberately used for augmented full covariance estimation

3 and 5 as the benchmark and compare the output SINR of optimized sparse arrays using two strategies. The snapshot number is changing from 10 to 1910 in a step of 100 and 1000 Monte Carlo simulations are run in each case. For each number, we collect the data of corresponding length either switching between sparse arrays 3 and 4 or using the nested array and augmenting the covariance. For more details on covariance augmentation, the readers can refer to [43] and reference therein. It is worth noting that the estimation accuracy of the full array covariance matrix significantly affect the selection of switched antennas. Hence, for different numbers of snapshots, Algorithm 2 is utilized to calculate the optimum adaptive sparse array with the input of estimated full array covariance. The curves of output SINR versus snapshot number in two cases of uncorrelated and correlated interfering signals are depicted in Figs. 5.12 and 5.13, respectively. For the latter, the correlation among different interferences are generated randomly. The proposed RCAS strategy can return the optimum sparse array that is array 5 when the snapshot number increases. Although the configuration

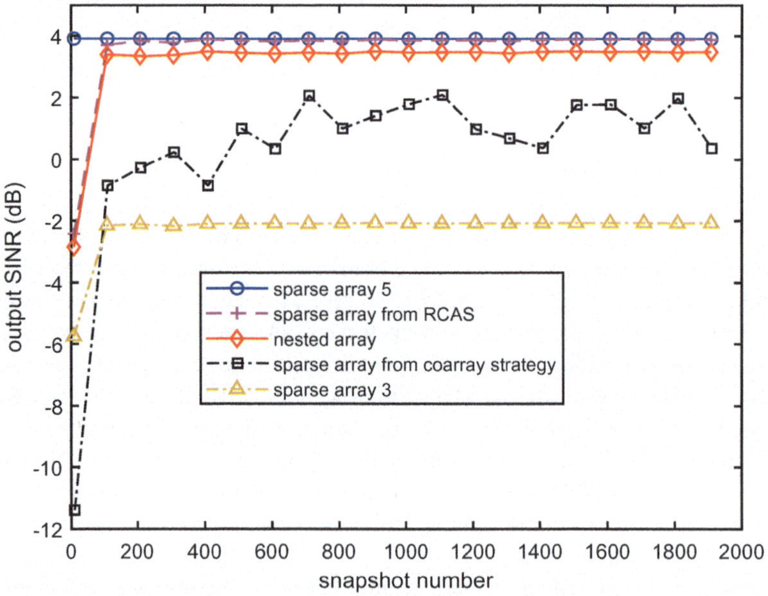

Fig. 5.12 Output SINR versus snapshot number in the case of uncorrelated interfering signals

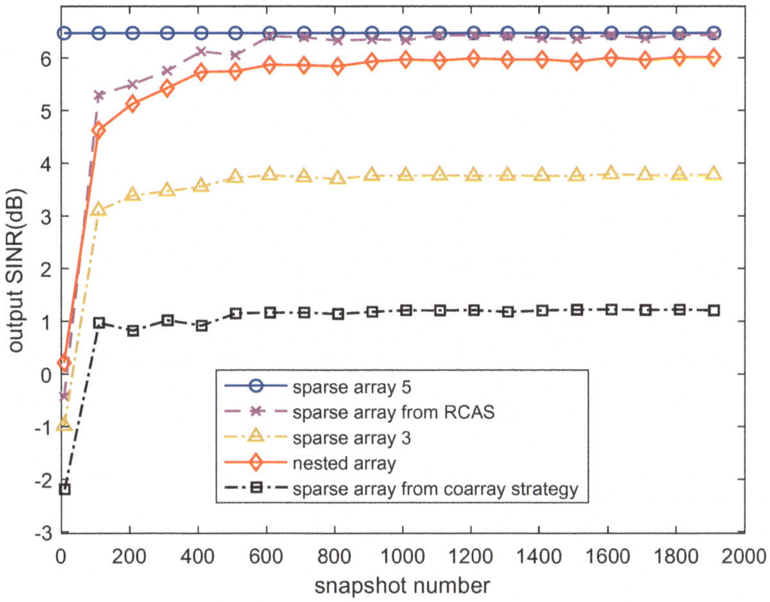

Fig. 5.13 Output SINR versus snapshot number in the case of correlated interfering signals

of nested array is not restricted to regularized antenna positions, its output SINR does not exhibit superiority over sparse array 5. The sparse array configured based on the environmental information obtained from augmented covariance matrix does not exhibit satisfactory performance especially when the snapshot number is small and the impinging interferences are correlated. This attributes to the fact that coarray-based signal processing usually requires a large number of snapshots and is restricted to dealing with uncorrelated signals.

We continue to examine an example of dynamic environment, which is described in Fig. 5.14. The total observing time is set as $100T$ sampling intervals with $T = 500$. Suppose that the target is coming from broadside, and there are two interferences coming from $-28°$ and $25°$ in the first period of the observing time. At the time instant of 30T, the scenario has changed to that of four interferences coming from $-31°$, $-12°$, $10°$ and $50°$. At the time instant of 60T, the scenario has changed again and four interferences are coming from $-28°$, $-12°$, $10°$ and $25°$, respectively. The INR of all interferences is 20 dB. We consider three strategies, those are fixed-array strategy, the proposed RCAS strategy and coarray strategy. In the first strategy, a fixed sparse array, that is the optimal sparse array configured for combined beamforming in scenario 1 using Eq. (5.56), is employed during the observing period. In the proposed RCAS strategy, complementary sparse arrays 3 and 4 are utilized for environment sensing and data collection, and adaptive sparse arrays are then configured based on the full array covariance, as indicated in the middle row of Fig. 5.14. In the coarray strategy, the full array covariance is obtained by augmenting the spatial lags of the nested array

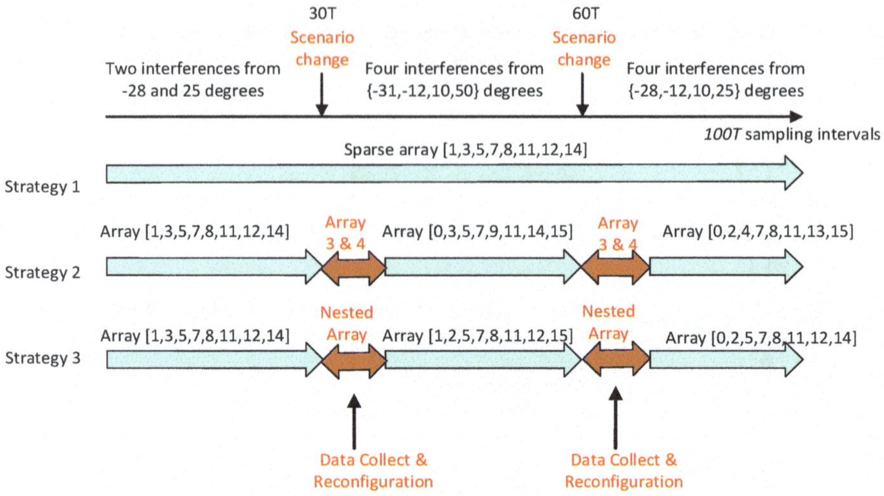

Fig. 5.14 Three strategies in dynamic environment and the antenna positions of arrays utilized in each strategy are indicated

shown in Fig. 5.11, according to the structure of Toeplitz matrix. The two adaptive sparse arrays are optimized based on the virtual covariance, as indicated in the bottom row of Fig. 5.14. We can see from Fig. 5.15 that the fixed sparse array experiences performance degradation at two durations of scenario changing. Though the nested array performs better than complementary sparse arrays during the second sensing stage, its configuration does not follow the rule of regularized antenna switching. The proposed RCAS strategy exhibits the best performance after adaptive sparse array reconfiguration and modest output SINR during the two environment sensing stages.

5.3 Cognitive Sparse Beamformer Design via Regularized Switching Network

In this section, we investigate cognitive sparse element-space and beam-space beamformer design via iterative antenna/beam replacement, respectively. Different from the existing works that either call for prior environmental information [197–201] or needs to obtain the covariance matrix of full array in advance [202, 203, 214], the proposed sparse beamformers are fully data-driven and eradicate any prior knowledge along with the assumption of static environment. Cognitive sparse beamformers can learn the environmental information through a continuous "perception-action" cycle, where perception is performed based on the collected data and "action" is to reconfigure the beamformer geometry and update excitation weights. The proposed regularized-switching design divides the full set of antennas/beams into groups and

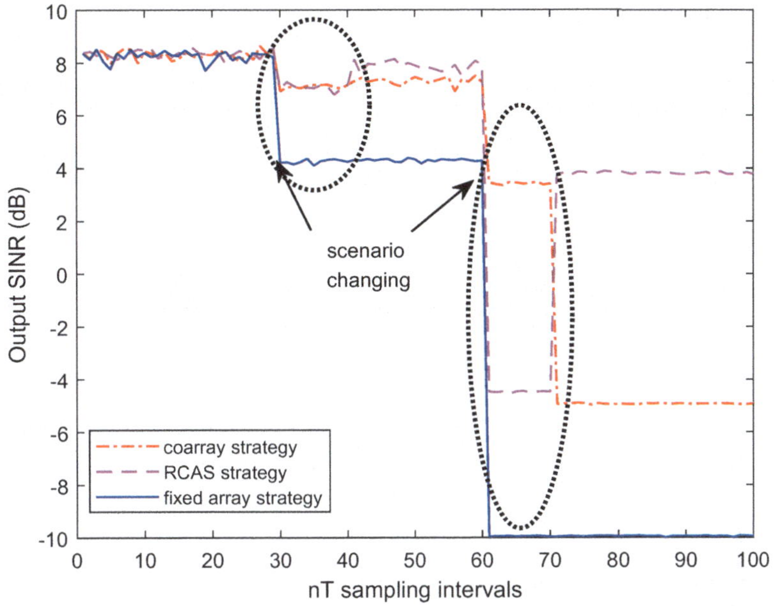

Fig. 5.15 Comparison of three strategies in dynamic environment. The environment has changed twice during the observation time as indicated by black arrows. The periods highlighted by black circles are stages for environmental sensing, data collection and array reconfiguration

precisely one antenna/beam is switched on in each group to compose the sparse beamformer. A deterministic sparse beamformer with a good quiescent beampattern is chosen for initialization, and then the currently employed antenna/beam is replaced with another one in the same group iteratively until achieving the maximum array gain (AG). Since each antenna/beam replacement along with the reception of new data corresponds to a different beamformer, calculating the variation of AG would require prohibitive computational complexity and memory usage. To achieve cognitive beamformer reconfiguration in real time, we derive closed-form formulas of AG variation, thus circumventing the repetitive cumbersome calculation of matrix inversion.

5.3.1 *Sparse Adaptive Beamforming*

We first derive adaptive beamforming in terms of two metrics of MPDR and LCMP, especially the expressions of array gain, which will be used as the metric for sparse beamformer design in the proposed work. And then the previous method of sparse beamformer design with a perfect prior information is briefly summarized for convenience of follow-on performance comparison.

5.3.1.1 Adaptive Beamforming

Adaptive beamforming extracts noise characteristics and intruding interference statistics from the received data and forms nulls towards the interferences automatically. Therefore, adaptive beamforming exhibits superior interference suppression performance, especially in a dynamic environment. The MPDR beamformer aims to minimize the array output power while maintaining a unit gain towards the look direction, which can be expressed as [204, 205],

$$\begin{aligned} \min_{\mathbf{w}} \ & \mathbf{w}^H \mathbf{R} \mathbf{w}, \\ \text{s.t.} \ & \mathbf{w}^H \mathbf{a}(\theta_0) = 1, \end{aligned} \tag{5.58}$$

where $\mathbf{w} \in \mathbb{C}^{N\times 1}$ is the beamforming weight vector and $\mathbf{a}(\theta_0)$ is the steering vector towards the direction of θ_0, that is,

$$\mathbf{a}(\theta_0) = [1, e^{jk_0 d \sin\theta_0}, \ldots, e^{jk_0 d(N-1)\sin\theta_0}]^T, \tag{5.59}$$

where the wavenumber is defined as $k_0 = 2\pi/\lambda$ with λ denoting wavelength and the steering direction θ_0 is measured relative to the array broadside.

In Eq. (5.58), $\mathbf{R}$ is the covariance matrix of the received data by the employed L-antenna sparse array and can be theoretically written as,

$$\begin{aligned} \mathbf{R} &= \sigma_s^2 \mathbf{a}(\theta_0)\mathbf{a}^H(\theta_0) + \mathbf{R}_n, \\ &= \sigma_s^2 \mathbf{a}(\theta_0)\mathbf{a}^H(\theta_0) + \sum_{j=1}^{J} \sigma_j^2 \mathbf{a}(\theta_j)\mathbf{a}^H(\theta_j) + \sigma_n^2 \mathbf{I}. \end{aligned} \tag{5.60}$$

The total number of interferences J and their respective arrival angles θ_j, $j = 1, \ldots, J$ are usually unknown. Neither are the signal power σ_s^2 and noise power σ_n^2. Thereby, we utilize the maximum likelihood estimate of the covariance to replace the true covariance $\mathbf{R}$ in practical applications, that is,

$$\hat{\mathbf{R}} = \frac{1}{T}\sum_{t=1}^{T} \mathbf{x}_t \mathbf{x}_t^H, \tag{5.61}$$

where $\mathbf{x}_t \in \mathbb{C}^L$ denotes the received corrupted data vector in the tth snapshot. No doubt that the snapshot number T affects the beamformer's performance as shown in the Sect. 5.3.4. The weight vector of MPDR beamformer is thus [215, 216],

$$\mathbf{w}_c = \frac{1}{\mathbf{a}^H(\theta_0)\hat{\mathbf{R}}^{-1}\mathbf{a}(\theta_0)} \hat{\mathbf{R}}^{-1}\mathbf{a}(\theta_0), \tag{5.62}$$

Utilizing Eq. (5.62), the array gain of MPDR beamformer is,

$$\mathcal{A}_c = \mathbf{a}^H(\theta_0)\bar{\mathbf{R}}_n^{-1}\mathbf{a}(\theta_0), \tag{5.63}$$

where $\bar{\mathbf{R}}_n = \mathbf{R}_n/\sigma_n^2$ is the normalized interference-plus-noise covariance matrix. Since $\bar{\mathbf{R}}_n$ is unavailable in practical applications, we utilize $\hat{\mathbf{R}}$ in place of $\bar{\mathbf{R}}_n$ as interferences are usually much stronger than white noise and the desired signal.

Although adaptive beamforming is superior to quiescent beamforming in terms of interference suppression, high sidelobe levels may constitute a potential threat in adaptive beamformer, especially when directional interferers are suddenly intruded on. As such, it is important to seek sparse arrays that maximize the AG with well-controlled sidelobes. The LCMP beamformer imposes beampattern constraints to control the sidelobes, that is [217],

$$\begin{aligned} \min_{\mathbf{w}} \ & \mathbf{w}^H\mathbf{R}\mathbf{w} \\ \text{s.t.} \ & \mathbf{w}^H\mathbf{C} = \mathbf{g}^T, \end{aligned} \tag{5.64}$$

where $\mathbf{C} = [\mathbf{a}(\theta_0), \mathbf{A}_s]$ contains the target steering vector and sidelobe manifold, $\mathbf{g} = [1, \delta, \ldots, \delta]^T$ and $\mathbf{A}_s$ denotes the array manifold matrix of the sidelobe region $\Omega_s = [\theta_1, \ldots, \theta_K]$, which is represented by K angular samples. The optimal weight vector is thus,

$$\mathbf{w}_l = \mathbf{R}^{-1}\mathbf{C}(\mathbf{C}^H\mathbf{R}^{-1}\mathbf{C})^{-1}\mathbf{g} = \mathbf{R}^{-1}\mathbf{C}\mathbf{U}^{-1}\mathbf{g}, \tag{5.65}$$

where $\mathbf{U} = \mathbf{C}^H\mathbf{R}^{-1}\mathbf{C}$. When the distortionless constraint is imposed and the signal is perfectly matched, the AG using the weight in Eq. (5.65) is,

$$\mathcal{A}_l = \frac{1}{\mathbf{g}^H[\mathbf{C}^H\hat{\mathbf{R}}^{-1}\mathbf{C}]^{-1}\mathbf{g}}. \tag{5.66}$$

Again, we utilize $\hat{\mathbf{R}}$ to replace the normalized interference-plus-noise covariance in this work.

Similarly, beamspace MPDR beamformer is expressed as,

$$\tilde{\mathbf{w}}_c = \frac{1}{\tilde{\mathbf{a}}^H(\theta_0)\tilde{\mathbf{R}}^{-1}\tilde{\mathbf{a}}(\theta_0)}\tilde{\mathbf{R}}^{-1}\tilde{\mathbf{a}}(\theta_0), \tag{5.67}$$

where $\tilde{\mathbf{R}} = \mathbf{B}^H\hat{\mathbf{R}}\mathbf{B}$, $\tilde{\mathbf{a}}(\theta_0) = \mathbf{B}^H\mathbf{a}(\theta_0)$ and $\mathbf{B} \in \mathbb{C}^{N\times N}$ is the beamspace transformation matrix, which can form a set of N beams and thus transform the element-space processing to beam-space domain. The ith column of $\mathbf{B}$ is defined as,

$$\mathbf{b}_i = \frac{1}{\sqrt{N}}[1, e^{jdk_0u_i}, \ldots, e^{j(N-1)dk_0u_i}]^T, i = 1, \ldots, N \tag{5.68}$$

which is the steering vector pointing towards direction u_i in u-space and $\{u_i, i = 1, \ldots, N\}$ is a set of uniformly spaced grid points with an interval of $2/N$, those are $\{-1, -1+2/N, \ldots, 1-2/N\}$. When the inter-element spacing $d = \lambda/2$, $\mathbf{B}$ becomes a unitary matrix, that is $\mathbf{B}^H\mathbf{B} = \mathbf{B}\mathbf{B}^H = \mathbf{I}$ and the set of N beams are orthogonal. Otherwise, the set of N beams are correlated.

The beamspace LCMP beamformer is then written as,

$$\tilde{\mathbf{w}}_l = \tilde{\mathbf{R}}^{-1}\tilde{\mathbf{C}}(\tilde{\mathbf{C}}^H\tilde{\mathbf{R}}^{-1}\tilde{\mathbf{C}})^{-1}\mathbf{g}, \tag{5.69}$$

where $\tilde{\mathbf{C}} = \mathbf{B}^H\mathbf{C}$. The AGs of both beamformers are thus,

$$\tilde{\mathcal{A}}_c = \tilde{\mathbf{a}}^H(\theta_0)\tilde{\mathbf{R}}^{-1}\tilde{\mathbf{a}}(\theta_0), \quad \tilde{\mathcal{A}}_l = \frac{1}{\mathbf{g}^H[\tilde{\mathbf{C}}^H\tilde{\mathbf{R}}^{-1}\tilde{\mathbf{C}}]^{-1}\mathbf{g}}. \tag{5.70}$$

To quantify the variation of the AG of each beamformer, $\tilde{\mathcal{A}}_c$ and $\mathcal{A}_c$ are chosen as the metric of MPDR beamformer reconfiguration, while inverse array gain (IAG) $\tilde{\mathcal{A}}_l^{-1}$ and $\mathcal{A}_l^{-1}$ are chosen for the LCMP beamformer due to the difficulty of reciprocal calculation.

5.3.1.2 Sparse Beamformer Design with Known Environmental Information

In the ideal case where the environmental information of interferences is known, such as the interference number and their DOAs, the AG of Capon beamformer can be approximated as [218],

$$\mathcal{A}_c \approx \mathbf{a}^H(\theta_0)\mathbf{a}(\theta_0) - \mathbf{a}^H(\theta_0)\mathbf{V}(\mathbf{V}^H\mathbf{V})^{-1}\mathbf{V}^H\mathbf{a}(\theta_0). \tag{5.71}$$

Here, the interferences are assumed to be much stronger than white noise. Utilizing the definition of the antenna selection vector $\mathbf{z} \in \{0, 1\}^{N\times 1}$, the optimum sparse MPDR beamformer can be constructed by solving the following optimization [218],

$$\begin{aligned} \max_{\mathbf{z},\gamma} \ & \gamma, \\ \text{s.t.} \ & \begin{bmatrix} \mathbf{V}^H\text{diag}(\mathbf{z})\mathbf{V} & \mathbf{V}^H\text{diag}(\mathbf{z})\mathbf{a}(\theta_0) \\ \mathbf{a}^H(\theta_0)\text{diag}(\mathbf{z})\mathbf{V} & \mathbf{a}^H(\theta_0)\text{diag}(\mathbf{z})\mathbf{a}(\theta_0) - \gamma \end{bmatrix} \succeq 0, \\ & \mathbf{z} \in \{0, 1\}^N, \\ & \mathbf{1}_M^T\mathbf{P}_l\mathbf{z} = 1, l = 1, \ldots, L, \end{aligned} \tag{5.72}$$

where $\mathbf{V} = [\mathbf{a}(\theta_1), \ldots, \mathbf{a}(\theta_J)]$ and $\mathbf{P}_l \in \{0, 1\}^{M\times N}$ is the group selection matrix. For the construction of sparse LCMP beamformer, we solve the following optimization problem,

$$\begin{aligned}
\min_{\mathbf{z},\mathbf{G},\gamma} \quad & \gamma, \\
\text{s.t.} \quad & \begin{bmatrix} \mathbf{G} & \mathbf{g} \\ \mathbf{g}^H & \gamma \end{bmatrix} \geq 0, \\
& \begin{bmatrix} \mathbf{V}^H \text{diag}(\mathbf{z})\mathbf{V} & \mathbf{V}^H \text{diag}(\mathbf{z})\mathbf{C} \\ \mathbf{C}^H \text{diag}(\mathbf{z})\mathbf{V} & \mathbf{C}^H \text{diag}(\mathbf{z})\mathbf{C} - \mathbf{G} \end{bmatrix} \geq 0, \\
& \mathbf{G} \geq 0; \\
& \mathbf{z} \in \{0, 1\}^N, \\
& \mathbf{1}_M^T \mathbf{P}_l \mathbf{z} = 1, l = 1, \ldots, L.
\end{aligned} \tag{5.73}$$

More derivation details can be found in our previous work [218]. For the construction of sparse beamspace beamformer, the same optimization in Eqs. (5.72) and (5.73) can be utilized with $\mathbf{a}(\theta_0)$ replaced by $\tilde{\mathbf{a}}(\theta_0)$, $\mathbf{V}$ replaced by $\tilde{\mathbf{V}}$ and $\mathbf{C}$ replaced by $\tilde{\mathbf{C}}$. However, the environmental information of interferences are usually not available in practice, thus data-independent sparse beamformer design is the core motivation of the following proposed work.

5.3.2 *Quiescent Beamformer Initialization*

When no environmental information is available at the start of operation, an optimum quiescent beamformer with a good-shaped beampattern is preferred due to its resilience against interferences from arbitrary directions. Quiescent beamformer design is essentially beampattern synthesis, which focuses on synthesizing beampatterns with mainbeam steering towards the target and reduced sidelobe levels.

5.3.2.1 Element-Space Quiescent Beamforming

Assume a large array of N uniformly spaced antennas and the inter-element spacing is d. The large array is divided into L groups and each group contains M antennas, that is $N = LM$, as shown in Fig. 5.16. Although the group division in Fig. 5.16 is continuous and uniform, arbitrary grouping is applicable too. There is one front-end processing channel in each group, and one RF switch can connect any antenna in this group with the corresponding front-end in a sufficiently small time via the predesigned circuitry. We investigate the sparse element-space beamformer design by regularized antenna switching.

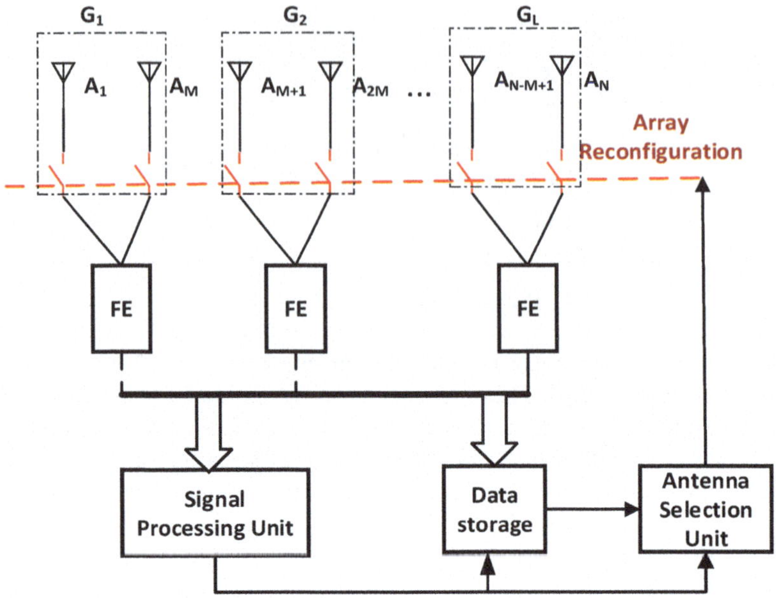

Fig. 5.16 The illustration of element-space cognitive beamformer by regularized antenna switching

The sparse quiescent element-space beamformer design can be formulated as,

$$\begin{aligned} \min_{\mathbf{w}} \ & \delta, \\ \text{s.t. } & \mathbf{w}^H \mathbf{a}(\theta_0) = 1, \\ & |\mathbf{w}^H \mathbf{A}_s| \leq \delta, \\ & \|\mathbf{P}_l \mathbf{w}\|_0 = 1, l = 1, \ldots, L, \end{aligned} \tag{5.74}$$

where δ is the upper bound of the synthesized sidelobe level. The set of cardinality constraints in the fourth line of Eq. (5.74) are utilized to select exactly one antenna in each group. Taking the continuous grouping in Fig. 5.16 as an example, then the group selection matrix $\mathbf{P}_l \in \{0, 1\}^{M \times N}$ are all zeros except those entries of $P_l(1, 1 + (l-1)M)$, $P_l(2, 2 + (l-1)M), \ldots, P_l(M, lM)$ being one, where $l = 1, \ldots, L$.

Clearly, the cardinality constraints in Eq. (5.74) render the problem difficult to solve. To proceed, we define an auxiliary binary variable $\mathbf{z} \in \{0, 1\}^N$ to transform the complex cardinality constraint into the real domain. As a result, Eq. (5.74) can be rewritten as,

$$\begin{aligned}
\min_{\mathbf{w}} \quad & \delta, \\
\text{s.t.} \quad & \mathbf{w}^H \mathbf{a}(\theta_0) = 1, \\
& |\mathbf{w}^H \mathbf{A}_s| \leq \delta, \\
& |\omega_i| \leq z_i, \; i = 1, \ldots, N, \\
& \mathbf{z} \in \{0, 1\}^N, \\
& \mathbf{1}_M^T \mathbf{P}_l \mathbf{z} = 1, l = 1, \ldots, L.
\end{aligned} \tag{5.75}$$

where ω_i and z_i are the ith element of $\mathbf{w}$ and $\mathbf{z}$ respectively.

We can see that variable $\mathbf{z}$ is actually a selection vector, which represents the sparse support of the beamforming weight vector $\mathbf{w}$. The binary constraints on $\mathbf{z} \in \{0, 1\}^N$ are difficult to solve and we resort to an iterative reweighting method to promote the binary sparsity of the selection vector. That is,

$$\begin{aligned}
\min_{\mathbf{w}} \quad & \delta + \rho \mathbf{c}^T \mathbf{z}, \\
\text{s.t.} \quad & \mathbf{w}^H \mathbf{a}(\theta_0) = 1, \\
& |\mathbf{w}^H \mathbf{A}_s| \leq \delta, \\
& |\omega_i| \leq z_i, i = 1, \ldots, N, \\
& |z_i| \leq \tau, i = 1, \ldots, N, \\
& \mathbf{1}_M^T \mathbf{P}_l \mathbf{z} = \tau, l = 1, \ldots, L,
\end{aligned} \tag{5.76}$$

where ρ is a trade-off parameter between the sidelobe level and the vector sparsity, and $\mathbf{c}$ is a reweighting vector for promoting binary sparsity. Here, we change the binary constraint on $\mathbf{z}$ to a latent box constraint $\mathbf{z} \in [0, \tau]$ with $0 < \tau \leq 1$ being the preset upper bound. The updating formula of $\mathbf{c}$ is first introduced in our previous work [219], and is given as follows,

$$c_i^{(k+1)} = (1 - (\frac{z_i^{(k)}}{\tau}))^\alpha - (\frac{z_i^{(k)}}{\tau})^\beta, \tag{5.77}$$

where α and β are two predefined parameters that control the shape of the reweighting curve. In such way, $c_i^{(k+1)}$ imposes a larger positive penalty when $z_i^{(k)}$ is closer to "0" and a smaller negative excitation when $z_i^{(k)}$ is closer to "τ" in the next iteration, in turn promoting the binary property of $\mathbf{z}/\tau$.

5.3.2.2 Beam-Space Quiescent Beamforming

Although antenna selection in element-space beamformer can significantly reduce the hardware cost and computational complexity, it inevitably exhibits performance degradation compared with the full array system. Alternatively, we consider another strategy, sparse beamspace beamformer, where a network of analogue phase shifters

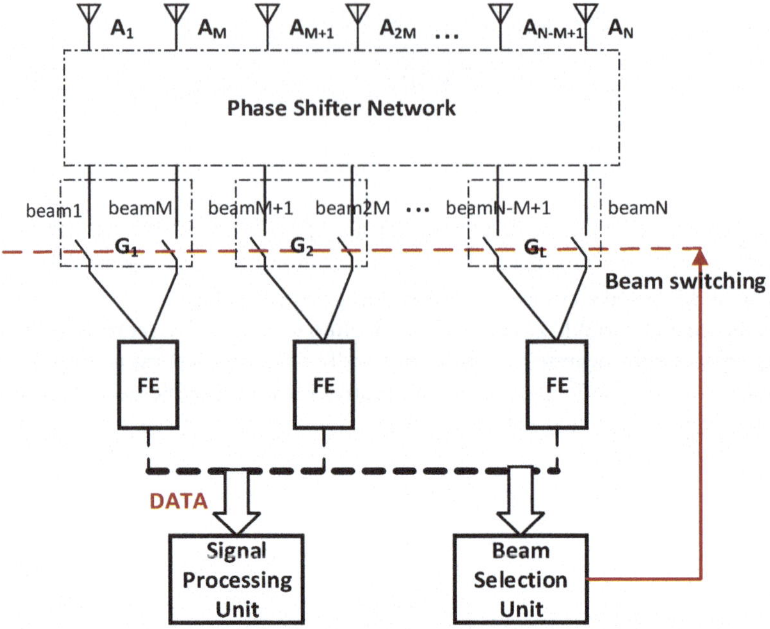

Fig. 5.17 The illustration of beamspace cognitive beamformer by regularized beam switching

are utilized to synthesize a set of N beams and an optimum subset of L beams are then selected to connect with the following front-end channels for further processing. To promote the practical implementation of cognitive sparse beamformers, regularized switching is employed as well, that is the N beams are divided into L groups and each group share one front-end processing channel as illustrated in Fig. 5.17. Again, grouping model is not restricted to the continuous one per Fig. 5.17 and different grouping models affect the beamformers' performance as shown in Sect. 5.3.4. The phase shifter network can be mathematically described by the beamspace transformation matrix $\mathbf{B} \in \mathbb{C}^{N \times N}$ described in Sect. 5.3.1.1.

To ensure a satisfactory performance of the beamformer, at least one beam close to the desired signal should be selected. Thus, the problem of regularized beam selection for quiescent beampattern synthesis is formulated as,

$$\begin{aligned} \min_{\mathbf{w}} \ & \delta, \\ \text{s.t.} \ & \mathbf{w}^H \tilde{\mathbf{a}}(\theta_0) = 1, \\ & |\mathbf{w}^H \tilde{\mathbf{A}}_s| \leq \delta, \\ & \|\mathbf{P}_d \mathbf{w}\|_0 \geq 1, \\ & \|\mathbf{P}_l \mathbf{w}\|_0 = 1, l \leq 1, \ldots, L, \end{aligned} \tag{5.78}$$

where $\mathbf{P}_d \in \{0, 1\}^{2\times N}$ is a beam selection matrix whose two rows correspond to the two beams adjacent to the direction of the desired signal θ_0, and $\tilde{\mathbf{A}}_s = \mathbf{B}^H \mathbf{A}_s$. The third constraint is imposed to select at least one beam neighbouring the desired signal for protection. The beam selection unit in Fig. 5.17 will calculate Eq. (5.78) using a similar procedure per Eqs. (5.75)–(5.76) and then determine the optimum subset of beams to connect with the following front-end processing channels. The signal processing unit then implements adaptive beamforming in a reduced dimensional beamspace.

5.3.3 Cognitive Sparse Beamformer Design

Initializing with the quiescent configuration, the cognitive beamformer begins to collect the data containing the environmental information and perform spatial filtering. Since the quiescent beamformer is not optimum in the specific environment, the antenna/beam switching unit, as shown in Fig. 5.18, will reconfigure to the optimum beamformer.

5.3.3.1 Cognitive Sparse Element-Space Beamformer Design

The design of cognitive sparse element-space beamformer comprises two consecutive operations, those are antenna removal and antenna addition. As shown in Fig. 5.18, the optimization within one group is an inner loop, where we remove the currently selected antenna in the examined group and then decide which antenna to be added in the same group with the removed one. The optimization of all groups constitutes an outer loop. To finally determine which antenna is switched on in each group, we need one removal calculation and M addition calculation. The entire calculation is repeated until the beamformer geometry converges. Since each antenna/beam on or off corresponds to a new beamformer, we derive closed-formed formulas to efficiently calculate the array gain variation to avoid computationally expensive matrix inversion. Next, we delineate these two calculations.

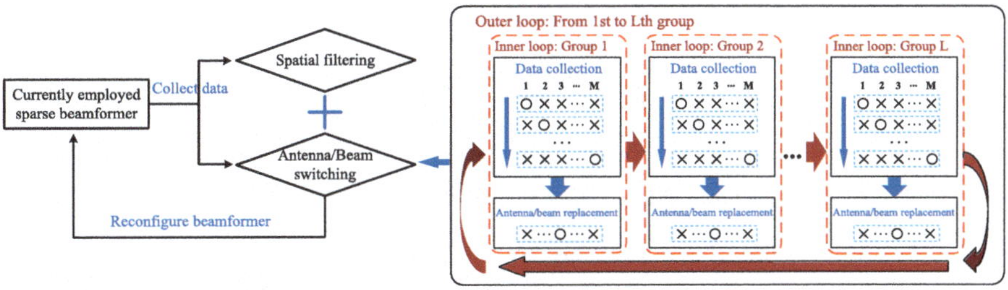

Fig. 5.18 The framework of the proposed cognitive sparse beamformer design

(a) Antenna Removal

The initialized quiescent sparse array receives the data and calculates the covariance $\hat{\mathbf{R}}$ and the optimum weight vector for spatial filtering can be calculated according to Eq. (5.65). Assume that we are examining the lth group and the antenna indexed by n in the lth group is to be removed. Let $\hat{\mathbf{R}}$ denote the covariance matrix of the currently employed L-antenna sparse array, then the covariance needs to be permuted such that the nth row and the nth column are both the last one. That means,

$$\mathbf{R}_{+n} = \mathbf{P}_n \hat{\mathbf{R}} \mathbf{Q}_n, \tag{5.79}$$

where $\mathbf{P}_n \in \{0, 1\}^{L \times L}$ and $\mathbf{Q}_n \in \{0, 1\}^{L \times L}$ are row permutation and column permutation matrices, which move the nth row and the nth column ($n = 1, \ldots, L$) to the last row and the last column respectively, without affecting the others. Similarly, we have that,

$$\mathbf{C}_{+n} = \mathbf{P}_n \mathbf{C}. \tag{5.80}$$

After row permutation, $\mathbf{C}_{+n}$ can be expressed as,

$$\mathbf{C}_{+n} = [\mathbf{C}_{-n}^T, \mathbf{c}_n]^T, \tag{5.81}$$

where $\mathbf{C}_{-n} \in \mathbb{C}^{(L-1)\times(K+1)}$ is the constraint matrix with $L-1$ antennas except for the nth antenna. And $\mathbf{c}_n = [e^{jk_0 dx_n \sin\theta_0}, e^{jk_0 dx_n \sin\theta_1}, \ldots, e^{jk_0 dx_n \sin\theta_K}]^T$ is the last row of the matrix $\mathbf{C}_{+n}$.

After permutation, the covariance matrix in Eq. (5.79) can be rewritten as,

$$\mathbf{R}_{+n} = \begin{bmatrix} \mathbf{R}_{-n} & \mathbf{r}_n \\ \mathbf{r}_n^H & r_{nn} \end{bmatrix}, \tag{5.82}$$

where $\mathbf{r}_n = E\{\mathbf{x}_{-n} x_n^*\}$ and $r_{nn} = E\{x_n^* x_n\}$ with $\mathbf{x}_{-n}$ and x_n denoting the received data by the $(L-1)$-element array excluding the nth antenna and by the nth antenna, respectively. And $\mathbf{R}_{-n} \in \mathbf{C}^{(L-1)\times(L-1)}$ is the covariance of the $(L-1)$-antenna array when the nth antenna is removed.

Utilizing block matrix inversion lemma, we can obtain that,

$$\begin{aligned} \mathbf{R}_{+n}^{-1} &= \begin{bmatrix} \mathbf{R}_{-n}^{-1} + \tau_n \mathbf{R}_{-n}^{-1} \mathbf{r}_n \mathbf{r}_n^H \mathbf{R}_{-n}^{-1} & -\tau_n \mathbf{R}_{-n}^{-1} \mathbf{r}_n \\ -\tau_n \mathbf{r}_n^H \mathbf{R}_{-n}^{-1} & \tau_n \end{bmatrix}, \\ &= \begin{bmatrix} \mathbf{D}_n & \tau_n \mathbf{d}_n \\ \tau_n \mathbf{d}_n^H & \tau_n \end{bmatrix}, \end{aligned} \tag{5.83}$$

where $\tau_n = 1/(r_{nn} - \mathbf{r}_n^H \mathbf{R}_{-n}^{-1} \mathbf{r}_n)$. From Eq. (5.83), we have that

$$\mathbf{D}_n = \mathbf{R}_{-n}^{-1} + \tau_n \mathbf{d}_n \mathbf{d}_n^H, \tag{5.84}$$

and

$$\mathbf{d}_n = -\mathbf{R}_{-n}^{-1}\mathbf{r}_n. \tag{5.85}$$

Since $\mathbf{R}_{+n}^{-1} = \mathbf{Q}_n^{-1}\hat{\mathbf{R}}^{-1}\mathbf{P}_n^{-1}$, we can easily obtain $\mathbf{D}_n$, $\mathbf{d}_n$ and τ_n by partitioning $\mathbf{R}_{+n}^{-1}$ according to Eq. (5.83). Furthermore, Eq. (5.84) implies that,

$$\mathbf{R}_{-n}^{-1} = \mathbf{D}_n - \tau_n \mathbf{d}_n \mathbf{d}_n^H. \tag{5.86}$$

Thus, utilizing the definition of $\mathbf{U}_{-n}^{-1}$ yields,

$$\mathbf{U}_{-n}^{-1} = [\mathbf{C}_{-n}^H \mathbf{R}_{-n}^{-1} \mathbf{C}_{-n}]^{-1}. \tag{5.87}$$

Note that the calculation of AG variation due to the removal of the nth antenna is not needed and we proceed to evaluate antenna addition next.

(b) Antenna Addition

Suppose the nth antenna in the lth group is removed, generating an $(L-1)$-element sparse array and the mth element in the lth group is added into the sparse array. The constraint matrix is changed to,

$$\mathbf{C}_{+m} = [\mathbf{C}_{-n}^T, \mathbf{c}_m]^T, \tag{5.88}$$

where $\mathbf{c}_m = [e^{jk_0 d x_m \sin\theta_0}, e^{jk_0 d x_m \sin\theta_1}, \ldots, e^{jk_0 d x_m \sin\theta_K}]^T$. Moreover, the covariance matrix of the new L-antenna sparse array can be expressed as,

$$\mathbf{R}_{+m} = \begin{bmatrix} \mathbf{R}_{-n} & \mathbf{r}_m \\ \mathbf{r}_m^H & r_{mm} \end{bmatrix}, \tag{5.89}$$

where $\mathbf{r}_m = E\{\mathbf{x}_{-n} x_m^*\}$ and $r_{mm} = E\{x_m^* x_m\}$, where $\mathbf{x}_{-n}$ is the received data vector of the $(L-1)$-antenna array with the nth antenna removed, x_m denoting the received data of the mth antenna and can be obtained by simply switching on the mth antenna with the lth front-end. In practice, both $\mathbf{r}_m$ and r_{mm} can be replaced by their maximum likelihood estimates. Following the same procedure of Eq. (5.83), we can obtain that,

$$\mathbf{R}_{+m}^{-1} = \begin{bmatrix} \mathbf{D}_m & \tau_m \mathbf{d}_m \\ \tau_m \mathbf{d}_m^H & \tau_m \end{bmatrix}, \tag{5.90}$$

where

$$\tau_m = 1/(r_{mm} - \mathbf{r}_m^H \mathbf{R}_{-n}^{-1} \mathbf{r}_m), \tag{5.91}$$

$$\mathbf{D}_m = \mathbf{R}_{-n}^{-1} + \tau_m \mathbf{d}_m \mathbf{d}_m^H, \tag{5.92}$$

and

$$\mathbf{d}_m = -\mathbf{R}_{-n}^{-1}\mathbf{r}_m. \tag{5.93}$$

Combining Eqs. (5.88) and (5.90), we have that,

$$\begin{aligned}\mathbf{U}_{+m} &= \mathbf{C}_{+m}^H \mathbf{R}_{+m}^{-1} \mathbf{C}_{+m}, \\ &= \mathbf{C}_{-n}^H \mathbf{D}_m \mathbf{C}_{-n} + \tau_m \mathbf{c}_m^* \mathbf{d}_m^H \mathbf{C}_{-n} \\ &+ \tau_m \mathbf{C}_{-n}^H \mathbf{d}_m \mathbf{c}_m^T + \tau_m \mathbf{c}_m^* \mathbf{c}_m^T, \\ &= \mathbf{U}_{-n} + \tau_m \mathbf{u}_m \mathbf{u}_m^H,\end{aligned} \tag{5.94}$$

where

$$\mathbf{u}_m = \mathbf{C}_{-n}^H \mathbf{d}_m + \mathbf{c}_m^* = -\mathbf{C}_{-n}^H \mathbf{R}_{-n}^{-1} \mathbf{r}_m + \mathbf{c}_m^*. \tag{5.95}$$

Applying Woodbury identity, we can obtain that

$$\mathbf{U}_{+m}^{-1} = \mathbf{U}_{-n}^{-1} - \frac{\mathbf{U}_{-n}^{-1}\mathbf{u}_m \mathbf{u}_m^H \mathbf{U}_{-n}^{-1}}{\tau_m^{-1} + \mathbf{u}_m^H \mathbf{U}_{-n}^{-1} \mathbf{u}_m}. \tag{5.96}$$

Based on Eq. (5.96), the IAG of the new L-antenna sparse array with the nth antenna removed and the mth antenna added can be written as,

$$\mathcal{A}_{+m}^{-1} = \mathbf{g}^H \mathbf{U}_{+m}^{-1} \mathbf{g} = \mathcal{A}_{-n}^{-1} - \frac{\mathbf{g}^H \mathbf{U}_{-n}^{-1} \mathbf{u}_m \mathbf{u}_m^H \mathbf{U}_{-n}^{-1} \mathbf{g}}{\tau_m^{-1} + \mathbf{u}_m^H \mathbf{U}_{-n}^{-1} \mathbf{u}_m}. \tag{5.97}$$

Thus, the IAG decrement by replacing the nth antenna with the mth one is,

$$\Delta \mathcal{A}_m^{-1} = \frac{\mathbf{g}^H \mathbf{U}_{-n}^{-1} \mathbf{u}_m \mathbf{u}_m^H \mathbf{U}_{-n}^{-1} \mathbf{g}}{\tau_m^{-1} + \mathbf{u}_m^H \mathbf{U}_{-n}^{-1} \mathbf{u}_m}. \tag{5.98}$$

Clearly, the larger the IAG decrement is, the more the AG increases by adding the mth antenna. Note that the matrices and vectors contained in $\mathbf{U}_{-n}^{-1}$, $\mathbf{u}_m$ and τ_m can all be calculated from the derivations above. The detailed calculation procedure of cognitive sparse element-space beamformer design is presented in Algorithm 3.

Algorithm 3: Cognitive Sparse Element-space Beamformer Design

Initialize with an L-antenna quiescent sparse array obtained from section II, collect the data and estimate covariance $\hat{\mathbf{R}}$ and inverse covariance $\hat{\mathbf{R}}^{-1}$.
repeat
  for *L groups of antennas* **do**
  - Find the selected antenna indexed by n in the lth group,
  - Permute $\hat{\mathbf{R}}^{-1}$ to obtain $\mathbf{R}_{+n}^{-1} = \mathbf{Q}_n \hat{\mathbf{R}}^{-1} \mathbf{P}_n$,
  - Decompose $\mathbf{R}_n^{-1}$ to obtain $\mathbf{D}_n$, τ_n, and $\mathbf{d}_n$,
  - Compute $\mathbf{R}_{-n}^{-1}$ per Eq. (5.86),
  - Compute $\mathbf{U}_{-n}^{-1}$ per Eq. (5.87).

  for *M antennas in the lth group* **do**
    - Switch on the mth antenna and collect data to obtain r_{mm} and $\mathbf{r}_m$,
    - Compose $\mathbf{C}_{+m}$ per Eqs. (5.81) and (5.88),
    - Calculate τ_m, $\mathbf{d}_m$, and $\mathbf{u}_m$,
    - Calculate the IAG variation per Eq. (5.98).

  end
  Seek $k = \max_{m=1,\ldots,M}\{\Delta\mathcal{A}_m^{-1}\}$,

  - Reconfigure the sparse array by switching the nth antenna off and the kth on;
  - Collect data with the new sparse array and obtain $\hat{\mathbf{R}}^{-1}$, construct $\mathbf{C}$;
  - Implement spatial filtering per Eq. (5.65).

  end
until *Sparse array configuration does not change*;

5.3.3.2 Cognitive Sparse Beam-Space Beamformer Design

As the spatial distribution of the target and the interferences is sparse, a much smaller number of beams are required for beamforming and thus beamspace processing can significantly reduce the computational complexity. Since neither the direction of the source nor those of interferences is known as a prior, the critical question is to select an optimum subset of beams for spatial filtering in terms of maximum array gain.

Assume that the phase shifter network can form a set of N beams and these beams are divided into L groups, with each group comprising M beams. A subset of L beams are selected to connect with the front-ends, each from one group. Similar to element-space beamformer design, we evaluate the L groups in the outer loop for beam removal and add the best beam in each group in the inner loop.

(a) Beam Removal

Suppose that a subset of L beams is currently employed for beamspace filtering, which can be represented by the beam space transformation matrix $\mathbf{B} \in \mathbb{C}^{L\times N}$. The beamspace covariance matrix is written as,

$$\tilde{\mathbf{R}} = \mathbf{B}^H\hat{\mathbf{R}}\mathbf{B} = \frac{1}{T}\sum_{t=1}^{T}\tilde{\mathbf{x}}_t\tilde{\mathbf{x}}_t^H, \tag{5.99}$$

where $\tilde{\mathbf{x}}_t = \mathbf{B}^H\mathbf{x}_t$ is the output of L beams at time instant t and $\mathbf{x}_t \in \mathbb{C}^N$ is the received data by the N-antenna array. Note that we do not have access to the element-space data $\mathbf{x}_t$, but the beamspace output $\tilde{\mathbf{x}}_t$ directly in beamspace processing.

The IAG variation of beamspace LCMP beamformer can be calculated in a similar way to that of elementspace LCMP. Here, we calculate the AG variation of MPDR beamformer as an explanation. Considering to remove the beam indexed by n in the lth group, we first permute the columns of the matrix $\mathbf{B}$ such that the nth beam is moved to the last column,

$$\mathbf{B}_n = \mathbf{B}\mathbf{Q}_n = [\mathbf{B}_{-n}, \mathbf{b}_n]. \tag{5.100}$$

where $\mathbf{b}_n$ denotes the nth beam and $\mathbf{B}_{-n}$ denotes the beamspace transformation matrix with the remaining $L-1$ beams after removing the nth beam. Using the permuted beamspace transformation matrix $\mathbf{B}_n$, the covariance matrix $\tilde{\mathbf{R}}_{+n} = \mathbf{B}_n^H\hat{\mathbf{R}}\mathbf{B}_n$ is rewritten as,

$$\begin{aligned}\tilde{\mathbf{R}}_{+n} &= \begin{bmatrix}\mathbf{B}_{-n}^H\hat{\mathbf{R}}\mathbf{B}_{-n} & \mathbf{B}_{-n}^H\hat{\mathbf{R}}\mathbf{b}_n\\ \mathbf{b}_n^H\hat{\mathbf{R}}\mathbf{B}_{-n} & \mathbf{b}_n^H\hat{\mathbf{R}}\mathbf{b}_n\end{bmatrix},\\ &= \begin{bmatrix}\tilde{\mathbf{R}}_{-n} & \mathbf{B}_{-n}^H\hat{\mathbf{R}}\mathbf{b}_n\\ \mathbf{b}_n^H\hat{\mathbf{R}}\mathbf{B}_{-n} & \mathbf{b}_n^H\hat{\mathbf{R}}\mathbf{b}_n\end{bmatrix}.\end{aligned} \tag{5.101}$$

Define $\tilde{\mathbf{r}}_n = \mathbf{B}_{-n}^H\hat{\mathbf{R}}\mathbf{b}_n$ and $\tilde{r}_{nn} = \mathbf{b}_n^H\hat{\mathbf{R}}\mathbf{b}_n$. The steering vector in the $(L-1)$-dimensional beamspace becomes,

$$\tilde{\mathbf{v}}_{s,-n} = \mathbf{B}_{-n}^H\mathbf{v}_s. \tag{5.102}$$

Utilizing the block matrix inversion lemma, the inverse covariance $\tilde{\mathbf{R}}_{+n}^{-1}$ is formulated as,

$$\begin{aligned}\tilde{\mathbf{R}}_{+n}^{-1} &= \begin{bmatrix}\tilde{\mathbf{R}}_{-n}^{-1} + \tilde{\tau}_n\tilde{\mathbf{R}}_{-n}^{-1}\tilde{\mathbf{r}}_n\tilde{\mathbf{r}}_n^H\tilde{\mathbf{R}}_{-n}^{-1} & -\tilde{\tau}_n\tilde{\mathbf{R}}_{-n}^{-1}\tilde{\mathbf{r}}_n\\ -\tilde{\tau}_n\tilde{\mathbf{r}}_n^H\tilde{\mathbf{R}}_{-n}^{-1} & \tilde{\tau}_n\end{bmatrix},\\ &= \begin{bmatrix}\tilde{\mathbf{D}}_n & \tilde{\tau}_n\tilde{\mathbf{d}}_n\\ \tilde{\tau}_n\tilde{\mathbf{d}}_n^H & \tilde{\tau}_n\end{bmatrix},\end{aligned} \tag{5.103}$$

where $\tilde{\tau}_n = 1/(\tilde{r}_{nn} - \tilde{\mathbf{r}}_n^H \tilde{\mathbf{R}}_{-n}^{-1} \tilde{\mathbf{r}}_n)$ and

$$\tilde{\mathbf{D}}_n = \tilde{\mathbf{R}}_{-n}^{-1} + \tilde{\tau}_n \tilde{\mathbf{d}}_n \tilde{\mathbf{d}}_n^H, \tag{5.104}$$

and

$$\tilde{\mathbf{d}}_n = -\tilde{\mathbf{R}}_{-n}^{-1} \tilde{\mathbf{r}}_n. \tag{5.105}$$

Also, $\tilde{\mathbf{R}}_{+n}^{-1}$ can be obtained by permuting the matrix $\tilde{\mathbf{R}}^{-1}$ by combining Eqs. (5.99) and (5.100), specifically,

$$\tilde{\mathbf{R}}_{+n}^{-1} = [\mathbf{B}_n^H \hat{\mathbf{R}} \mathbf{B}_n]^{-1} = [\mathbf{Q}_n^H \mathbf{B}^H \hat{\mathbf{R}} \mathbf{B} \mathbf{Q}_n]^{-1} = \mathbf{Q}_n^{-1} \tilde{\mathbf{R}}^{-1} \mathbf{Q}_n^{-H}. \tag{5.106}$$

Accordingly, $\tilde{\mathbf{D}}_n$, $\tilde{\mathbf{d}}_n$ and $\tilde{\tau}_n$ can be obtained by partitioning the matrix $\tilde{\mathbf{R}}_{+n}^{-1}$ per Eq. (5.103). Subsequently,

$$\tilde{\mathbf{R}}_{-n}^{-1} = \tilde{\mathbf{D}}_n - \tilde{\tau}_n \tilde{\mathbf{d}}_n \tilde{\mathbf{d}}_n^H. \tag{5.107}$$

(b) Beam Addition

Suppose that the mth beam in the lth group is going to be added. The covariance matrix becomes,

$$\tilde{\mathbf{R}}_{+m} = \mathbf{B}_{+m}^H \hat{\mathbf{R}} \mathbf{B}_{+m}, \tag{5.108}$$

where $\mathbf{B}_{+m} = [\mathbf{B}_{-n}, \mathbf{b}_m]$ and $\mathbf{b}_m$ denotes the mth beam. The beamspace steering vector becomes,

$$\tilde{\mathbf{v}}_{s,+m} = \mathbf{B}_{+m}^H \mathbf{v}_s. \tag{5.109}$$

The covariance $\tilde{\mathbf{R}}_{+m} = \mathbf{B}_{+m}^H \hat{\mathbf{R}} \mathbf{B}_{+m}$ can be partitioned as,

$$\begin{aligned} \tilde{\mathbf{R}}_{+m} &= \begin{bmatrix} \mathbf{B}_{-n}^H \hat{\mathbf{R}} \mathbf{B}_{-n} & \mathbf{B}_{-n}^H \hat{\mathbf{R}} \mathbf{b}_m \\ \mathbf{b}_m^H \hat{\mathbf{R}} \mathbf{B}_{-n} & \mathbf{b}_m^H \hat{\mathbf{R}} \mathbf{b}_m \end{bmatrix}, \\ &= \begin{bmatrix} \tilde{\mathbf{R}}_{-n} & \mathbf{B}_{-n}^H \hat{\mathbf{R}} \mathbf{b}_m \\ \mathbf{b}_m^H \hat{\mathbf{R}} \mathbf{B}_{-n} & \mathbf{b}_m^H \hat{\mathbf{R}} \mathbf{b}_m \end{bmatrix}. \end{aligned} \tag{5.110}$$

Define $\tilde{\mathbf{r}}_m = \mathbf{B}_{-n}^H \hat{\mathbf{R}} \mathbf{b}_m$ and $\tilde{r}_{mm} = \mathbf{b}_m^H \hat{\mathbf{R}} \mathbf{b}_m$. In practical applications, $\tilde{\mathbf{r}}_m$ and $\tilde{r}_{mm}$ are estimated by,

$$\tilde{\mathbf{r}}_m = \frac{1}{T} \sum_{t=1}^{T} \tilde{\mathbf{x}}_{-n,t} \tilde{x}_{m,t}^*, \tag{5.111}$$

and

$$\tilde{r}_{mm} = \frac{1}{T} \sum_{t=1}^{T} |\tilde{x}_{m,t}|^2, \tag{5.112}$$

where $\tilde{\mathbf{x}}_{-n,t}$ and $\tilde{x}_{m,t}$ denote the output of the $L-1$ beams with the nth beam removed and the mth beam in the tth snapshot, respectively.

Utilizing the block matrix lemma gain, then the inverse covariance is written by,

$$
\begin{aligned}
\tilde{\mathbf{R}}_{+m}^{-1} &= \begin{bmatrix} \tilde{\mathbf{R}}_{-n}^{-1} + \tilde{\tau}_m \tilde{\mathbf{R}}_{-n}^{-1} \tilde{\mathbf{r}}_m \tilde{\mathbf{r}}_m^H \tilde{\mathbf{R}}_{-n}^{-1} & -\tilde{\tau}_m \tilde{\mathbf{R}}_{-n}^{-1} \tilde{\mathbf{r}}_m \\ -\tilde{\tau}_m \tilde{\mathbf{r}}_m^H \tilde{\mathbf{R}}_{-n}^{-1} & \tilde{\tau}_m \end{bmatrix}, \\
&= \begin{bmatrix} \tilde{\mathbf{D}}_m & \tilde{\tau}_m \tilde{\mathbf{d}}_m \\ \tilde{\tau}_m \tilde{\mathbf{d}}_m^H & \tilde{\tau}_m \end{bmatrix},
\end{aligned} \tag{5.113}
$$

where $\tilde{\tau}_m = 1/(\tilde{r}_{mm} - \tilde{\mathbf{r}}_m^H \tilde{\mathbf{R}}_{-n}^{-1} \tilde{\mathbf{r}}_m)$, and

$$
\tilde{\mathbf{D}}_m = \tilde{\mathbf{R}}_{-n}^{-1} + \tilde{\tau}_m \tilde{\mathbf{d}}_m \tilde{\mathbf{d}}_m^H, \tag{5.114}
$$

and

$$
\tilde{\mathbf{d}}_m = -\tilde{\mathbf{R}}_{-n}^{-1} \tilde{\mathbf{r}}_m. \tag{5.115}
$$

Combining Eqs. (5.113) and (5.109), the AG of the new sparse beamspace beamformer is,

$$
\begin{aligned}
\mathcal{A}_m &= \tilde{\mathbf{v}}_{s,+m}^H \tilde{\mathbf{R}}_{+m}^{-1} \tilde{\mathbf{v}}_{s,+m}, \\
&= [\tilde{\mathbf{v}}_{s,-n}^H, \mathbf{v}_s^H \mathbf{b}_m] \begin{bmatrix} \tilde{\mathbf{D}}_m & \tilde{\tau}_m \tilde{\mathbf{d}}_m \\ \tilde{\tau}_m \tilde{\mathbf{d}}_m^H & \tilde{\tau}_m \end{bmatrix} \begin{bmatrix} \tilde{\mathbf{v}}_{s,-n} \\ \mathbf{b}_m^H \mathbf{v}_s \end{bmatrix}, \\
&= \tilde{\mathbf{v}}_{s,-n}^H \tilde{\mathbf{D}}_m \tilde{\mathbf{v}}_{s,-n} + \tilde{\tau}_m \mathbf{v}_s^H \mathbf{b}_m \tilde{\mathbf{d}}_m^H \tilde{\mathbf{v}}_{s,-n} \\
&\quad + \tilde{\tau}_m \tilde{\mathbf{v}}_{s,-n}^H \tilde{\mathbf{d}}_m \mathbf{b}_m^H \mathbf{v}_s + \tilde{\tau}_m \mathbf{v}_s^H \mathbf{b}_m \mathbf{b}_m^H \mathbf{v}_s, \\
&= \tilde{\mathbf{v}}_{s,-n}^H \tilde{\mathbf{R}}_{-n}^{-1} \tilde{\mathbf{v}}_{s,-n} + \tilde{\tau}_m |\tilde{\mathbf{v}}_{s,-n}^H \tilde{\mathbf{d}}_m + \mathbf{v}_s^H \mathbf{b}_m|^2.
\end{aligned} \tag{5.116}
$$

The beamspace AG increment by replacing the nth beam with the mth one is,

$$
\begin{aligned}
\Delta\mathcal{A}_m &= \tilde{\mathbf{v}}_{s,+m}^H \tilde{\mathbf{R}}_{+m}^{-1} \tilde{\mathbf{v}}_{s,+m} - \tilde{\mathbf{v}}_{s,-n}^H \tilde{\mathbf{R}}_{-n}^{-1} \tilde{\mathbf{v}}_{s,-n}, \\
&= \tilde{\tau}_m |\tilde{\mathbf{v}}_{s,-n}^H \tilde{\mathbf{d}}_m + \mathbf{v}_s^H \mathbf{b}_m|^2, \\
&= \tilde{\tau}_m |\mathbf{v}_s^H \mathbf{b}_m - \tilde{\mathbf{v}}_{s,-n}^H \tilde{\mathbf{R}}_{-n}^{-1} \tilde{\mathbf{r}}_m|^2, \\
&= \frac{|\mathbf{v}_s^H \mathbf{b}_m - \tilde{\mathbf{v}}_{s,-n}^H \tilde{\mathbf{R}}_{-n}^{-1} \tilde{\mathbf{r}}_m|^2}{\tilde{r}_{mm} - \tilde{\mathbf{r}}_m^H \tilde{\mathbf{R}}_{-n}^{-1} \tilde{\mathbf{r}}_m}.
\end{aligned} \tag{5.117}
$$

The detailed updating procedure of cognitive sparse beam-space beamformer design is presented in Algorithm 4.

Algorithm 4: Beam-space Sparse Beamformer Design

```
Collect the data from L channels and estimate beamspace covariance R̃ and inverse
  covariance R̃^{-1}.
repeat
  for L beams in the current beamspace do
    • Find the selected beam indexed by n
      in the lth group,
    • Permute R̃^{-1} to obtain R̃_{+n}^{-1} = Q_n^{-1} R̃^{-1} Q_n^{-H},
    • Partition R̃_{+n}^{-1} to get D̃_n, τ̃_n, d̃_n per Eq. (5.103),
    • Compute R̃_{-n}^{-1} per Eq. (5.107),
    • Compute ṽ_{s,-n} per Eq. (5.102).
    for M beams in the lth group do
      • Switch on the mth beam, collect data,
        obtain r̃_mm, r̃_m per Eqs. (5.111) and (5.112),
      • Calculate the mth beam b_m per Eq. (5.68),
      • Calculate the AG variation per Eq. (5.117).
    end
    Seek k = max_{m=1,...,M}{ΔA_m},
    • Reconfigure the beamspace by switching the
      nth beam off and the kth on,
    • Collect data and obtain R̃^{-1},
    • Implement spatial filtering per Eq. (5.67).
  end
until Sparse array configuration does not change;
```

Remark 5.1 For both proposed Algorithm 1 and Algorithm 2, spatial filtering is performed by signal processing unit based on the currently composed beamformers concurrently with the beamformer reconfiguration by the antenna selection unit.

Remark 5.2 For both proposed Algorithm 1 and Algorithm 2, compared with the conventional direct matrix inversion based method, instead of calculating an $L \times L$-dimensional matrix inversion for each antenna/beam (a total of $N = ML$ times in each outer loop), they only require to calculate an $(L-1) \times (L-1)$-dimensional matrix inversion for each group of antennas/beams (a total of L times in each outer loop). In addition, we utilize the properties of block matrix to obtain the inverse of the $(L-1) \times (L-1)$-dimensional matrix, which reduces the computational complexity from $O(L^3)$ to $O(L^2)$. Therefore, the proposed method can significantly reduce the computational complexity resulted from matrix inversion. Suppose that we sample T groups of receive data to construct the covariance matrix for each optimization and the number of constraints for LCMP is P. Then the computational complexities of the two proposed algorithms, the direct LCMP method in element space and the direct MPDR method in beam space are provided in Table 5.3.

Table 5.3 Comparison of computational complexity

Method		Computational complexity (Number of complex multiplications)	Simplification	
			P is larger than L and M	P is small
Element space	Algorithm 1	$\frac{2}{3}P^3L + PL^3 + P^2L^2 + 2ML^3 + ML^2P + P^2ML + TML^2$	$O(P^3L + P^2L^2 + PL^3 + ML^3 + MPL^2 + MP^2L + TML^2)$	$O(ML^3 + TML^2)$
	LCMP	$\frac{2}{3}ML^4 + PML^3 + P^2ML^2 + \frac{2}{3}P^3ML + P^2ML + PML + TML^3$	$O(ML^4 + PML^3 + P^2ML^2 + P^3ML + TML^3)$	$O(ML^4 + TML^3)$
Beam space	Algorithm 2	$TML^2 + 2ML^3 + M^2L^2$	$O(TML^2 + ML^3 + M^2L^2)$	
	MPDR	$M^2L^3 + TML^3 + (2/3)ML^4 + ML^3 + ML^2$	$O(TML^3 + ML^4 + M^2L^3)$	

The variation curves of computational complexity versus different parameters are provided in Fig. 5.19. We separately change the number of groups L and the number of antennas/beams M in each group, the corresponding computational complexity curves in one outer loop is depicted in Fig. 5.19a–b. It can be seen from the results that compared with the direct matrix inversion based methods, the proposed algorithms can greatly reduce the computational complexity. Among all the parameters, the number of groups L has the greatest impact on the complexity. Besides, the computational complexity of beam space is slightly larger than that of element space. In addition, we fix the total number of antennas/beams N to 100, and change the number of groups L. In this case, the number of antennas/beams in each group is $M = N/L$ and the curve of computational complexity is given in Fig. 5.19c. Figure 5.19d indicates that for a given number of total antennas/beams, the more groups they are divided into, the higher the computational complexity will be. This is because more groups means the beamformer will select more antennas/beams, which results in better beamforming performance but higher computational complexity.

5.3.4 Numerical Analysis

Extensive simulation results are presented in this section to validate the theoretical part of this section

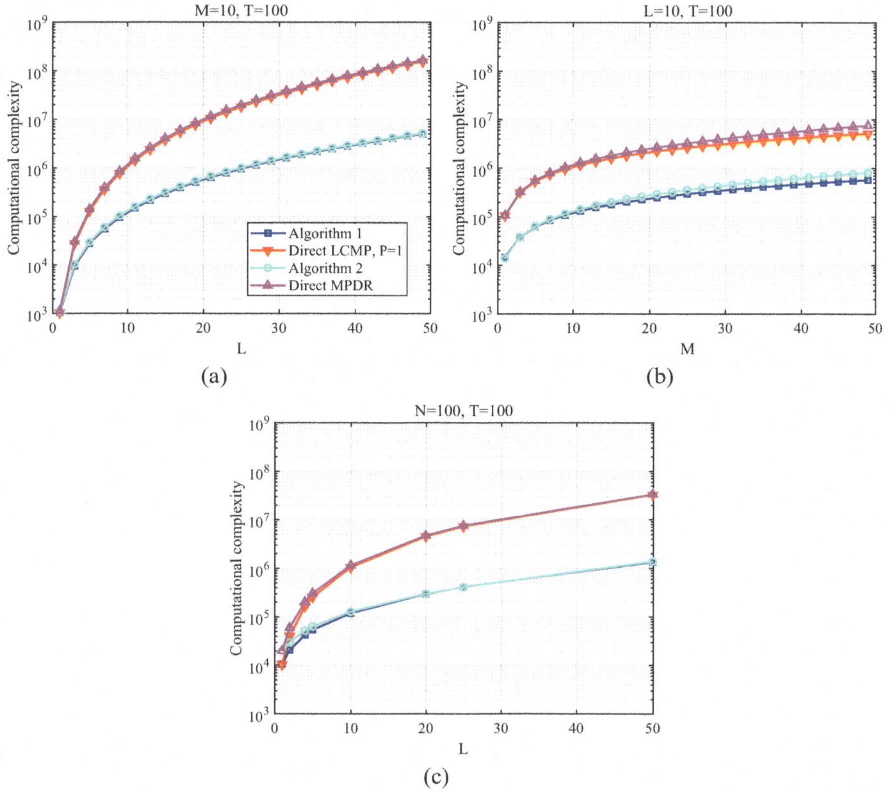

Fig. 5.19 Computational complexity versus **a** the number of groups L; **b** the number of antennas M in each group. **c** Computational complexity versus the number of groups L when the total number of antennas N is fixed to 100

5.3.4.1 Initialization of Element-Space Quiescent Beamformer

First, we verify the performance of the regularized antenna switching method for quiescent beamformer design proposed in section II through a comparative experiment. Consider a uniform linear array (ULA) composed of 20 antennas with an inter-element spacing of a quarter wavelengths. This linear array is continuously divided into 10 groups and each group contains two consecutive antennas. There are totally 10 front-ends and each front-end is responsible to connect with the two antennas in one group. A quiescent sparse beamformer with 10 antennas is composed through regularized antenna switching. For the beampattern requirement, the main beam points towards 0° and the sidelobe angular region is set as $[-90°, -10°] \cup [10°, 90°]$. We run the proposed method in Eq. (5.76) and obtain a sparse array, named as sparse array 1, as shown in Fig. 5.20a. For selecting ten antennas from 10 separate groups, there are a total of 1024 different choices. We enumerate all these choices and find the sparse array that achieves the lowest peak sidelobe level (PSLL), named as sparse

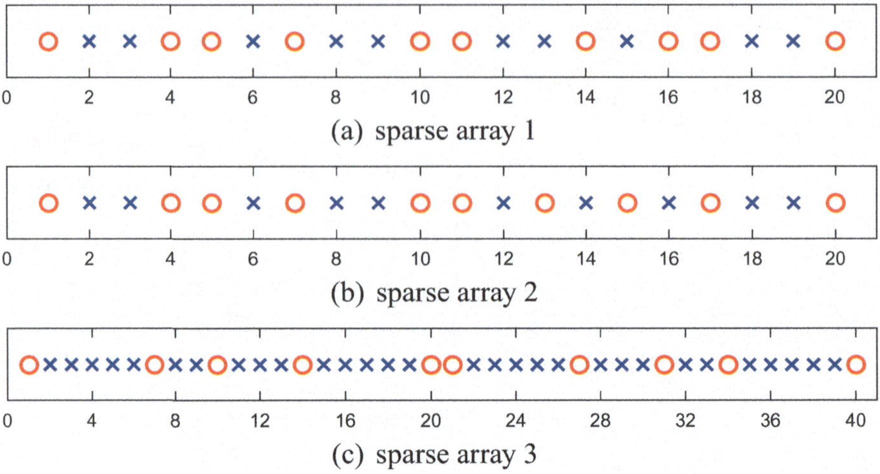

Fig. 5.20 The optimum quiescent sparse arrays obtained by **a** the proposed method from 20 elements; **b** enumeration from 20 elements; **c** the proposed method from 40 elements. "Circle" indicates selected antennas and "cross" discarded

array 2, in Fig. 5.20b. We plot the quiescent beampatterns of array 1 and array 2 in Fig. 5.21, where the PSLL of array 1 is –17.1 dB and that of array 2 is –17.2 dB. Through comparison, the sparse array obtained by the proposed selection method exhibits only 0.1 dB PSLL degradation, which manifests the effectiveness of the proposed method for securing a good sub-optimal solution.

As enumeration is restricted to small-scaled problems, we continue to consider a relatively large array. Suppose the ULA consists of 40 antennas with an inter-element spacing of 1/8 wavelengths. It is divided into 10 groups and each group is composed of 4 consecutive antennas. We choose one antenna from each group and the constraints of the beampattern is the same as the former experiment. The sparse array obtained by the proposed selection method and its corresponding quiescent beampattern are given in Figs. 5.20c and 5.21 respectively. The PSLL of array 3 is –17.9 dB, which is slightly lower than sparse arrays (1) and (2), and all exhibit good quiescent beampattern.

5.3.4.2 Cognitive Element-Space Beamformer Design

We begin with verifying the performance of Algorithm 1 in a static environment. Assuming that the desired signal comes from the direction of 0° with a signal-to-noise ratio (SNR) of 0dB, and there are seven interferences from the directions of −80°, −70°, −10°, 10°, 30°, 65° and 80° respectively, with an interference-to-noise ratios (INR) of 20 dB. We consider two cases described in Sect. 5.3.4.1, those are a ULA composed of 20 antennas and initialized with sparse array 1 in case 1, and a ULA composed of 40 antennas and initialized with sparse array 3 in case 2. Ideally, if

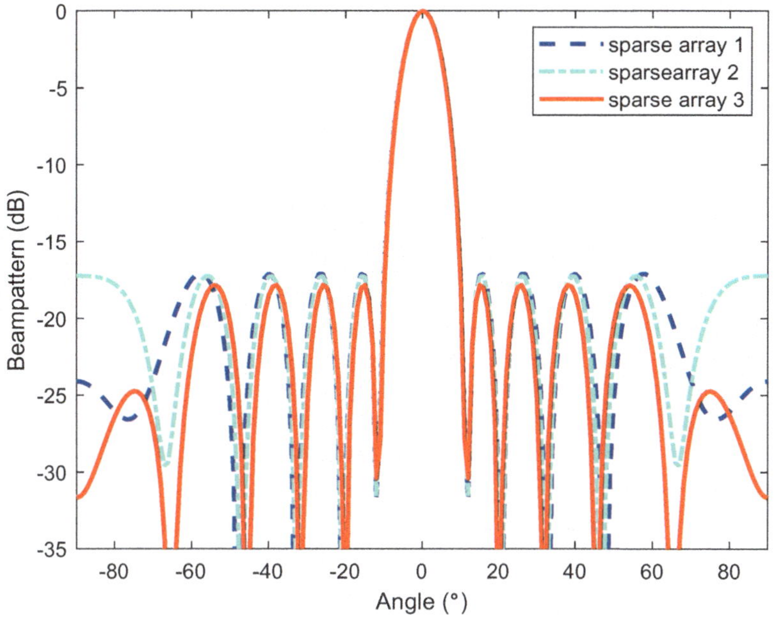

Fig. 5.21 The quiescent beampatterns of three sparse arrays shown in Fig. 5.20

the information of interferences is known, we can obtain the optimum sparse arrays according to Eqs. (5.72) and (5.73) in both cases. The corresponding two arrays are defined as sparse array 4 and sparse array 5, and are shown in Fig. 5.22a and b. Then we implement Algorithm 1 to achieve iterative sparse array reconfiguration in the two cases. We define the sparse arrays after convergence in case 1 and case 2 as sparse array 6 and sparse array 7 respectively, which are given in Fig. 5.22c and d. The adaptive beampatterns of sparse arrays 6 and 7 are depicted in blue solid lines in Fig. 5.23, where we can see that both sparse arrays can form deep nulls towards the interferences while preserving the desired signal. To further reduce the PSLL of adaptive beampatterns, we rerun Algorithm 1 in the metric of LCMP by imposing some constraints on the sidelobe region. The two LCMP sparse arrays are termed as sparse array 8 and sparse array 9 in two cases, respectively and plotted in Fig. 5.22e and f. Their corresponding beampatterns are depicted in red dash lines in Fig. 5.23. It can be observed from Fig. 5.23 that the PSLL of LCMP sparse beamformers is nearly 5 dB lower than that of MPDR sparse beamformers.

The convergence performance of the proposed Algorithm 1 is demonstrated in Fig. 5.24. As defined in Fig. 5.18, we refer to the antenna replacement in each group as an inner iteration, and the updating of all groups as an outer iteration. Thereby, each outer iteration contains 10 inner iterations in the examined example. From Fig. 5.24, we can argue that the convergence speed of the proposed design is fast with no more than three outer iterations. In Fig. 5.24a and b, the two dashed lines depict the upper

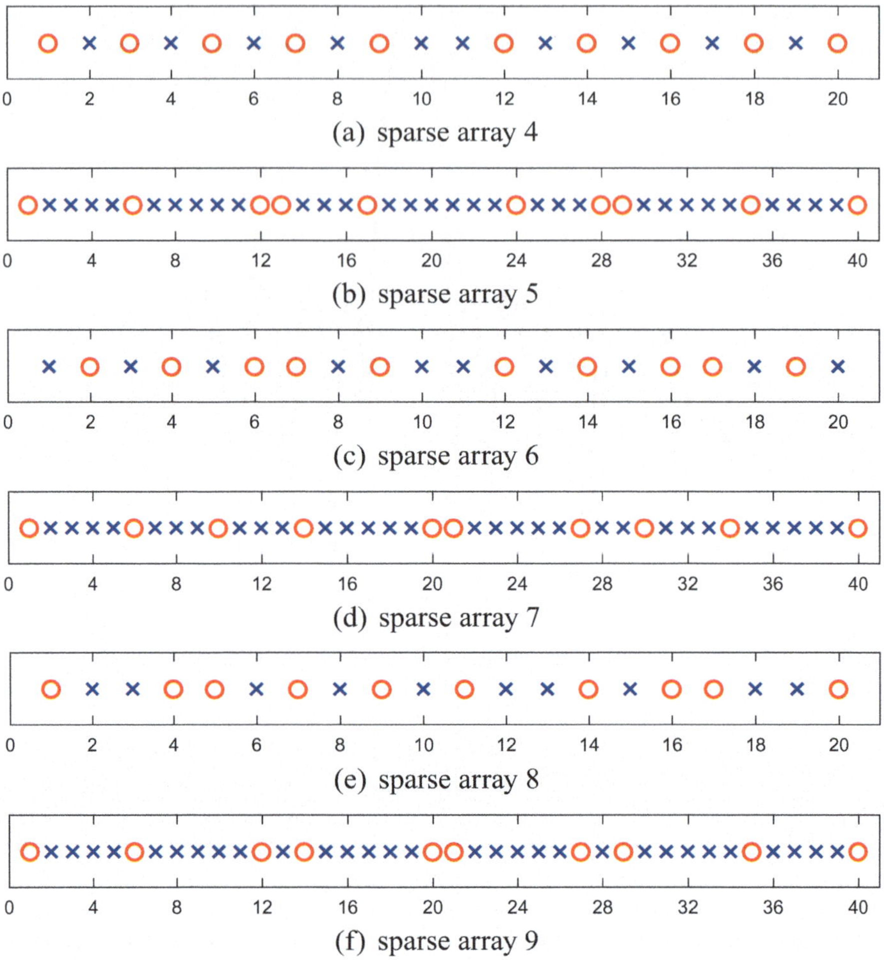

Fig. 5.22 Optimum sparse arrays obtained by Eq. (5.72) in **a** case 1 **b** case 2. Optimum MPDR sparse arrays in **c** case 1 **d** case 2. Optimum LCMP sparse arrays in **e** case 1 **f** case 2

bounds of AG in the two cases using sparse arrays 4 and 5, and the two dashed-dot lines depict the array gains of the optimal sparse array obtained by the model-based method described in Section II-B, which assumes the environmental prior knowledge is known. As can be seen from Fig. 5.24, the AG of configured beamformers gradually improves with increased iterations. When the algorithm converges, both beamformers can achieve an AG higher than 9.4dB, which is less than 0.1 dB under the upper bound and no more than 0.05 dB compared with the model based method in [218]. Although the finally converged sparse array only exhibits two different antenna positions compared with the initial quiescent beamformer, the AG of reconfigured beamformer in case 1 is improved by 2.8dB, while 5.1 dB in case 2. Taking case

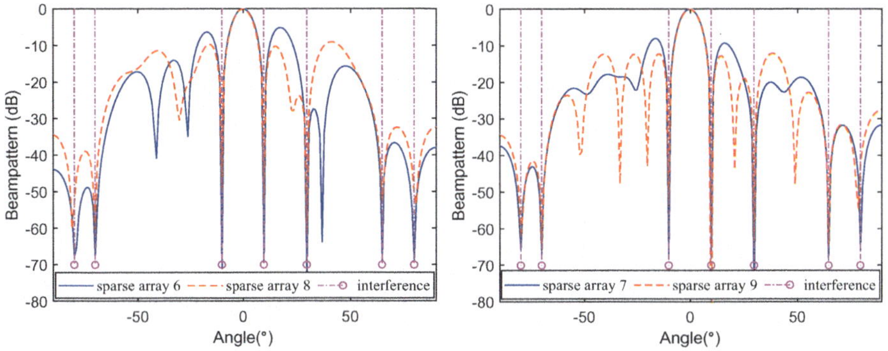

(a) (b)

Fig. 5.23 Adaptive beampatterns in **a** case 1 **b** case 2

Fig. 5.24 Array gain versus iteration number in two cases of array 6 and 7

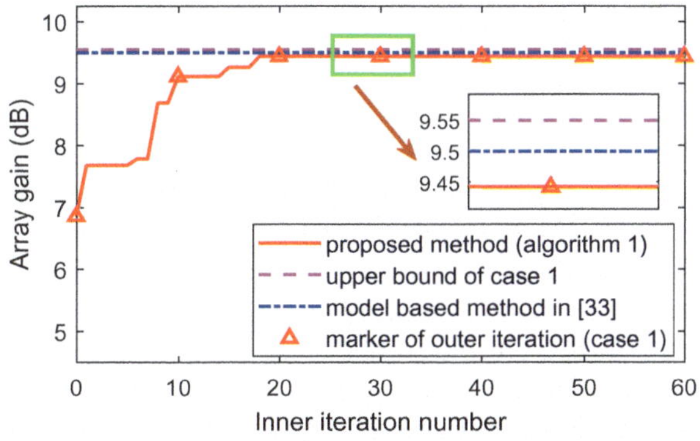

(a) case 1

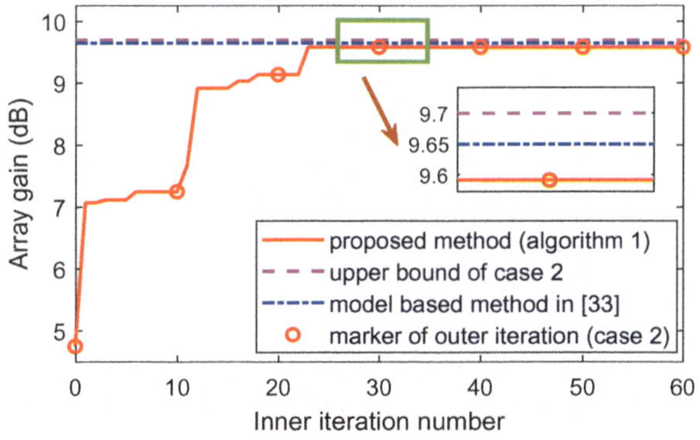

(b) case 2

Table 5.4 Iterative sparse array reconfiguration for case 2

Iteration no.		Antenna index									
Outer	Inner										
0	0	1	7	10	14	20	21	27	31	34	40
1	1	**3**	7	10	14	20	21	27	31	34	40
	3	3	7	**12**	14	20	21	27	31	34	40
	6	3	7	12	14	20	**23**	27	31	34	40
2	1	**1**	7	12	14	20	23	27	31	34	40
	2	1	**6**	12	14	20	23	27	31	34	40
	6	1	6	12	14	20	**21**	27	31	34	40
	8	1	6	12	14	20	21	27	**30**	34	40
3	3	1	6	**10**	14	20	21	27	30	34	40

* The red number indicates the antenna that has changed compared with the former iteration

2 as an example, the detailed iterations of sparse array reconfiguration are enlisted in Table 5.4. In this table, we only present the iterations with antenna replacement. The algorithm converges after the 3rd inner iteration of the 3rd outer iteration. We notice that the discarded antennas in the previous iteration will be switched on again with the updating of other groups. Therefore, the optimality of antenna positions is relative with respect to the status of other antennas.

We continue to investigate the influence of snapshot number on the output performance of Algorithm 1. For a data-dependent adaptive beamformer, the number of snapshots affects the estimation accuracy of covariance matrix, and thus influencing the output performance. A different beamformer is generated once antenna replacement is conducted and new data is received. For different number of snapshots, Algorithm 1 will optimize the sparse beamformer based on the estimated covariance from a given number of received data. We change the number of snapshots, T, from 10 to 1910 in an interval of 100 and run 1000 Monte Carlo simulations for each number. Note that different snapshot number leads to differed beamformer configuration caused by the estimation accuracy of covariance matrix. We apply Algorithm 1 in both cases 1 and 2, and plot the AG value after convergence versus snapshot number, as shown in Fig. 5.25. We can see that, with the increase of snapshot number, the output performance of the beamformer improves and tends to be stable. When the number of snapshots is larger than 100, the two sparse beamformers exhibit good performance, whose AG approached the upper bound obtained using sparse arrays 4 and 5. We also plot the AG of sparse arrays 1 and 3 without reconfiguration for comparison. Clearly, the reconfigured adaptive beamformer gives a performance improvement of more than 2.5 and 5.2 dB in case 1 and case 2, respectively. This demonstrates the necessity of beamformer reconfiguration for performance improvement.

To further illustrate the effectiveness of the proposed design, we conduct simulations in a dynamic environment. We consider three different scenarios, where the direction of the desired signal remains the same (0° and 0dB), but the interferences are changing. We start from the same scenario as that in the static environment. In

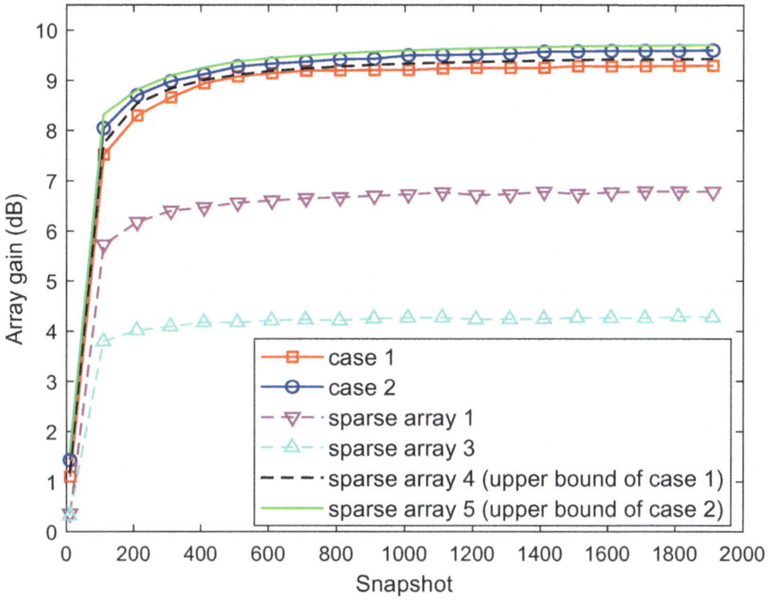

Fig. 5.25 Array gain versus snapshot number in different cases

scenario 2, the seven interferences are changing directions to $-70°$, $-50°$, $-10°$, $10°$, $15°$, $35°$ and $80°$, and in scenario 3, changing to $-60°$, $-50°$, $-10°$, $10°$, $25°$, $35°$ and $80°$. The INRs of all interferences are 20 dB. We examine the two cases of arrays described in Sect. 5.3.4.1 in this dynamic environment and the results are given in Fig. 5.26. For each antenna replacement, the number of snapshots we received is $T = 1000$ and we defined this duration as one sampling interval. As shown in Fig. 5.26, at time instants $50T$ and $150T$, the environment changes from scenario 1 to scenario 2 and from scenario 2 to scenario 3 respectively. When the environment changes, the output performance of the current beamformer decreases abruptly. And during the followed reconfiguration period (marked by the green dash ellipses), the beamformer is cognizant of performance degradation and apply the "perception-action" process to adapt to the new environment. Consequently, the reconfigured beamformer can quickly recover its performance against the interferences.

Furthermore, we explore the influence of various group sizes on the results when the total number of antennas is fixed. We still utilize the 40 antennas described in case 2, the difference is that we change the group sizes. We change the number of groups among $T = \{8, 10, 20, 40\}$, and accordingly, the number of antennas in each group is $M = \{5, 4, 2, 1\}$. In particular, $T = 40$ means the whole array is selected. The performance of Algorithm 1 for various group sizes is depicted in Fig. 5.27. Besides, the running times of one outer iteration for different group sizes are provided in Table 5.5. From the results, we can see that with the increase of the number of groups, the arrray gain becomes higher. This is because more groups means more

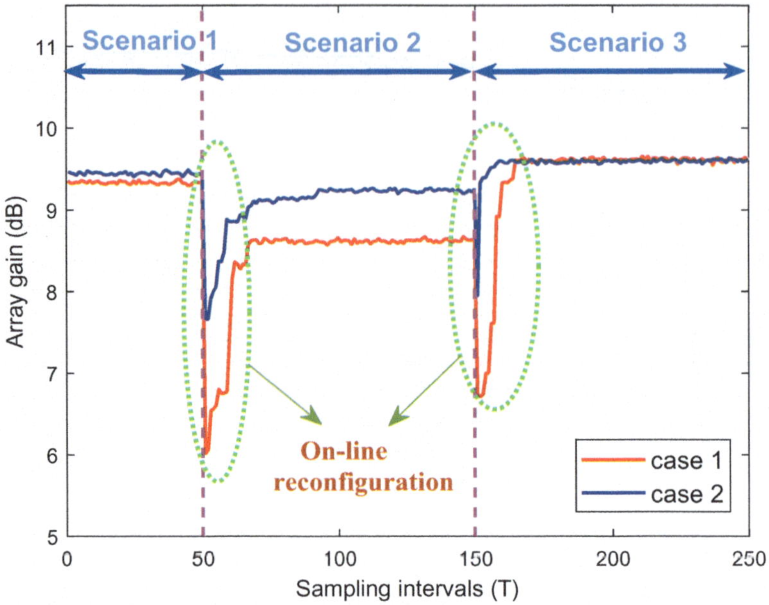

Fig. 5.26 Performance of Algorithm 1 in dynamic environment

Fig. 5.27 Array gain versus iteration number for various group size

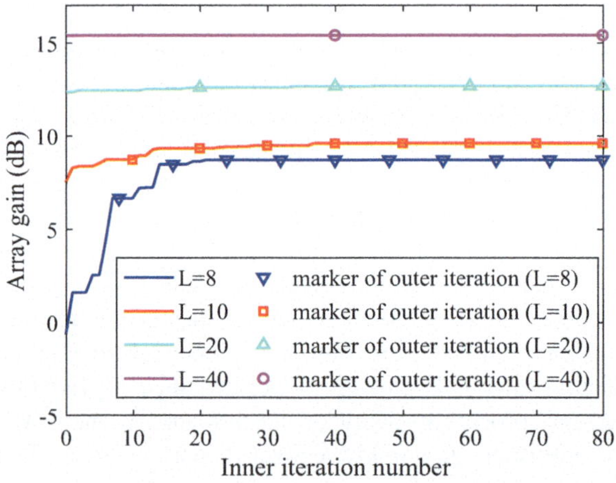

antennas will be selected for data processing. However, more groups will bring higher hardware costs. And in terms of running time, more groups will also bring greater computational complexity, thus increase the response time of the system. Therefore, we need to consider the trade-off between array gain and computational complexity to reasonably choose the size of each group in practical application. In addition, from the comparison in Table 5.5, we can see that the computational complexity of

Table 5.5 The running time of one outer iteration for different group sizes

	L = 8 (s)	L = 10 (s)	L = 20 (s)	L = 40 (s)
Algorithm 1	0.0460	0.0589	0.1180	0.2661
Direct LCMP	0.3756	0.5586	2.2865	9.2085

Algorithm 1 is significantly improved compared with the direct LCMP method. And with the increase of L, the improvement is more significant.

Last but not least, we examine the impact of different group update sequences. We randomly generate 200 different group update orders, and apply Algorithm 1 to compare the obtained sparse array performance with that of sequential update (that is, updating from the 1st to the Lth group sequentially). The simulation results are shown in Fig. 5.28, which indicate that the array gain of the cognitive sparse beamformer is varying under different group update orders, but not n an obvious way. Especially for case 2, when the total number of antennas is large and the inter-element spacing is small (which implies an increased continuity of the design), a satisfactory array gain can be achieved under an arbitrary group update order. In practical applications, in order to further ensure the array gain of the cognitive beamformer, on the premise of neglecting the computational complexity, we can apply different update orders and select the best sparse array beamformers.

5.3.4.3 Initialization of Beam-Space Quiescent Beamformer

In this subsection, we examine the beam-space beamformer initialization proposed in section II-B. Consider a ULA composed of 40 antennas with an inter-element spacing of half wavelength, which can form 40 mutually orthogonal beams. We divided these 40 beams into 10 groups with 4 beams in each group and select one beam from each group to compose the beam-space beamformer. For the beampattern requirement, the main beam points to 0° and the sidelobe region is set as $[-90, -5] \cup [5, 90]$. In this experiment, we also consider two cases of different grouping models, continuous grouping and non-continuous grouping, shown in Fig. 5.29. For continuous grouping, one group is composed of four adjacent beams; while for non-continuous grouping, four beams separated by 10 beams in between form one group. For the former model, more than three continuous beams cannot be selected simultaneously; while the latter model enables the selection of continuous beams. In these two models, we use Eq. (5.78) to design the quiescent beamformer for initialization. The results are depicted in Fig. 5.30 and the selected beams are given in Table 5.6. From Fig. 5.30, we can see that the PSLL of the continuous grouping is –25 dB, while that of the non-continuous grouping is –39 dB. Although different grouping models affect the performance of the configured beamformers, the investigation of optimum grouping models is out of the scope of this work.

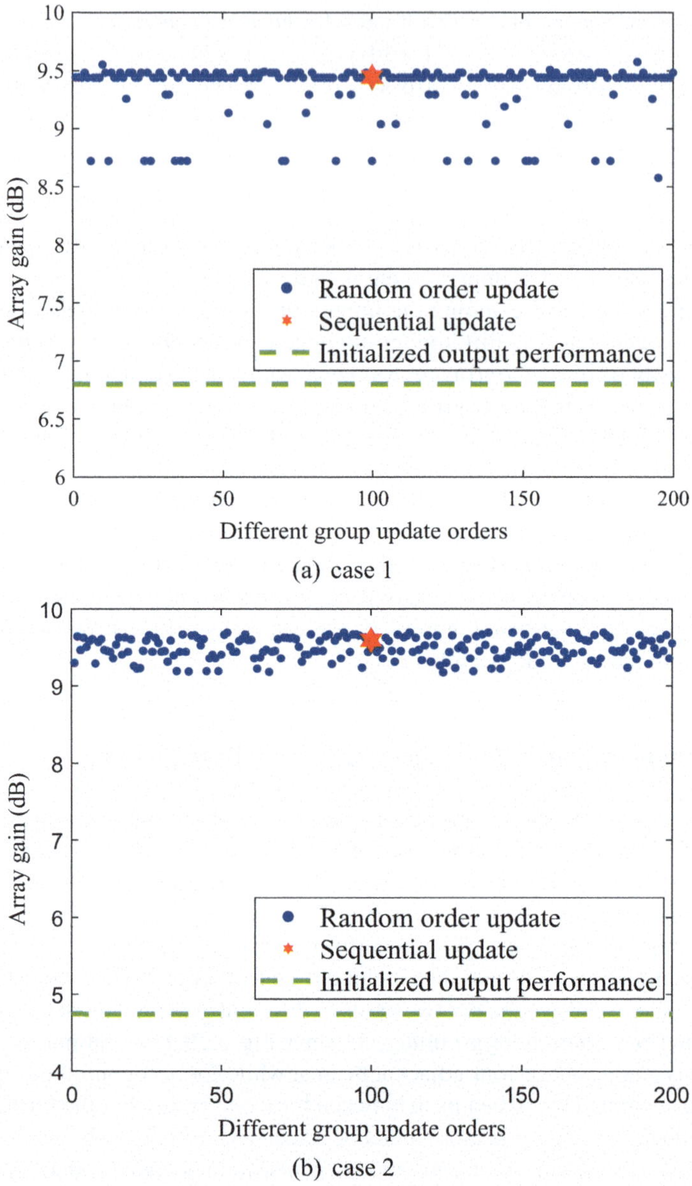

Fig. 5.28 Array gain versus different group update orders in two cases

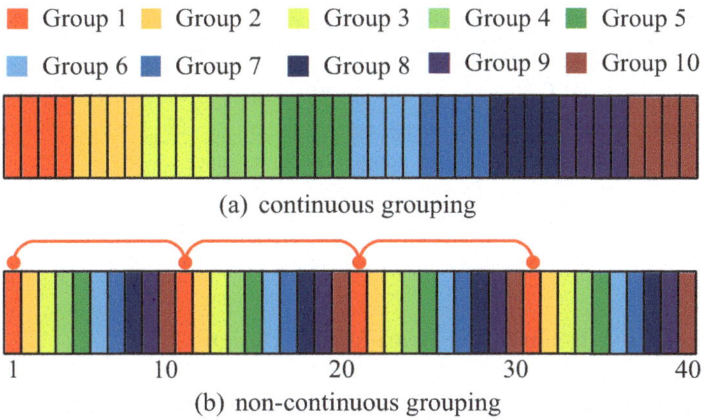

Fig. 5.29 Different grouping models of beam channels

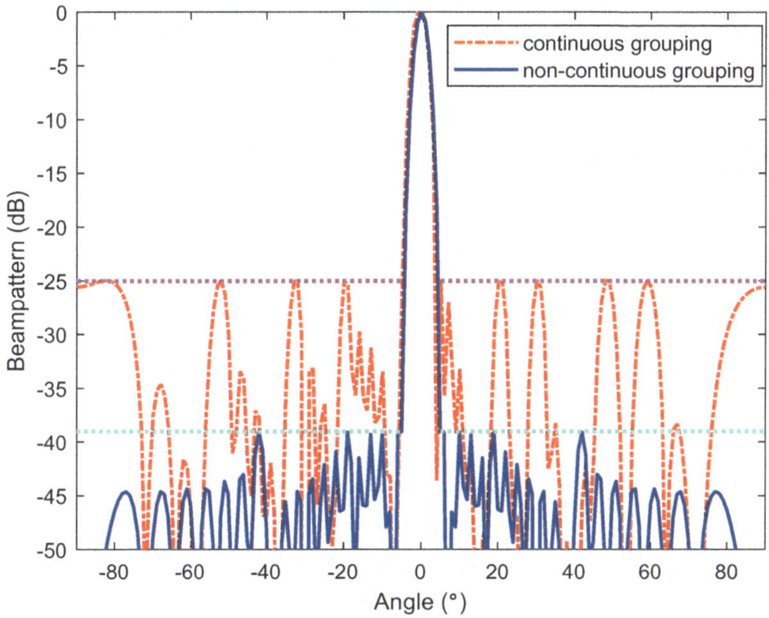

Fig. 5.30 The quiescent beampatterns under two grouping models

Table 5.6 The optimum beams choosen under two modes

Mode	Selected beams of each group after initialization								
Continuous	1	5	10	14	20	21	28	31	36	38
Non-continuous	21	22	23	34	15	6	27	8	19	20
Grouping	Selected beams of each group after convergence									
Continuous	1	8	12	16	20	21	25	29	33	37
Non-continuous	21	2	13	14	15	16	17	18	19	40

5.3.4.4 Cognitive Beam-Space Beamformer Design

We continue to validate the proposed iterative beam replacement in Algorithm 2. Assume that the desired signal comes from 0° with a SNR of 0 dB and there are 7 interferences coming from −80°, −22°, −17°, −12°, −7°, 30° and 80° with INRs of 40 dB. We apply Algorithm 2 to achieve on-line beam replacement under the two models defined in Fig. 5.29. The curves of AG versus inner iteration number are given in Fig. 5.31, which shows that the output performance of the beamformer gradually improves through iterations. For continuous grouping, the AG is increased by 2.2 dB and finally reaches 14.4 dB. And for non-continuous grouping, the AG is increased by 1.8 dB and finally reaches 15.9 dB. Compared with the continuous model, the performance of the non-continuous one is significantly better in terms of both beampattern and AG. Specifically, the output AG of non-continuous model can gradually approach the upper bound of the full adaptive beamformer ($10 * \log_{10} 40 = 16$dB), which is depicted by the black dash line in Fig. 5.31. In addition, the convergence speed of Algorithm 2 is very fast and both models can converge within two outer iterations. The selected beams after convergence of the two models are given in Table 5.6 and the corresponding adaptive beampatterns are given in Fig. 5.32. The

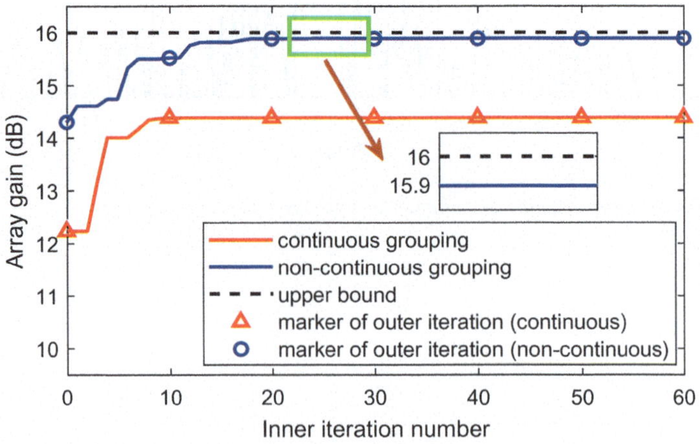

Fig. 5.31 Array gain versus iteration number in the cases of two modes

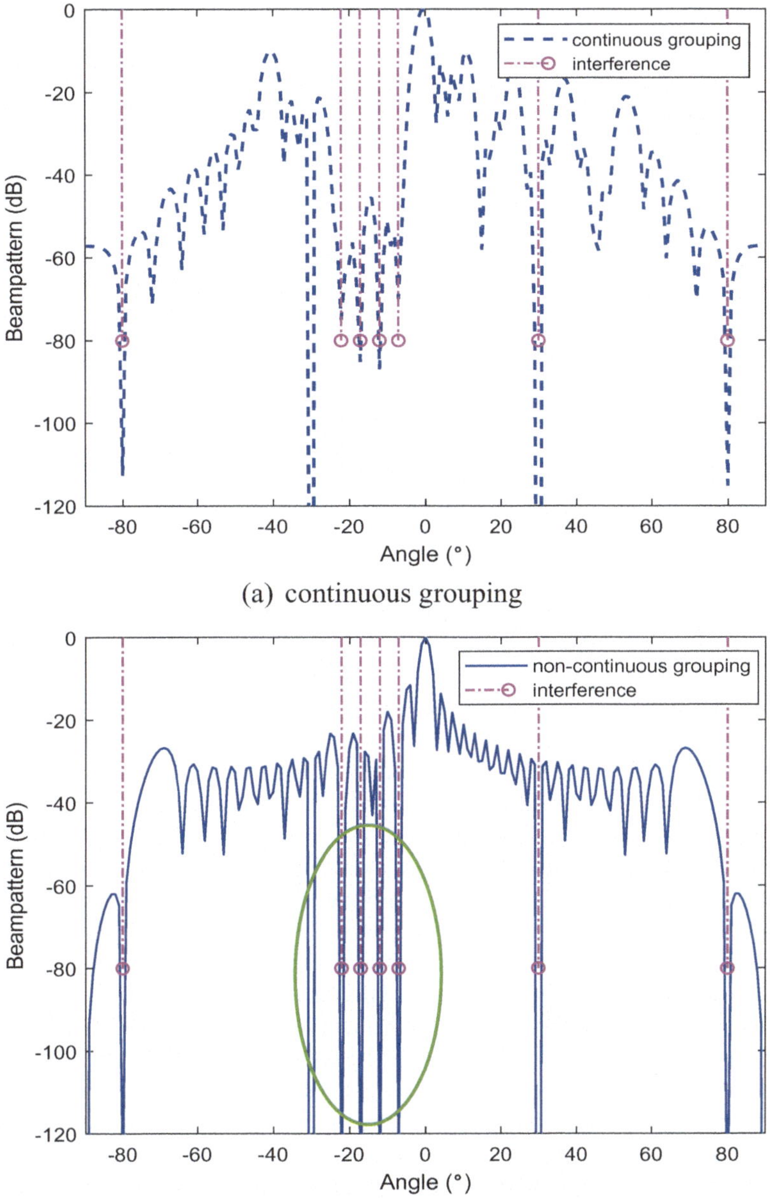

(a) continuous grouping

(b) non-continuous grouping

Fig. 5.32 Adaptive beampatterns of the converged beams under two grouping models

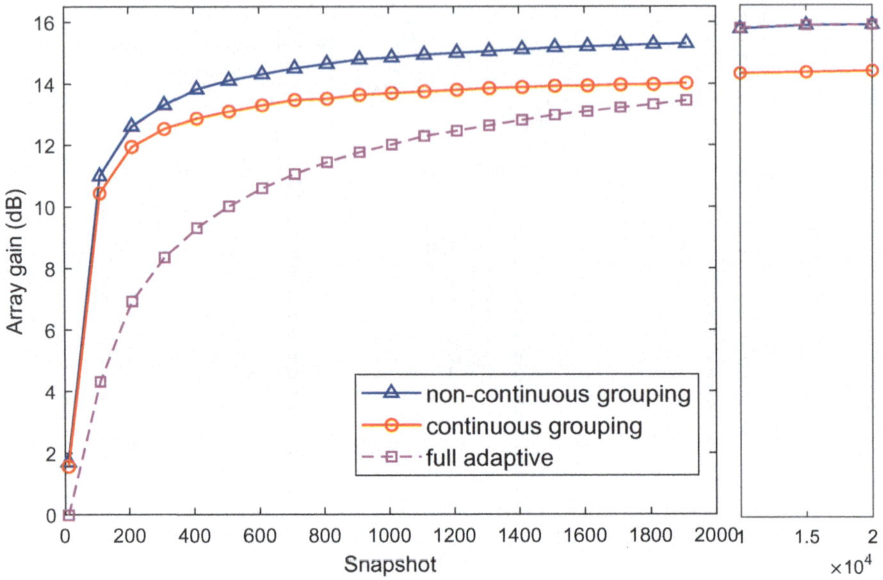

Fig. 5.33 Array gain versus the number of snapshots for different grouping models

beampatterns indicate that nulls can be formed in the directions of the interferences for both models, while the sparse beamformer in the non-continuous model can form a deeper null compared with that in the continuous grouping, especially for the four interferences marked in the green ellipse. Meanwhile, the sidelobe level of non-continuous grouping is lower as well.

In the following example, we change the snapshot number to study the performance of Algorithm 2. The snapshot number is changing from 10 to 1910 with an interval of 100. For each number of snapshots, we apply the on-line beam replacement process to the non-continuous grouping and continuous grouping, and conduct 1000 Monte Carlo simulations. In addition, we simulate the case of full beam adaptation, that is, all beams are used for adaptive beamforming. The curves of AG versus snapshot number are shown in Fig. 5.33. As can be seen from Fig. 5.33, when the number of snapshots is small, the performance of sparse adaptive beamformer is significantly higher than that of full adaptive beamformer, as sparse beamformer requires less training data to accurately estimate the covariance matrix. This proves the necessity of sparse beamformer design in dynamic environment. When the number of snapshots is sufficiently large (larger than 1×10^4 per the right part of Fig. 5.33), the performance of non-continuous grouping is as good as that of full adaptation, which is more than 3 dB higher than that of continuous model.

In the last example, we conduct a simulation in a dynamic environment. Similar to the final example in Sect. 5.3.4.2, we consider three scenarios with different interference environments. Scenario 1 remains the same as the one in the beginning. For scenario 2, there are seven interferences coming from the directions of $-70°$, $-20°$,

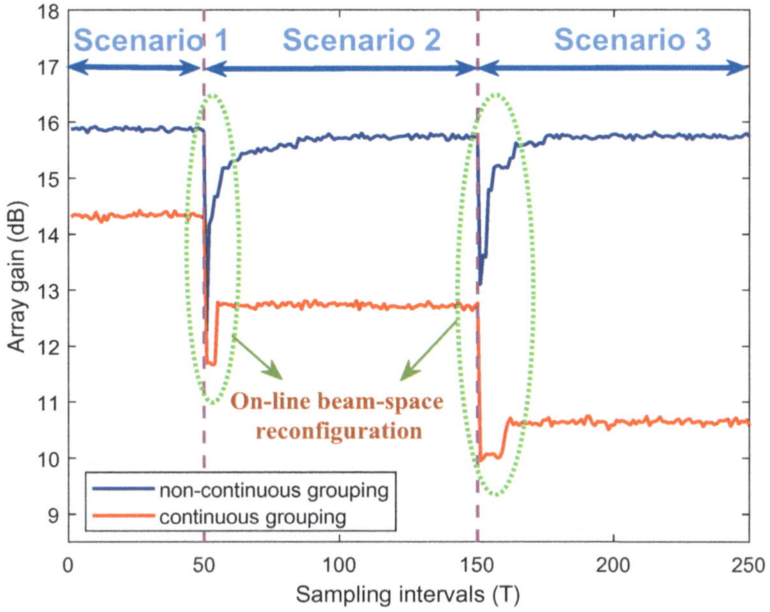

Fig. 5.34 Performance of Algorithm 2 in dynamic environment

$-7°$, $7°$, $15°$, $20°$, $80°$ and $-80°$, $-17°$, $-12°$, $-7°$, $7°$, $12°$, $60°$ for scenario 3. The INRs of all interferences in three scenarios are 40 dB and the desired signal remains unchanged (0° and 0dB). We examine both cases of continuous and non-continuous grouping and the results are given in Fig. 5.34. For each inner iteration, we sample $T = 1000$ snapshots. It can be seen from Fig. 5.34 that when the environment changes at instants $50T$ and $150T$, both models can quickly adapt to the new environment by on-line beam reconfiguration. After convergence, the AG of sparse beam-space beamformers in the non-continuous grouping model can reach more than 15.7 dB.

5.4 Summary

In this chapter, we introduced the cognitive-driven optimization of sparse sensing.

In Sect. 5.2, a complementary sparse array switching (RCAS) strategy was proposed for adaptive sparse array beamformer design, which aimed to swiftly adapt intertwined array configuration and excitation weights in accordance to dynamic environment. The work in Chap. 4 usually assumed either known or estimated environmental information, which was regarded as an impediment to practical implementation. The RCAS works in two steps. First, a set of deterministic complementary sparse arrays, all with good quiescent beampatterns, was designed and a full data collection was conducted by switching among them. Then, the adaptive sparse array was configured for the specific environment, based on the data collected in the first step. Deterministic and adaptive sparse arrays in both design steps were restricted to regularized antenna switching for increased practicability. The RCAS was devised as an exclusive cardinality-constrained optimization and an iterative algorithm was proposed to solve it effectively. We conducted a rigorous theoretical analysis and proved that the proposed algorithm is an equivalent transformation to the original cardinality-constrained optimization. Extensive simulation results validated the effectiveness of proposed RCAS strategy.

In Sect. 5.3, we proposed a novel cognitive sparse beamformer design in the framework of regularized antenna switching. The proposed beamformer optimization did not require any priori environmental information and was capable of perceiving the environment from the real data and swiftly adapting entwined beamformer geometry and excitation weights according to environmental dynamics through a "perception-action" cycle. Specifically, the proposed beamformer replaced one antenna or beam with other candidates in the same group in the metric of maximizing the array gain iteratively. To achieve efficient on-line beamformer reconfiguration, we derived closed-form formulas to quantify the array gain variation in both metrics of MPDR and LCMP. Extensive numerical results manifested the effectiveness of the proposed cognitive sparse beamformers, especially in dynamic environment.

Chapter 6
Sparse Sensing for MIMO Array Radar

Multi-Input Multi-Output (MIMO) array radar is a new type of radar system that adopts waveform diversity technology at the transmitter. In essence, different transmit antennas can transmit differed orthogonal or partially correlated signals. In this regard, MIMO radar offers more degrees of freedom (DoFs) compared with the phased array counterparts. This provides opportunities for improved radar functions, such as target detection, beamforming and direction of arrival (DOA) estimation. In the previous chapters, sparse receiver array design was delineated, which enhances the radar performance by fully utilizing the spatial DoFs and reduce hardware cost. Recently, with the increased demands of cost saving and improved performance, the optimization of sparse transceiver for MIMO radar has become particularly important. In this chapter, we examine the sparse optimization of MIMO radar transceiver for different tasks. In Sect. 6.1, we briefly introduce the principle of MIMO array transceiver and review the state of the art in sparse array design. In Sect. 6.2, the sparse MIMO array transceiver design for enhanced adaptive beamforming is presented under the assumption of known environmental conditions. We examine the active sparse array design enabling the maximum signal to interference plus noise ratio (MaxSINR) beamforming at the MIMO radar receiver through successive convex approximation (SCA) incorporating the two dimensional group sparsity promoting regularization. In Sect. 6.3, the cognitive-driven optimization of sparse array transceiver for MIMO radar beamforming is introduced, which further eliminates the prerequisite of prior information. We propose a cognitive-driven MIMO array design where both the beamforming weights and the transceiver configuration are adaptively and concurrently optimized in dynamic operating environment via a "perception-action" cycle. Finally, concluding remarks are provided in Sect. 6.4.

X. Wang et al., *Sparse Sensing and Sparsity Sensed in Multi-sensor Array Applications*, https://doi.org/10.1007/978-981-99-9558-5_6

6.1 Introduction

In the early 21st century, around 2003 and 2004, the idea of MIMO system and MIMO signal processing was introduced into radar target detection, which results in a new system, i.e., MIMO radar inspired from the MIMO wireless communication system. Unlike a standard phased-array radar, which transmits scaled versions of a single waveform, MIMO radar simultaneously transmits diverse (possibly independent or partially correlated) waveforms via multiple antennas and utilizes multiple antennas to receive the reflected signals as well. Like MIMO communications, MIMO radar offers a new paradigm for signal processing research. MIMO radar possesses significant potentials for fading mitigation, resolution enhancement, and interference and jamming suppression. Fully exploiting these potentials can result in significantly improved target detection, parameter estimation, as well as target tracking and recognition performance [55, 220–223].

There are two basic regimes of MIMO array model considered in the current literature. In the first regime, the transmit array and the receive array are broadly spaced, providing independent scattering responses for each antenna pairing, sometimes referred to as distributed MIMO radar. The distributed MIMO radar, different from the conventional multistatic radar system, enables each transmit antenna to detect the radar target from different directions by utilizing spatial diversity and increasing the inter-element spacing of the transmit array. The received echo signal can be considered as the superposition of independent fading signals. Combining these echo signals from different independent observation paths can obtain an approximately constant radar cross section (RCS) of the target, so as to suppress the RCS flicker of the target and obtain a large spatial diversity gain. In the second regime, the transmit array and the receive array are closely spaced so that the target is in the far field of the transmit-receive array, sometimes referred to as coherent MIMO radar. In terms of transmit signals, phased array radar can be regarded as a special case of colocated MIMO radar.

In this chapter, we consider the colocated MIMO array, which can significantly improve angle estimation performance. The estimation performance improvement can be dramatic when optimized sparse arrays are used. In some sense, the performance of the MIMO systems can be characterized by a virtual array constructed by the convolution of the locations of the transmit and receive antenna locations. Consequently, a filled virtual array can be constructed by using sparse constituent arrays. In principle, this virtual array can be much larger than the array of an equivalent traditional system; thus, the MIMO system will have much better intrinsic resolution.

Waveform diversity of MIMO array radar enables different transmit waveforms from different antennas. The transmit waveform of the colocated MIMO array can be divided into two categories: orthogonal and partially correlated. When the transmit waveforms are orthogonal, the transmit beampattern is omni-directional. In this case, the virtual aperture expansion can be realized by signal separation at the receive end, so as to obtain high angular resolution. When the transmit waveforms are partially correlated, the transmit beampattern can be optimized to any desired shape and direction. In this case, transmit beampatterns can be designed so that the transmit energy can be reasonably allocated to the target area of interest.

Consider a colocated MIMO narrow-band radar system equipped with a ULA composed of N_t antennas and N_r receivers. The system transmits N_t non-coherent waveforms, denoted by $\mathbf{S} = [\mathbf{s}(1), \ldots, \mathbf{s}(N)] \in \mathbb{C}^{N_t \times N}$ with N being the snapshot number in each transmit pulse. Assume that there is a point target locates in the direction of θ in the far field of the transmitting and receiving arrays. Then, the received signal matrix in a single pulse can be described by,

$$\mathbf{Y} = \beta \mathbf{a}_R(\theta)\mathbf{a}_T^T(\theta)\mathbf{S} + \mathbf{N}, \tag{6.1}$$

where $\mathbf{a}_T(\theta)$ and $\mathbf{a}_R(\theta)$ are the transmit and receive steering vector towards the direction θ, respectively, and $\mathbf{N}$ represents the complex noise and interferences, whose each column is independent and its elements follows the identically distributed complex Gaussian model.

After performing the pulse compression using $\mathbf{S}^H(\mathbf{S}\mathbf{S}^H)^{-1/2}$ on the data matrix $\mathbf{Y}$, the output data matrix is given by,

$$\mathbf{Z} = \tilde{\beta} \mathbf{a}_R(\theta)\mathbf{a}_T^T(\theta)\hat{\mathbf{R}}^{1/2} + \tilde{\mathbf{N}}, \tag{6.2}$$

where $\tilde{\beta} = \sqrt{N}\beta$, $\hat{\mathbf{R}} = \frac{1}{N}\mathbf{S}\mathbf{S}^H$ is the signal covariance matrix, and $\tilde{\mathbf{N}}$ is the corresponding interference-pluse-noise data matrix. When the transmit waveforms are orthogonal to each other, the signal covariance matrix $\hat{\mathbf{R}} = \mathbf{I}$.

Vectorizing the above received data matrix $\mathbf{Z}$ yields that,

$$\mathbf{z} = \text{vec}(\mathbf{Z}) = \tilde{\beta}\mathbf{A}\mathbf{v}(\theta) + \mathbf{z}_n, \tag{6.3}$$

where $\mathbf{z}_n = \text{vec}(\tilde{\mathbf{N}})$, $\mathbf{A} = \hat{\mathbf{R}}^{1/2} \otimes \mathbf{I}_{N_r}$ and $\mathbf{v}(\theta) = \mathbf{a}_T(\theta) \otimes \mathbf{a}_R(\theta) \in \mathbb{C}^{N_t N_r \times 1}$ denotes the virtual steering vector coming from the transmit and receive arrays towards the direction θ.

It is seen from (6.3) that the obtained vector $\mathbf{z}$ can be regarded as the received signal of $N_t \times N_r$ virtual antennas composed of the transmit and receive arrays, as shown in Fig. 6.1. This virtual array is termed as sum-coarray, which enables the

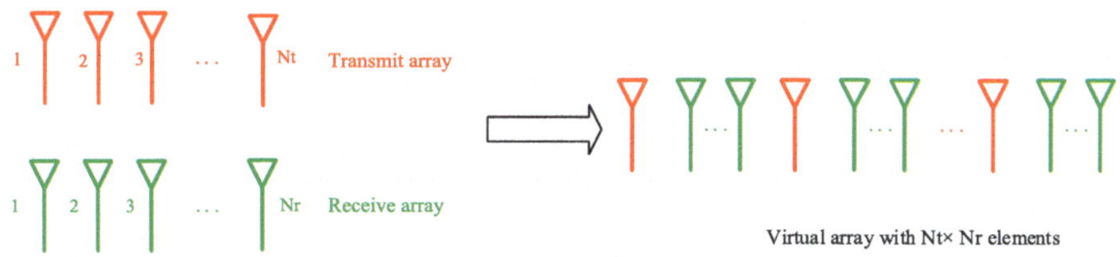

Fig. 6.1 Construction of MIMO virtual array

signal processing from the perspective of co-array. Equation (6.3) mathematically implies that MIMO radar own more spatial degrees of freedom than traditional phase-array radars and thus can enhance the parameter identification ability owing to the larger aperture and more antenna elements.

The demand for higher angular resolution in radar and communications applications has spurred the use of larger arrays. Yet, increasing the number of antenna elements leads to an increase in hardware cost and complexity, as well as higher power consumption. Therefore, various strategies have been developed to exploit the redundancy that is inherent to fully populated arrays in order to reduce the number of radio frequency (RF) front-ends while preserving the desirable array performance. Thus, array reconfigurability through antenna selection was exploited to mitigate interference and maximize the signal to interference plus noise ratio (SINR) [187, 200]. Sparse array design, aided by emerging fast antenna switching technologies, can lower the overall system cost by reducing the number of expensive front-end processing channels.

Benefiting from the larger aperture generated by the virtual arrays of MIMO radar, higher estimation accuracy and better beamforming performance can be achieved. If the sparse configuration of the MIMO transceiver is optimized, the performance of MIMO radar can be further enhanced. As mentioned above, the conventional MIMO radar has a sparse transceiver array. The inter-element spacing of the receive array is half wavelength, whereas that of the transmit array is multiple wavelengths. Hence, the MIMO sum coarray is a compact ULA with a large virtual aperture that enables high spatial resolution [224–226]. This configuration, however, may not render an optimum beamforming in terms of maximizing the output SINR (MaxSINR) for a given environment [227, 228]. Sparse MIMO array beamforming design can seek to optimize the transmit sensor locations and the corresponding waveform correlation matrix [229, 230]. The optimal design of sparse MIMO transceiver can be divided into two types: structured and unstructured sparse arrays.

The environment-independent, structured sparse array design seeks to increase the number of the spatial autocorrelation lags and maximize the contiguous segment of the coarray aperture for a limited number of sensors. The main task, therein, is to enable DOA estimation involving more sources than the physical sensors [231–234]. For beamforming applications, the environment-independent design criteria typically strives to achieve desirable beampattern characteristics such as broad main lobe and minimum sidelobe levels and frequency invariant beampattern for wide-band design [235]. However, the environment-dependent, unstructured sparse receive beamformer, which strives to achieve MaxSINR, can potentially provide sparse configurations that improve target detection and estimation accuracy for the operating environment [187, 202, 236–239].

In Sect. 6.2, the unstructured sparse MIMO array transceiver design for MaxSINR with known environmental information is discussed. Non-uniform sparse arrays have emerged as an effective technology which can be deployed in various active and passive sensing modalities, such as acoustics and radio frequency (RF) applications, including GPS. Fundamentally, sparse array design seeks optimum system performance, under both noise and interference, while being cognizant of the limitations on cost and aperture. Within the RF beamforming paradigm, maximizing the signal to interference plus noise ratio (SINR) amounts to concurrently optimizing the array configuration and beamforming weights. Equivalently, the problem can be cast as continuously selecting the best antenna locations to deliver MaxSINR under dynamic environment. In lieu of constantly placing antennas to new positions, a more practical and feasible approach is to have a uniform full array and then switch among antennas with a fixed number of front-end chains, yielding the MaxSINR sparse beamformer for a given environment.

In Sect. 6.3, we consider the cognitive-driven optimization of sparse MIMO array transceiver for adaptive beamforming in dynamic environment. When the environment information is unavailable, we need to learn the environmental information automatically from the received data. MIMO radars enable superior capabilities compared with standard phased arrays radar. Cognitive radar continuously interacts with the environment and update the radar parameters through the acquired knowledge [207]. In MIMO radar, a "perception-action" cognition approach typically entails adaptive optimizations of the transmit waveforms and power allocation [240–242]. In addition to these parameters, array reconfiguration (hardware) can significantly improve system performance beyond that achieved with fixed antenna positions [243, 244].

6.2 Sparse MIMO Transceiver Design for MaxSINR

Sparse array design aided by emerging fast switching technologies can lower the overall system overhead by reducing the number of expensive transceiver chains. In this section, we assume a perfect knowledge of the environment and examine the active sparse array design enabling the MaxSINR beamforming at the MIMO radar receiver. The proposed approach entails an entwined design, i.e., jointly selecting the optimum transmit and receive sensor locations for accomplishing MaxSINR receive beamforming. Specifically, we consider a colocated MIMO radar platform with orthogonal transmit waveforms, and examine antenna selections at the transmit and receive arrays. The optimum active sparse array transceiver design problem is formulated as successive convex approximation (SCA) alongside the two-dimensional group sparsity promoting regularization. Several examples are provided to demonstrate the effectiveness of the proposed approach in utilizing the given transmit/receive array aperture and degrees of freedom for achieving MaxSINR beamforming.

The MaxSINR MIMO beamformer design jointly selects the optimum transmit and receive sensor locations for implementing an efficient receive beamformer. We assume that the transmit sensors emit orthogonal waveforms. The MIMO radar implementing sensor-based orthogonal transmit waveforms does not benefit from coherent transmit processing gain achieved when using directional beamforming. In this context, the proposed approach is fundamentally different from the parallel sparse array MIMO beamforming designs, which primarily rely on optimizing the transmit sensor locations and the corresponding correlation matrix of the transmit waveform sequence [245–247]. The proposed approaches, therein, maximize the transmitted signal power towards the perspective target locations while minimizing the cross-correlation of the target returns to achieve efficient receive beamforming characteristics. Moreover, existing design schemes essentially pursue an uncoupled transmit/receive design, which is in contrast to the approach considered in this chapter, which incorporates sparse receiver design in conjunction with transmit array optimization.

The proposed transceiver beamforming design problem, in essence, amounts to configuring the transmit/receive array with the corresponding optimum receiver beamforming weights. To this end, we pursue a data dependent approach that jointly optimizes the beamforming sensor positions and weights. The optimization requires the knowledge of the perspective desired source locations and some prior knowledge of the interfering signals as well. We optimally select M_t from M transmitters and N_r from N receivers. Such selection is a binary optimization problem, and is NP-hard. In order to avoid extensive computations associated with enumeration of all possible array configurations, we apply convex relaxation. The design problem is posed as SCA with two-dimensional reweighted mixed $l_{1,2}$-norm penalization to jointly invoke sparsity in the transmit and receive dimensions.

6.2.1 Problem Formulation

We consider a MIMO radar with M transmitters and N receivers illuminating the scene through omni-directional beampattern. This is typically achieved through emissions of orthogonal waveforms across the transmit array. Specifically, we pursue a co-located MIMO arrangement with the transmit and receive arrays in close vicinity. As such, a far-field source is presented by equal angles of departure and arrival. Consider K target sources arriving from $\{\theta_{s,k}\}$. The received baseband data $\mathbf{x}(n) \in \mathbb{C}^{MN\times 1}$ after matched filtering at the N element uniformly spaced receiver array at the time instant n is given by,

$$\mathbf{x}(n) = \sum_{k=1}^{K} s_k(n)\mathbf{b}(\theta_{s,k}) + \sum_{l=1}^{L_c} c_l(n)\mathbf{b}(\theta_{j,l}) + \sum_{q=1}^{Q} \mathbf{i}_q(n) + \mathbf{v}(n), \tag{6.4}$$

where, $s_k(n) \in \mathbb{C}$ is the kth reflected target signal. The extended steering vector $\mathbf{b}(\theta)$ of the virtual array is $\mathbf{b}(\theta) = \mathbf{a}_T(\theta) \otimes \mathbf{a}_R(\theta)$. In the case of uniform transmit and receive linear arrays with a respective inter-element spacing of d_t and d_r, the transmit and receive steering vectors are given by,

$$\mathbf{a}_T(\theta) = [1 \quad e^{j2\pi(d_t/\lambda)cos\theta} \quad \ldots \quad e^{j2\pi(M-1)(d_t/\lambda)cos\theta}]^T, \tag{6.5}$$

and

$$\mathbf{a}_R(\theta) = [1 \quad e^{j2\pi(d_r/\lambda)cos\theta} \quad \ldots \quad e^{j2\pi(N-1)(d_r/\lambda)cos\theta}]^T. \tag{6.6}$$

The variance of additive Gaussian noise $\mathbf{v}(n) \in \mathbb{C}^{MN \times 1}$ is σ_v^2 at the receiver output. There are L_c interferences $c_l(n)$ mimicking target reflected signal and Q narrowband interferences $\mathbf{i}_q(n) \in \mathbb{C}^{MN \times 1}$. The latter is defined as the Kronecker product of the receiver steering vector $\mathbf{a}_R(\theta_{i,q})$ and the matched filtering output of the interference $\mathbf{j}_q(n)$ such that $\mathbf{i}_q(n) = \mathbf{j}_q(n) \otimes \mathbf{a}_R(\theta_{i,q})$. The received data vector $\mathbf{x}(n)$ is then linearly combined to maximize the output SINR. The output signal $y(n)$ of the optimum beamformer for MaxSINR is given by [185],

$$y(n) = \mathbf{w}^H \mathbf{x}(n), \tag{6.7}$$

where $\mathbf{w}$ is the beamformer weight at the receiver. The optimal solution $\mathbf{w}_o$ can be obtained by solving the optimization problem that seeks to minimize the interference power at the receiver output while preserving the desired signal. The constraint minimization problem can be cast as,

$$\begin{aligned} \underset{\mathbf{w} \in \mathbb{C}^{MN}}{\text{minimize}} \quad & \mathbf{w}^H \mathbf{R}_{\mathbf{x}} \mathbf{w}, \\ \text{s.t.} \quad & \mathbf{w}^H \mathbf{R}_s \mathbf{w} = 1, \end{aligned} \tag{6.8}$$

where the source correlation matrix is $\mathbf{R}_s = \sum_{k=1}^{K} \sigma_{s,k}^2 \, \mathbf{b}(\theta_{s,k}) \mathbf{b}^H(\theta_{s,k})$, with $\sigma_{s,k}^2 = E\{s_k(n) s_k^H(n)\}$ denoting the average received power from the kth target return. The data correlation matrix, $\mathbf{R}_{\mathbf{x}} \approx (1/T) \mathbf{x}(n) \mathbf{x}(n)^H$, is directly estimated from the T received data snapshots. The solution to the optimum weights in (6.8) is given by $\mathbf{w}_o = \mathscr{P}\{\mathbf{R}_{\mathbf{x}}^{-1} \mathbf{R}_s\}$, with the operator $\mathscr{P}\{.\}$ representing the principal eigenvector of the input matrix. This optimum solution yields the MaxSINR, SINR_o, given by [185],

$$\text{SINR}_o = \Lambda_{max}\{\mathbf{R}_x^{-1} \mathbf{R}_s\}, \tag{6.9}$$

which is the MaxSINR given by the maximum eigenvalue (Λ_{max}) of the product of the two matrices, the inverse of interference plus noise correlation matrix and the desired source correlation matrix. It is clear that the resulting solution holds irrespective of the array configuration, whether the array is uniform or sparse. For the latter, the performance of MaxSINR beamformer is intrinsically tied to the array configuration. The sparse optimization of the above formulation is explained in Sect. 6.2.2.

6.2.2 Sparse Array Transceiver Design

The separate sensor selection problem for a joint transmit and receive design is a combinatorial optimization problem and can't be solved in polynomial time. We formulate the sparse array design problem by exploiting the structure of the received signal model and solve it by applying sequential convex approximation. To exploit the sparse structure of the joint transmit and receive sensor selection, we introduce a two dimensional $l_{1,2}$-mixed norm regularization to recover group sparse solutions. One dimension pertains to the transmitter sparsity, whereas the other sparsity pattern is associated with the receiver side. Moreover, this two-dimensional sparsity pattern is entwined and coupled, meaning that when one transmitter is discarded, all N receiving data pertaining to this transmitter is zero. Similarly, the receiver is discarded only when its received data corresponding to all the M transmitters is zero. The structure of the optimal sparse beamforming weight vector $\mathbf{w} \in \mathbb{C}^{MN}$ is elucidated in (6.10), where $\mathbf{w}(i,\ j)$ denotes the entries of $\mathbf{w}$ indexed from i to j. Also, $\checkmark$ denotes a sensor location activated, and $\times$ denotes a sensor not activated.

$$\overbrace{\underbrace{\begin{pmatrix}\checkmark\\ \times\\ \times\\ \checkmark\\ \vdots\\ \times\end{pmatrix}}_{\text{(Tx 1 active)}}}^{\mathbf{w}(1,N)}\ \overbrace{\underbrace{\begin{pmatrix}\checkmark\\ \times\\ \times\\ \checkmark\\ \vdots\\ \times\end{pmatrix}}_{\text{(Tx 2 active)}}}^{\mathbf{w}(N+1,2N)}\ \overbrace{\underbrace{\begin{pmatrix}\times\\ \times\\ \times\\ \times\\ \vdots\\ \times\end{pmatrix}}_{\text{(Tx 3 inactive)}}}^{\mathbf{w}(2N+1,3N)}\ \cdots\ \overbrace{\underbrace{\begin{pmatrix}\checkmark\\ \times\\ \times\\ \checkmark\\ \vdots\\ \times\end{pmatrix}}_{\text{(Tx } M \text{ active)}}}^{\mathbf{w}(N(M-1)+1,MN)} \tag{6.10}$$

In (6.10), each column vector denotes the receive beamformer weights for a fixed transmit location. It is noted that the optimal sparse beamformer, corresponding to the active transmit and receive locations, follows a group sparse structure along the transmit and receive dimensions (the consecutive N sensors are discarded vertically or the consecutive M sensors are discarded horizontally. It is evident that the missing transmit sensor at position 3, for example, translates to the sparsity along all the corresponding N entries of $\mathbf{w}$ (N consecutive $\times$ vertically in (6.10)). Similarly, the group sparsity is also invoked across the received signals corresponding to all transmitters (M consecutive $\times$ horizontally in (6.10)).

6.2.2.1 Group Sparse Solutions via SCA

The problem in (6.8) can equivalently be rewritten by swapping the objective and constraint functions as follows,

$$\begin{aligned} \underset{\mathbf{w}\in\mathbb{C}^{MN}}{\text{minimize}} \quad & \mathbf{w}^H\bar{\mathbf{R}}_s\mathbf{w} \\ \text{s.t.} \quad & \mathbf{w}^H\mathbf{R}_{\mathbf{x}}\mathbf{w} \le 1 \end{aligned} \tag{6.11}$$

where $\bar{\mathbf{R}}_s = -\mathbf{R}_s$. The sparse MIMO configuration of uniformly spaced receivers and transmitters with a respective inter-element spacing of d_r and $d_t = Nd_r$ is employed for data collection. The covariance matrix $\mathbf{R}_{\mathbf{x}}$ of the full receive virtual array can be then obtained by the matrix completion method.

The beamforming weight vectors are generally complex valued, whereas the quadratic functions are real. This observation allows expressing the problem with only real variables which is typically accomplished by replacing the correlation matrix $\bar{\mathbf{R}}_s$ by $\tilde{\mathbf{R}}_s$ and concatenating the beamforming weight vector accordingly,

$$\tilde{\mathbf{R}}_s = \begin{bmatrix} \text{real}(\bar{\mathbf{R}}_s) & -\text{imag}(\bar{\mathbf{R}}_s) \\ \text{imag}(\bar{\mathbf{R}}_s) & \text{real}(\bar{\mathbf{R}}_s) \end{bmatrix}, \tag{6.12}$$

$$\tilde{\mathbf{w}} = \begin{bmatrix} \text{real}(\mathbf{w}) \\ \text{imag}(\mathbf{w}) \end{bmatrix}$$

Similarly, the received data correlation matrix $\mathbf{R_x}$ is replaced by $\tilde{\mathbf{R}}_\mathbf{x}$. The problem in (6.11) then becomes,

$$\begin{aligned} \underset{\tilde{\mathbf{w}} \in \mathbb{R}^{2MN}}{\text{minimize}} \quad & \tilde{\mathbf{w}}' \tilde{\mathbf{R}}_s \tilde{\mathbf{w}}, \\ \text{s.t.} \quad & \tilde{\mathbf{w}}' \tilde{\mathbf{R}}_\mathbf{x} \tilde{\mathbf{w}} \leq 1, \end{aligned} \tag{6.13}$$

where $'$ denotes transpose operation. After expressing the constraint in terms of real variables, we convexify the objective function by utilizing the first order approximation iteratively,

$$\begin{aligned} \underset{\tilde{\mathbf{w}} \in \mathbb{R}^{2MN}}{\text{minimize}} \quad & \mathbf{m}^{(k)\prime} \tilde{\mathbf{w}} + b^{(k)}, \\ \text{s.t.} \quad & \tilde{\mathbf{w}}' \tilde{\mathbf{R}}_\mathbf{x} \tilde{\mathbf{w}} \leq 1, \end{aligned} \tag{6.14}$$

where, $\mathbf{m}$ and b, updated at the $k+1$th iteration, are given by $\mathbf{m}^{(k+1)} = 2\tilde{\mathbf{R}}_s \tilde{\mathbf{w}}^{(k)}$, $b^{(k+1)} = -\tilde{\mathbf{w}}^{(k)\prime} \tilde{\mathbf{R}}_s \tilde{\mathbf{w}}^{(k)}$, respectively. Finally, to invoke sparsity in the beamforming weight vector, the re-weighted mixed $l_{1,2}$-norm is adopted primarily for promoting group sparsity.

$$\underset{\tilde{\mathbf{w}}, \mathbf{c}, \mathbf{r}}{\text{minimize}} \quad \mathbf{m}^{(\mathbf{k})\prime} \tilde{\mathbf{w}} + b^{(k)} + \alpha(\mathbf{p}' \mathbf{c}) + \beta(\mathbf{q}' \mathbf{r}) \tag{6.15}$$

$$\text{s.t.} \quad \tilde{\mathbf{w}}' \tilde{\mathbf{R}}_\mathbf{x} \tilde{\mathbf{w}} \leq 1, \tag{6.15a}$$

$$||\mathbf{P}_i \odot \tilde{\mathbf{w}}||_2 \leq c_i, \tag{6.15b}$$

$$0 \leq c_i \leq 1, \quad i = 1, \ldots, M \tag{6.15c}$$

$$||\mathbf{Q}_j \odot \tilde{\mathbf{w}}||_2 \leq r_j, \tag{6.15d}$$

$$0 \leq r_j \leq 1, \quad j = 1, \ldots, N \tag{6.15e}$$

$$\mathbf{1}_M' \mathbf{c} = M_t, \tag{6.15f}$$

$$\mathbf{1}_N' \mathbf{r} = N_r, \tag{6.15g}$$

and

$$\mathbf{P}_i = [\overbrace{0...0...0}^{\substack{N\ elements \\ 1st\ group}} \ldots \overbrace{1...1...1}^{\substack{N\ elements \\ ith\ group}} \ldots \overbrace{1...1...1}^{\substack{N\ elements \\ (M+i)th\ group}} \ldots \overbrace{0...0...0}^{\substack{N\ elements \\ 2Mth\ group}}]', \tag{6.16}$$

$$\mathbf{Q}_j = [\overbrace{\underbrace{0\ldots0}_{(j-1)\ 0s}\ 1\ 0\ldots0}^{\substack{N\ elements\ of \\ the\ 1st\ group}} \ldots \overbrace{\underbrace{0\ldots0}_{(j-1)\ 0s}\ 1\ 0\ldots0}^{\substack{N\ elements\ of \\ the\ 2Mth\ group}}]'. \tag{6.17}$$

Here, $\odot$ means the element-wise product, $\mathbf{c} \in [0, 1]^M$ and $\mathbf{r} \in [0, 1]^N$ are two auxiliary selection vectors taking values between 0 and 1, $\mathbf{P}_i \in \{0, 1\}^{2MN}$ in (6.15b) is the transmission selection vector, which is used to select the real and imaginary parts of the N weights corresponding to the ith transmitter, as shown in Eq. (6.16). Equations (6.15c) and (6.15f) indicate that we can select M_t transmitters at most. Similarly, $\mathbf{Q}_j$ in (6.15d) is the receiver selection vector, as shown in (6.17). Equations (6.15e) and (6.15g) indicate that we can select N_r receivers at most. The two parameters α and β are used to control the sparsity of the transmitters and the receivers, $\mathbf{p}$ and $\mathbf{q}$ are the reweighting coefficient vectors for the transmitters and receivers respectively and the detailed update method is given as follows.

6.2.2.2 Reweighting Update

A common method of updating the reweighting coefficient is to take the reciprocal of $|\mathbf{w}|$ [213]. However, this can't control the number of elements to be selected. Thus, similar to [219], in order to make the number of selected antennas controllable, we update the weights using the following formula,

$$p_i^{(k)} = \frac{1 - c_i^{(k)}}{1 - e^{-\beta_0 c_i^{(k)}} + \epsilon} - (\frac{1}{\epsilon})(c_i^{(k)})^{\alpha_0} \tag{6.18}$$

$$q_i^{(k)} = \frac{1 - r_i^{(k)}}{1 - e^{-\beta_0 r_i^{(k)}} + \epsilon} - (\frac{1}{\epsilon})(r_i^{(k)})^{\alpha_0} \tag{6.19}$$

Here, α_0 and β_0 are two parameters that control the shape of the curve, and the parameter ϵ avoids the unwanted explosive case. The essence of this re-weight update method is to take a large positive penalty for a entry close to zero and a small negative reward for the entry close to 1. As a result, the entries of the two selection vectors $\mathbf{c}$ and $\mathbf{r}$ tend to be either 0 or 1. Through iterative regression, we can select M_t transmitters and N_r receivers. In addition, by controlling the values of α and β, we can obtain different sparsities of transmitters and receivers [248].

6.2.3 Simulations

In this subsection, we demonstrate the effectiveness of our proposed sparse MIMO radar from the perspective of output SINR. We compare the performance of the optimal MIMO array transceiver configured according to the proposed algorithm with randomly configured MIMO arrays. In practice, the interference may be caused by co-existence in the same bandwidth or being deliberately positioned at some angles transmitting the same waveform as targets of interest. We consider both kinds of interferences.

6.2.3.1 Example 1

In this example, we fix the direction of the interferences, and change the arrival angle of the target from 0° to 90°. We consider a full uniform linear MIMO array consisting of 8 transmitters (M = 8) and 8 receivers (N = 8). We select $M_t = 4$ transmitters and $N_r = 4$ receivers. For the receiver, we set the minimum spacing of the sensors to $\lambda/2$, while for the transmitter, we set the minimum spacing of the sensors to $N\lambda/2$. Suppose there are two interferences with an interference to noise-ratio (INR) of 13dB and arrival angles of $\theta_q = [40°, 70°]$. The SNR of the desired signal is fixed at 20dB. We simulate co-existing interferences and deliberate interferences. The output SINR versus target angle is shown in Fig. 6.2. In this figure, case 1 corresponds to two deliberate interferences, whereas case 2 represents two co-existing interferences. It can be seen that, in both cases, the optimal sparse MIMO arrays exhibit better performance than randomly selected sparse arrays. Figure 6.3, depicts the configuration of the optimal MIMO array for a target at 80°.

6.2.3.2 Example 2

In this example, we consider the scenario where the interferences are spatially close to the target which causes a great adverse impact on the array performance. We change the angle of the target from 0° to 90°. For each target angle, two interferences are generated from the proximity of ±5° away from the target. We consider a full uniform linear MIMO array consisting of 20 transmitters (M = 20) and 20 receivers (N = 20). We select 5 transmitters and 5 receivers to compose the sparse MIMO array. The other simulation parameters remain the same as those in example 1. Again, we plot the output SINR versus target angle, as shown in Fig. 6.4. In case 1, the two interferences are deliberate, whereas in case 2, both interferences are co-existing. It can be observed that when the interferences are spatially close to the target, the proposed optimal MIMO sparse array exhibits more superiority than randomly configured sparse MIMO arrays.

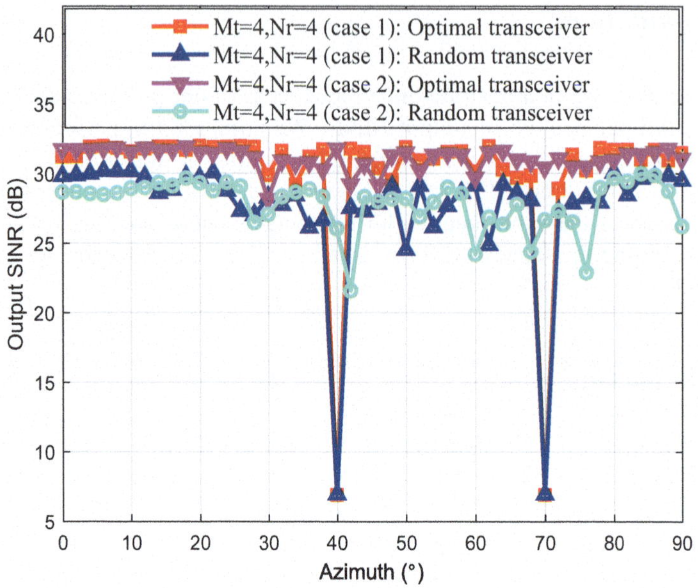

Fig. 6.2 Relationship between output SINR and target angle with the directions of interferences being fixed

Fig. 6.3 Optimal MIMO transceiver configuration when the target is from 80°

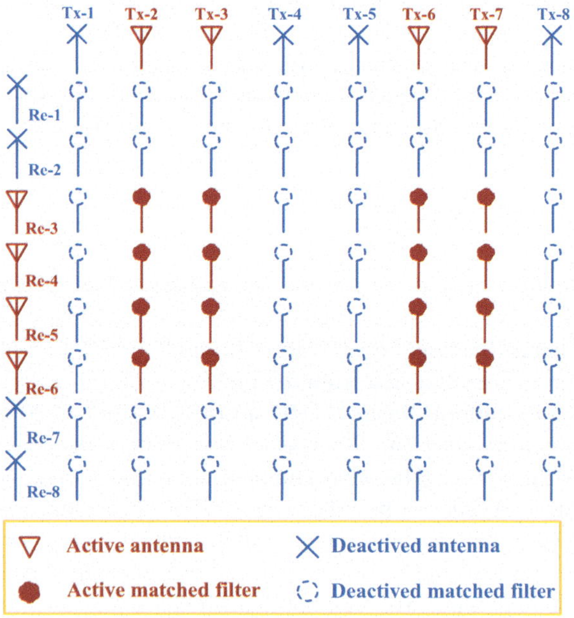

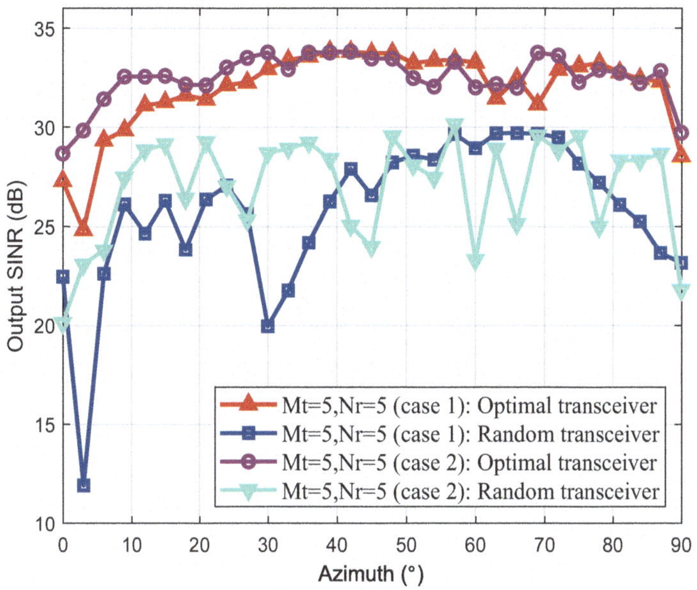

Fig. 6.4 Relationship between the output SINR and the target angle when the interferences are closed to the target

6.3 Cognitive-Driven Optimization of Sparse MIMO Beamforming

In the previous section, the optimum sparse MIMO array transceiver design is performed assuming that the environmental information is completely known. This may pose challenging problems for practical implementation, especially in a dynamic operating environment. In such environment, a cognitive MIMO radar would be capable of adjusting system parameters adaptively by sensing and learning. Beamforming performance of MIMO radar is guided by both beamforming weight coefficients and the transceiver configuration. In this section, we examine a cognitive-driven MIMO array design where both the beamforming weights and the transceiver configuration are adaptively and concurrently optimized under different environmental conditions. The perception-action cycle involves data collection of full virtual array, covariance reconstruction and joint design of the transmit and receive arrays by antenna selections. The optimal transceiver array design is realized by promoting two-dimensional group sparsity via iteratively minimizing reweighted mixed $l_{2,1}$-norm, with constraints imposed on transceiver antenna spacing for proper transmit/receive isolation. Simulations are provided to demonstrate the "perception-action" capability of the cognitive sparse MIMO array in achieving enhanced beamforming and anti-jamming in dynamic target and interference environment.

MIMO radars enable superior capabilities compared with standard phased arrays radar. Cognitive radar continuously interacts with the environment and update the radar parameters through the acquired knowledge [207]. In MIMO radar, "perception-action" cognition approach typically entails adaptive optimizations of the transmit waveforms and power allocation [240–242]. In addition to these parameters, array reconfiguration (hardware) can significantly improve system performance beyond that achieved with fixed antenna positions [243, 244]. Motivated by this fact, we propose a cognitive-driven sparse MIMO array design method that incorporates sparse transceiver array optimization. The schematic system diagram is shown in Fig. 6.5, where a fully-connected switching network is employed for cognitive array reconfiguration. The task is to continuously and cognitively select different subsets of antennas from a large array to deliver the best performance under time-varying environment.

Sparse array design, aided by emerging fast antenna switching technologies, can lower the overall system cost by reducing the number of expensive front-end processing channels. The conventional MIMO radar has a sparse transceiver array. The inter-element spacing of the receive array is half wavelength, whereas that of the transmit array is multiple wavelengths. Hence, the MIMO sum coarray is a compact uniform linear array (ULA) with a large virtual aperture that enables high spatial resolution [224–226]. This configuration, however, may not render an optimum beamforming in terms of MaxSINR [227, 228]. Sparse MIMO array beamforming design can achieve optimization of the transmit sensor locations and the corresponding waveform correlation matrix for synthesizing a desired beampattern [229, 230]. This approach, however, essentially pursues an decoupled transmit/receive design. In this chapter, we pursue the coupled design approach and seek optimum sparse transceiver, configuring the MaxSINR beamformer at the MIMO radar receiver. We assume a colocated MIMO radar platform with orthogonal transmit waveforms. The adaptive beamforming is implemented on the virtual array after matched filtering. In the proposed design, the transceiver array configuration and beamforming weights are concurrently optimized to seek the best output SINR performance. This requires a cognitive operation where the present sensed system information is utilized to determine the next system parameters.

In the first step of the proposed cognitive beamforming, the MIMO radar senses the environment by sequentially switching to different sets of antennas for data collection, based on which a full covariance matrix of the large virtual array is constructed. In the second step of learning, a constrained design optimization problem is formulated and solved to simultaneously provide the configurations of both the transmit and receive arrays associated with MaxSINR. The radar will then act by switching on the selected transmit and receive antennas, and applying the resultant optimum beamforming weights after matched filtering. It is noted that cycle is triggered each time the output SINR degrades from a certain threshold performance, thereby making antenna selection a cognitive operation. A reweighted mixed $l_{2,1}$-norm minimization is employed to promote a two-dimensional group sparsity for MIMO transceiver design. Moreover, for practical consideration, the optimum cog-

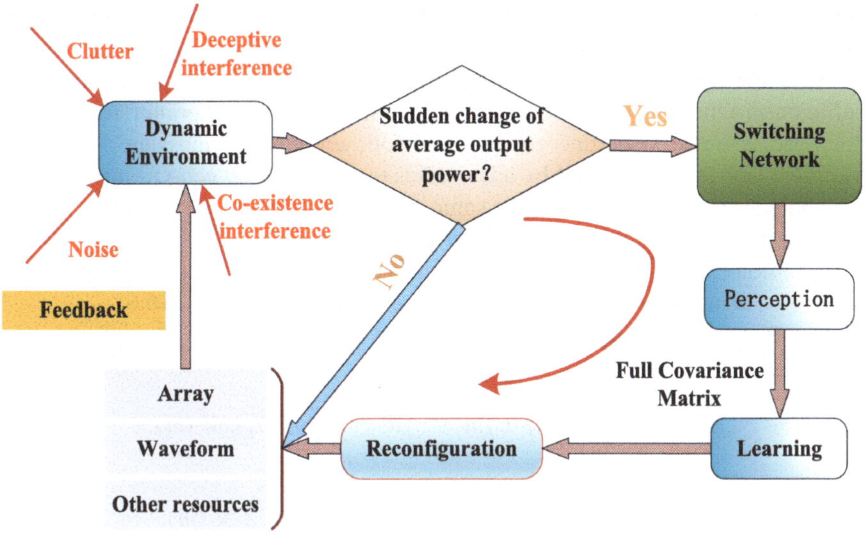

Fig. 6.5 Schematic diagram of cognitive MIMO Radar

nitive MIMO radar is designed under antenna isolation constraints to control the minimum spacing between transmitter and receiver antennas.

6.3.1 Full Covariance Construction

In the first step, the MIMO radar senses the RF environment by sequentially switching to different sets of sparse arrays. This action enables data collection and a follow-on construction of the full covariance matrix corresponding to the virtual transceiver after matched filtering. Consider a cognitive MIMO radar with a total of M antennas. Later, only M_t transmit antennas and N_r (in this section, we assume that $M_t = N_r$) receive antennas will be optimally selected from the M available antennas. The full covariance matrix corresponding to co-located M-antenna transmit and receive arrays is constructed as follows.

The received signal at the nth time instant is given by,

$$\mathbf{x}(n) = \eta_0(n)\mathbf{a}_R(\theta_0)\mathbf{a}_T^T(\theta_0)\mathbf{s}(n) + \sum_{l1=1}^{L_1} j_{l_1}(n)\mathbf{a}_R(\theta_{j,l_1}) + \sum_{l2=1}^{L_2} \eta_{l_2}(n)\mathbf{a}_R(\theta_{j,l_2})\mathbf{a}_T^T(\theta_{j,l_2})\mathbf{s}(n) + \mathbf{v}(n), \tag{6.20}$$

where $\mathbf{s}(n) \in \mathbb{C}^{M\times 1}$ represents the M orthogonal transmit signals, θ_0 denotes the target angle, $j_{l_1}(n)$ represents the l_1-th co-existing interference whose arrival angle relative to the receiver is θ_{j,l_1}. The variables $\eta_0(n)$ and $\eta_{l_2}(n)$ represent the reflection coefficients of the target and the l_2-th deceptive interference, or spoofer, respectively, which are assumed to be uncorrelated and follow complex Gaussian distribution. Deceptive interference is used by adversaries to mimic target echo to interfere with target detection and thus exhibits the same waveform with radar transmit signal. The arrival angle of the l_2th deceptive interference is θ_{j,l_2}, $\mathbf{v}(n) \in \mathbb{C}^{M\times 1}$ is additive white Gaussian noise. Also, $\mathbf{a}_T(\theta)$ and $\mathbf{a}_R(\theta)$ represent the steering vectors of the transmit and receive arrays which are defined in the same way as (6.5) and (6.6).

After matched filtering, we obtain the data matrix $\mathbf{Y}(m) \in \mathbb{C}^{M\times M}$ for the mth radar pulse.

$$\begin{aligned}\mathbf{Y}(m) &= \eta_0(m)\mathbf{a}_R(\theta_0)\mathbf{a}_T^T(\theta_0)\mathbf{R} + \sum_{l1=1}^{L_1} \mathbf{a}_R(\theta_{j,l_1})\mathbf{i}_{l_1}^T(m) \\ &+ \sum_{l2=1}^{L_2} \eta_{l_2}(m)\mathbf{a}_R(\theta_{j,l_2})\mathbf{a}_T^T(\theta_{j,l_2})\mathbf{R} + \mathbf{V}'(m),\end{aligned} \tag{6.21}$$

where $\mathbf{R} = \sigma_s^2\mathbf{I}$ is the auto-correlation matrix of M orthogonal transmit signals with equal power of σ_s^2, $\mathbf{i}_{l_1}(m) = \sum_n j_{l_1}(n)\mathbf{s}^*(n)$ is the output of matched filtering of the l_1th interference. Since the interference is not related to the transmit waveforms, $\mathbf{i}_{l_1}(m)$ still follows Gaussian distribution. We assume that the reflection coefficients obey the Swerling II target model, i.e., they change from one pulse repetition period to another. Vectorizing the data matrix $\mathbf{Y}(m)$, we generate the data vector received by the virtual array,

$$\begin{aligned}\mathbf{y}(m) &= \text{vec}(\mathbf{Y}(\mathbf{m})) \\ &= \eta_0(m)\sigma_s^2\mathbf{b}(\theta_0) + \sum_{l1=1}^{L_1} \mathbf{a}_R(\theta_{j,l_1}) \otimes \mathbf{i}_{l_1}(m) \\ &+ \sum_{l2=1}^{L_2} \eta_{l_2}(m)\sigma_s^2\mathbf{b}(\theta_{j,l_2}) + \mathbf{v}'(m),\end{aligned} \tag{6.22}$$

where the definition of $\mathbf{b}(\theta_i)$ is given in (6.4). The covariance matrix of the full virtual array is given by,

$$\mathbf{R}_x = E\left\{\mathbf{y}(m)\mathbf{y}^H(m)\right\} \approx \frac{1}{T}\sum_{m=1}^{T} \mathbf{y}(m)\mathbf{y}^H(m) \tag{6.23}$$

where a total of T pulses is assumed for sensing. From Eqs. (6.21), (6.22) and (6.23), we can observe that in order to obtain the covariance matrix of the full virtual array, we need to find the data matrix $\mathbf{Y}(m)$.

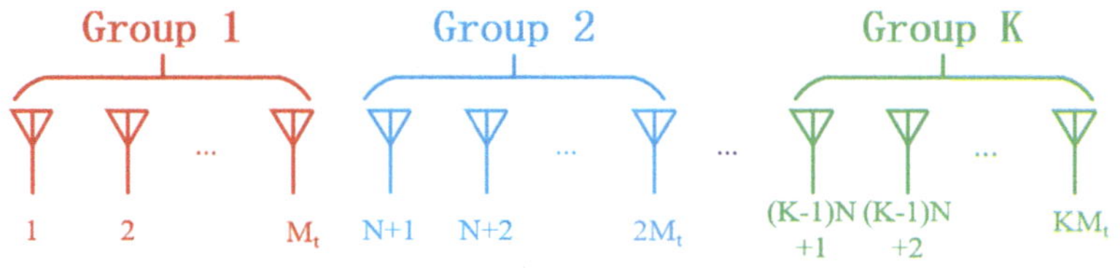

Fig. 6.6 Division of the large M-antenna uniform linear array into K groups

As the data matrix $\mathbf{Y}$ corresponds to the full virtual array, it is a challenging problem to construct $\mathbf{Y}$ using a sparse array with a small number of antennas. In the sequel, we propose a time-multiplexing method to sequentially switch on different sets of antennas for data collection and followed full covariance construction. Matrix $\mathbf{Y}$ is a Hankel matrix, that is, the elements along each anti sub-diagonal are equal. Thereby, we only need to estimate the elements on the first row and the last column to infer all elements of the full matrix.

Suppose we select M_t transmitters and M_t receivers from the M-antenna array, respectively. In this case, the M-antenna array is divided into K subarrays and each subarray comprises M_t consecutive antennas as shown in Fig. 6.6; thus $M = KM_t$. Accordingly, we divide the matrix $\mathbf{Y}$ into K^2 sub-matrices with the same size of $M_t \times M_t$ as follows,

$$\mathbf{Y} = \begin{bmatrix} \mathbf{Y}_{11} & \mathbf{Y}_{12} & \dots & \mathbf{Y}_{1K} \\ \mathbf{Y}_{21} & \mathbf{Y}_{22} & \dots & \mathbf{Y}_{2K} \\ \dots & \dots & \dots & \dots \\ \mathbf{Y}_{(K-1)1} & \mathbf{Y}_{(K-1)2} & \dots & \mathbf{Y}_{(K-1)K} \\ \mathbf{Y}_{K1} & \mathbf{Y}_{K2} & \dots & \mathbf{Y}_{KK} \end{bmatrix} \tag{6.24}$$

Utilizing the Hankel matrix structure, if the $2K - 1$ submatrices on the first row and the Kth column are estimated, as highlighted in color in Eq. (6.24), we can then recover the entire matrix $\mathbf{Y}$, and subsequently obtain the covariance matrix using Eqs. (6.22) and (6.23). The $2K - 1$ submatrices can be estimated via multiplexing to different sets of antennas according to the following procedure within one radar pulse.

- For submatrix $\mathbf{Y}_{1i} (2 \le i \le K)$, highlighted in blue, we use the first M_t antennas as the receiving array, and use the ith group containing from $[(i-1)M_t + 1]$th to (iM_t)th antennas to transmit M_t orthogonal waveforms. After matched filtering at the receiver, we can estimate matrix $\mathbf{Y}_{1i}$, and the other matrices in the first row by sequentially multiplexing to next transmit array.
- For submatrix $\mathbf{Y}_{jK} (2 \le j \le K-1)$, highlighted in green, we use the $[(K-1)M_t + 1]$th to (KM_t)th antennas as the transmit array, and switch the receive array sequentially from $(j-1)M_t + 1$ to jM_t. After matched filtering at the receiver, we can estimate matrix $\mathbf{Y}_{jK}$, and the other matrices in the Kth column.

- For submatrix $\mathbf{Y}_{11}$, highlighted in red, we take the 1st to M_tth antennas as the receiving array and the $M_t + 1$th to $(2M_t)$th antennas as the transmitting array. The set of orthogonal transmit waveforms are then phase rotated by $e^{j(2\pi/\lambda)(M_t d)cos\theta_0}$, which is equivalent to the selection of the transmit array composed of the 1st to M_tth antennas. After matched filtering at the receiver, we can estimate matrix $\mathbf{Y}_{11}$. The matrix $\mathbf{Y}_{KK}$ can be estimated in the same way.

6.3.2 *Optimal Transceiver Design*

After obtaining the covariance matrix of the full virtual array, the next step for cognitive MIMO radar is to determine the optimal transceiver, including both the beamforming weights and array configurations.

6.3.2.1 Beamforming for MIMO Radar

When the M antennas are used for both the transmit and receive arrays, as the case in Sect. 6.3.1, the output of the MIMO receiver is given by,

$$z(m) = \mathbf{w}^H \mathbf{y}(m). \tag{6.25}$$

where beamforming weight vector $\mathbf{w} \in \mathbb{C}^{M^2 \times 1}$ is applied to the virtual array at the receiver after matched filtering. The optimal of MaxSINR beamformer is referred to as Capon beamformer, which can be obtained by minimizing the noise and interference power without weakening the desired signal [185, 202, 249]. That is,

$$\begin{aligned} \underset{\mathbf{w} \in \mathbb{C}^{M^2 \times 1}}{\text{minimize}} \quad & \mathbf{w}^H \mathbf{R}_\mathbf{x} \mathbf{w}, \\ \text{s.t.} \quad & \mathbf{w}^H \mathbf{b}(\theta_0) = 1, \end{aligned} \tag{6.26}$$

where s.t. represents "subject to". The solution of the above constrained optimization problem is given by,

$$\mathbf{w}_o = [\mathbf{b}^H(\theta_0)\mathbf{R}_x^{-1}\mathbf{b}(\theta_0)]^{-1}\mathbf{R}_x^{-1}\mathbf{b}(\theta_0) \tag{6.27}$$

6.3.2.2 Sparse Transceiver Design

For MIMO radar, beamforming in the receiver is implemented on the M^2-sensor virtual array after matched filtering, as shown in Fig. 6.7. There are M virtual sensors associated with each transmit or receive physical sensor. If a sensor is not activated, it implies that none of the M virtual sensors associated with this sensor is selected.

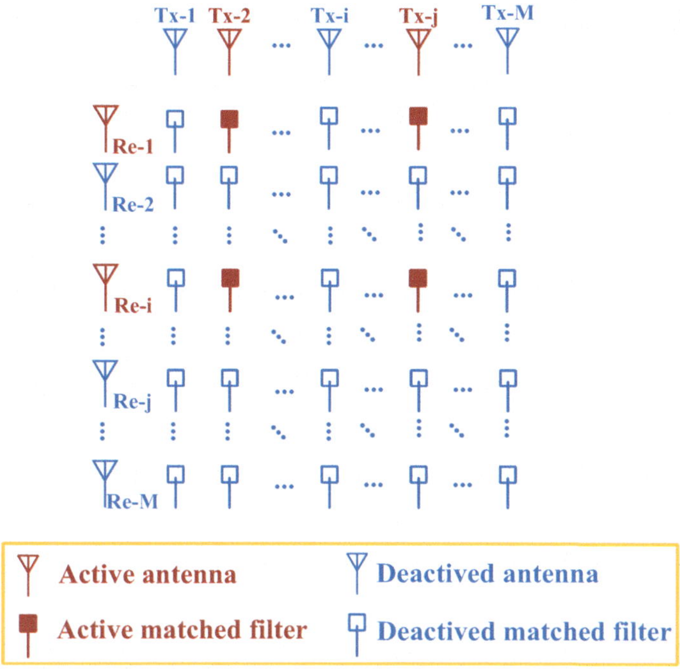

Fig. 6.7 Schematic diagram of virtual sensor for cognitive MIMO radar and the illustration of two-dimensional group sparsity

When we leverage the sparsity of beamforming weight vector to design sparse arrays, it suggests that the M consecutive weights vertically or horizontally are all zeros. For example, in Fig. 6.7, the first sensor is not activated for waveform transmission, so none of the virtual sensors in the first column are selected. Similarly, the second sensor is not activated for receiving signals, so all the virtual sensors in the first row are discarded.

It can be inferred from the above analysis that a two-dimensional group sparsity promoting method can be employed to design the sparse transceiver configuration of MIMO radar. Suppose we choose M_t out of the M sensors as the transmit array and another M_t different sensors as the receiving array. When a sensor is activated, there is at least one non-zero weight coefficient related to the corresponding M virtual sensors. On the contrary, when a sensor is not activated, the weights of the corresponding M virtual sensors all assume zero values. Thus, the sensor selection scheme needs to warrant that the weights of the entire row or column of virtual sensors in Fig. 6.6 are either all zero or have at least one non-zero entry.

The beamforming weight vectors are generally complex valued, whereas the quadratic function in Eq. (6.26) are real. This observation allows expressing the problem with only real variables, which is typically accomplished by converting the problem from the complex domain to the real domain and concatenating the vectors accordingly,

$$\tilde{\mathbf{R}}_x = \begin{bmatrix} \text{real}(\mathbf{R}_x) & -\text{imag}(\mathbf{R}_x) \\ \text{imag}(\mathbf{R}_x) & \text{real}(\mathbf{R}_x) \end{bmatrix}, \tilde{\mathbf{w}} = \begin{bmatrix} \text{real}(\mathbf{w}) \\ \text{imag}(\mathbf{w}) \end{bmatrix},$$
$$\tilde{\mathbf{b}}(\theta_0) = \begin{bmatrix} \text{real}(\mathbf{b}(\theta_0)) \\ \text{imag}(\mathbf{b}(\theta_0)) \end{bmatrix}, \tag{6.28}$$

where $\tilde{\mathbf{R}}_x \in \mathbb{R}^{2M^2 \times 2M^2}$, $\tilde{\mathbf{w}} \in \mathbb{R}^{2M^2 \times 1}$ and $\tilde{\mathbf{b}}(\theta_0) \in \mathbb{R}^{2M^2 \times 1}$ are the real domain vectors of $\mathbf{R}_x$, $\mathbf{w}$ and $\mathbf{b}(\theta_0)$, respectively. Therefore, the sensor selection problem of sparse array construction can be described as the following optimization,

$$\underset{\tilde{\mathbf{w}},\mathbf{c},\mathbf{r}}{\text{minimize}} \quad \tilde{\mathbf{w}}\tilde{\mathbf{R}}_x\tilde{\mathbf{w}} \tag{6.29}$$
$$\text{s.t.} \quad \tilde{\mathbf{w}}^H\tilde{\mathbf{b}}(\theta_0) = 1, \tag{6.29a}$$
$$||\mathbf{P}_i \odot \tilde{\mathbf{w}}||_2 = c_i, \quad i = 1, \ldots, M \tag{6.29b}$$
$$||\mathbf{c}||_0 = M_t, \tag{6.29c}$$
$$||\mathbf{Q}_j \odot \tilde{\mathbf{w}}||_2 = r_j, \quad j = 1, \ldots, M \tag{6.29d}$$
$$||\mathbf{r}||_0 = M_t, \tag{6.29e}$$
$$\mathbf{c}^H\mathbf{r} = 0, \tag{6.29f}$$

where $\odot$ denotes the element-wise product, $\mathbf{P}_i$ and $\mathbf{Q}_j$ are the transmit selection matrix and the receive selection matrix respectively which are defined in (6.16) and (6.17). Constraints (6.29c) and (6.29e) indicate that M_t sensors are selected for each transmitting and receiving array, and $||.||_0$ denotes the l_0 norm. Constraint (6.29f) implies that the transmitting array and receiving array do not share any sensors.

6.3.2.3 Reweighted $l_{2,1}$-Norm Minimization

The l_0-norm constraints in (6.29c) and (6.29e) are non-convex, which renders the above optimization problem difficult to solve. We utilize the l_1-norm to approximate the l_0-norm [250]. To further promote sparsity, an iterative reweighted l_1-norm was proposed in [218]. Based on this idea, we employ an iterative reweighted mixed $l_{2,1}$-norm to promote two-dimensional group sparsity, which can be described as follows,

$$\underset{\tilde{\mathbf{w}},\mathbf{c},\mathbf{r}}{\text{minimize}} \quad \tilde{\mathbf{w}}\tilde{\mathbf{R}}_x\tilde{\mathbf{w}} + \alpha(\mathbf{p}^H\mathbf{c}) + \beta(\mathbf{q}^H\mathbf{r}) \tag{6.30}$$
$$\text{s.t.} \quad \tilde{\mathbf{w}}^H\tilde{\mathbf{b}}(\theta_0) = 1, \tag{6.30a}$$
$$||\mathbf{P}_i \odot \tilde{\mathbf{w}}||_2 \le c_i, \tag{6.30b}$$
$$0 \le c_i \le 1, \quad i = 1, \ldots, M \tag{6.30c}$$
$$||\mathbf{Q}_j \odot \tilde{\mathbf{w}}||_2 \le r_j, \tag{6.30d}$$

$$0 \leq r_j \leq 1, \quad j = 1, \ldots, M \tag{6.30e}$$

$$\mathbf{1}_M^H \mathbf{c} = M_t, \quad \mathbf{1}_M^H \mathbf{r} = M_t, \tag{6.30f}$$

$$\mathbf{c}^H \mathbf{r} = 0, \tag{6.30g}$$

where α and β are two parameters to balance between the array sparsity and output SINR of the transceiver. A common method of updating the reweighting coefficient is to take the reciprocal of the absolute weight value. However, this fails to control the number of elements to be selected. Thus, similar to Sect. 6.2.2.2, in order to control the number of selected antennas, we update the reweighting coefficient $\mathbf{p}$ and $\mathbf{q}$ in the same way as (6.18) and (6.19) for each iteration.

In practical implementation, a close distance between the transmit and receiving antennas will cause power leakage and unwanted coupling effect. To achieve high isolation between the transmit and receive channels, which is always desirable, the selected transmit and receive antennas should be sufficiently spatially separated. Therefore, for the transceiver array design, the two positions neighbouring the selected transmitting antenna should be vacant and not hosting any receive antenna. To satisfy this requirement, we replace the constraint in (6.30g) with the following new constraints,

$$\begin{aligned} &c_i + r_{i-1} + r_i + r_{i+1} \leq 1, \quad i = 2, \ldots, M-1, \\ &c_1 + r_1 + r_2 \leq 1, \\ &c_M + r_M + r_{M-1} \leq 1. \end{aligned} \tag{6.31}$$

6.3.3 New Transceiver Reconfiguration

We obtain the optimal transceiver configuration in the current environment through sensing and learning. In the action phase, and according to the two vectors $\mathbf{c}$ and $\mathbf{r}$ from Eq. (6.30), we switch on the selected M_t antennas indicated by $\mathbf{c}$ to transmit the M_t orthogonal waveforms, and switch on the selected M_t receiving antennas indicated by $\mathbf{r}$ to receive data. After matched filtering, we perform beamforming with the weights $\mathbf{w}$ obtained from Eq. (6.30). Consequently, a new transceiver array will be reconfigured to adapt to the new environment for performance improvement. When the environment changes, the output SINR decreases, instigating the beginning of another "perception-action" cycle.

6.3.4 Simulations

In this subsection, we demonstrate the effectiveness of the proposed cognitive-driven sparse transceiver design for MIMO radar in different scenarios.

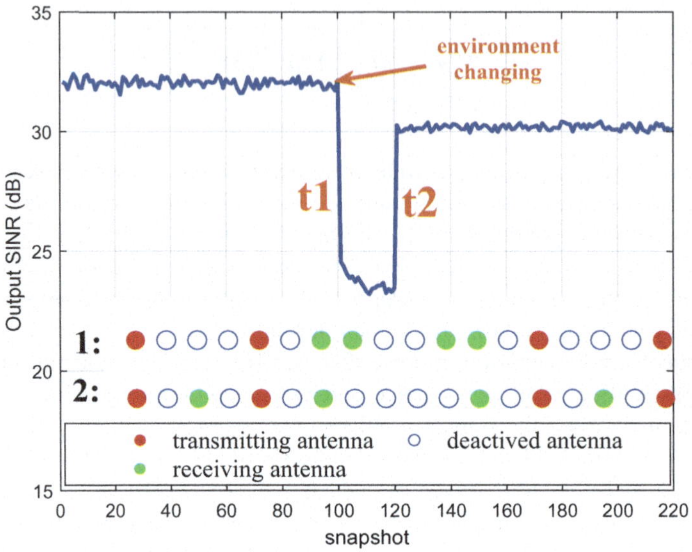

Fig. 6.8 Response of cognitive MIMO radar to environmental changes

6.3.4.1 Example 1

In this example, we simulate the work flow of the proposed cognitive MIMO radar and evaluate its performance. We assume that there are a total of 18 antennas with an inter-element spacing of $\lambda/2$, from which $M_t = 4$ antennas are selected for transmitting and another $M_t = 4$ antennas for receiving. In the first part of the simulation, that is, before time t_1, the target is at angle 65° with the signal-to-noise ratio (SNR) is 20 dB. There are two deceptive interferences with an interference-to-noise ratio (INR) of 15 dB located at angle 50° and 60°, respectively. At a certain moment t_1, the environment suddenly changes, and another interference close to the target at 63° is switched on. We can observe that the output SINR decreases abruptly starting from time instant t_1, as shown in Fig. 6.8. This triggers the cognitive mechanism of the MIMO radar, which re-initiates sensing and then learning from time t_1 to time t_2. It reconfigures the optimal transceiver arrays responding to the environmental change at time t_2. Accordingly, the output SINR is now maximized under the new environment. The two optimum transceiver configurations 1 and 2 before and after the environmental change are shown in Fig. 6.8. Evidently, the selection of transmit antennas remains the same, while the configuration of receiving array has significantly changed. In this experiment, the output SINR of the new environment is slightly lower than that of the initial environment, which is due to the additional interference closed to the expected signal.

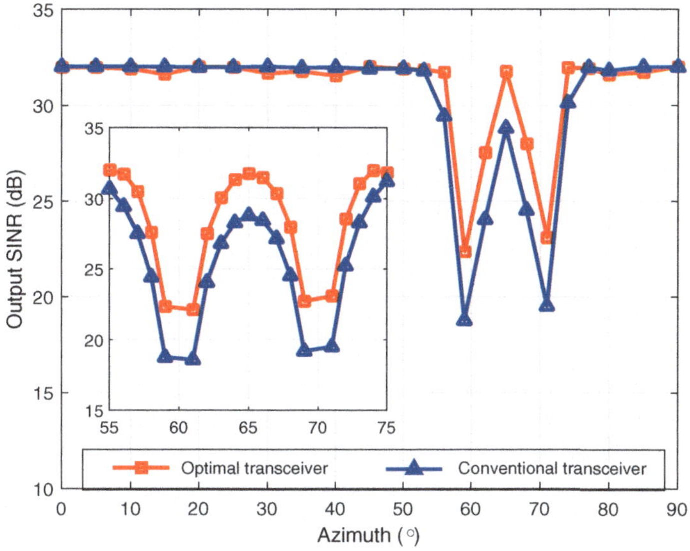

Fig. 6.9 Relationship between output SINR and target angle with the directions of interferences being fixed

6.3.4.2 Example 2

In this example, we compare the performance of the optimal transceiver with that of conventional transceiver that comprises antennas indexed by 1, 5, 9 and 13 for transmitting and antennas indexed by 15 16, 17 and 18 for receiving. Two deceptive interferences are impinging on the array from 60° and 70°, respectively, with an INR of 15dB. The scanning angle is changing from 0° to 90° and the SNR is set as 20dB. Other parameters remain the same as in example 1. We configure an optimal transceiver for each scanning angle. The curve of output SINR versus scanning angle is plotted in Fig. 6.9, where we can observe that when the interferences are widely separated from the target, the superiority of the proposed optimal transceiver is not obvious. On the other hand, when the interferences move closer to the target, the performance of the optimal transceiver exhibits noticeable improvement compared to that of conventional MIMO array.

6.3.4.3 Example 3

From example 2, we can argue that the optimal cognitive MIMO radar behaves well when the interferences are close to the target, which is difficult to deal with in practice. Thus, in this example, we focus on this specific scenario. For each scanning angle, we investigate two cases, where the incident angles of two deceptive interference are $\pm 5°$ relative to the target in case 1 and then $\pm 3°$ in case 2. Again, we plot the curve of

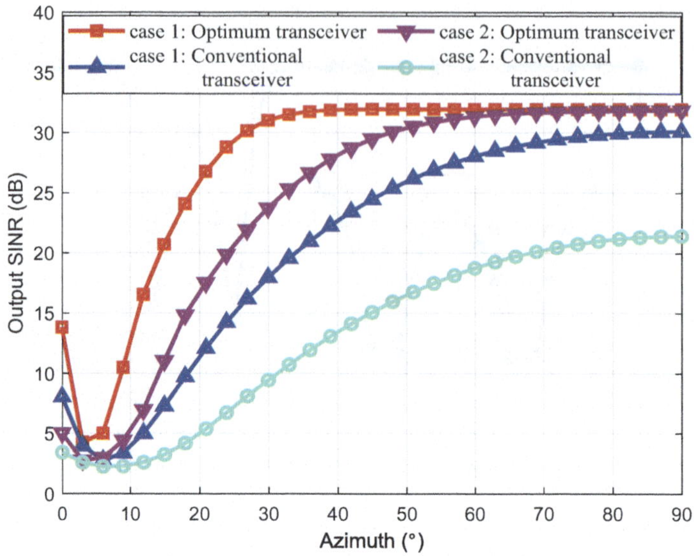

Fig. 6.10 Relationship between the output SINR and the target angle when the interferences are close to the target

output SINR versus the scanning angle, as shown in Fig. 6.10. It can be observed that the optimal transceiver array selected by the optimal cognitive MIMO radar clearly improves the performance. Moreover, the closer the interference is relative to the target, the higher the improvement. Additionally, there is an interesting phenomenon as seen from Fig. 6.10, that the linear MIMO transceiver behaves worse in the endfire direction compared with the broadside direction. The reason is that the beampatterns of linear arrays are broadened in the endfire direction, thus limiting their spatial resolution.

6.3.4.4 Example 4

In this example, we examine the effect of the number of candidate antennas and the antenna spacing on the output performance of our optimal sparse transceivers. The output SINR versus the scanning angle is depicted in Fig. 6.11. Again, the linear array exhibits degraded performance when steering in the endfire direction. This phenomenon can be ameliorated by increasing the total array aperture, as proved by the blue curve. We can see that when the sensor spacing is $\lambda/2$, the performance of the optimal sparse transceiver improves with increased number of candidate antennas. For a fixed number of antennas, increasing the inter-element spacing improves the performance of the optimal sparse transceiver at lower scanning angle. However, the performance becomes unstable when scanning to broadside direction.

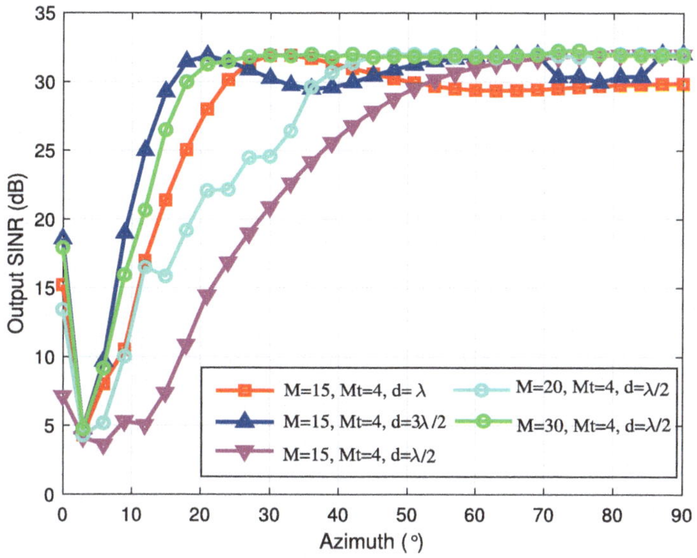

Fig. 6.11 The influence of the total number of sensors and the antenna spacing on the output SINR of optimal sparse transceiver array

6.4 Summary

In this chapter, we examined the sparse MIMO array transceiver design for different tasks and under different criteria. The problem of sparse transceiver design for MIMO radar was first considered under the assumption of prior knowledge of the source/interference environment. For a given number of transmitting and receiving antennas, a sparse MIMO array structure in terms of MaxSINR was jointly designed. It was shown by simulations that the proposed sparse MIMO transceiver selection algorithm provides superior interference suppression performance over randomly designed MIMO transceiver arrays. We then proposed a novel cognitive MIMO radar which adaptively optimizes both beamforming weights and transceiver array configuration for MaxSINR beamforming in a dynamic environment. This cognitive-driven optimization framework can learn the environment automatically, thus eliminating the assumption of any prior knowledge. From the same large array, a given number of antennas were selected to construct the transmit and receive arrays jointly. The two arrays were not allowed to have overlapping or adjacent antennas for improved isolation. During perception in the underlying perception-action cycle, the data is collected by sequentially switching among different sets of antennas to construct the covariance matrix of a full virtual array. The simulations showed that the proposed cognitive-driven sparse MIMO array design can adaptively reconfigure the transceiver to maximize the beamforming performance with dynamically changing environments.

Chapter 7
Sparse Sensing for Target Detection

In this chapter, we investigate the problem of slow target detection in heterogeneous clutter in the context of radar space-time adaptive processing (STAP). Traditional STAP approaches require a large number of training data to estimate the clutter covariance matrix, which prohibits their practical applications. In order to address the issue of limited training data especially in the heterogeneous scenarios, we propose a novel thinned STAP via selecting an optimum subset of antenna-pulse pairs in the metric of maximum output signal-to-clutter-plus-noise ratio (SCNR). The proposed thinned STAP strategy defines a new parameter, named spatial spectrum correlation coefficient (S^2C^2), to analytically characterize the effect of space-time configuration on STAP performance and reduce the dimensionality of traditional STAP. Three algorithms are proposed to solve the antenna-pulse pair selection problem. The effectiveness of the proposed strategy is confirmed by both simulations and expreriments, especially by employing the MCARM data set. In Sect. 7.2, we review the clutter model and introduce the parameter of S^2C^2. In Sect. 7.3, we propose three approaches to solve the antenna-pulse pair selection problem. Section 7.3.4 gives simulation and experimental results that validate the performance of the proposed strategy. Finally, some conclusions are drawn in Sect. 7.4.

7.1 Introduction

Space-time adaptive processing (STAP) involves combining signals linearly from an array of antennas and multiple pulses to improve airborne radar detection performance in severe clutter and jamming environments [65, 251, 252]. The optimum STAP processor employs the clutter-plus-noise covariance matrix (CCM) to whiten the received data prior to the application of a matched-filter detector. As the CCM is usually unknown, practical STAP approaches such as the sample matrix inversion (SMI) method obtain the maximum likelihood estimate of the CCM from secondary cells. The number of independent and identically distributed (IID) training data,

X. Wang et al., *Sparse Sensing and Sparsity Sensed in Multi-sensor Array Applications*, https://doi.org/10.1007/978-981-99-9558-5_7

required by the SMI to ensure an average signal-to-clutter-plus-noise ratio (SCNR) loss within 3dB of the optimum processor, is twice the number of degrees of freedom (DoFs) of the detector. This is typically on the order of several hundreds for common STAP applications and can far exceed the available data measurements [253]. For non-homogeneous environment, the SMI may incur a significant performance degradation due to lack of sufficient IID training data [254].

In addition to clutter heterogeneity and limited sample support, practical implementations of STAP continue to face other challenges including high computational cost in real-world scenarios. A radar array with N antennas and M coherent pulses involves the inversion of an $NM \times NM$ matrix, requiring an order of $(NM)^3$ operations. The fact that the algorithm must be executed for each range gate, as well as angle and Doppler bins exacerbates the problem. Fortunately, most interference suppression problems in airborne radar systems are rank deficient in nature, that is, they require fewer adaptive DoFs compared to those offered by the array. By appropriate design, the redundant DoFs can be discarded and partial STAP can be applied [255]. For instance, preprocessing can be used to select statistically representative training data in order to mitigate clutter heterogeneity [256–260]. Knowledge-aided (KA) STAP incorporates a priori knowledge into the estimation process to accelerate the convergence of the CCM [261–264]. Test data only algorithms such as the D^3 algorithm [265] and the MLED [266] do away with the need for training data. An image processing-based STAP technique was proposed in [267]. Principle Component Analysis (PCA) projects the data onto a lower dimensional subspace [175, 268] to reduce the high sample support. Some recent work has considered utilizing sparse recovery (SR) techniques of the clutter spectrum in the angle-Doppler plane for moving target indication [269, 270]. However, these SR approaches are computationally more expensive than conventional STAP because they require a full-dimensional matrix inversion and the recovery procedure calls for additional computations. The work in [271] assumes the sparse property of the STAP filter weights instead of the clutter spectrum and utilized the l_1-norm to promote sparsity.

In this chapter, we propose a novel thinned STAP approach consisting of an antenna-pulse selection strategy where we select a subset of optimal antenna-pulse pairs in each range gate prior to applying STAP and target detection. Traditional selection strategies viewed the antenna array separately from the pulse train and examined the selection on the array only in airborne radar to reject ground clutter [272]. Whereas in this chapter we view the interleaved antenna-pulse data as a virtual large array where each sensor is specified by the antenna number and pulse number. We carry out the "thinning" in both space and time through joint antenna-pulse selection to obtain a sub-configuration comprising the optimal sensor elements and consequently the optimal subset of antenna-pulse pairs.

The chapter is organized as follows: In Sect. 7.2, we formulate the parameter, referred to as spatial spectrum correlation coefficient (S^2C^2), to characterize the separation between the target and the clutter subspace and use it for analytically expressing the output SCNR. In Sect. 7.3, we propose three approaches to solve the antenna-pulse selection problem based on the three formulations of the S^2C^2. In addition to extensive simulations, the performance of the proposed antenna-pulse selection strategy is evaluated using monostatic Multi-Channel Airborne Radar Measurement (MCARM) data in Sect. 7.3.4. Finally, some conclusions are drawn in Sect. 7.4.

7.2 Definition of Spatial Spectral Correlation Coefficient

In this section, we introduce a new parameter S^2C^2, named spatial spectrum correlation coefficient, to analytically characterize the effect of space-time configuration on STAP performance and reduce the dimensionality of traditional STAP. It is worth to note that the parameter S^2C^2 is essentially an extension of the parameter SCC defined in chapter 4 from the spatial domain to the joint spatial-spectral domain.

7.2.1 Signal Model

The geometry of a sidelooking airborne array radar of N uniformly spaced antennas with inter-element spacing d in a Cartesian Coordinate system is shown in Fig. 3.2 in [65]. Let $\mathcal{P}$ be a scatterer patch on the ground having elevation angle θ and azimuth angle ϕ with respect to the array center. Consider an antenna with a distance $nd, n = 0, \cdots, N-1$ relative to the array origin. The signal received by the antenna from the stationary scatterer patch $\mathcal{P}$ is phase shifted relative to the origin by,

$$n2\pi f_s = n\frac{2\pi}{\lambda}d\cos\phi\cos\theta, \tag{7.1}$$

where $f_s = (d/\lambda)\cos\phi\cos\theta \in [-1/2, 1/2]$ is the normalized spatial frequency under the condition $d = \lambda/2$ and λ is the wavelength. The spatial steering vector can be expressed as,

$$\mathbf{a}_s = [1, e^{j2\pi f_s}, \cdots, e^{j(N-1)2\pi f_s}]^T. \tag{7.2}$$

In pulse Doppler radar, the Doppler frequency is measured by phase comparison between echo signals due to a transmitted coherent pulse train with pulse repetition interval (PRI) $\tilde{T}$. The unit phase shift incurred by the platform moving with the velocity v_p is given by

$$2\pi f_d = 2\pi\tilde{T}\frac{2v_p}{\lambda}\cos\phi\cos\theta. \tag{7.3}$$

Here we denote $f_d = (2v_p\tilde{T}/\lambda)\cos\phi\cos\theta$ as the normalized Doppler frequency. Then the temporal steering vector with M coherent pulses can be written as,

$$\mathbf{a}_t = [1, e^{j2\pi f_d}, \cdots, e^{j(M-1)2\pi f_d}]^T. \tag{7.4}$$

Thus, the interleaved space-time steering vector, $\mathbf{a}(\theta, \phi) \in \mathbb{C}^{NM \times 1}$, is

$$\mathbf{a}(\theta, \phi) = \mathbf{a}_s \otimes \mathbf{a}_t, \tag{7.5}$$

where $\otimes$ denotes Kronecker product. Consider a single range increment, specified by the elevation angle θ, the total clutter echo is an integral over the various contributions from all ground scatterers in azimuth,

$$\mathbf{c}(\theta) = \int_{\phi=0}^{2\pi} \mathcal{A} D(\phi, \theta) G(\phi, \theta) \mathbf{a}(\theta, \phi) d\phi, \tag{7.6}$$

where $\mathcal{A}$ stands for ground reflectivity that is assumed to be a circular complex Gaussian variable, $D(\phi, \theta)$ and $G(\phi, \theta)$ are the receive and transmit directivity patterns respectively. The trajectory of clutter spectrum in the angle-Doppler (f_s-f_d) plane is very important for extracting the information on clutter subspace. For a sidelooking array, the clutter trajectory is,

$$f_d = \kappa f_s, \tag{7.7}$$

which is a straight line in the f_d-f_s plane with a slope $\kappa = 2v_p\tilde{T}/d$. This implies that most of the clutter energy is concentrated in a ridge and effective clutter rank is much smaller than NM. In fact, for a sidelooking ULA, Brennan's rule gives the effective rank of the CCM to be, [251, 273],

$$N_e = \text{int}\{N + \beta(M-1)\} \tag{7.8}$$

where int{} denotes the next integer number. This rule states that there are at most N_e clutter eigenvalues that are larger than the noise floor σ_n^2. In other words, the clutter subspace is spanned by the N_e eigenvectors corresponding to the dominant eigenvalues. And then the clutter covariance matrix can be expressed as,

$$\mathbf{Q}_c = \sum_{i=1}^{N_e} \sigma_i^2 \mathbf{e}_i \mathbf{e}_i^H = \sum_{j=1}^{N_e} P_j \mathbf{v}_j \mathbf{v}_j^H, \tag{7.9}$$

where $\mathbf{e}_i$ and σ_i^2 are the ith eigenvector and its corresponding eigenvalue of clutter covariance matrix $\mathbf{Q}_c$ respectively. We call $\mathbf{e}_i, i = 1, \cdots, N_e$ as clutter eigenbasis in the following. As shown in [274], for large enough clutter-to-noise ratio (CNR), the clutter subspace is also spanned by N_e Fourier basis vectors $\mathbf{v}_j, j = 1, \cdots, N_e$ with power coefficients P_j. The definition of Fourier basis $\mathbf{v}_j$ is same with that of the interleaved space-time steering vector $\mathbf{a}$ in Eq. (7.5) with its corresponding spatial and Doppler frequencies. Since the two sets of N_e basis vectors span the same clutter subspace, i.e. span($\mathbf{e}_i, i = 1, \cdots, N_e$) = span($\mathbf{v}_j, j = 1, \cdots, N_e$), each Fourier basis vector can be expressed as a linear combination of eigenbasis, i.e.,

$$\mathbf{v}_j = \sum_{i=1}^{N_e} \mu_i^j \mathbf{e}_i. \tag{7.10}$$

Thus, the center points of the resolution grids along the clutter trajectory can be used as a set of approximate Fourier basis vectors of the clutter subspace, which circumvents the need for the computationally expensive eigenvalue decomposition. There is no specific rule that governs this choice and the simplest approach is to choose the cells where the clutter power is largest. Better clutter representation is achieved by analysing a zero-padding Fourier spectrum and including more basis vectors at the expense of increased complexity.

The detection problem is usually treated as a hypothesis test to determine the presence of a target. The model of the received signal, $\mathbf{x} \in \mathbb{C}^{NM}$, in a single range increment is,

$$H_0 : \mathbf{x} = \mathbf{c} + \mathbf{n}, \tag{7.11}$$

where H_0 denotes the null hypothesis. For simplicity of notation, we put $\mathbf{c}$ instead of $\mathbf{c}(\theta)$ and $\mathbf{n}$ denotes the Gaussian white noise with power σ_n^2. The alternative hypothesis is given by,

$$H_1 : \mathbf{x} = \alpha\mathbf{t} + \mathbf{c} + \mathbf{n}, \tag{7.12}$$

where α denotes the complex amplitude of target signal. The target space-time steering vector $\mathbf{t} \in \mathbb{C}^{NM\times 1}$ can be similarly constructed according to Eqs. (7.2), (7.4) and Eq. (7.5) with $f_s = (d/\lambda)\cos\phi_t \cos\theta_t$ and $f_d = 2v_t\tilde{T}/\lambda$ given a target in elevation θ_t and azimuth ϕ_t with radial velocity v_t relative to the platform.

We assume that all components of $\mathbf{x}$ are mutually independent. The CCM, $\mathbf{Q}$, is then a sum of clutter and noise covariance matrices as follows,

$$\mathbf{Q} = E\{\mathbf{x}\mathbf{x}^H\} = \sigma_n^2 \mathbf{I}_{NM} + \mathbf{Q}_c, \tag{7.13}$$

where $\mathbf{Q}_c$ stands for the clutter covariance matrix and is usually rank-deficient. For a sidelooking linear array, Brennan's rule gives the clutter rank N_e per Eq. (7.8).

As the clairvoyant CCM is not available in practice, we adopt the Adaptive Matched Filter (AMF) detector in this work [98], which is formulated as

$$\frac{|\mathbf{v}^H \hat{\mathbf{Q}}^{-1} \mathbf{x}|^2}{\mathbf{v}^H \hat{\mathbf{Q}}^{-1} \mathbf{v}} \lessgtr_{H_1}^{H_0} \mathcal{T}, \tag{7.14}$$

where $\mathcal{T}$ is the threshold value and $\mathbf{v}$ is the scanning steering vector over the angle-Doppler plane. The CCM $\mathbf{Q}$ is estimated by $\hat{\mathbf{Q}} = \frac{1}{L}\sum_{l=1}^{L} \mathbf{x}(l)\mathbf{x}^H(l)$ given L homogeneous training data under the hypothesis H_0. Clearly, the maximum value of Eq. (7.14) occurs when $\mathbf{v} = \mathbf{t}$ in the cell under test (CUT).

The theoretical false alarm rate P_{FA} of the AMF for a standard ULA with K antenna-pulse pairs and K_t training data, is

$$P_{\mathrm{FA}} = \int_0^1 \left(1 + \frac{\gamma \eta}{K_t}\right)^{-L} f_\beta(\eta; L+1, K-1) d\eta, \tag{7.15}$$

where γ is the threshold value of the AMF detector and $L = K_t - K + 1$. And η is a type I beta distributed variable with parameters $L + 1$ and $K - 1$. The detection probability P_{D}, on the other hand, is

$$P_{\mathrm{D}} = 1 - \int_0^1 h(\eta) f_\beta(\eta; L+1, K-1) d\eta, \tag{7.16}$$

where

$$h(\eta) = \left(1 + \frac{\gamma \eta}{K_t}\right)^{-L} \sum_{l=1}^{L} \binom{L}{l} \left(\frac{\gamma \eta}{K_t}\right)^{l} G_l\left(\frac{\eta K \rho}{1 + \frac{\gamma \eta}{K_t}}\right), \tag{7.17}$$

and

$$G_l(y) = e^{-y} \sum_{n=0}^{l} \frac{y^n}{n!}, \tag{7.18}$$

where ρ is the signal to noise ratio (SNR). The number K of selected antenna-pulse pairs can be determined from Eq. (7.16) by setting P_{FA}, P_{D} and SNR to the desired values.

7.2.2 *Definition of Spatial Spectral Correlation Coefficient*

Based on the signal model introduced above, we define the parameter, Spatial Spectral Correlation Coefficient, in three different expressions, max-cosine expression, matrix-vector expression and determinant expression.

7.2.2.1 Max-Cosine Expression

First consider the simplest case where the rank of the clutter is one $N_e = 1$, we define the $S^2C^2$of the full configuration as,

$$\alpha = \frac{\mathbf{s}^H \mathbf{c}(f_s, f_d)}{\sqrt{(\mathbf{s}^H \mathbf{s})(\mathbf{c}(f_s, f_d)^H \mathbf{c}(f_s, f_d))}} = \frac{\mathbf{s}^H \mathbf{c}(f_s, f_d)}{NM}. \tag{7.19}$$

Clearly that the value of S^2C^2 is within the range [0, 1] and represents the cosine value of the angle between the target and clutter steering vectors. The smaller the value of the S^2C^2 is, the more separable the target and clutter are, thereby implying better adaptive processing performance. In the general case of $N_e > 1$, the $S^2C^2$is defined as the maximum cosine value between the target and the N_e clutter steering vectors. That is,

$$\alpha = \max_{i=1,\ldots,N_e} \frac{\mathbf{s}^H \mathbf{c}_i(f_s, f_d)}{NM}. \tag{7.20}$$

We can argue that the smaller the $S^2C^2$is, the more separable beteen the target and the clutter subspace, thus implying the higher output SCNR.

7.2.2.2 Matrix-Vector Expression

Proceeding from Eq. (7.9), we arrange the N_e Fourier basis vectors into a matrix $\mathbf{V}_c \in \mathbb{C}^{MN \times N_e}$,

$$\mathbf{V}_c = [\mathbf{v}_1, \mathbf{v}_2, \cdots, \mathbf{v}_{N_e}]. \tag{7.21}$$

Defining $\mathbf{P} = \text{diag}[P_1, \cdots, P_{N_e}]$, we rewrite the CCM in Eq. (7.13) as,

$$\mathbf{Q} = \sigma_n^2 \mathbf{I}_{NM} + \mathbf{V}_c \mathbf{P} \mathbf{V}_c^H. \tag{7.22}$$

Applying the Woodbury matrix identity to the inverse CCM, $\mathbf{Q}^{-1}$, yields,

$$\mathbf{Q}^{-1} = \frac{1}{\sigma_n^2}(\mathbf{I}_{NM} - \mathbf{V}_c(\sigma_n^2 \mathbf{P}^{-1} + \mathbf{V}_c^H \mathbf{V}_c)^{-1} \mathbf{V}_c^H). \tag{7.23}$$

We assume the clutter is much stronger than white noise, i.e. $P_1 > \cdots > P_{N_e} \gg \sigma_n^2$, then Eq. (7.23) can be further simplified as

$$\mathbf{Q}^{-1} \approx \frac{1}{\sigma_n^2}(\mathbf{I}_{NM} - \mathbf{V}_c(\mathbf{V}_c^H \mathbf{V}_c)^{-1} \mathbf{V}_c^H). \tag{7.24}$$

From the above equation it is clear that when the CNR is high, $\mathbf{Q}^{-1}$ approximates the clutter nullspace. Accordingly, the optimum STAP weight vector, [65, 98],

$$\mathbf{w}_{\text{opt}} = \eta \mathbf{Q}^{-1}\mathbf{t} \approx \frac{\eta}{\sigma_n^2}(\mathbf{I}_{NM} - \mathbf{V}_c(\mathbf{V}_c^H \mathbf{V}_c)^{-1} \mathbf{V}_c^H)\mathbf{t}, \tag{7.25}$$

becomes the interference eigencanceler proposed in [175]. Here $\eta = (\mathbf{t}\mathbf{Q}^{-1}\mathbf{t})^{-1/2}$ does not affect the output SCNR.

The steering vector of the target signal $\mathbf{t}$ can be decomposed into two orthogonal components: the clutter subspace component $\mathbf{t}_c$ and the nullspace component $\mathbf{t}_\perp$, i.e., $\mathbf{t} = \mathbf{t}_c + \mathbf{t}_\perp$ with $\mathbf{t}_c = (\mathbf{V}_c(\mathbf{V}_c^H \mathbf{V}_c)^{-1} \mathbf{V}_c^H)\mathbf{t}$ and $\mathbf{t}_\perp = (\mathbf{I}_{NM} - \mathbf{V}_c(\mathbf{V}_c^H \mathbf{V}_c)^{-1} \mathbf{V}_c^H)\mathbf{t}$ respectively. The STAP weight vector $\mathbf{w}_{\text{opt}}$, is along the clutter nullspace direction $\mathbf{t}_\perp$. We define the S^2C^2 as the absolute value of the cosine of the angle between the target signal $\mathbf{t}$ and clutter subspace component $\mathbf{t}_c$, i.e.,

$$|\alpha| = |\cos\vartheta| = \left|\frac{\mathbf{t}^H\mathbf{t}_c}{\|\mathbf{t}\|_2\|\mathbf{t}_c\|_2}\right|. \tag{7.26}$$

Here the length of $\mathbf{t}$ is $\|\mathbf{t}\|_2 = \sqrt{MN}$ under the assumption of isotropic antennas. As the SCNR_{out} is directly related to the squared value of the S^2C^2, we substitute the expression of $\mathbf{t}_c$ into Eq. (7.26) and take the squared value yields,

$$\begin{aligned}|\alpha|^2 &= \frac{|\mathbf{t}^H\mathbf{V}_c(\mathbf{V}_c^H\mathbf{V}_c)^{-1}\mathbf{V}_c^H\mathbf{t}|^2}{MN\|\mathbf{V}_c(\mathbf{V}_c^H\mathbf{V}_c)^{-1}\mathbf{V}_c^H\mathbf{t}\|_2^2},\\ &= \frac{1}{MN}\mathbf{t}^H\mathbf{V}_c(\mathbf{V}_c^H\mathbf{V}_c)^{-1}\mathbf{V}_c^H\mathbf{t}.\end{aligned} \tag{7.27}$$

Finally, combining Eqs. (7.24) and (7.27), the output SCNR (SCNR_{out}) becomes

$$\begin{aligned}\text{SCNR}_{\text{out}} &= \sigma_t^2\mathbf{t}^H\mathbf{Q}^{-1}\mathbf{t}\\ &\approx \frac{\sigma_t^2}{\sigma_n^2}\mathbf{t}^H(\mathbf{I}_{NM} - \mathbf{V}_c(\mathbf{V}_c^H\mathbf{V}_c)^{-1}\mathbf{V}_c^H)\mathbf{t}\\ &\approx \text{SNR}\cdot MN(1-|\alpha|^2),\end{aligned} \tag{7.28}$$

where σ_t^2 denotes the power of the target signal and $\text{SNR} = \sigma_t^2/\sigma_n^2$. Eq. (7.28) shows that the SCNR_{out} depends on two factors: the number of available DoFs (which is MN for the full configuration) and the squared S^2C^2 value $|\alpha|^2$. When the number of available DoFs is fixed, the performance can be improved by changing the space-time configuration to reduce the S^2C^2 value. Thus the S^2C^2 characterizes the effect of the space-time geometry on the adaptive filtering performance and, in this respect, is an effective metric for optimum antenna-pulse selection.

7.2.2.3 Determinant Expression

The matrix-vector expression of the S^2C^2 given in Eq. (7.27) is not a convenient form for antenna-pulse selection. Below we derive a compact formula of the S^2C^2 based on Eq. (7.27) in terms of matrix determinants.

Let the clutter cross-correlation matrix $\mathbf{D}_c \in \mathbb{C}^{N_e\times N_e}$ be

$$\begin{aligned}\mathbf{D}_c &= \mathbf{V}_c^H\mathbf{V}_c\\ &= \begin{bmatrix}\rho_{11} & \rho_{12} & \cdots & \rho_{1N_e}\\ \rho_{21} & \rho_{22} & \cdots & \rho_{2N_e}\\ \cdots & \cdots & \cdots & \cdots\\ \rho_{N_e1} & \rho_{N_e2} & \cdots & \rho_{N_eN_e}\end{bmatrix},\end{aligned} \tag{7.29}$$

where the entry $\rho_{ij} = \mathbf{v}_i^H\mathbf{v}_j$ for $i, j = 1, ..., N_e$. Note that the estimated Fourier basis vectors are not orthogonal in practical applications, thus we do not simplify $\mathbf{D}_c$

as $\mathbf{I}_{N_e}$ for generality. Define $\mathbf{V}_t = [\mathbf{t}, \mathbf{v}_1, \mathbf{v}_2, \cdots, \mathbf{v}_{N_e}]$, the target plus clutter cross-correlation matrix $\mathbf{D}_t \in \mathbb{C}^{(N_e+1)\times(N_e+1)}$ is

$$\begin{aligned}\mathbf{D}_t &= \mathbf{V}_t\mathbf{V}_t^H \\ &= \begin{bmatrix} \rho_{tt} & \rho_{t1} & \cdots & \rho_{tN_e} \\ \rho_{1t} & \rho_{11} & \cdots & \rho_{1N_e} \\ \cdots & \cdots & \cdots & \cdots \\ \rho_{N_et} & \rho_{N_e1} & \cdots & \rho_{N_eN_e} \end{bmatrix}, \\ &= \begin{bmatrix} MN & \mathbf{t}^H\mathbf{V}_c \\ \mathbf{V}_c^H\mathbf{t} & \mathbf{D}_c \end{bmatrix}, \end{aligned} \tag{7.30}$$

where the entry $\rho_{tj} = \rho_{jt}^* = \mathbf{t}^H\mathbf{v}_j$ for $j = 1, ..., N_e$ and $\rho_{tt} = \mathbf{t}^H\mathbf{t} = MN$ assuming MN antenna-pulse pairs. Utilizing the determinant formula of a block matrix in Eq. (7.30) yields,

$$|\mathbf{D}_t| = |\mathbf{D}_c|(MN - \mathbf{t}^H\mathbf{V}_c\mathbf{D}_c^{-1}\mathbf{V}_c^H\mathbf{t}). \tag{7.31}$$

Thus,

$$\mathbf{t}^H\mathbf{V}_c\mathbf{D}_c^{-1}\mathbf{V}_c^H\mathbf{t} = MN - \frac{|\mathbf{D}_t|}{|\mathbf{D}_c|}. \tag{7.32}$$

Substituting Eqs. (7.29) and (7.32) into Eq. (7.27), the expression of the S^2C^2 can be rewritten in terms of the ratio of two matrix determinants,

$$|\alpha|^2 = \frac{1}{MN}\mathbf{t}^H\mathbf{V}_c\mathbf{D}_c^{-1}\mathbf{V}_c^H\mathbf{t} = 1 - \frac{|\mathbf{D}_t|}{MN|\mathbf{D}_c|}. \tag{7.33}$$

In the case of single interference as in [104], i.e. when $N_e = 1$, the two cross-correlation matrices in Eqs. (7.29) and (7.30) reduce to $\mathbf{D}_c = \rho_{11} = MN$ and

$$\mathbf{D}_t = \begin{bmatrix} MN & \rho_{t1} \\ \rho_{1t} & MN \end{bmatrix}, \tag{7.34}$$

respectively. Substituting $|\mathbf{D}_c| = MN$ and $|\mathbf{D}_t| = M^2N^2 - |\rho_{t1}|^2$ into Eq. (7.33) yields $|\alpha|^2 = |\rho_{t1}|^2/(M^2N^2)$. Thus this generalized formula is in accordance with the formula of the single interference case given by Eq. (11) in [104].

Now, we proceed to substitute Eq. (7.33) into Eq. (7.28),

$$\text{SCNR}_{\text{out}} \approx \text{SNR} \cdot MN(1 - |\alpha|^2) \approx \text{SNR} \cdot \frac{|\mathbf{D}_t|}{|\mathbf{D}_c|}. \tag{7.35}$$

Note that Eq. (7.35) is equivalent to Eq. (7.28), but there is no explicit linear dependence between the performance and the number of selected antenna-pulse pairs in Eq. (7.35). This further demonstrates the non-linear relationship between the SCNR_{out}

and the number of DoFs. In the next section, we implement antenna-pulse selection in terms of maximizing the $\mathrm{SCNR}_{\mathrm{out}}$ in Eq. (7.35) directly.

7.3 Thinned STAP via Antenna-Pulse Selection

In this section, we propose three selection algorithms in terms of maximizing the output SCNR based on the definitions in Sects. 7.2.2.1, 7.3.2 and 7.3.3. Simulation results are provided in Sects. 7.3.4 and experimental results are presented in Sect. 7.3.5.

7.3.1 Iterative Min-Max Algorithm

We propose a Min-Max algorithm that minimizes the maximum absolute $\mathrm{S}^2\mathrm{C}^2$ between the target steering vector and selected N_e clutter steering vectors, i.e. $\min_{\mathbf{z}} \max\{|\mathbf{z}^T \mathbf{w}_{cs}^i|, i = 1, \cdots, N_r\}$, which is convex optimization. It is clear that the optimum configuration includes the boundary antenna-pulse pairs, which can guarantee the largest aperture and high resolution. But, the optimum configuration that consists of the boundary antenna-pulse pairs typically exhibits high sidelobes [209, 275]. In order to solve this problem, we place constraints on the $\mathrm{S}^2\mathrm{C}^2$ values of the resolution grids off the clutter ridge in the Spatial-Doppler plane. Let us denote $\mathbf{f} = [f_s, f_d]$ with $\mathbf{f}_{i,j} = [f_s^i, f_d^j] \in [-0.5, 0.5] \times [-0.5, 0.5], i, j = 1, ..., L_1(L_2)$ be the specified resolution grids in Spatial-Doppler plane with the constrained sidelobe level. Then the correlation steering vector $\mathbf{w}_{i,j}$ is defined as, [104],

$$\mathbf{w}_{i,j} = \mathbf{s}^* \odot (e^{j2\pi \mathbf{n} f_s^i} \otimes e^{j2\pi \mathbf{m} f_d^j}), i, j = 1, ..., L_1(L_2). \tag{7.36}$$

The problem, incorporated with constrained sidelobe level, is formulated as,

$$\begin{aligned} \min_{\mathbf{z},t} \ & t, \\ \text{s.t.} \ & |\mathbf{z}^T \mathbf{w}_{cs}^i| \leq t, i = 1, \cdots, N_r \\ & \mathbf{z} \in \{0, 1\}^{NM}, \\ & \mathbf{1}^T \mathbf{z} = K, \\ & |\mathbf{z}^T \mathbf{w}_{i,j}| \leq \delta_{i,j}, i, j = 1, ..., L_1(L_2), \\ & t > 0. \end{aligned} \tag{7.37}$$

where $\mathbf{w}_{cs}^i$ is the correlation steering vector corresponding to the ith clutter Fourier basis and $\mathbf{1}$ is the vector of ones. Also $\delta_{i,j}$ is the desired normalized sidelobe level with respect to the mainlobe and can be set to $0.5K$ [176].

The optimization problem in Eq. (7.37) is convex except for the binary constraint $\mathbf{z} \in \{0, 1\}^{NM}$, which renders it an NP-hard combinatorial problem. To overcome this difficulty, we use the difference of convex sets (DCS) method, [104], to reformulate the optimization.

Finally, a sequential convex programming based on the first order Taylor decomposition is adopted to solve Eq. (7.37) iteratively. The kth iteration of the Min-Max algorithm then becomes,

$$
\begin{aligned}
\min_{\mathbf{z},t} \quad & t + \mu(\mathbf{1} - 2\mathbf{z}_{k-1})^T \mathbf{z}; \\
\text{s.t.} \quad & |\mathbf{z}^T \mathbf{w}_{cs}^i| \leq t, \quad i = 1, \cdots, N_r \\
& \mathbf{z} \in [0, 1]^{NM}, \\
& \mathbf{1}^T \mathbf{z} = K, \\
& |\mathbf{z}^T \mathbf{w}_{i,j}| \leq \delta_{i,j}, i, j = 1, ..., L_1(L_2), \\
& t > 0.
\end{aligned} \tag{7.38}
$$

where μ is a regularization parameter that balances between the solution sparsity and S^2C^2minimization. Note that Eqs. (7.37) and (7.38) are equivalent and have the same minimum objective value for large enough μ. Moreover, Eq. (7.38) has a Linear Programming (LP) structure and is solvable in reasonable time through some LP solvers [276].

7.3.2 D.C. Programming

Define the binary selection vector $\mathbf{z} \in \{0, 1\}^{MN}$ with "one" meaning the corresponding antenna-pulse pair is selected and "zero" meaning discarded. Then, the two cross-correlation matrices of the selected subarray can be expressed as

$$
\mathbf{D}_c(\mathbf{z}) = \mathbf{V}_c^H \text{diag}(\mathbf{z}) \mathbf{V}_c, \ \mathbf{D}_t(\mathbf{z}) = \mathbf{V}_t^H \text{diag}(\mathbf{z}) \mathbf{V}_t, \tag{7.39}
$$

with the two matrices $\mathbf{V}_c$ and $\mathbf{V}_t$ defined in Eqs. (7.21) and (7.30). Note that $\mathbf{D}_c(\mathbf{z})$ is not an identity matrix after implementing selection although the N_e clutter basis vectors of the full configuration are mutually orthogonal. This is the second reason why we could not simplify $\mathbf{D}_c$ as $\mathbf{I}_{N_e}$ in Eq. (7.29). Correspondingly, the SCNR_{out} of the selected sub-configuration can be written as

$$
\text{SCNR}_{\text{out}} = \text{SNR} \cdot \frac{|\mathbf{D}_t(\mathbf{z})|}{|\mathbf{D}_c(\mathbf{z})|} = \text{SNR} \cdot \frac{|\mathbf{V}_t^H \text{diag}(\mathbf{z}) \mathbf{V}_t|}{|\mathbf{V}_c^H \text{diag}(\mathbf{z}) \mathbf{V}_c|}. \tag{7.40}
$$

The antenna-pulse selection problem in terms of maximizing the SCNR_{out} is formulated as,

$$
\begin{aligned}
\min_{\mathbf{z}} \quad & \frac{|\mathbf{D}_c(\mathbf{z})|}{|\mathbf{D}_t(\mathbf{z})|}, \\
\text{s.t.} \quad & \mathbf{z} \in \{0, 1\}^{MN}.
\end{aligned} \tag{7.41}
$$

Note that the objective function in Eq. (7.41) is equivalent to $\max_{\mathbf{z}} \mathrm{SCNR}_{\mathrm{out}}$. It is clear that the global minimizer is a vector of all ones, when no restriction is placed on the number of selected entries. As it is critical in STAP that the number of DoFs be matched to the available training data, we constrain the number of selected antenna-pulse pairs to be K, given $L = 2K$ training data. The optimization problem becomes,

$$
\begin{aligned}
\min_{\mathbf{z}} \quad & \frac{|\mathbf{D}_c(\mathbf{z})|}{|\mathbf{D}_t(\mathbf{z})|}, \\
\text{s.t.} \quad & \mathbf{z} \in \{0, 1\}^{MN}, \\
& \mathbf{1}^T \mathbf{z} = K.
\end{aligned} \tag{7.42}
$$

Define the feasible set $\mathcal{S} = \{\mathbf{z} : \mathbf{z} \in \{0, 1\}^{MN}\}$, which comprises the extreme points of the polytope $\mathcal{D} = \{\mathbf{z} : 0 \preceq \mathbf{z} \preceq 1\}$, and relax the constraint $\mathbf{z} \in \mathcal{S}$ to $\mathbf{z} \in \mathcal{D}$. Since both $\mathbf{D}_c(\mathbf{z})$ and $\mathbf{D}_t(\mathbf{z})$ are positive definite and the logarithm function is monotonically increasing, Eq. (7.42) is relaxed into the following problem,

$$
\begin{aligned}
\min_{\mathbf{z}} \quad & \log(|\mathbf{D}_c(\mathbf{z})|) - \log(|\mathbf{D}_t(\mathbf{z})|), \\
\text{s.t.} \quad & \mathbf{1}^T \mathbf{z} = K, \\
& \mathbf{z} \in \mathcal{D}.
\end{aligned} \tag{7.43}
$$

The objective function of Eq. (7.43) is a difference of two concave functions and belongs to D.C. Programming [277–279]. A convex-concave procedure (CCP) is adopted to solve it [133], which is proven to be able to converge to a stationary point in [280] with a feasible initial point. Although the stationary point may not be necessarily the local minimum, it was proven to converge to the global optimum under certain conditions in [279]. Note that the DCA (d.c. algorithm) in [277, 278] is reduced to the CCP for differentiable optimization problems. In the following, we give a detailed implementation of the CCP.

The concave function $f(\mathbf{z}) = \log(|\mathbf{D}_c(\mathbf{z})|)$ is approximated iteratively by its first-order Taylor decomposition in the $(k+1)$th iteration as

$$
f(\mathbf{z}) \approx \hat{f}(\mathbf{z}) = f(\mathbf{z}^{(k)}) + \nabla f(\mathbf{z}^{(k)})^T (\mathbf{z} - \mathbf{z}^{(k)}). \tag{7.44}
$$

The jth entry of the gradient $\nabla f(\mathbf{z}^{(k)})$ is

$$
\nabla f_j(\mathbf{z}^{(k)}) = \mathrm{tr}\left\{\mathbf{D}_c^{-1}(\mathbf{z}^{(k)})(\mathbf{v}_{c,j}\mathbf{v}_{c,j}^H)\right\}, \tag{7.45}
$$

where the operator tr{•} takes the trace of the matrix • and $\mathbf{v}_{c,j} \in \mathbb{C}^{N_e \times 1}$ is the jth row vector of $\mathbf{V}_c$. Substituting Eqs. (7.44), (7.45) into Eq. (7.43) and dropping the constant terms, the $(k+1)$th iteration is formulated as,

$$\begin{aligned} \min_{\mathbf{z}} \quad & \nabla f(\mathbf{z}^{(k)})^T \mathbf{z} - \log(|\mathbf{D}_t(\mathbf{z})|), \\ \text{s.t.} \quad & \mathbf{1}^T \mathbf{z} = K, \\ & \mathbf{z} \in \mathcal{D}. \end{aligned} \tag{7.46}$$

According to [131, 137], the global optimum solution of a D.C. programming is on the edge of the polytope $\mathcal{D}$, which is sparse but not necessarily binary. Solving DC problems most often relies on branch and bound methods or cutting plane methods, whereas these global methods often prove slow in practice, requiring many partitions or cuts. Therefore, we are instead concerned with local heuristic that can find local optimum binary solutions rapidly.

Inspired by the effective Gaussian randomization approach in [281], we modify the randomization procedure to obtain a binary solution. We assume a random vector ξ with each component $\xi_i \sim \mathcal{N}(\hat{\mathbf{z}}_i, \varepsilon_i)$, where the mean $\hat{\mathbf{z}}$ is the optimum solution of Eq. (7.46) and ε_i denotes the variance of the ith entry ξ_i. After obtaining a random vector ξ by sampling $\mathcal{N}(\hat{\mathbf{z}}, \text{diag}(\varepsilon))$, we then set the first K largest entries to one and others to zero to generate a feasible point. Moreover, we repeat this random sampling multiple times and choose the one that yields the best objective. As shown in [281], randomization provides an effective approximation for a small number of randomizations. The same conclusion applies here. In order to preserve the one and zero entries of $\hat{\mathbf{z}}$ while varying the non-binary entries, we set $\varepsilon_i = \hat{\mathbf{z}}_i(1 - \hat{\mathbf{z}}_i)$ with the highest variance 0.25 for the most ambiguous entry 0.5. We point out that D.C. Programming can also be adapted to find the sparsest configuration with the required performance. This is achieved by swapping the output SCNR term in the objective function with the number constraint in the second line of Eq. (7.46).

7.3.3 Modified Correlation Measurement

Since the CCP requires solving a sequence of convex optimization problems with computational complexity of order $O((MN)^{3.5})$ for each iteration in the worst case scenario, D.C. Programming is computationally expensive although with theoretical convergence guarantee. We devise a simple alternative search method, named Modified Correlation Measurement (MCM) based on [104]. The MCM belongs to greedy search algorithm which is effective in solving combinatorial optimization problems although without complete mathematical proof up to now [282]. The detailed implementation procedure of the MCM is presented in Table 7.1. Basically, all antenna pulse pairs are selected initially and a backward search is utilized to discard the antenna-pulse pair that gives the smallest objective value in Eq. (7.42) in each iteration until K antenna-pulse pairs remain.

Table 7.1 Modified Correlation Measurement Method

Step 1	Select all antenna-pulse pairs, i.e. $\mathbf{z} = \mathbf{1}$, iteration number $k = 1$, index set $\chi^{(1)} = [1, \cdots, MN]$;
Step 2	For $l = 1 : MN - k + 1$ Set $\tilde{\mathbf{z}} = \mathbf{z}$ and $\tilde{\mathbf{z}}(\chi^{(k)}(l)) = 0$, Calculate $r(l) = \frac{\lvert \mathbf{D}_c(\tilde{\mathbf{z}}) \rvert}{\lvert \mathbf{D}_t(\tilde{\mathbf{z}}) \rvert}$, End;
Step 3	Obtain $i = \arg\min_{l=1,\ldots,MN-k+1} r(l)$;
Step 4	Set $\mathbf{z}(\chi^{(k)}(i)) = 0$, and update the index set $\chi^{(k+1)} = \chi^{(k)} \setminus \chi^{(k)}(i) = \{n \in \chi^{(k)}, n \neq \chi^{(k)}(i)\}$;
Step 5	$k = k + 1$, If $k = MN - K$, terminate; otherwise go back to Step 2.

In practice we may need to obtain the sparsest space-time configuration which satisfies the performance requirement. Since the output SCNR is monotonically increasing with the number of selected antenna-pulse pairs, the MCM can be adapted to achieve this aim. Instead of enforcing constraints on the number of selected antenna-pulse pairs, the termination condition in Step 5 can be the reduction of the output SCNR.

Now Let us calculate the computational complexity of the MCM method. As there are in total $MN - K$ steps in MCM and we need to calculate the determinant of a matrix with dimension $N_e \times N_e$ in each step, the computational complexity of the MCM is thus of the order $O(N_e^3 \times (MN - K))$. Taking the filter weight calculation into account, i.e. implementing STAP with K antenna-pulse pairs, the total computational load is of the order $O(N_e^3 \times MN + K^3)$, which is smaller than that of the traditional STAP of complexity $O(M^3N^3)$. The reason is that the effective clutter rank $N_e \ll MN$, such as $N_e = M + N$ for the sidelooking array. More importantly, the proposed selection strategy is most advantageous for detecting slow-moving targets with reduced training data, which means it is not necessary to implement selection for each angular-Doppler bin.

7.3.4 Simulations

We now validate the proposed thinned STAP using simulated (Sects. 7.3.4.1–7.3.4.4) and experimental (Sect. 7.3.5) data. We utilize the MCM method to calculate optimum sub-configurations and the CNR is fixed at 20 dB throughout this section. We take the sidelooking array as an example in the whole simulation, even though the proposed antenna-pulse selection strategy has no restriction on the array orientation. We use a radar platform with 11 antennas and 20 pulses from Sects. 7.3.4.2–7.3.4.4. Without special notice, the spatial frequency of the target is set as zero.

7.3.4.1 Validation of Two Proposed Methods

In order to demonstrate the effectiveness of the proposed two antenna-pulse selection methods, we use a small radar system which permits us to enumerate all possible sub-configurations. Thus the radar has 4 antennas and 4 pulses and we select 10 antenna-pulse pairs out of 16. We inject a target at a broadside, that is $f_s = 0$. The Doppler frequency is swept over the range $[-0.5, 0.5]$. An optimum subset of 10 antenna-pulse pairs is selected through both proposed selection methods as well as enumeration. In Fig. 7.1, we plot the normalized distance D_e between the evaluated $\widetilde{\text{SCNR}}_{\text{out}}$ and the true value SCNR_{out} as a function of f_d, i.e.,

$$D_e = \frac{\text{SCNR}_{\text{out}} - \widetilde{\text{SCNR}}_{\text{out}}}{\text{SCNR}_{\text{out}}}, \tag{7.47}$$

where $\widetilde{\text{SCNR}}_{\text{out}}$ is the output SCNR evaluated at the obtained solution and SCNR_{out} denotes the true optimum value through enumeration. We can see that both methods are capable of finding the global optimum configuration for slow-moving targets. In other scenarios the sub-optimum solutions can also produce acceptable performance through empirical simulations.

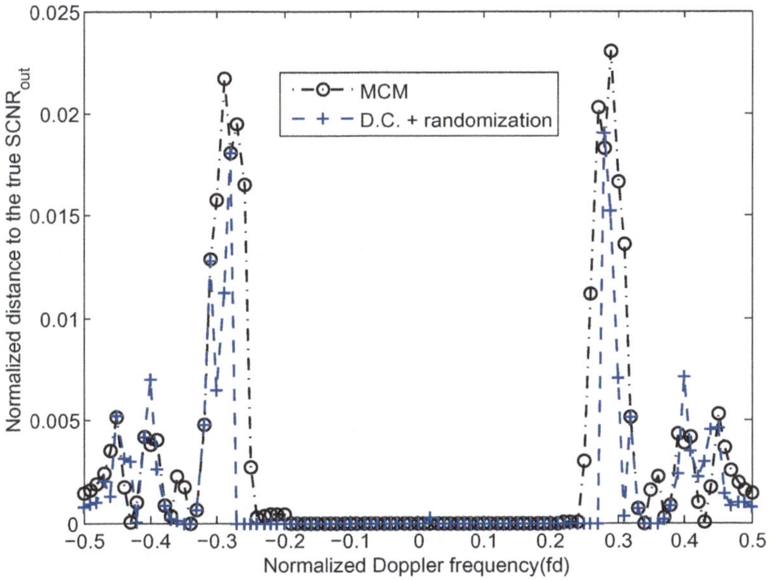

Fig. 7.1 Validation of the MCM, D.C. programming and randomization

7.3.4.2 Redundancy of Full Configuration

A large array of uniformly spaced elements often exhibits significant redundancy, meaning that dropping the redundant elements hardly affects performance. To show this, we examine the relationship between the detection probability and the number of selected antenna-pulse pairs K. Again, we run the simulation for both the Fourier and eigenbasis in Fig. 7.2. We consider two scenarios where the target normalized Doppler frequency is $f_d = 0.05$ and $f_d = 0.1$ respectively. The output SCNR value is set to -5 dB for the full configuration in the two scenarios and homogeneous clutter is assumed. For each K, the optimum configuration of K antenna-pulse pairs is calculated. We utilize $L = 2K$ IID training data to calculate the CCM in the outer plot and fix $L = 220$ in the inner plot. We can see that the optimum sub-configuration achieves the same detection performance as the full one with a much smaller number of DoFs from the outer plot, only about 80 antenna-pulse pairs when $f_d = 0.05$. The number increases to 120 when $f_d = 0.1$. When there is a fixed number of training data, such as $L = 220$ in the inner plot, the best performance is achieved in the neighbourhood of $K = L/2$. An optimum sub-configuration with 50 antenna-pulse pairs is sufficient when $f_d = 0.05$, whereas the detection probability is only 65% when $f_d = 0.1$. These results demonstrate the significant redundancy of the full configuration, especially for slow-moving targets where the clutter is the dominant interfering source. The reason exists in Eq. (7.35), which illustrates the SCNR loss

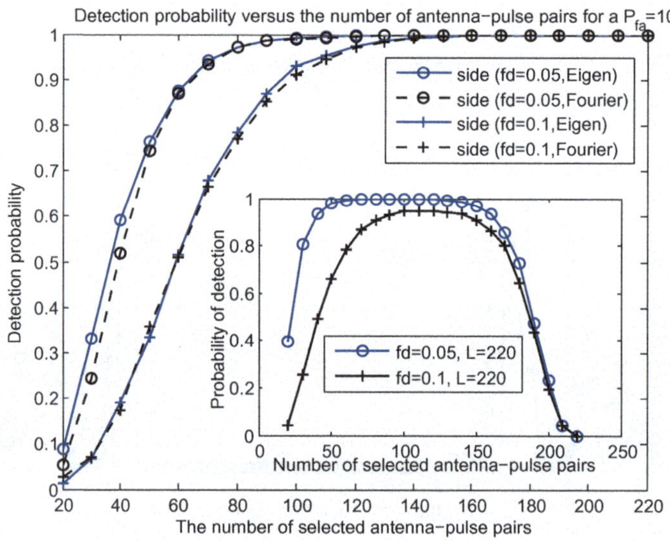

Fig. 7.2 The detection probability versus different numbers of antenna-pulse pairs. The number of training data keeps twice of that of antenna-pulse pairs, i.e. $L = 2K$ in the outer plot; whereas the number of training data is fixed as $L = 220$ in the inner. Each point is averaged over 10^5 Monte-Carlo runs

caused by selection can be significantly compensated by the reduction of S^2C^2, whereas $SCNR_{out}$ is linearly proportional to the number of antenna-pulse pairs K for fast targets. Also observe that there is slight performance difference between the eigenbasis based selection and that of the Fourier basis, which justifies the use of the Fourier basis especially that it is computationally much simpler. The antenna-pulse selection is carried out on the Fourier basis in what follows.

7.3.4.3 Normalized SCNR Loss

We still assume homogeneous clutter returns in this simulation. The number of IID training data required by the SMI algorithm for the full configuration to achieve an SCNR loss within 3 dB of the optimum is $2NM = 440$. But we set the number of selected antenna-pulse pairs to $K = 100$, meaning that only $2K = 200$ training data snapshots are required for the sub-configurations. The performance of clutter suppression strategies in STAP is commonly assessed using the normalized SCNR loss, [252],

$$L_s = \frac{|\mathbf{v}^H \hat{\mathbf{Q}}^{-1} \mathbf{v}|^2}{|\mathbf{v}^H \hat{\mathbf{Q}}^{-1} \mathbf{Q} \hat{\mathbf{Q}}^{-1} \mathbf{v}||\mathbf{v}^H \mathbf{Q}^{-1} \mathbf{v}|}, \tag{7.48}$$

where $\hat{\mathbf{Q}} = \frac{1}{L} \sum_{i=1}^{L} \mathbf{x}_i \mathbf{x}_i^H$ is the sample CCM and $\mathbf{v}$ is the scanning steering vector. Firstly, we examine the CCM convergence rate in the upper plot of Fig. 7.3. The normalized Doppler frequency of the target is set to $f_d = 0.05$, where the corresponding optimum sub-configuration is shown in Fig. 7.4. The number of homogeneous training data is changing from 100 for the sub-configuration (220 for the full configuration to avoid singularity) to 440 in steps of 10. We can see that the CCM of the full configuration requires at least 300 IID training data to converge, which is much slower than the sub-configuration. We then examine the relationship between L_s and the normalized Doppler frequency in the lower plot of Fig. 7.3. The normalized Doppler frequency of the target sweeps over the range $[-0.5, 0.1]$ and we calculate the optimum subset of 100 antenna-pulse pairs corresponding to each frequency. We compare the optimum configuration, which we denote by "sub_{opt}" using 200 training data, to the full configuration, denoted by "full" using 440 training data. The performances of the full and sub-configuration with 300 training data are also included for reference. We can observe the considerable performance degradation for the full configuration without sufficient training data.

7.3.4.4 Robustness Against Heterogeneity

Finally we illustrate the improvement of the detection probability that the selection strategy provides in heterogeneous clutter in Fig. 7.5. In the simulation, the probability of false alarm rate is set to $P_{\text{fa}} = 10^{-3}$. As we are primarily interested in slow moving targets, we assume the normalised target Doppler frequency to be uniformly

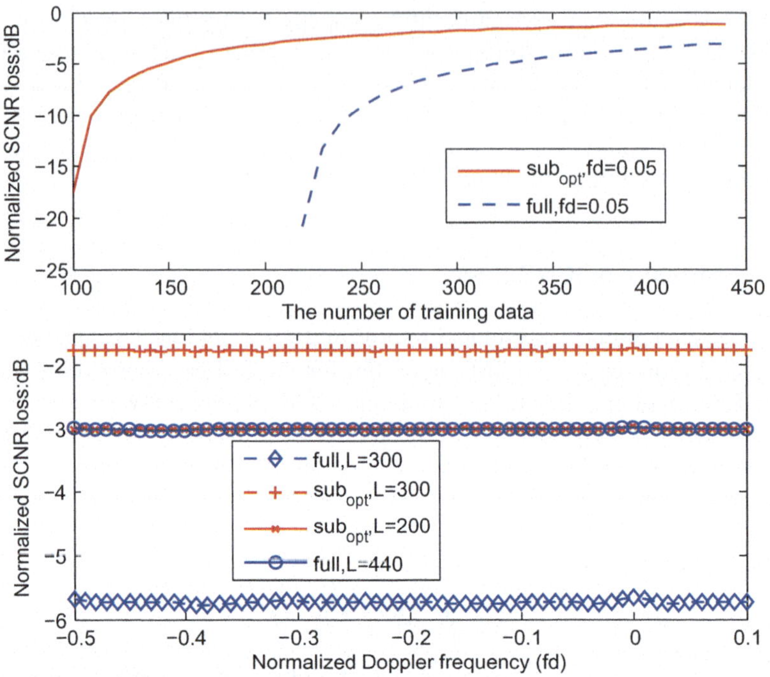

Fig. 7.3 Normalized SCNR loss versus the number of training data with $f_d = 0.05$ (upper plot) and versus normalised Doppler frequency with $L = 2K \setminus 300$ (lower plot), each point is averaged over 10^4 Monte-Carlo runs

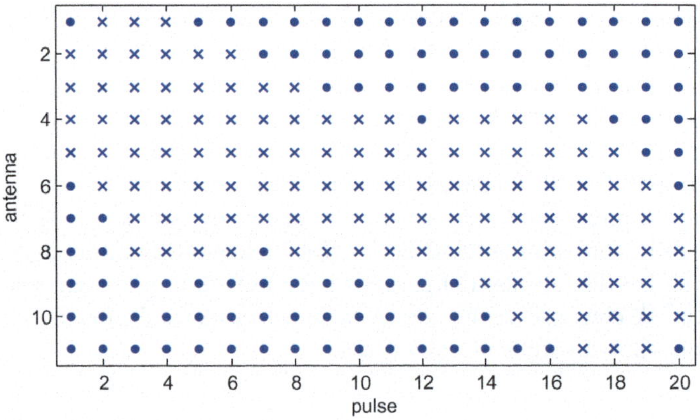

Fig. 7.4 100 selected antenna-pulse pairs in the cases of $f_d = 0.05$. The filled dots stand for selected and the cross means discarded

distributed over the range $[-0.1, 0.1]$. We use the label "homo" to refer to curves for a homogeneous environment, leaving all others that show the performance under heterogeneity without a label. In the homogeneous case, we obtain the detection curves for the full configuration when the CCM is estimated from 440 IID training data snapshots. In the heterogeneous scenario, we simulate the clutter heterogeneity by inserting 60 high-amplitude, mainbeam discrete targets into various range cells with normalized spatial and Doppler frequencies uniformly distributed within $[-0.5, 0.5]$, [256]. The required training data snapshots for both sub-configurations are selected using the generalized inner product (GIP)-based non-homogeneity detector (NHD) of [256]. Specifically, we sort the GIP values and retain the $2K = 200$ realizations corresponding to smallest GIP value. Moreover, the labels $2K$ and $2MN$ indicate the number of snapshots used for the CCM estimation and the curve labelled "Ext" is the result of using all the snapshots that are returned by the NHD with GIP value below a predetermined threshold. In order to better exhibit the superiority of the proposed selection method, we compare its detection performance with the iterative min-max algorithm [153], denoted as "min-max". Note that the sub_{non} is the sub-configuration with the first 100 antenna-pulse pairs selected. We make the following observations:

- The sub-configurations combined with the NHD retain practically the same detection performance as the homogeneous case. We do not plot the curves for the sub-configurations in the homogeneous case for simplicity.

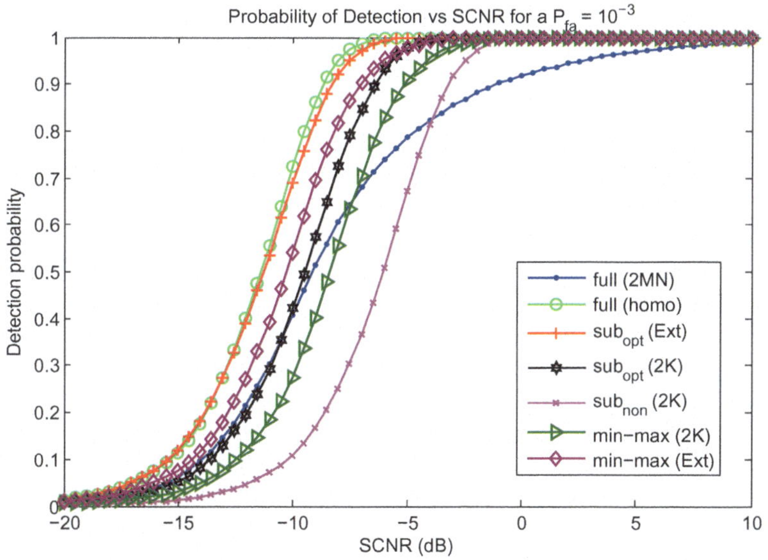

Fig. 7.5 Detection probability curve versus SCNR in both homogeneous and heterogeneous clutter, each point is averaged over 10^5 Monte-Carlo runs

- The proposed subspace based selection method outperforms the iterative min-max algorithm, as the generalized S^2C^2 directly relates the output SCNR with the selected antenna-pulse pairs;
- The optimum sub-configuration exhibits better performance than the non-optimum and min-max configurations in both homogeneous and heterogeneous clutter;
- In heterogeneous clutter, the optimum sub-configuration combined with the NHD outperforms the full array, as it requires fewer IID training data than the full one.
- Interestingly, we find that the optimum sub-configuration with extended IID training data exhibits similar performance with the full configuration in the homogeneous clutter, which further illustrates the redundancy of the full configuration in the scenarios where the target is close to the clutter trajectory.

7.3.5 Experimental Results

The experimental radar data we used is the MCARM database [283]. The radar platform was travelling at around 100 ms^{-1} over an area that includes a variety of terrain types as well as interfering discretes. The radar was comprised of a rectangular array of two rows in elevation and 11 sensors in azimuth and collected a coherent pulse interval of length 128 pulses. The pulse repetition frequency was 1984 Hz. The examples presented here are based on the acquisition 575 from flight 5. Since the number of available range cells is only 630, we employ the upper 11 uniformly spaced antennas and the first 20 pulses which satisfies the sample size requirement for the CCM estimation. The clutter spectrum in the angle-Doppler plane is estimated in the left plot of Fig. 7.6. The clutter ridge is shifted right in the Doppler axis due to the small crab angle.

Using a similar approach to [284], we inject one target into the range gate 290, which is located at broadside and Doppler bin 3 (corresponding to a Doppler frequency of 46.5 Hz) with a magnitude 0.003. The target location is also marked in Fig. 7.6. The selected optimum sub-configuration with 100 antenna-pulse pairs is shown in Fig. 7.8. The clutter spectrum corresponding to the optimum sub-configuration is presented in the right plot of Fig. 7.6. We can see that the clutter trajectory is stretched and narrowed, which increases the separation between the target and the clutter. The normalized AMF statistics versus range cells from 250 to 400 are shown in Fig. 7.7. The thresholds for both full and sub-configurations obtained from the simulations above are also shown. Although one cannot ascertain whether all the peaks above the threshold are valid targets or false alarms, it is highly unlikely that they would all be valid targets. It is certainly fair to say the optimum sub-configuration produces a much more reasonable detection performance than both the full configuration and the min-max solution, showing a significant reduction in the number of false alarms. The reason is that the maximum number of homogeneous training data is around 300 for acquisition 575 in flight 5 [283], whereas the full configuration requires 440 IID snapshots (Fig. 7.8).

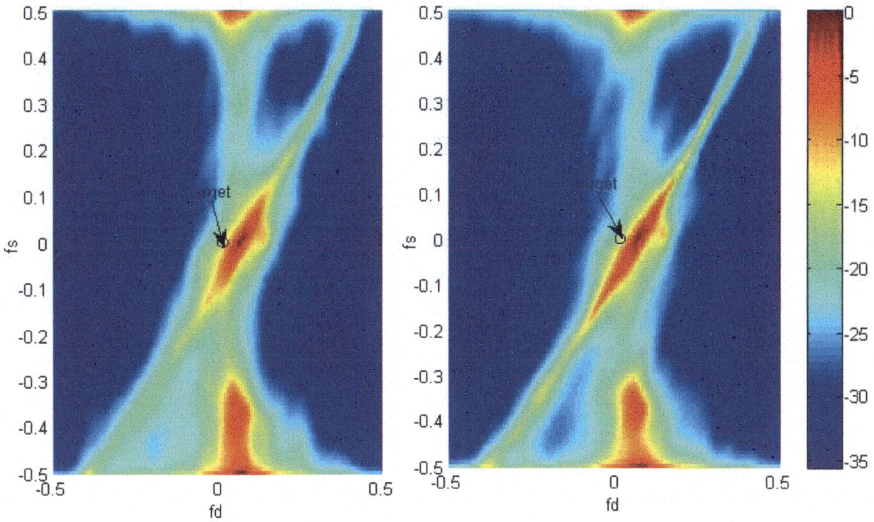

Fig. 7.6 Clutter spectrum in angle-Doppler plane for both full configuration (left) and the selected optimum sub-configuration (right)

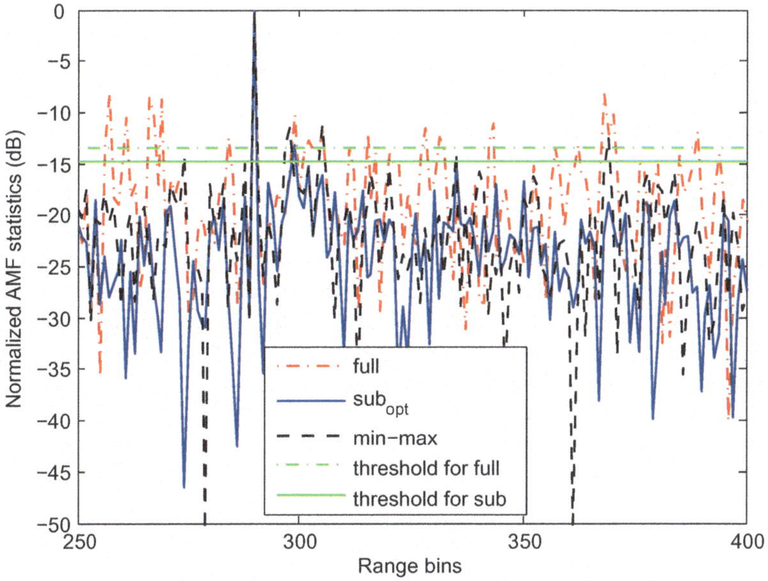

Fig. 7.7 The normalized AMF statistics versus range bins for MCRAM data at Doppler bin 3 and broadside

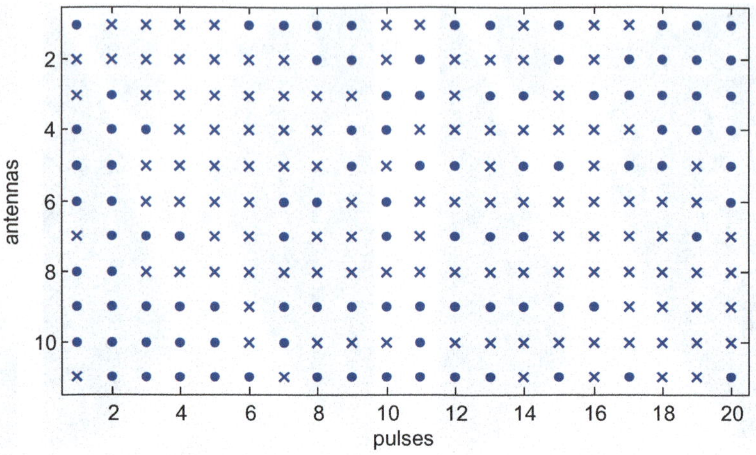

Fig. 7.8 The selected optimum sub-configuration with 100 antenna-pulse pairs

7.4 Summary

In this chapter, we introduced a new perspective, where we viewed the interleaved antenna-pulse data as forming a single large sensor array, and investigated a thinned STAP by carrying out selection to this large sensor array. Each sensor, specified by the antenna number and pulse number, corresponds to an antenna-pulse pair. We reformulated Spatial and Spectral Correlation Coefficient (S^2C^2) to characterize the space-time separation between the target and clutter subspace. Based on the S^2C^2, we implemented the antenna-pulse selection as a maximization of the output SCNR and presented three algorithms to select the optimum antenna pulse pairs based on the three definitions of S^2C^2. The performance of the proposed antenna-pulse selection strategy was validated using simulations and MCARM experimental data, which demonstrated its effectiveness on robust radar target detection while reducing the computational load and addressing the problems of clutter heterogeneity and limited sample support.

Chapter 8
Sparse Sensing for Dual-Functional Radar Communications

The ongoing intensive research has been developing multi-function solutions to the coexistence of radar and communications in the increasingly congested radio frequency (RF) spectral environment. Dual-function radar communications (DFRC) systems have recently been proposed to enable the coexistence of radar and wireless communications. The nominal array configuration for existing DFRC systems is uniform and of fixed-structured, however, it is not necessarily optimum in every sense, and ignores the additional degrees of freedom (DoFs) provided by the flexibility in configuring the antenna array. In this chapter, we introduce the sparse sensing for dual-functional radar communications.

In Sect. 8.2, we investigate the optimum sparse array design for dual-function radar communications. We solve the problem of sparse transmit array design for DFRC systems by antenna selection where same or different antennas are assigned to different functions. We consider three different types of DFRC systems which implement different simultaneous beamformers associated with single and different sparse arrays with shared aperture. We utilize the array configuration as an additional spatial degree of freedom (DoF) to suppress the cross-interference and facilitate the cohabitation of the two system functions.

In Sect. 8.3, we consider the sparse array optimization for spatial index modulation. We propose a new technique for communication information embedding into the emission of multiple-input multiple-output (MIMO) radar using sparse antenna array configurations. We show that by reconfiguring sparse transmit array through antenna selection and reordering waveform-antenna paring, a data rate of megabits per second can be achieved for a moderate number of transmit antennas.

X. Wang et al., *Sparse Sensing and Sparsity Sensed in Multi-sensor Array Applications*, https://doi.org/10.1007/978-981-99-9558-5_8

8.1 Introduction

In recent years, radio frequency (RF) spectrum is becoming increasingly congested with an exponentially growing demand by end-consumers. Consequently, defense applications are losing spectrum to commercial communications, and have to operate in congested environments. Emerging research in multi-function platforms aims at using common or shared aperture and frequency spectrum between radar, electronic warfare, and military communications [285–289], whose coexistence benefits from common transmit platform, and in turn moving away from independent systems [290–292]. In order to enable usage or sharing of spectrum resources and platform hardware, a dual-function radar communications (DFRC) system utilizing waveform diversity in tandem with amplitude/phase control of the radar beam was introduced in [293–298], where radar is considered as the primary function and presents itself as a system of opportunity to secondary communication functions. In dual-function paradigm, identical signals, same carrier frequency and bandwidth, and common antenna array are deployed to fulfill the objectives of both radar and communication operations. In DFRC system, secondary communications strive to embed a sequence of binary data during each radar pulse which can be achieved through scaling or modulation of either the radar beam or the radar waveform or both. This signal embedding, however, should be accomplished with no, or minimum, alterations to primary radar function, whether it is detection, tracking, or estimation.

One signaling strategy for embedding information into the radar pulsed emissions uses sidelobe amplitude modulation (AM) and changes the sidelobe level (SLL), according to the information message, towards the intended communication user direction [299]. In lieu of AM, a coherent phase-modulation (PM)-based method was proposed in [300, 301] to embed one symbol into the radar emission by controlling the phase of complex transmit array pattern. The benefits of decomposing the radar pulse into different waveforms were demonstrated in [301] where a communication symbol is embedded as a phase rotation between a pair of transmitted orthogonal waveforms. A multi-waveform amplitude shift keying (ASK) strategy was introduced in [302] to embed one binary bit with each orthogonal waveform through bilevel sidelobe control.

Similar to the offering of multi-waveforms, the dual functions of the radar communications system can be improved by properly utilizing the multi-sensor transmit/receive array configurations [303–305]. Although the nominal array configuration for existing DFRC systems is uniform and of fixed-structured, it is not necessarily optimum in every sense, and ignores the additional degrees of freedom (DoFs) provided by the flexibility in configuring the antenna array [43, 156, 209, 244, 306]. Non-uniform sparse arrays have attracted increased attention in multi-sensor transmit/receive systems as an effective solution to reduce the system's complexity and cost, yet retain desired performance [307–309]. Sparse array design is often cast as optimally placing a given number of antennas on a larger number of possible uniform grid points. In so doing, we are able to span large aperture without introducing unwanted high sidelobes, thus improving spatial resolution. Sparse array design becomes an "antenna selection" problem when the number of antennas is equal to the number of grid points, but with fewer RF units. In this case, antenna selection amounts, in essence, to assigning antennas to RF units. In transmit antenna selection, the number of expensive RF chains, which consist of digital-to-analog converter, up-converter, filters and power amplifiers, is smaller than the number of available transmit antenna elements [163, 310]. Thus sparse arrays can, undoubtedly, alleviate pressures on the resource management and efficiency requirements on power amplifiers.

In Sect. 8.2, we introduce the sparse array design to promote dual-function radar communications. We consider the problem of sparse transmit array design for DFRC systems by antenna selection where same or different antennas are assigned to different functions. We consider three different types of DFRC systems which implement different simultaneous beamformers associated with single and different sparse arrays with shared aperture. We utilize the array configuration as an additional spatial degree of freedom (DoF) to suppress the cross-interference and facilitate the cohabitation of the two system functions. It is shown that the use of sparse arrays adds to improved angular resolution with well-controlled sidelobes on DFRC system paradigm. The utilization of sparse arrays in DFRC systems is validated using simulation examples.

In Sect. 8.3, we consider the sparse array optimization for spatial index modulation. We propose a new technique for communication information embedding into the emission of multiple-input multiple-output (MIMO) radar using sparse array configurations. The phases induced by antenna displacements in a sensor array are unique, which makes array configuration feasible for symbol embedding. We also exploit the fact that in a MIMO radar system, the association of independent waveforms with the transmit antennas can change over different pulse repetition periods without impacting the radar functionality. We show that by reconfiguring sparse transmit array through antenna selection and reordering waveform-antenna paring, a data rate of megabits per second can be achieved for a moderate number of transmit antennas. To counteract practical implementation issues, we propose a regularized antenna selection based signaling scheme. The possible data rate is analyzed and the symbol/bit error rates are derived. Simulation examples are provided for performance evaluations and to demonstrate the effectiveness of proposed DFRC technique.

8.2 Optimum Sparse Array Design for DFRC

In this section, we add to DFRC system paradigm by introducing a new co-existence approach based on antenna selection. Specifically, we examine the problem of sparse array beampattern synthesis with a fixed number of transmit antennas under the framework of dual functional system design. Three different types of DFRC systems are considered, namely, a system that implements (a) single sparse array with one set of weights, i.e., a single beamformer; (b) single sparse array but with different beamformers; (c) multiple sparse but complementary arrays with different beamformers. The latter case is referred to as "shared aperture", where the combined sparse arrays span the given system aperture. The main advantages of utilizing sparse arrays in DFRC systems are manifested by simulation results in the suppression of cross-interference between the two functions and improved hardware efficiency.

8.2.1 System Configuration and Signal Model

We consider a joint radar communications platform equipped with a reconfigurable transmit antenna array through an antenna selection network, as shown in Fig. 8.1. There are K transmit antennas uniformly spaced with an inter-element spacing of d and M front-ends available for waveform transmitting. The antenna selection network comprises K RF switches which connect/disconnect antennas with front-ends. Suppose a transmit array is configured with M selected antennas located at $p_m d, m = 1, \ldots, M$ with $p_m \in \mathbb{N}$ being non-negative integers. The radar receiver, on the other hand, employs an array of N receive antennas with an arbitrary linear configuration. Without loss of generality, a single-element communication receiver is assumed to be located in the far field at direction θ_c, which is known to the

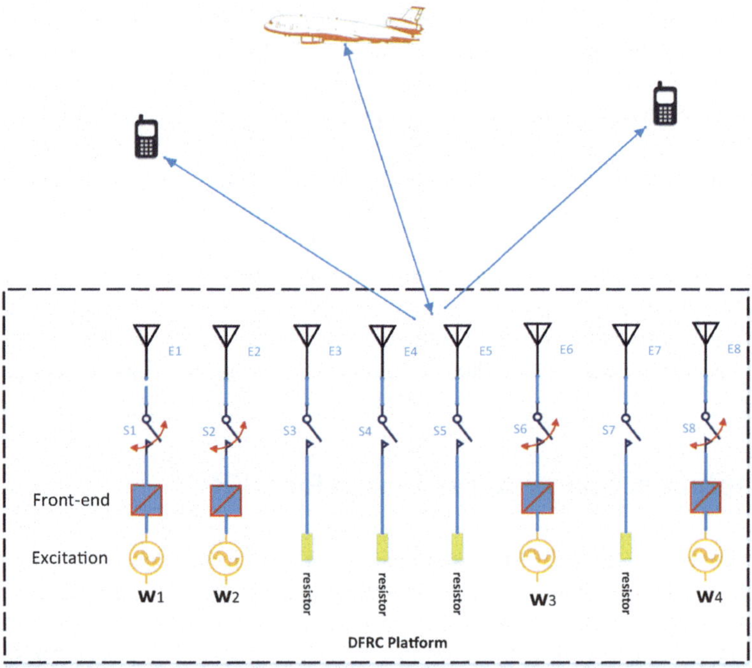

Fig. 8.1 Joint platform of dual function radar communications with antenna selection network

transmitter. Let $\Psi_l(t), l = 1, \ldots, L$ be a set of L orthogonal waveforms, each occupying the same bandwidth. In other words, the spectral contents of all waveforms fully overlap in the frequency domain. Each waveform is normalized to have unit power, i.e., $\int_T |\Psi_l(t)|^2 dt = 1$, with T and t denoting the radar pulse duration and the fast time index, respectively. It is further assumed that the orthogonality condition $\int_T \Psi_l(t)\Psi_{l'}^*(t)dt = 0$ is satisfied for $l \neq l'$, where $()^*$ stands for the complex conjugate.

Let $\mathbf{s}(t;\tau)$ be the $M \times 1$ baseband transmit signal vector during the τth radar pulse. Assume that Q far-field targets at directions $\theta_q, q = 1, \ldots, Q$, located within the radar main beam, are observed in the background of strong clutter and interference. The $N \times 1$ baseband representation of the signals at the output of the radar receive antenna array is given by,

$$\mathbf{x}(t;\tau) = \sum_{q=1}^{Q} \beta_q(\tau)\mathbf{s}^H(t;\tau)\mathbf{a}(\theta_q)\mathbf{b}(\theta_q) + \mathbf{n}(t;\tau), \tag{8.1}$$

where $\beta_q(\tau)$ is the qth target reflection coefficient which obeys the Swerling-II target model [299], i.e., they remain constant during the entire pulse duration, but vary independently from pulse to pulse. The vector $\mathbf{a}(\theta)$ is the steering vector of the transmitting array, defined as,

$$\mathbf{a}(\theta) = [e^{jk_0 p_1 d \sin\theta}, \ldots, e^{jk_0 p_M d \sin\theta}]^T, \tag{8.2}$$

where $k_0 = 2\pi/\lambda$ is the wavenumber. The steering vector of the receiving array, $\mathbf{b}(\theta)$, can be defined in a similar way as $\mathbf{a}(\theta)$. The vector $\mathbf{n}(t;\tau)$ is of $N \times 1$ dimension, representing the unwanted clutter, interference and white noise in the τth radar pulse.

The baseband signal received by the communication antenna can be expressed as,

$$x_c(t;\tau) = \beta_c(\tau)\mathbf{s}^H(t;\tau)\mathbf{a}(\theta_c) + n_c(t;\tau), \tag{8.3}$$

where $\beta_c(\tau)$ is the channel coefficient of the received signal that summarizes the propagation environment between the transmit array and the communication receiver during the τth pulse, and $\mathbf{a}(\theta_c)$ is the steering vector of the transmit array toward the communication direction θ_c. In addition, $n_c(t;\tau)$ is the noise signal. We can observe from Eqs. (8.1) and (8.3) that the array configuration, via the non-uniform inter-element spacing, plays a pivotal role in determining the beamforming performance, which can be considered as an additional DoF designated for performance enhancement. We elaborate on the design of sparse arrays for DFRC systems in the following sections, under different constraints of common and shared system resources.

8.2.2 Design of Common Array with Single Beamformer

This section considers a common transmit array with a single beamformer for both radar and communication functions, and examines the sparse array design for different signaling strategies in phased array radar.

8.2.2.1 Revisit of Different Signaling Schemes

The general expression of transmit signal for the phased-array radar can be written as,

$$\mathbf{s}(t;\tau) = \Psi(t)\mathbf{w}(\tau), \tag{8.4}$$

where the $M \times 1$ beamforming weight vector $\mathbf{w}(\tau)$, $\|\mathbf{w}(\tau)\|_2 = 1$, is varied in PRIs and required to satisfy a certain desired transmit power radiation pattern. Substituting Eq. (8.4) into Eq. (8.3), the transmit beamforming complex gain towards the communication receiver becomes $\mathbf{w}^H\mathbf{a}(\theta_c) = G_c e^{j\phi_c}$ with G_c and ϕ_c denoting the magnitude and phase, respectively. During each radar pulse, communications can be simultaneously achieved with radar functions by embedding information into either the amplitude G_c or the phase ϕ_c. The associated methods amount to radar beam modulations, and are termed AM- and PM-based methods, respectively.

The AM-based method is to embed information into the radar emission via modulating the complex gain amplitude G_c towards the intended communication direction. To satisfy the primary radar operation requirements, the radar mainlobe is kept unchanged during the entire CPI. Thus, the AM-based method can only enable information delivery to a communication receiver located within the sidelobe region. Accordingly, a narrower radar mainlobe implies a wider angular sector where the communication function can take place. The N_b information bits are mapped into a dictionary of $J = 2^{N_b}$ sidelobe levels (SLLs) denoted as $G_c \in \{\Delta_1, \ldots, \Delta_J\}$. Therefore, the implementation of this method requires a single radar waveform $\Psi(t)$ and J beamforming weight vectors associated with distinct SLLs. The PM-based method, on the other hand, embeds information by controlling the phase ϕ_c of the beam radiated towards the communication receiver. The N_b sequence of binary bits are mapped into a dictionary of J phase symbols denoted as $\phi_c \in \{\phi_1, \ldots, \phi_J\}$. For coherent communications, this method amounts to using one waveform and J beamforming weights associated to deliver J distinct phases.

Matched filtering the received communication signal in Eq. (8.3) with the waveform $\Psi(t)$ yields,

$$y_c(\tau) = \int_T x_c(t,\tau)\Psi(t)dt = \beta_c(\tau)G_c e^{j\phi_c} + n_c(\tau), \tag{8.5}$$

where $n_c(\tau) = \int_T n_c(t;\tau)\Psi(t)dt$ is the additive noise term after integration. The embedded communication symbol can be estimated as,

$$\hat{G}_c = \left|\frac{y_c(\tau)}{\beta_c(\tau)}\right| \quad \text{and} \quad \hat{\phi}_c = \text{angle}(y_c(\tau)) - \text{angle}(\beta_c(\tau)), \tag{8.6}$$

where $|\cdot|$ stands for absolute value and angle$(\cdot)$ is the angle of the argument. The actual embedded binary message can be decoded by comparing the complex gain estimate with the pre-defined J-dimensional dictionary. Both the AM and PM methods can achieve a data rate of $R_b = N_b \cdot \text{PRF}$ in bits per second, with PRF denoting the pulse repetition frequency. The hybrid amplitude phase modulation method, such as quadrature amplitude modulation (QAM), can certainly achieve a doubled data rate by choosing the amplitude Δ_k and the phase ϕ_k from the respective pre-defined dictionary G_c and ϕ_c, compared with single AM or PM beam modulation.

In addition, two other signaling strategies for information embedding using waveform diversity in tandem with beamforming gain control were proposed in [302]. The multi-waveform two-level ASK (2ASK) and binary phase shift keying (BPSK) methods employ multiple waveforms in conjunction with two transmit beamforming weight vectors, denoted as $\mathbf{w}_H$ and $\mathbf{w}_L$ ($\mathbf{w}_P$ and $\mathbf{w}_Q$ for BPSK). In order to embed N_b bits per radar pulse, N_b orthogonal waveforms are simultaneously transmitted with each waveform delivering one information bit to the communication receiver. During each radar pulse, the waveform $\Psi_n(t), n = 1, \ldots, N_b$ is radiated either via $\mathbf{w}_H$ ($\mathbf{w}_P$) for $b_n = 1$ or $\mathbf{w}_L(\mathbf{w}_Q)$ for $b_n = 0$. Accordingly, the transmit signals in Eq. (8.4) can be rewritten as,

$$\mathbf{s}(t) = \sum_{n=1}^{N_b} \sqrt{\frac{P_t}{N_b}} \left(b_n \mathbf{w}_i + (1 - b_n)\mathbf{w}_j\right) \Psi_n(t), i = H, P; j = L, Q. \tag{8.7}$$

Here, the power budget P_t is assumed to be equally distributed among the N_b orthogonal waveforms. Matched filtering the received signal with each orthogonal waveform yields the $N_b \times 1$ data vector $y_{c,n}, n = 1, \ldots, N_b$, defined as

$$y_{c,n}(\tau) = \sqrt{\frac{P_t}{N_b}} \beta_c(\tau)[(1 - b_n)\mathbf{w}_i + b_n \mathbf{w}_j]^H \mathbf{a}(\theta_c) + n_{c,n}(\tau), \tag{8.8}$$
$$n = 1, \ldots, N_b, i = H, P; j = L, Q.$$

where $n_{c,n}(\tau) = \int_T n_c(t; \tau)\Psi_n(t)dt, n = 1, \ldots, N_b$. The embedded communication symbol can then be estimated according to Eq. (8.6). Clearly, in this case, the data rate is $N_b \times$ PRF bits per second.

8.2.2.2 Design of Common Array with Single Beamformer

For a given transmit array configuration, the design of the beamforming weight vector $\mathbf{w}$ embedding the kth communication symbol $\Delta_k e^{j\phi_k}$ can be formulated in two different cases. In the case of desired focused beampattern for the radar function, the weight vector $\mathbf{w}$ is designed to maintain the power radiation in the sidelobe region $\bar{\Theta}$ under the specified level ρ, while maintaining a unit gain towards the radar target direction θ_t. The problem formulation can be written as,

$$\begin{aligned} \textbf{Focused Pattern:} \quad & \min_{\mathbf{w},\alpha} \ \alpha, \\ \text{subject to} \quad & \mathbf{w}^H \mathbf{a}(\theta_t) = e^{j\mu(\theta_t)}, \\ & \left|\mathbf{w}^H \mathbf{a}(\theta_l)\right| \le \rho + \alpha, \ \theta_l \in \bar{\Theta}, l = 1, \ldots, L_s \\ & \mathbf{w}^H \mathbf{a}(\theta_c) = \Delta_k e^{j\phi_k}, \ k = 1, \ldots, J, \end{aligned} \tag{8.9}$$

where $\theta_k, k = 1, \ldots, L_s$ are L_s samples of the sidelobe region $\bar{\Theta}$, and $\mu(\theta_t)$ denotes the phase of unit complex gain. Here, α is an auxiliary variable for controlling sidelobe levels and its maximum value is 0. In the case of desired flat-top beampattern for the radar function, the main function of $\mathbf{w}$ is to concentrate the transmit power within a certain angular sector $\Theta = [\theta_{\min}, \theta_{\max}]$. The mainlobe ripples are required to be less than a specified level ϵ and the power radiation level corresponding to the sidelobe region $\bar{\Theta}$ is well-controlled under the specified value ρ. This problem can be formulated as,

$$\begin{aligned} \textbf{Flat-top Pattern:} \quad & \min_{\mathbf{w},\alpha} \ \alpha, \\ \text{subject to} \quad & \left|\mathbf{w}^H \mathbf{a}(\theta_i) - e^{j\mu(\theta_i)}\right| \le \epsilon, \ \theta_i \in \Theta, i = 1, \ldots, L_m, \\ & \left|\mathbf{w}^H \mathbf{a}(\theta_l)\right| \le \rho + \alpha, \ \theta_l \in \bar{\Theta}, l = 1, \ldots, L_s \\ & \mathbf{w}^H \mathbf{a}(\theta_c) = \Delta_k e^{j\phi_k}, \ k = 1, \ldots, J, \end{aligned} \tag{8.10}$$

where $\theta_i, i = 1, \ldots, L_m$ are samples of the mainlobe region Θ, and $\mu(\theta_i), i = 1, \ldots, L_m$ stands for the mainlobe phase profile. Different from the formulation in [301], the mainlobe phase profile $\mu(\theta_i), i = 1, \ldots, L_m$ is adjustable in lieu of assuming fixed values. We use the phase profile of the desired beampattern, $e^{j\mu(\theta)}$, as a free parameter in the optimization in order to achieve better approximation to the desired beampattern. In this respect, we utilize an alternating descent method that iteratively shifts between the weight vector $\mathbf{w}$ and the phase profile $e^{j\mu(\theta)}$ [104]. In the $(l+1)$th iteration, we compute the weight vector $\mathbf{w}^{(l+1)}$ by solving Eq. (8.10) based on the phase profile $e^{j\mu^{(l)}(\theta)}$ from the previous lth iteration. Then, we update the phase profile as follows,

$$e^{j\mu^{(l+1)}(\theta_i)} = \frac{\mathbf{w}^{(l+1)H}\mathbf{a}(\theta_i)}{|\mathbf{w}^{(l+1)H}\mathbf{a}(\theta_i)|}, i = 1, \ldots, L_m, \theta_i \in \Theta. \tag{8.11}$$

It was proved in [104] that the combination of phase profile updating formula in Eq. (8.11) and alternating descent method is capable of converging to the ideal power pattern synthesis within the mainlobe angular region, that is $-\epsilon \le |\mathbf{w}^H \mathbf{a}(\theta_i)| - 1 \le \epsilon, i = 1, \ldots, L_m$ with ϵ denoting the maximum allowable ripple.

The beamforming weight design in Eqs. (8.9) and (8.10) often assumes a uniform transmit array configuration with M antennas. It is difficult to design a large-aperture uniform linear array (ULA) due to increased number of antennas and associated hardware cost, which may, in turn, limit the angular region where the communication function takes place. The optimally designed sparse array is capable of achieving high angular resolution without introducing unwanted high sidelobes.

The sparse array transmit pattern synthesis considered in this work involves the integrated design of sparse array configuration and the corresponding beamforming weights. Define an antenna selection vector $\mathbf{r} \in \{0, 1\}^M$ with entry "1" denoting the respective antenna selected and entry "0" discarded. The positions of "1" entries in the vector $\mathbf{r}$ determine the transmit array configuration. The beamforming weight vector $\mathbf{w}$ is mandated to exhibit the same sparse structure as the selection vector $\mathbf{r}$. The design of a common sparse array with a single beamformer should satisfy both radar and communication functions. Taking the focused beampattern synthesis in Eq. (8.9) as an example, the sparse array design for dual-functional systems can be formulated as follows,

$$
\begin{aligned}
\min_{\mathbf{w},\mathbf{r},\alpha} \quad & \alpha, && (8.12a)\\
\text{subject to} \quad & \mathbf{w}^H\mathbf{a}(\theta_t) = e^{j\mu(\theta_t)}, \\
& \left|\mathbf{w}^H\mathbf{a}(\theta_l)\right| \le \rho + \alpha, \ \theta_l \in \bar{\Theta}, l = 1, \ldots, L_s \\
& \mathbf{w}^H\mathbf{a}(\theta_c) = \Delta_k e^{j\phi_k}, && (8.12b)\\
& |w_m|^2 \le r_m, m = 1, \ldots, K && (8.12c)\\
& \mathbf{r} \in \{0, 1\}^M, \ \mathbf{1}^T\mathbf{r} = M, && (8.12d)
\end{aligned}
$$

where the constraint in Eq. (8.12b) is used to embed the communication symbol $\Delta_k e^{j\phi_k}$. The constraint in Eq. (8.12c) is imposed to couple the two variables $\mathbf{w}$ and $\mathbf{r}$, and promotes the same sparsity of the beamforming weight vector $\mathbf{w}$ as the selection vector $\mathbf{r}$. The constraint in Eq. (8.12d) restrains the number of selected antennas to be exactly M.

The problem in Eq. (8.12a) belongs to notorious combinatorial optimization. In order to eliminate the boolean constraint in Eq. (8.12d), the formulation in Eq. (8.12a) can be rewritten as,

$$
\begin{aligned}
\min_{\mathbf{w},\mathbf{r},\alpha} \quad & \alpha + \gamma\mathbf{r}^T(\mathbf{1} - \mathbf{r}), && (8.13)\\
\text{subject to} \quad & \mathbf{w}^H\mathbf{a}(\theta_t) = e^{j\mu(\theta_t)}, \\
& \left|\mathbf{w}^H\mathbf{a}(\theta_l)\right| \le \rho + \alpha, \ \theta_l \in \bar{\Theta}, l = 1, \ldots, L_s \\
& \mathbf{w}^H\mathbf{a}(\theta_c) = \Delta_k e^{j\phi_k}, \\
& |w_m|^2 \le r_m, m = 1, \ldots, K \\
& 0 \le \mathbf{r} \le 1, \ \mathbf{1}^T\mathbf{r} = M.
\end{aligned}
$$

Here, the boolean constraint on $\mathbf{r}$ is tantamount to the combination of the box constraint $0 \leq \mathbf{r} \leq 1$ and the second part of the objective function $\min_{\mathbf{r}} \mathbf{r}^T(\mathbf{1}-\mathbf{r})$. Additionally, γ represents a trade-off parameter which compromises between the peak sidelobe level (PSL) and the sparseness of the selection vector $\mathbf{r}$. The two formulations in Eqs. (8.12a) and (8.13) become equivalent when γ tends to infinity [131, 137].

The second part of the objective function $\mathbf{r}^T(\mathbf{1}-\mathbf{r})$ is concave, and thus it is difficult to minimize Eq. (8.13) directly. A sequential convex programming (SCP) based on iteratively linearizing the second concave function is utilized to reformulate the non-convex problem to a series of convex subproblems, each of which can be optimally solved using convex programming [133]. The integrated sparse array design and beampattern synthesis in the (k+1)th iteration can be formulated based on the solution from the kth iteration $\mathbf{r}^{(k)}$ as,

$$\begin{aligned}
\min_{\mathbf{w},\mathbf{r},\alpha} \quad & \alpha + \gamma[\mathbf{r}^T(\mathbf{1}-2\mathbf{r}^{(k)}) + \mathbf{r}^{(k)T}\mathbf{r}^{(k)}], \qquad (8.14)\\
\text{subject to} \quad & \mathbf{w}^H\mathbf{a}(\theta_t) = e^{j\mu(\theta_t)},\\
& \left|\mathbf{w}^H\mathbf{a}(\theta_l)\right| \leq \rho + \alpha,\ \theta_l \in \bar{\Theta}, l = 1, \ldots, L_s\\
& \mathbf{w}^H\mathbf{a}(\theta_c) = \Delta_k e^{j\phi_k},\\
& |w_m|^2 \leq r_m, m = 1, \ldots, K\\
& 0 \leq \mathbf{r} \leq 1,\ \mathbf{1}^T\mathbf{r} = M.
\end{aligned}$$

Eq. (8.14) can be formulated into a standard second order cone programming (SOCP) and effectively solved by various types of software packages, such as CVX [142]. Note that Eq. (8.14) is a convex relaxation of the original combinatorial optimization problem in Eq. (8.12a), thus the SCP is a local heuristic and the obtained solution is a sub-optimum local minimizer. In order to find a sufficiently good sub-optimum sparse array, the typical remedy is to initialize the algorithm with several feasible points $\mathbf{r}^{(0)}$ and find the one with the minimum objective value over the different runs. To accelerate the convergence rate of the algorithm and increase its robustness against initial search points, we propose a new method to update the selection vector $\mathbf{r}$ iteratively. The updating rule of the selection vector in the $(k+1)$th iteration is expressed as,

$$\mathbf{r}^{(k+1)} = \mathbf{r}^{(k)} \oplus \Delta\mathbf{r}, \qquad (8.15)$$

where $\oplus$ denotes modulus-two addition and $\Delta\mathbf{r} \in \{0, 1\}^K$ is a K-dimensional vector with all zeros except for two entries with value one. The positions $p_i, i = 1, 2$ of the two "one" entries are determined by $\mathbf{w}^{(k+1)}$ and $\mathbf{r}^{(k)}$ together as follows,

$$\begin{aligned}
p_1 &= \mathrm{argmax_n}\{|\mathrm{w}_\mathrm{n}^{(\mathrm{k}+1)}| : \mathrm{r}_\mathrm{n}^{(\mathrm{k})} = 0\}, \qquad (8.16)\\
p_2 &= \mathrm{argmin_n}\{|\mathrm{w}_\mathrm{n}^{(\mathrm{k}+1)}| : \mathrm{r}_\mathrm{n}^{(\mathrm{k})} = 1\}.
\end{aligned}$$

Table 8.1 The detailed description of sparse array design for DFRC systems

Step 0	Initialize the threshold value δ and phase profile $\mu(\theta)$; Set the maximum iteration number Q
Step 1	generate a random feasible starting point $\mathbf{r}^{(0)}$
Step 2	**WHILE:** Run the optimization in Eq. (8.14) based on $\mathbf{r}^{(k)}$ to obtain $\mathbf{w}^{(k+1)}$
Step 3	Update $\mathbf{r}^{(k+1)}$ according to Eqs. (8.15) and (8.16) based on $\mathbf{w}^{(k+1)}$ and $\mathbf{r}^{(k)}$
Step 3	Update the phase profile $\mu^{(k+1)}(\theta)$ according to Eq. (8.11)
Step 4	If $\|\mathbf{w}^{(k+1)} - \mathbf{w}^{(k)}\|_2 \geq \delta$ and $k \leq Q$, set $k = k + 1$ and go to Step 2
	If $\|\mathbf{w}^{(k+1)} - \mathbf{w}^{(k)}\|_2 \leq \delta$, store the obtained sparse beamformer and terminate
Step 5	If $\|\mathbf{w}^{(k+1)} - \mathbf{w}^{(k)}\|_2 > \delta$ and $k > Q$, go to Step 1. **END**

The integrated design of sparse array configuration $\mathbf{r}$ and the associated beamforming weight $\mathbf{w}$ comprises two main stages. In the first stage, a sparse array $\mathbf{r}$ and the associated beamformer $\mathbf{w}_1$ are obtained for embedding the first communication symbol from Eq. (8.14). In the second stage, all other beamformers $\mathbf{w}_k$, $k = 2, \ldots K$ for embedding the remaining symbols based on the obtained sparse array are calculated from Eq. (8.9). Thus, it is not necessary to perform antenna selection and reconfigure the array at each PRI. Note that an alternating descent algorithm is deployed in both stages to synthesize the desired beampattern by iteratively updating the mainlobe phase profile. The phase profile $\mu(\theta_i)$, $\theta_i \in \Theta$ is initialized according to the following formula,

$$\mu(\theta_t) = 1, \text{ for focused beam}; \quad \mu(\theta_i) = -2\pi \sin(\theta_i), \theta_i \in \Theta, \text{ for flat-top beam.} \tag{8.17}$$

The detailed description of the sparse array design for DFRC systems is summarized in Table 8.1. To reduce the effect of initial search points, in addition to the proposed updating rule, a new starting point is selected whenever the iteration number exceeds the maximum number Q.

8.2.3 *Design of Common Array with Multiple Beamformers*

Now, we assume that both radar and communication functions deploy the same sparse transmit array while being associated with different beamformers. This strategy is illustrated in Fig. 8.2, where $\mathbf{w}_r$ and $\mathbf{w}_c$ are weight vectors corresponding to two beamformers for radar and communications, respectively. The cross-interference between different beamformers can be mitigated by utilizing orthogonal waveforms and spatial filtering. The composite transmit signal is written as,

$$\mathbf{s}(t; \tau) = \mathbf{w}_r \Psi_1(t) + \mathbf{w}_c \Psi_2(t), \tag{8.18}$$

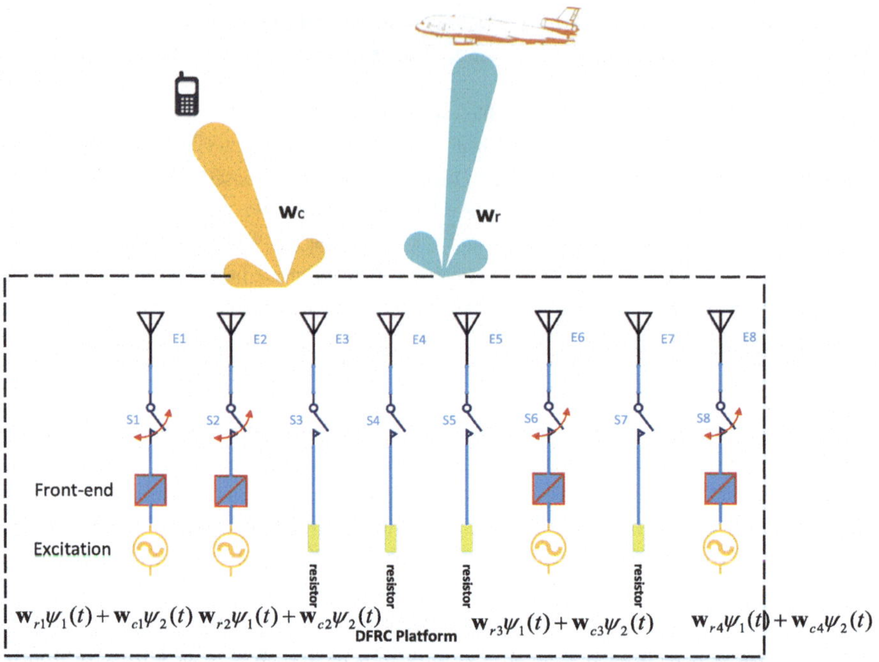

Fig. 8.2 Joint platform of DFRC systems with a common sparse array and multiple beamformers

Matched filtering the radar received signal in Eq. (8.1) with the waveform $\Psi_1(t)$ yields,

$$\mathbf{x}(\tau) = \sum_{q=1}^{Q} \beta_q(\tau)\mathbf{w}_r^H \mathbf{a}(\theta_q)\mathbf{b}(\theta_q) + \mathbf{n}(\tau). \tag{8.19}$$

Similarly, matched filtering the communication received signal in Eq. (8.3) with the waveform $\Psi_2(t)$ yields,

$$x_c(\tau) = \beta_c(\tau)\mathbf{w}_c^H \mathbf{a}(\theta_c) + n_c(\tau). \tag{8.20}$$

Thus, the radar and communication functions do not interfere with each other by utilizing the waveform diversity. As such, the beamformer $\mathbf{w}_r$ is responsible to both synthesize the desired beampattern for radar applications and mitigate the interference to communications caused by the radar function. On the other hand, the beamformer $\mathbf{w}_c$ is utilized to provide directive gain towards the communication receiver. The common sparse array with two respective beamformers for radar and communications can be designed as follows,

$$\min_{\mathbf{w}_r, \mathbf{w}_c, \alpha} \quad \alpha + \gamma \mathbf{r}^T(\mathbf{1} - \mathbf{r}), \tag{8.21a}$$

$$\begin{aligned}
\text{subject to} \quad & \mathbf{w}_r^H \mathbf{a}(\theta_t) = e^{j\mu(\theta_t)}, \\
& \left|\mathbf{w}_r^H \mathbf{a}(\theta_l)\right| \le \rho + \alpha, \ \theta_l \in \bar{\Theta}, l = 1, \ldots, L_s \\
& \left|\mathbf{w}_c^H \mathbf{a}(\theta_l)\right| \le \rho + \alpha, \ \theta_l \in \bar{\Theta}_c, l = 1, \ldots, L_c & (8.21\text{b}) \\
& \mathbf{w}_r^H \mathbf{a}(\theta_c) = 0, & (8.21\text{c}) \\
& \mathbf{w}_c^H \mathbf{a}(\theta_c) = 1; & (8.21\text{d}) \\
& |w_{i,m}|^2 \le r_m, m = 1, \ldots, K, i = \{r, c\} & (8.21\text{e}) \\
& 0 \le \mathbf{r} \le 1, \ \mathbf{1}^T \mathbf{r} = M, & (8.21\text{f})
\end{aligned}$$

where $\bar{\Theta}_c$ is the sidelobe angular region pre-defined for communications, which contains the radar mainlobe sector Θ. The constraint in Eq. (8.21b) restrains the impact of communications on the radar function to be less than ρ. The constraint in Eq. (8.21c) imposes orthogonality between the radar beamforming weights and the communication steering vector, such that the radar function does not interfere the communication function. The constraint $\mathbf{w}_c^H \mathbf{a}(\theta_c) = 1$ provides unit gain towards the communication receiver. The two constraints in Eq. (8.21e) guarantee the common sparse array for two respective beamformers associated with radar and communication functions. We do not restrain the communication beamformer to put a null against the radar target, as the reflected communication signal may be utilized by the radar for detection performance improvement. It is worth noting that communications and radar functions can be implemented independently and concurrently due to different beamformers and orthogonal waveforms. In this respect, it is unnecessary to embed communication symbols into the radar pulses by utilizing the signaling strategies described in Sect. 8.2.2.1.

8.2.4 Design of Intertwined Subarrays with Shared Aperture

In this section, we consider a separated antenna deployment strategy which partitions the K available antennas into two sparse subarrays, one for radar and the other for communications. Assume that M_r out of K antennas are allocated for radar and the remaining $M_c = K - M_r$ antennas for communications. The two corresponding subarrays are entwined and together form a contiguous filled aperture. The M_r-antenna radar transmit subarray is capable of concentrating the transmit power towards the target and a deep null against the communication receiver to mitigate the cross-interference. On the contrary, the remaining M_c-antenna communication subarray is required to point the beam towards the communication receiver and maintains well-controlled sidelobes in other angular regions. The formulation of this problem is similar to Eq. (8.21a), with the only difference that the constraints in Eqs. (8.21e) and (8.21f), respectively, change to

$$|w_{r,m}|^2 \leq r_m, \; |w_{c,m}|^2 \leq 1 - r_m, m = 1, \ldots, K \tag{8.22}$$

and

$$0 \leq \mathbf{r} \leq 1, \; \mathbf{1}^T \mathbf{r} = M_r, \tag{8.23}$$

which implies that the two beamformers exhibit complementary sparse structures, instead of a common structure as per Eq. (8.21a). Note that the sparse supports of two beamformers $\mathbf{w}_r$ and $\mathbf{w}_c$ are interleaved as restricted in Eq. (8.22), and combined beamformer sparse arrays span the entire array aperture. The sparse array configuration remains unchanged with respect to PRI for different communication symbols. Reconfiguration is required when the electromagnetic environment changes, such as the direction of the communication receiver. Various off-the-shelf software packages have been developed to effectively and efficiently solve the above SOCP problem [140, 311, 312], which facilitates the real-time sparse array reconfiguration in time-varying environment (Fig. 8.3).

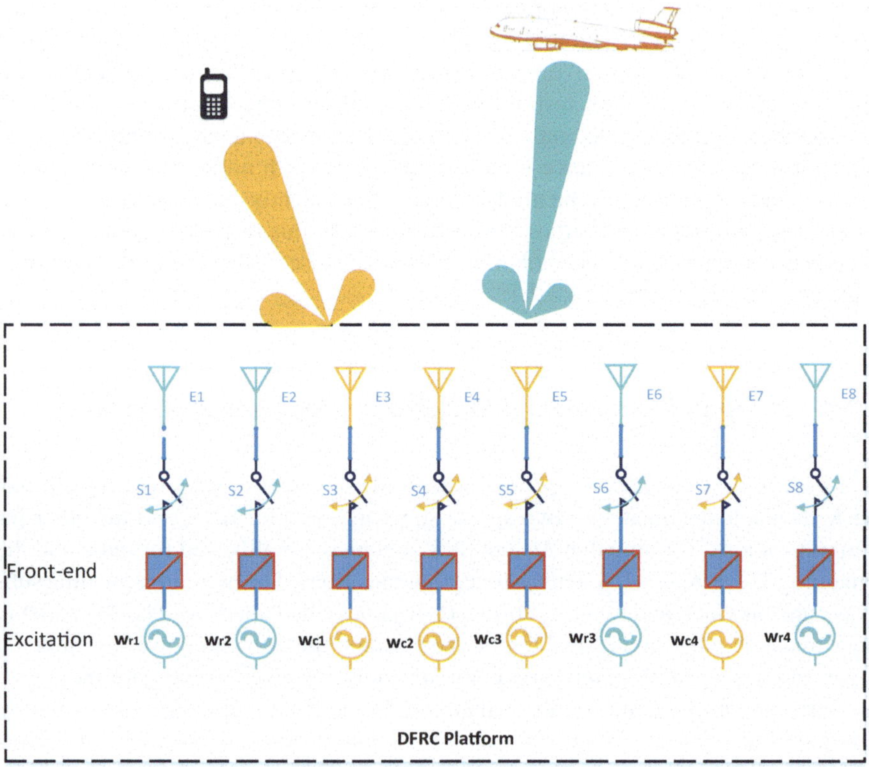

Fig. 8.3 Joint platform of DFRC systems with intertwined subarrays for radar and communications

8.2.5 Simulations

In the simulations, we consider a radar platform with $K = 40$ antennas arranged in a ULA with an inter-element spacing of 0.25λ. The antenna selection network is deployed to select a subset of M antennas in Examples 1 and 2, and select two entwined subarrays of 20 antennas in Example 3. We evaluate the performance of sparse arrays in DFRC systems by plotting the power pattern and the bit error rate (BER) curves in different scenarios.

8.2.5.1 Example 1: Common Array Design with Single Beamformer

We first investigate the common sparse array design with a single beamformer for DFRC systems. We assume that the radar target is arriving from the direction $\theta_t = 0°$ in the focused beam case and within the angular sector $\Theta = [-10°, 10°]$ in the flat-top beam case. A single communication receiver is assumed at direction $\theta_c = -40°$. Two sparse arrays of 10 antennas are selected for two cases of focused and flat-top beams, respectively. The communication information is embedded during each radar pulse via both the AM and PM signaling schemes. The four communication symbols for the AM and PM modulations are pre-defined as $\Delta = \{0.05, 0.03, 0.01, 0.001\}$ and $\phi_c = \{-\pi/2, 0, \pi/2, \pi\}$, respectively. The two selected 10-antenna sparse arrays are denoted as array (a) and array (b), and plotted in Fig. 8.4. The beampatterns of the sparse arrays (a) and (b) with four modulated SLLs towards the communication receiver are depicted in Figs. 8.5 and 8.6. Clearly, the selected sparse arrays do not utilize the available full aperture, as the communication receiver is located far from the radar target and the beampattern mainlobe is relatively wide. The PSL is set as −20dB for both cases and the maximum peak-to-peak mainlobe ripples are set as 0.8dB. Both the focused and flat-top beampatterns exhibit almost the same shape in the entire angular region excluding the communication receiver direction. The radiation power level of both sparse arrays in the radar sidelobe areas is at least 23dB lower than the mainlobe.

Figure 8.7 shows the BER curves versus signal-to-noise ratio (SNR) for variable data rates using the sparse array shown in Fig. 8.4a as well as a 10-antenna ULA with a half-wavelength inter-element spacing. The data rates used are 1, 2, 4, and 8 bits per pulse. The SNR is defined as $10\log_{10}(P_t/\sigma_n^2)$ with P_t and σ_n^2 denoting the transmit power and noise power, respectively. Information embedding is performed using 2ASK and orthogonal waveforms, i.e., eight orthogonal waveforms are used for the case of 8 bits per pulse. The figure shows that all BER curves decrease as the SNR increases. However, since the total power budget P_t is fixed, the BER curve associated with the 2 bits per pulse is shifted by 3dB to the right on the SNR axis as compared to the BER curve associated with the 1 bit per pulse case. This is attributed to the fact that the total power is divided equally between the two orthogonal waveforms used for the case of 2 bits per pulse while the total power is assigned to a single waveform for the case of 1 bit per pulse. As the data rate increases, the total power

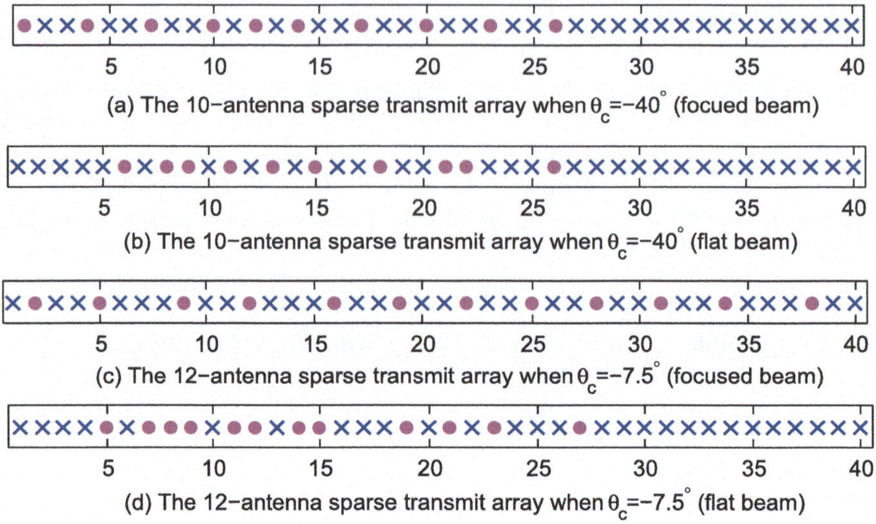

Fig. 8.4 Configurations of proposed sparse arrays: **a** 10-antenna sparse array when $\theta_c = -40°$ for focused beam; **b** 10-antenna sparse array when $\theta_c = -40°$ for flat-top beam; **c** 12-antenna sparse array when $\theta_c = -7.5°$ for focused beam; **d** 12-antenna sparse array when $\theta_c = -7.5°$ for flat-top beam. Filled circles represents selected and cross for discarded

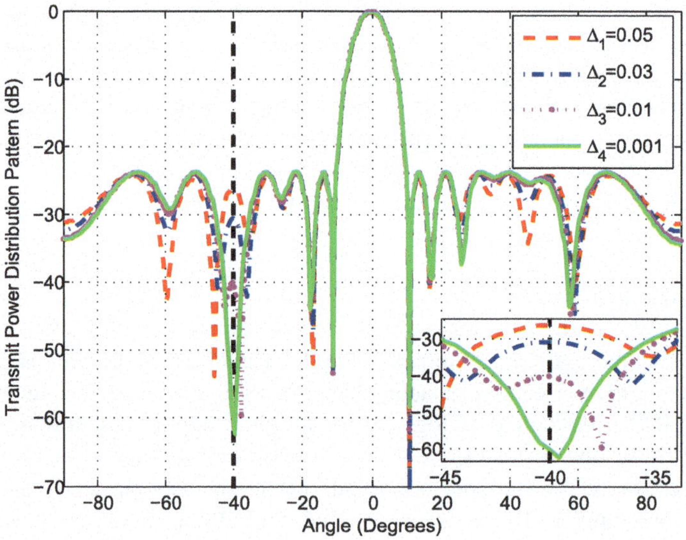

Fig. 8.5 Synthesized focused beampatterns of 10-antenna sparse array **a** for communication symbols $\Delta = \{0.05, 0.03, 0.01, 0.001\}$

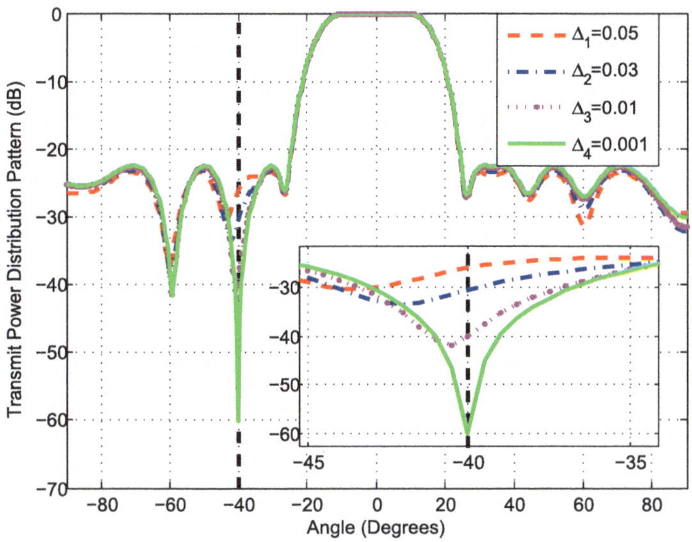

Fig. 8.6 Synthesized flat-top beampatterns of 10-antenna sparse array **b** for communication symbols $\Delta = \{0.05, 0.03, 0.01, 0.001\}$

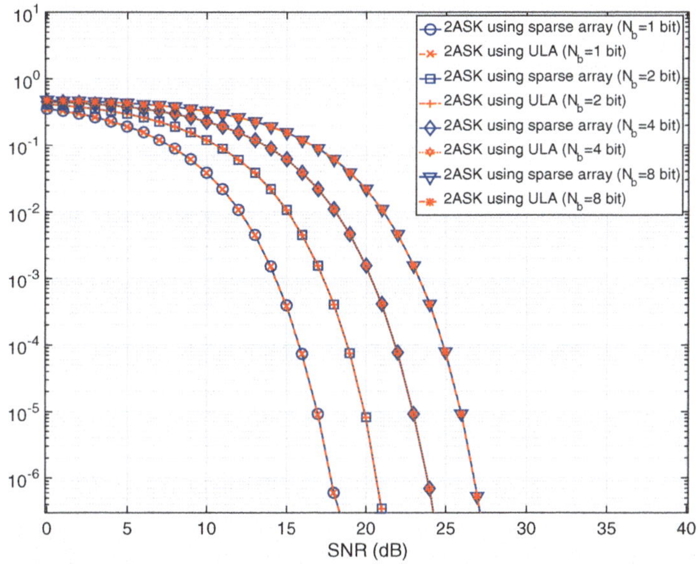

Fig. 8.7 BER versus SNR using 10-antenna sparse array **a** and 10-antenna ULA. Communication receiver locates at direction $\theta_c = -40°$

is divided amongst more waveforms, causing the respective BER curves to further shift to the right on the SNR axis. The figure also shows that both the 10-antenna sparse array and the 10-antenna ULA yield the same BER performance since the communication direction is well separated from the mainlobe region and the SLLs used for communications are controlled to be the same for both arrays. Although the communications performance is the same, the sparse array configuration has a narrower mainlobe which enables higher angular resolution for the radar operation as compared to the ULA configuration.

We then change the direction of the communication receiver to be $\theta_c = -7.5°$, which is close to the radar target. The number of selected antennas is increased to 12 and the configurations of two selected 12-antenna sparse arrays are depicted in Fig. 8.4c and d. The sparse array in Fig. 8.4c aims at synthesizing a focused-shape beampattern pointing towards the radar direction $\theta_t = 0°$ while ensuring that the communication direction $\theta_c = -7.5°$ is sitting in the sidelobe region. We also use for comparison a 12-antenna ULA with inter-element spacing of half wavelength. The beampatterns of the sparse array (c) and ULA with four modulated SLLs are plotted in Figs. 8.8 and 8.9, respectively. We can observe that the selected 12-antenna sparse array (c) occupies almost the full array aperture. As a result, the synthesized beampatterns exhibit high angular resolution compared with that of the 12-antenna ULA. Again, the power patterns of the sparse array (c) exhibit almost the same shape with a constant PSL of −20dB regardless of different communication symbols, whereas the power level of the ULA reaches −12.9dB at the communication angle of −7.5°. The figures also show that the communication direction is located in the

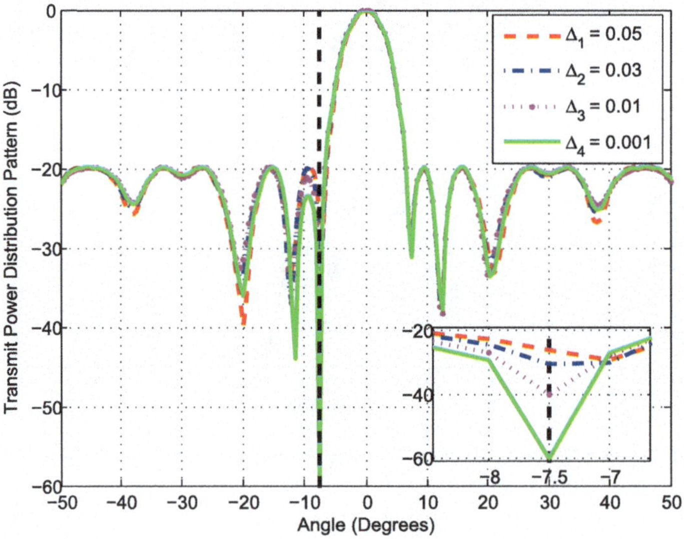

Fig. 8.8 Synthesized focused beampatterns of sparse array **c** for communication symbols $\Delta = \{0.05, 0.03, 0.01, 0.001\}$

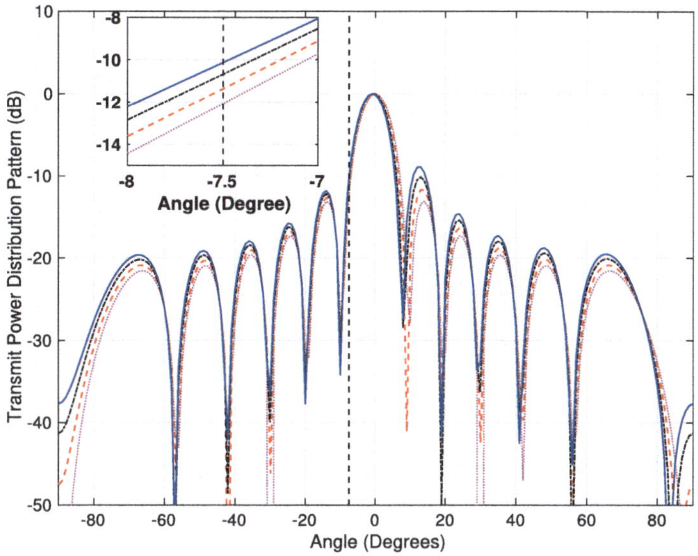

Fig. 8.9 Synthesized focused beampatterns of 12-antenna ULA for communication symbols $\Delta = \{0.3, 0.28, 0.26, 0.24\}$

sidelobe region in the case of sparse array (c) while it overlaps with the mainlobe for the ULA. The sparse array (d) is used to synthesize a flat-top beampattern with the communication receiver located in the mainlobe. Figure 8.10 shows that the power pattern of the sparse array (d) exhibits a sharper transition band compared to that of the ULA, which enables better transmit power concentration and improved robustness. The phase profiles around the direction of the communication receiver for each QPSK symbol are plotted in Fig. 8.11. The figure shows that the phase patterns for sparse array (d) change linearly with the arrival angle, which implies constant phase differences between different QPSK symbols. Thus, robust communication performance against the communication receiver angle deviation can be achieved.

Figure 8.12 shows the BER curves versus SNR for information embedding towards a communication receiver located in the direction $\theta_c = -7.5°$. We use multi-waveform 2ASK information embedding for the sparse array and ULA associated with Figs. 8.8 and 8.9. Note that the direction $\theta_c = -7.5°$ overlaps with the mainlobe of the ULA while it is well separated from the mainlobe of the sparse array thanks to the narrow mainlobe property of the sparse array. Since the radar operation requires the mainlobe to remain the same during the entire dwell time of the radar operation, the ULA is designed such that the transmit gain associated with binary bit "0" equals 0.9 times of the transmit gain associated with binary bit "1", i.e., the maximum variation of the transmit gain within the mainlobe of the radar is kept within 10% of its maximum value. A larger variation between the two beampatterns is undesirable and can cause disturbance to the radar operation. Figure 8.12 shows that the BER performance for the sparse array outperforms that associated with the

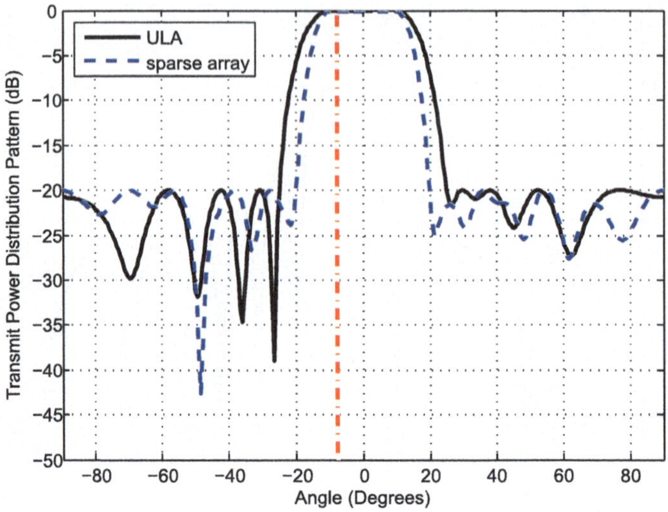

Fig. 8.10 Synthesized flat-top beampatterns of 12-antenna sparse array **d** and 12-antenna ULA for PM-based signaling scheme

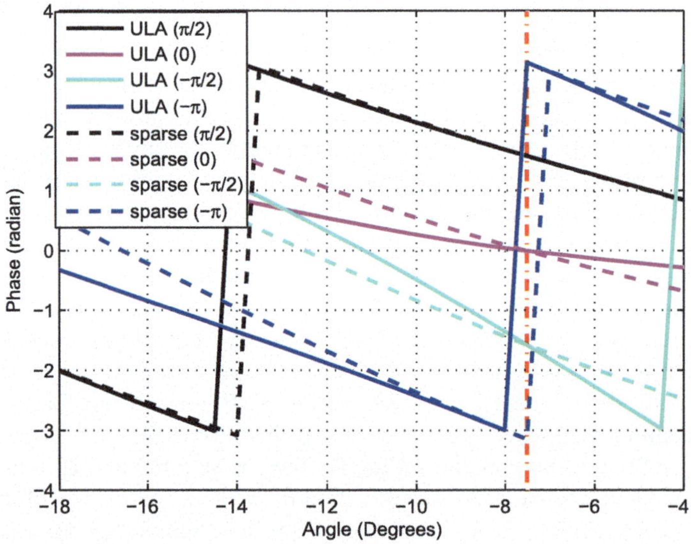

Fig. 8.11 The phase profiles around the direction of communication receiver for each QPSK symbol $\{-\pi/2, 0, \pi/2, \pi\}$. The phases are within the range $[-\pi.\pi]$ radian

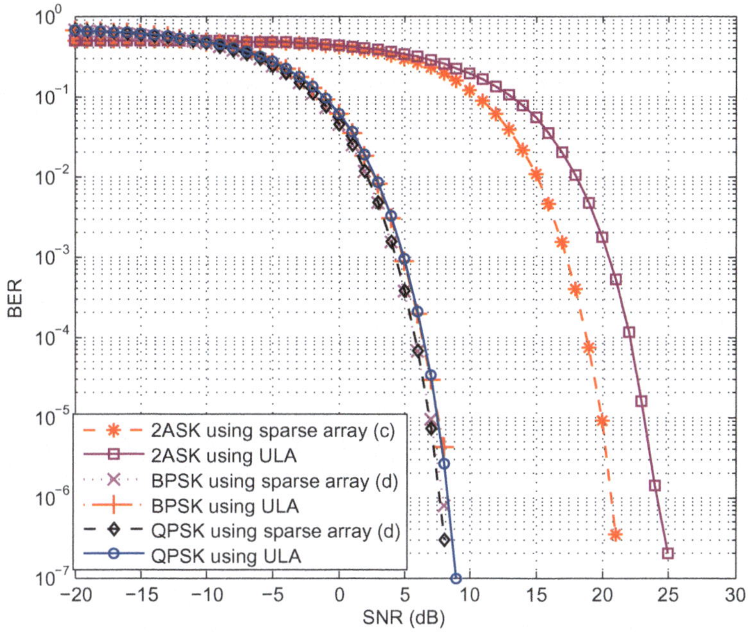

Fig. 8.12 BER versus SNR using 12-antenna sparse arrays **c** and **d** and 12-antenna ULA. Communication receiver locates at direction $\theta_c = -7.5°$

ULA. This can be attributed to the fact that the communication direction is located outside the mainlobe of the sparse array and, therefore, the SLLs used can be well separated from each other. It is worth noting that, if the radar operation requires the mainlobe for the ULA to be exactly same as that of the sparse array, then 2ASK communication technique completely fails.

As the PM signaling scheme enables the concurrent communications within the radar mainlobe direction, we calculate the BER curves versus SNR utilizing the BPSK and QPSK modulations for the sparse array shown in Fig. 8.4d, compared with the 12-antenna ULA. Two orthogonal waveforms and two phase symbols $\{0, -\pi\}$ are used for the BPSK embedding scheme, whereas one orthogonal waveform and four phase symbols $\{-\pi, -\pi/2, 0, \pi/2\}$ are used to test the QPSK signaling scheme. The results are plotted in Fig. 8.12. We can observe that the sparse array exhibits a slightly better communication performance utilizing the two PM signaling schemes.

The BER is also computed versus angle of the communication receiver using the 2ASK signaling schemes for the sparse array and the ULA associated with the focused transmit beams shown in Figs. 8.8 and 8.9. The SNR is fixed to 20 dB. The BER curves for the sparse and the ULA arrays are shown in Fig. 8.13. The figure demonstrates that the use of sparse array enables communications delivery towards the intended communication direction and does not enable eavesdroppers located at other directions to intercept the embedded data. The figure also shows that the

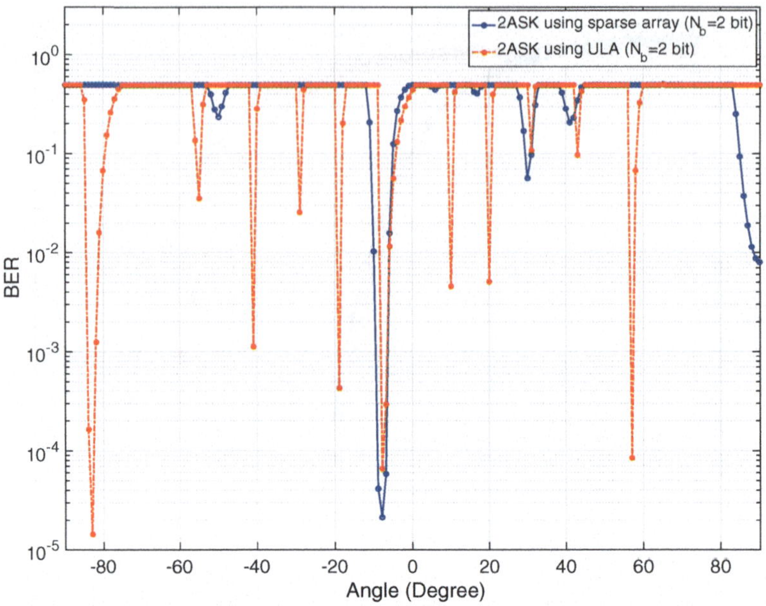

Fig. 8.13 BER versus direction of the communication receiver using 12-antenna sparse array **c** and 12-antenna ULA. SNR=20 dB is used

use of ULA allows eavesdroppers located at several directions in the sidelobe region to detect the data. Therefore, the use of sparse arrays for information embedding guarantees more security as compared to ULA.

Finally, we analyze the robustness of the proposed sparse array design method against the choice of initial search points. The communication receiver is assumed to locate at $-7.5°$. We compare the sensitivity of the proposed updating rule shown in Eqs. (8.15) and (8.16) with the traditional method in both cases of focused and flat-top beampattern synthesis. We conduct 100 Monte-Carlo trials with randomly chosen initial search points. The PSL of synthesized beampatterns with different starting points is summarized in the histograms of Fig. 8.14. We can see that the proposed updating rule can design a sparse array with a satisfied beampattern, regardless of the initial search point, thus exhibiting much better robustness and shorter convergence time compared with the traditional method.

8.2.5.2 Example 2: Common Array Design with Multiple Beamformers

We then consider the shared sparse array design with respective beamformers for radar and communications. The radar target and a single communication receiver are assumed at direction $\theta_t = 0°$ and $\theta_c = -5°$, respectively. The sparse array of 15 antennas is selected and associated with two designed beamformers. One beamformer

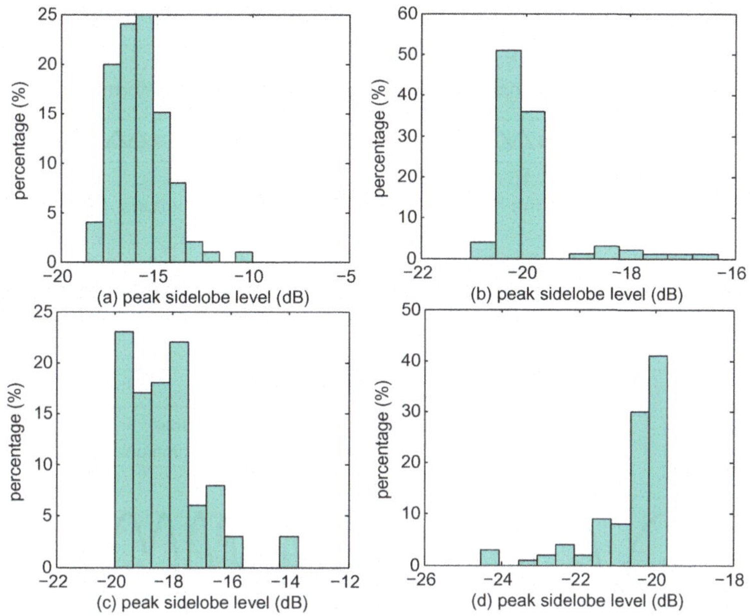

Fig. 8.14 Histograms: **a** PSL of traditional updating rule in the case of focused beam. **b** PSL of proposed updating rule in the case of focused beam. **c** PSL of traditional updating rule in the case of flat-top beam. **d** PSL of proposed updating rule in the case of flat-top beam

aims at concentrating the radiation power towards the radar target and forming a complete null towards the communication receiver. The other beamformer strives to minimize the power level within the sidelobe angular sector while maintaining the unit array again towards the communication receiver. As the communication receiver is close to the radar target, a desired beampattern with the half power beam width (HPBW) of around 3° is required. The structure of the 15-antenna sparse array is depicted in Fig. 8.15 (e), which spans the full aperture length. The power patterns associated with the two beamformers are depicted in Fig. 8.16 with a peak SLL of -15dB. The 15-antenna ULA with an inter-element spacing of half wavelength fails to synthesize a narrow mainlobe towards the target while simultaneously forming a deep null in the communication direction due to its low spatial resolution. We then plot the power patterns of a 20-antenna ULA with inter-element spacing of $\lambda/2$ in Fig. 8.17 for comparison. The source efficiency of deploying sparse arrays for DFRC systems are clearly manifested from the comparable performance of the 20-antenna ULA with that of the 15-antenna sparse array. It is worth noting that plotting BER curves becomes unnecessary for performance validation here, as communication symbols are no longer needed to embed into the radar pulse transmitting due to independent beamformers and waveform diversity.

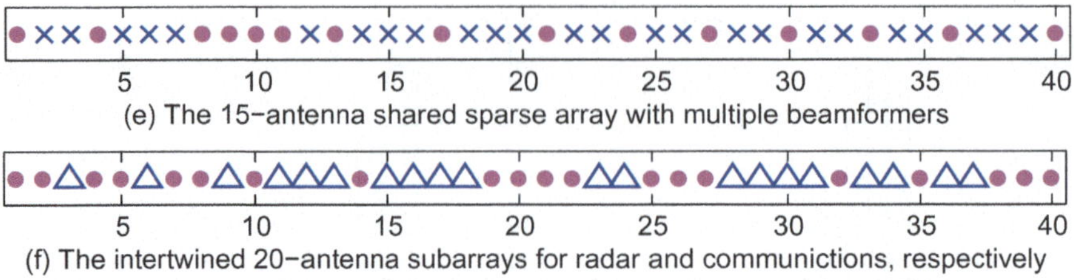

Fig. 8.15 **e** Proposed 15-antenna sparse array for multiple beamformers when $\theta_c = -5°$. **f** The two intertwined 20-antenna subarrays for radar and communications, respectively. The subarray indicated by filled circles is for radar function and that indicated by triangles is for communications

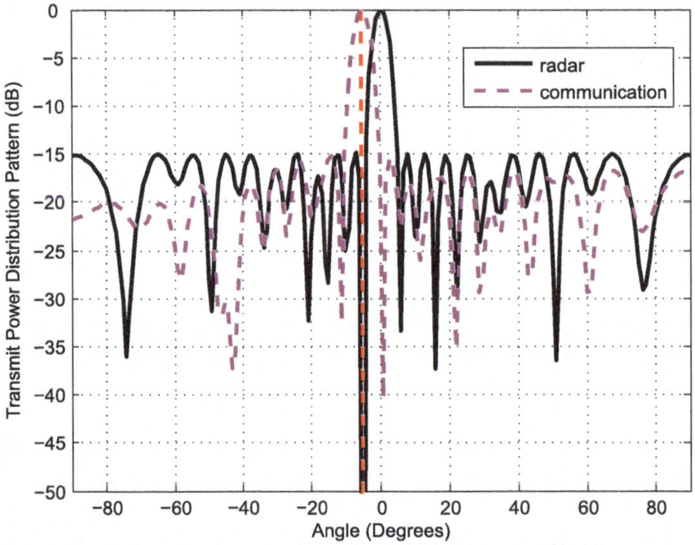

Fig. 8.16 Transmit power patterns of common 15-antenna sparse array associated with multiple beamformers

8.2.5.3 Example 3: Intertwined Subarray Design with Shared Aperture

In this example, we proceed to investigate the intertwined sparse array design with shared aperture for radar and communication functions. The available 40 antennas are partitioned into two sparse subarrays: one for radar function and the other for communications. The optimum subarrays will most likely be entwined and not necessarily separated, or split. The two subarrays are plotted in Fig. 8.15 (f) and their corresponding beampatterns are depicted in Fig. 8.18. We can observe that the subarray for radar function forms a narrow beam towards the target and a deep null towards the communication receiver. The subarray for communications remains a constant -15dB power level in the sidelobe angular region including the radar target direction. We also plot the power patterns of two split subarrays in Fig. 8.19 for

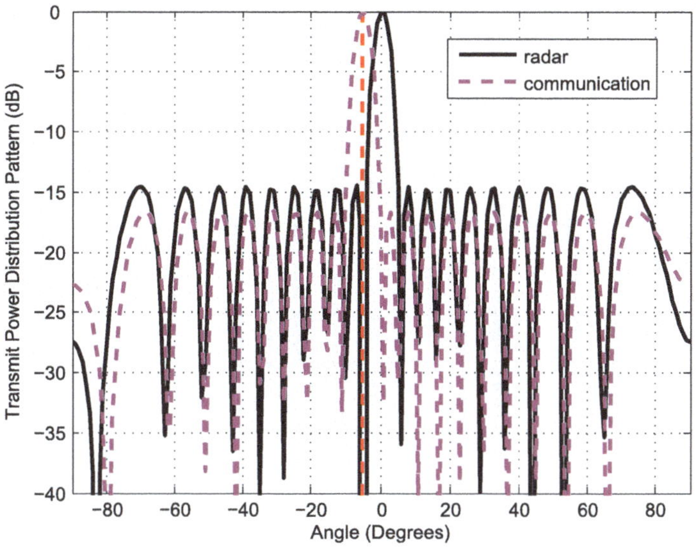

Fig. 8.17 Transmit power patterns of common 20-antenna ULA associated with multiple beamformers

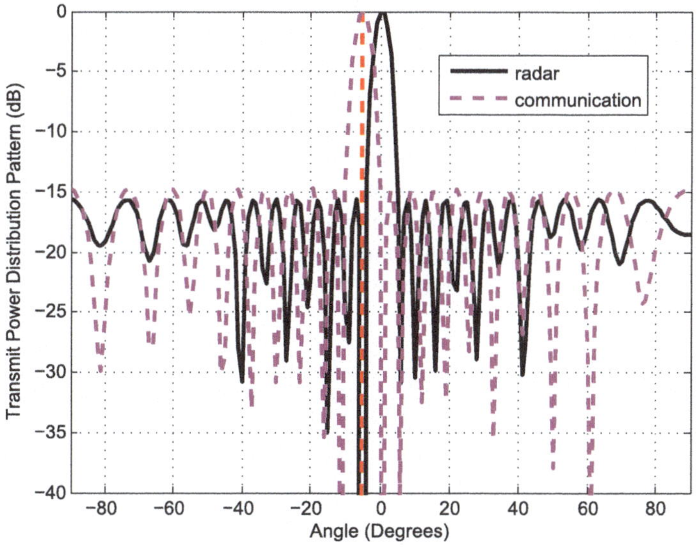

Fig. 8.18 Transmit power patterns of intertwined 20-antenna subarrays for radar and communications

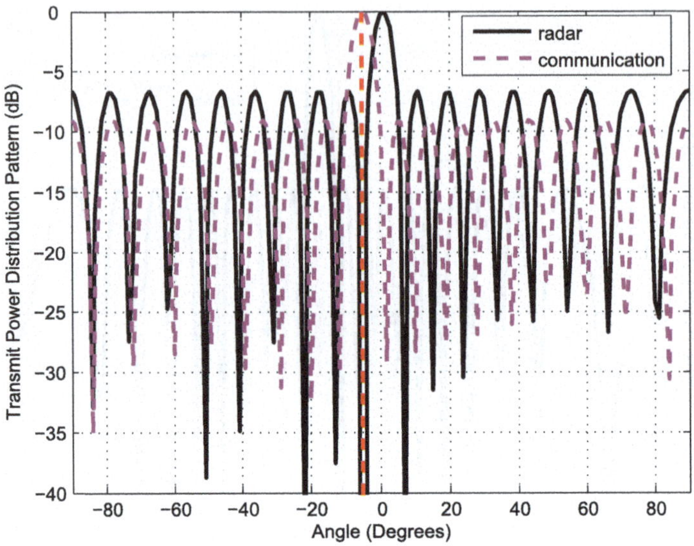

Fig. 8.19 Transmit power patterns of split 20-antenna subarrays for radar and communications

comparison, where the first 20 antennas compose a subarray for radar function and the remaining 20 antennas compose the other subarray for communications. We can observe that the radiation patterns of two split subarrays exhibit high SLLs and wide mainlobes due to its low spatial resolution caused by limited aperture length, which will undoubtedly affect the normal radar function.

We calculate the aperture efficiency of sparse arrays (a), (c), (e), (f) and associated weights in Table 8.2. The aperture efficiency is defined as the ratio between the directivity of sparse array and that of uniformly excited ULA, which determines the directivity loss due to the non-uniform tapering. It is defined as $\eta = \frac{G_s}{G_0}$, with

$$G_0 = \frac{M^2}{M + 2\sum_{m=1}^{M-1}(M-m)\text{sinc}(mk_0 d)}, \tag{8.24}$$

and

$$G_s = \frac{|\sum_m^M w_m|^2}{\sum_{n=1}^M \sum_{m=1}^M w_n w_m^* \text{sinc}((n-m)k_0 d)}. \tag{8.25}$$

We can see from Table 8.2 that the aperture efficiency of proposed sparse arrays are relatively high.

Table 8.2 Aperture efficiency of sparse arrays (a), (c), (e), (f)

Array (a)	Array (c)	Array (e)	Array (f)
0.95	0.91	0.86	0.87

8.3 Sparse Array Reconfiguration for Spatial Index Modulation

As antenna array technology progresses, sophisticated antenna selection schemes through RF switches and array reconfiguration methods that were previously infeasible begin to become possible. Sparse antenna arrays with non-uniform inter-element spacing attract increased attention in multi-sensor transmit/receive systems as an effective solution to reduce the system's complexity and cost, yet retain multifaceted benefits [104]. Taking the notion of sparse arrays further, in this section, we propose a technique utilizing array configuration for reliable communication symbol embedding concurrently with MIMO radar operation through antenna selection. We investigate the problem of expressing sparse array configurations and their association with independent waveforms as unique communication symbols. In spectrum sharing perspective, the deployment of reconfigurable sparse arrays by antenna selection can, undoubtedly, alleviate pressures on the resource management and efficiency requirements. Simulation results show that the versatility of sparse array configurations facilitates the realization of multiple functions on the same system and the performance enhancement.

8.3.1 System Configuration and Signal Model

We consider a joint MIMO-radar communications platform equipped with a reconfigurable transmit antenna array through an antenna selection network as shown in Fig. 8.20. This joint system can simultaneously detect radar targets of interest while sending communication symbols to downlink users. There are M transmit antennas uniformly located in the platform with an inter-element spacing of d and K $(K<M)$ front-ends installed for waveform transmitting. The antenna selection network comprises M RF switches and their on/off status can be changed to connect/disconnect the corresponding antennas with the following front-ends. Note that only K antennas are switched on for waveform transmitting during each pulse repetition interval (PRI) and the remaining $M-K$ antennas are either switched off or connected to resistors. Suppose a transmit array is configured with K selected antennas located at $p_k d, k = 1, \ldots, K$ with $p_k \in \{0, \ldots, M-1\}$. The radar receiver employs an array of N receive antennas with an arbitrary linear configuration. It is assumed that both the transmit and receive arrays are closely spaced such that a target in the far-field would be seen from the same direction by both arrays. Without loss of generality,

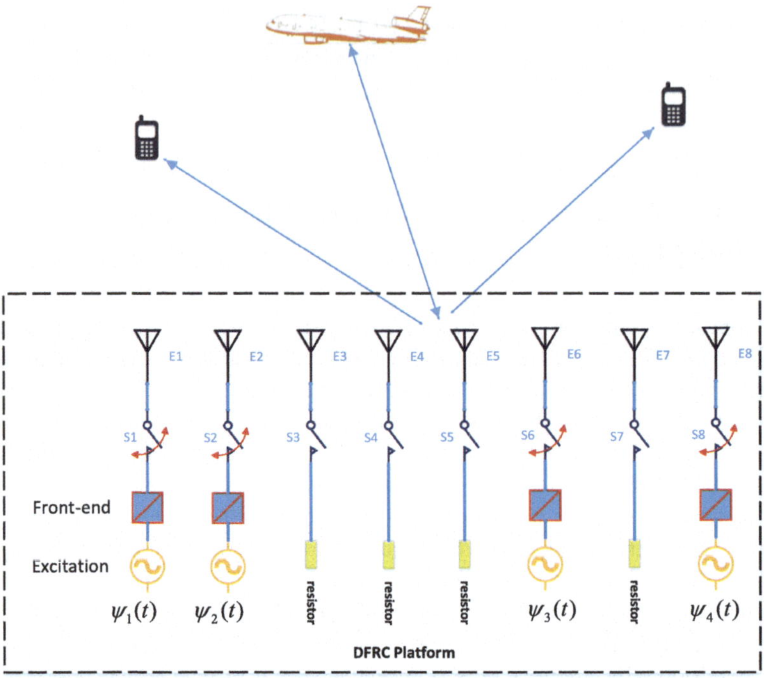

Fig. 8.20 Joint platform of a DFRC system with antenna selection network

a single-element communication receiver is assumed to be located in direction θ_c, which is exactly known to the transmitter.

Let $\Psi_k(t), k = 1, \ldots, K$ be a set of K orthogonal waveforms, each occupying the same bandwidth. In other words, the spectral contents of all waveforms overlap in the frequency domain. Assume that each waveform is normalized to have unit power, i.e., $\int_T |\Psi_k(t)|^2 dt = 1$, with T and t denoting the waveform duration and the fast time index, respectively. It is further assumed that the orthogonality condition $\int_T \Psi_k(t)\Psi_{k'}^*(t)dt = 0$ is satisfied for $k \neq k'$, where $()^*$ stands for the complex conjugate. Assume that Q far-field targets of interest arriving from the directions $\theta_q, q = 1, \ldots, Q$, located within the radar main beam, are observed in the background of strong clutter and interferences, such as television, radio and signals from other commercial communication services as well as deliberate jammers. The $N \times 1$ baseband representation of the signals at the output of the radar receive antenna array is given by,

$$\mathbf{x}(t;\tau) = \sum_{q=1}^{Q} \beta_q(\tau) \left(\tilde{\mathbf{a}}^T(\theta_q)\Psi(t)\right) \mathbf{b}(\theta_q) + \mathbf{n}(t;\tau), \tag{8.26}$$

where τ is the pulse number, $\beta_q(\tau)$ is the qth target reflection coefficient,[1] $\tilde{\mathbf{a}}^T(\theta_q)$ and $\mathbf{b}(\theta_q)$ denote the steering vectors of the sparse transmit array and the receive array, respectively, $(\cdot)^T$ stands for the transpose, $\Psi(t) \triangleq [\Psi_1(t), \ldots, \Psi_K(t)]^T$ is the $K \times 1$ vector of orthogonal waveforms, and $\mathbf{n}(t;\tau)$ is the $N \times 1$ vector of zero mean summarizing the unwanted clutter, interferences and white noise during the τth radar pulse. In (8.26), the sparse transmit array steering vector, $\tilde{\mathbf{a}}(\theta)$, can be defined as,

$$\tilde{\mathbf{a}}(\theta) = [e^{jk_0 p_1 d \sin\theta}, \ldots, e^{jk_0 p_K d \sin\theta}]^T, \tag{8.27}$$

where $k_0 = 2\pi/\lambda$ is the wavenumber and $p_k \in \{0, \ldots, M-1\}, k = 1, \ldots, K$. The steering vector of the MIMO radar receive array, $\mathbf{b}(\theta)$, can be defined in a similar way as that of $\tilde{\mathbf{a}}(\theta)$.

The signal at the output of the communication receiver can be modelled as

$$x_c(t;\tau) = \alpha_{\text{ch}}(\tau)\tilde{\mathbf{a}}^T(\theta_c)\Psi(t) + n_c(t;\tau), \tag{8.28}$$

where $\alpha_{\text{ch}}(\tau)$ is the channel coefficient of the received signal that summarizes the propagation environment between the transmit array and the communication receiver during the τth pulse and $\tilde{\mathbf{a}}(\theta_c)$ is the steering vector of the selected transmit array toward the communication direction θ_c. In addition, $n_c(t;\tau)$ is the noise signal interfering the communication process in the τth radar pulse. We assume that the channel coefficient α_{ch} is known or accurately estimated and remains unchanged during the entire coherent processing interval of the DFRC system. Therefore, for the rest of this paper, we remove the dependency of the channel coefficient on the pulse index τ.

At the MIMO radar receiver, the received signal components associated with the individual transmitted waveforms can be obtained using matched filtering to Eq. (8.26). The signals observed at the output of the radar receiver are the $KN \times 1$ extended vector of virtual data, that is,

$$\begin{aligned} \mathbf{y}_r(\tau) &= \text{vec}\left(\int_T \mathbf{x}(t;\tau)\Psi^H(t)dt\right), \\ &= \sum_{q=1}^{Q} \beta_q(\tau)[\tilde{\mathbf{a}}(\theta_q) \otimes \mathbf{b}(\theta_q)] + \mathbf{n}(\tau), \end{aligned} \tag{8.29}$$

where $\text{vec}(\cdot)$ is the operator that stacks the columns of a matrix into one column vector, $\otimes$ denotes the Kronecker product, ^{H} stands for the Hermitian transpose, and $\mathbf{n}(\tau) = \text{vec}\left(\int_T \mathbf{n}(t;\tau)\Psi^H(t)dt\right)$ is the $KN \times 1$ additive noise term after matched filtering.

The communication receiver is assumed to have perfect knowledge of the orthogonal waveforms $\Psi_k(t), k = 1, \ldots, K$. Moreover, it is assumed that the phase syn-

[1] The target reflection coefficients are assumed to obey the Swerling-II target model [299], i.e., they remain constant during the entire pulse duration, but vary independently from pulse to pulse.

chronization between the transmit array and the communication receiver is adjusted. Matched filtering the received data in Eq. (8.28) to each waveform $\Psi_k(t)$ yields,

$$\begin{aligned} y_{c,k}(\tau) &= \int_T x_c(t;\tau)\Psi_k^*(t)dt, \\ &= \alpha_{\mathrm{ch}}\tilde{\mathbf{a}}_k(\theta_c) + n_{c,k}(\tau), \quad k = 1, \ldots, K \end{aligned} \tag{8.30}$$

where $\tilde{\mathbf{a}}_k = e^{jk_0 p_k d \sin\theta_c}$ denotes the kth entry of the selected transmit array steering vector $\tilde{\mathbf{a}}(\theta_c)$ and $n_{c,k}(\tau) = \int_T n_c(t;\tau)\Psi_k^*(t)dt, k = 1, \ldots, K$ are additive noise terms after matched filtering. Array configurations denote the spatial DoFs and can be combined with waveform design in temporal domain to embed communication symbols concurrently with MIMO radar functions. We elaborate on the deployment of sparse arrays for symbol embedding in DFRC systems in Sects. 8.3.2–8.3.4.

8.3.2 *Antenna Selection Based Spatial Index Modulation*

There are totally M antennas installed in the common transmit platform and an antenna selection network is deployed to select K out of M antennas. We deploy K orthogonal waveforms, $\Psi_1(t), \ldots, \Psi_K(t)$, and transmit them via the selected K antennas. It is clear that the steering vectors of selected sparse arrays can be estimated by the communication receiver after matched filtering and utilized to embed communication symbols from Eq. (8.30).

8.3.2.1 Information Embedding Scheme

The steering vector of the M-antenna full transmit array is denoted as $\mathbf{a}(\theta)$ and can be expressed as

$$\mathbf{a}(\theta) = [e^{jk_0 p_1 d \sin\theta}, \ldots, e^{jk_0 p_M d \sin\theta}]^T. \tag{8.31}$$

Denote the $K \times M$ selection matrix during the τth radar pulse as $\mathbf{P}(\tau) \in \{0, 1\}^{K \times M}$, where there is only one entry being "1" in each row and in the kth column corresponding to the kth selected antenna, $k \in \{1, \ldots, M\}$. Applying the selection matrix $\mathbf{P}(\tau)$ to the steering vector $\mathbf{a}(\theta)$ of the full transmit array yields the $K \times 1$ steering vector of the selected subarray, that is,

$$\tilde{\mathbf{a}}(\theta;\tau) = \mathbf{P}(\tau)\mathbf{a}(\theta). \tag{8.32}$$

The K orthogonal waveforms are transmitted via the selected K antennas, and the $N \times 1$ complex vector of the radar received observations in Eq. (8.26) can be rewritten as,

$$\mathbf{x}(t,\tau) = \sum_{q=1}^{Q} \beta_q(\tau)[\mathbf{a}^T(\theta_q)\mathbf{P}^T(\tau)\Psi(t)]\mathbf{b}(\theta_q) + \mathbf{n}(t,\tau). \tag{8.33}$$

Let $\mathcal{P} = \{1, \ldots, M\}$ label the full set of antennas installed in the transmit platform. During each radar pulse, a subset $\mathcal{S}$ of K antennas are selected from the full set $\mathcal{P}$ for waveform transmitting. Such a selection is essentially a combinatorial problem. There are totally $L = C_M^K = \frac{M!}{K!(M-K)!}$ different subsets, $\mathcal{S}_l \subset \mathcal{P}, l = 1, \ldots, L$, and each subset $\mathcal{S}_l$ corresponds to a unique selection matrix $\mathbf{P}(\tau)$ with "1" entries located in the columns indicated by $\mathcal{S}_l$, which in turn corresponds to a unique steering vector $\tilde{\mathbf{a}}(\theta;\tau)$. For each subarray $\mathcal{S}_l$, a communication symbol consisting of N_b bits can be defined.

Assume that the communication receiver knows its direction θ_c relative to the stationary MIMO transmit platform. The signal at the output of the communication receiver antenna in Eq. (8.28) is remodelled as,

$$x_c(t,\tau) = \alpha_{\text{ch}}\mathbf{a}^T(\theta_c)\mathbf{P}^T(\tau)\Psi(t) + n_c(t,\tau), \tag{8.34}$$

Matched filtering the received data with the set of K orthogonal waveforms yields,

$$\begin{aligned}\mathbf{y}_c(\tau) &= \text{vec}\left\{\int_T x_c(t,\tau)\Psi^H(t)dt\right\}, \\ &= \alpha_{\text{ch}}\mathbf{P}(\tau)\mathbf{a}(\theta_c) + \mathbf{n}_c(\tau), \\ &= \alpha_{\text{ch}}\tilde{\mathbf{a}}(\theta_c;\tau) + \mathbf{n}_c(\tau).\end{aligned} \tag{8.35}$$

where $\mathbf{y}_c(\tau) = [y_{c,1}(\tau), \ldots, y_{c,K}(\tau)]^T$ and $\mathbf{n}_c(\tau) = [n_{c,1}(\tau), \ldots, n_{c,K}(\tau)]^T$. Thus, the communication receiver signal at the output of the matched-filter is a scaled and noisy selection of the full steering vector $\mathbf{a}(\theta_c)$, meaning that the selected sparse array $\mathcal{S}_l$ can be recovered from the received vector $\mathbf{y}_c(\tau)$. We propose to utilize the steering vector of the sparse transmit array $\mathcal{S}_l$ as codes to embed communication symbols.

8.3.2.2 Detection of Communication Symbols

As mentioned above, the steering vector of the selected sparse array $\mathcal{S}_l, l = 1, \ldots, L$ can be utilized to embed communication symbols. Thus, a dictionary of L unique symbols can be constructed as

$$\mathbb{D} = \{\tilde{\mathbf{a}}_1(\theta_c), \ldots, \tilde{\mathbf{a}}_L(\theta_c)\}, \tag{8.36}$$

where $\tilde{\mathbf{a}}_l(\theta_c)$ is the steering vector corresponding to the selected sparse array $\mathcal{S}_l \subset \mathcal{P}$ towards the communication receiver. The cardinality L of the dictionary determines the capacity of the communication system and is usually not a power of 2. The fact that only a subset of L available symbols is required offers flexibility in the design of the actual system as well as improved noise immunity. A compressive study of the selection of symbol subset will be discussed in the Sect. 8.3.2.5.

Let us assume that the channel is estimated accurately. In practice, training sequences can be periodically transmitted to update the channel estimate and adjust phase synchronization between the transmit array and the communication receiver. During each radar pulse, an N_b-bit communication information is first converted into the corresponding decimal number n_d. The K antennas comprised by the sparse array $\mathcal{S}_{n_d}$ are then selected to transmit the K orthogonal waveforms. The steering vector of the selected transmit sparse array during the τth radar pulse can be estimated as,

$$\hat{\mathbf{a}}(\theta_c;\tau) = (1/\alpha_{\text{ch}})\mathbf{y}_c(\tau). \tag{8.37}$$

The communication receiver then calculates the distance between the estimated vector $\hat{\mathbf{a}}(\theta_c;\tau)$ and each element of the dictionary $\mathbb{D}$, that is $D^l = \|\hat{\mathbf{a}}(\theta_c;\tau) - \tilde{\mathbf{a}}_l(\theta_c)\|_2, l = 1,\ldots,L$. The embedded communication symbol can be found with the smallest distance $\min_l D^l$ and then converted into the corresponding binary sequence. The detection of each communication symbol requires LK complex multiplications, thus computational complexity increases proportionally with the data rate.

Given that the radar transmits one symbol per pulse, the symbol rate of the communication system is identical to the pulse repetition frequency (f_{PRF}). The number of bits that can be transmitted per symbol is

$$N_b = \lfloor \log_2 L \rfloor, \tag{8.38}$$

where $\lfloor\cdot\rfloor$ stands for the largest integer that is no greater than the argument. Thus, the resulting data rate of the antenna selection based dual-functional systems is $N_b \times f_{\text{PRF}}$ bit per second (bps).

8.3.2.3 Angular Ambiguities

Note that the steering vector comprises the phase terms $\phi_k = k_0 p_k d \sin\theta_c, k = 1,\ldots,K, p_k \in 0,\ldots,M-1$ produced by the displacement of the K selected antennas. The mapping between the phase term $\phi_m = k_0 p_m d \sin\theta_c, m = 1,\ldots,M$ and the antenna position p_m is not one-to-one, as the phase is periodic with a period of 2π. That means there may exist multiple antennas in the array producing the same phase term, giving rise to the problem of angular ambiguity. The trivial ambiguity happens when $\theta_c = 0$, that is when the communication receiver is at broadside direction. The phases are zero regardless of antenna positions and all entries of $\mathbf{a}(\theta_c)$ are one. Thus, no information can be embedded via antenna selection. When $|\theta_c|$ is small, the phases produced by all the M antennas are equally spaced on the unit circle with values $k_0 m d \sin\theta_c, m = 0,\ldots,M-1$. As θ_c increases, the phase difference between two adjacent antennas increases and the Mth antenna reaches an angle for which we have the largest spread around the unit circle, that is,

$$\phi_M = k_0(M-1)d\sin\theta_c = 2\pi - k_0 d\sin\theta_c. \tag{8.39}$$

Solving Eq. (8.39) yields $\theta_{cm} = \sin^{-1}\left(\frac{2\pi}{Mk_0 d}\right)$. We refer to this angle as the maximal spread angle. When $\theta_c > \theta_{cm}$, it is likely that two or more antennas exhibit the same phase value. This happens when their phases are equal modulo 2π. In general, angular ambiguities happen when $(M-m)k_0 d\sin\theta_c = 2\pi$, giving $\theta_c = \sin^{-1}\frac{2\pi}{(M-m)k_0 d}$ for $m = 1, \ldots, M-1$. It is important to note, however, that when the arrival angle of communication receiver θ_c is small, the performance may be poor if the full number of bits is used. Therefore, it is advantageous to transmit at the maximal spread angle θ_{cm} which gives the best performance. We then describe a scheme to mitigate the ambiguities and steer the performance of the maximal spread angle to any receiver spatial angle.

The ambiguities described above can be mitigated by introducing additional phase rotation to each transmit antenna. Denote the vector of phase rotations assigning to the M antennas as $\mathbf{u} = [e^{j\varphi_1}, \ldots, e^{j\varphi_M}]^T$. We pre-multiply element-wise at the transmitter the vector of orthogonal waveforms, $\Psi(t)$, by the selected phase vector $\mathbf{P}(\tau)\mathbf{u}$. That means once an antenna is selected, the corresponding phase rotation is multiplied. Then the vector of phase-shifted waveforms become $\tilde{\Psi}(t) = \text{diag}(\mathbf{P}(\tau)\mathbf{u})\Psi(t)$, with diag$(\cdot)$ denoting a diagonal matrix with the vector $\cdot$ populating along the diagonal. The set of rotated waveforms $\tilde{\Psi}(t)$ still preserve the orthogonality, which is proved as follows,

$$\begin{aligned}\int_0^T \tilde{\Psi}(t)\tilde{\Psi}^H(t)dt &= \text{diag}(\mathbf{P}(\tau)\mathbf{u})\int_0^T \Psi(t)\Psi^H(t)dt\ \text{diag}(\mathbf{P}(\tau)\mathbf{u})^H, \quad (8.40)\\ &= \text{diag}(\mathbf{P}(\tau)\mathbf{u})\text{diag}(\mathbf{P}(\tau)\mathbf{u}^*),\\ &= \mathbf{I},\end{aligned}$$

Thus, the phase-rotated waveforms $\tilde{\Psi}(t)$ does not affect the normal operation of radar functions. The matched-filtered signal at the communication receiver in Eq. (8.35) becomes,

$$\mathbf{y}_c(\tau) = \alpha_{\text{ch}}\text{diag}(\mathbf{P}(\tau)\mathbf{u})\tilde{\mathbf{a}}(\theta_c;\tau) + \mathbf{n}_c(\tau). \tag{8.41}$$

The received signal vector now has phases $\tilde{\phi}_k = \phi_k + \varphi_k = k_0 p_k d\sin\theta_c + \varphi_{i_k}, k = 1, \ldots, K, i_k \in \{1, \ldots, M\}$. Thus, we can deduce a specific phase rotation for each transmit antenna, such that the phases of all M antennas are uniformly distributed around the unit circle at the spatial angle θ_c of the communication receiver. That means, $\tilde{\phi}_m = 2\pi(m-1)/M, m = 1, \ldots, M$. Then, the phase rotation for the mth antenna can be calculated as,

$$\varphi_m = \frac{2\pi(m-1)}{M} - k_0 p_m d\sin\theta_c, m = 1, \ldots, M. \tag{8.42}$$

In this manner, not only are we able to mitigate the ambiguities, but also to deliver the best symbol dictionary to any receiver.

8.3.2.4 Symbol Error Rate

Let us assume without loss of generality that the transmitted sparse array is $\mathcal{S}_i$, whose corresponding steering vector $\tilde{\mathbf{a}}_i$ comprises K phases of value $\tilde{\phi}_k = 2\pi(i_k - 1)/M, i_k = \{1, \ldots, M\}$. It is worth noting that the dependence of the steering vector $\tilde{\mathbf{a}}_i$ on the angle θ_c of the communication receiver is suppressed due to the additional phase rotations. The symbols in the dictionary defined in Eq. (8.36) change to

$$\tilde{\mathbf{a}}_l = [e^{j2\pi(l_1-1)/M}, \ldots, e^{j2\pi(l_K-1)/M}]^T, l_k = \{1, \ldots, M\}. \tag{8.43}$$

Let us define the distance between the estimated steering vector $\hat{\mathbf{a}}_i$ and each code $\tilde{\mathbf{a}}_l$ in the dictionary as $D^l = \|\hat{\mathbf{a}}_i - \tilde{\mathbf{a}}_l\|_2$. Then the probability of a correct symbol detection is given by

$$P_d = P\left(D^i < D^l, \forall l = 1, \ldots, L, l \neq i\right). \tag{8.44}$$

It is worth noting that a symbol error may not occur even if the noise places some phase $\tilde{\phi}_k = 2\pi(i_k - 1)/M$ of the received signal closer to another constellation, $2\pi(l_k - 1)/M, i_k, l_k \in \{1, \ldots, M\}$, such that $l_k \neq i_k, k = 1, \ldots, K$, provided that $D^i < D^l$. Thus, for each symbol $\mathcal{S}_i$, we have

$$P\left(D^i < D^l\right) \geq \Pi_{k=1}^{K} P(D_k^i < D_k^l), \tag{8.45}$$

where $P(D_k^i < D_k^l)$ denotes the probability of a correct detection of the kth phase term. Detecting each phase of the steering vector is similar to the M-ray phase-shift keying (M-ary PSK) scenario, where every phase $\tilde{\phi}_k$ is taken out of M uniformly distributed signal constellations around a unit circle with an angular separation of $\gamma = 2\pi/M$. The average probability of symbol error for M-ary PSK modulation with sufficiently high signal-to-noise ratio (SNR) is [313],

$$Q(\rho, \gamma) = \text{erfc}\left(\sqrt{\rho}\sin(\frac{\gamma}{2})\right), \tag{8.46}$$

where ρ stands for the SNR and erfc denotes the complementary error function. Thus, we have that

$$P\left(D_k^i < D_k^l\right) = 1 - Q(\rho, \gamma), k = 1, \ldots, K. \tag{8.47}$$

Substituting Eq. (8.47) into Eq. (8.45) yields the lower bound of the detection probability,

$$P_d \geq [1 - Q(\rho, \gamma)]^K. \tag{8.48}$$

The upper bound of symbol error rate (SER) is then obtained by,

$$P_e = 1 - P_d \leq 1 - [1 - Q(\rho, \gamma)]^K. \tag{8.49}$$

The embedded symbol is detected by comparing the distance between the estimated steering vector and each code in the dictionary. Thus, it is preferred that the distance between any two code vectors in the dictionary be maximized. For radar modalities, target detection is the main objective. The detection performance is directly related to the efficacy of clutter cancellation. As the same set of K orthogonal waveforms $\Psi(t)$ are transmitted during each PRI, there is no attendant Doppler coherency degradation as existed in waveform modulation scheme proposed in [314]. However, the transmit array configuration affects the radar detection performance significantly. Thereby, the selection of symbol subset should consider two criteria together, the performance of communication functions and a satisfying radar transmit beampattern.

8.3.2.5 Selection of Constellation Symbols

We consider the first criterion of communication performance, that is selecting a subset of $L_b = 2^{N_b}$ symbols from L candidates, such that the distance between any two symbols in the dictionary is maximized. Without loss of generality, the total number M of installed antennas is assumed to be even. As all the M antennas are uniformly distributed around the unit circle with a phase difference of $2\pi/M$, it is intuitive that the two symbols with the largest distance are $\tilde{\mathbf{a}}_1 = \{1, e^{j2\pi/M}, \ldots, e^{j2\pi(K-1)/M}\}$ and $\tilde{\mathbf{a}}_2 = \{-1, e^{j2\pi(M/2+1)/M}, \ldots, e^{j2\pi(M/2+K-1)/M}\}$. That means each pair of antennas in symbols $\tilde{\mathbf{a}}_1$ and $\tilde{\mathbf{a}}_2$ are center-symmetrically distributed in the upper and lower half circles, respectively. The largest distance can be calculated as $\|\tilde{\mathbf{a}}_1 - \tilde{\mathbf{a}}_2\|^2 = 4K$. Initialize the symbol subset as $\mathbb{D}_c = \{\tilde{\mathbf{a}}_1, \tilde{\mathbf{a}}_2\}$, and $\mathbf{z}_l$ are selection vectors corresponding to $\tilde{\mathbf{a}}_l$ such that $\tilde{\mathbf{a}}_l = \mathbf{a}(\mathbb{Z}(\mathbf{z}_l)), l = 1, 2$, where $\mathbb{Z}(\mathbf{z})$ denotes the sparse support of vector $\mathbf{z}$. The remaining $L_b - 2$ symbols can be found as follows:

$$\begin{aligned} \max_{\mathbf{z}, \nu} \quad & \nu, \\ \text{subject to} \quad & \|\text{diag}(\mathbf{z})\mathbf{a} - \text{diag}(\mathbf{z}_l)\mathbf{a}\|_2^2 \geq \nu, l = 1, \ldots, |\mathbb{D}_c|, \\ & \mathbf{z} \in \{0, 1\}^M, \ \mathbf{1}^T\mathbf{z} = K, \end{aligned} \tag{8.50}$$

where $\mathbf{a} = [1, e^{j2\pi/M}, \ldots, e^{j2\pi(M-1)/M}]^T$ is the steering vector of the full array after phase rotation and the vector comprised of the non-zero entries of diag($\mathbf{z}_l$)$\mathbf{a}$ is a symbol in $\mathbb{D}_c$. The selection variable $\mathbf{z}$ is binary with entry one denoting the corresponding antenna selected and entry zero discarded, $|\mathbb{D}_c|$ stands for the cardinality of the symbol subset $\mathbb{D}_c$. The constraint $\mathbf{1}^T\mathbf{z} = K$ controls the number of selected antennas to be exactly K.

As explained in [209], the binary property of the selection variable $\mathbf{z} \in \{0, 1\}^M$ is tantamount to

$$\max_{\mathbf{z}} \ \mathbf{z}^T(\mathbf{z} - \mathbf{1}) \text{ subject to } 0 \leq \mathbf{z} \leq 1, \tag{8.51}$$

where the inequality is the constraint applying to each entry in vector $\mathbf{z}$. Combining Eqs. (8.51) with (8.50) yields the following formulation,

$$\max_{\mathbf{z},\nu} \ \nu + \mu[\mathbf{z}^T(\mathbf{z}-\mathbf{1})], \tag{8.52}$$
$$\text{subject to } \ \mathbf{a}^H\text{diag}(\mathbf{z})\mathbf{a} - 2\mathbf{a}^H\text{diag}(\mathbf{z})\text{diag}(\mathbf{z}_l)\mathbf{a} + K \leq \nu, l = 1, \ldots, |\mathbb{D}_c|,$$
$$0 \leq \mathbf{z} \leq 1, \ \mathbf{1}^T\mathbf{z} = K,$$

where the trade-off parameter μ is used to control the emphasis between symbol distance and the boolean property of the selection vector $\mathbf{z}$. Equal importance can achieved by setting $\mu = 1$.

Next, we consider the second criterion of radar transmit pattern synthesis and the optimum dictionary is denoted as $\mathbb{D}_r$. As shown in Eq. (8.29), the virtual extended signal vector of the MIMO radar is the Kronecker product between transmit and receive array steering vectors. That is,

$$\mathbf{c}(\theta) = \tilde{\mathbf{a}}(\theta) \otimes \mathbf{b}(\theta). \tag{8.53}$$

Assume that the beamforming weight vector is $\mathbf{w}$, the overall beampattern of MIMO radar can be expressed as

$$B(\theta) = \left|\mathbf{w}^H\mathbf{c}(\theta)\right| = \left|\mathbf{w}^H[\tilde{\mathbf{a}}(\theta) \otimes \mathbf{b}(\theta)]\right|. \tag{8.54}$$

We can see that the shape of overall beampattern is affected by both transmit and receive array configurations. Since the structure of receive array is fixed, it is preferred that sparse transmit array configurations satisfy a certain desired power radiation pattern, when combined with the given receive array. The main function of MIMO radar is to concentrate the transmit power within a certain angular sector $\Theta = [\theta_{\min}, \theta_{\max}]$, where the radar signal may come from. The beampattern corresponding to the sidelobe region $\bar{\Theta}$ is required to be less than a pre-defined sidelobe level ϵ. The selection of symbol subset satisfying the criterion of radar function can be formulated as follows:

$$\max_{\mathbf{z},\mathbf{w},\rho} \ \rho + \mu[\mathbf{z}^T(\mathbf{z}-\mathbf{1})], \tag{8.55}$$
$$\text{s.t. } \ |\mathbf{w}^H\mathbf{c}(\theta_i) - e^{j\mu(\theta_i)}| \leq \rho, \ \theta_i \in \Theta, i = 1, \ldots, L_m,$$
$$|\mathbf{w}^H\mathbf{c}(\theta_k)| \leq \epsilon, \ \theta_k \in \bar{\Theta}, k = 1, \ldots, L_s$$
$$|\mathbf{J}_m\mathbf{w}| \leq z_m, m = 1, \ldots, M$$
$$0 \leq \mathbf{z} \leq 1, \mathbf{1}^T\mathbf{z} = K,$$

where $\theta_i, i = 1, \ldots, L_m$ and $\theta_k, k = 1, \ldots, L_s$ are L_m and L_s samples of the mainlobe region Θ and sidelobe region $\bar{\Theta}$, respectively, and $\mu(\theta)$ is the user-defined mainlobe phase profile, ρ denotes the allowable maximum mainlobe ripple. The weight vector $\mathbf{w}$ exhibits a block sparsity with $N - K$ blocks of M entries being zero. In addition, the matrix $\mathbf{J}_m \in \{0, 1\}^{N \times MN}$ is utilized to extract the $[(m-1)N+1] \sim (mN)$th entries of the weight vector $\mathbf{w}$. The matrix has "one" entry in each row and in the $[(m-1)N+1] \sim (mN)$th columns, and all other entries being zero. The constraints

$|\mathbf{J}_m \mathbf{w}| \leq z_m, m = 1, \ldots, M$ are used to promote the same group sparsity of weight vector $\mathbf{w}$ as the selection variable $\mathbf{z}$.

Clearly, the objective functions in Eqs. (8.52) and (8.55) are concave, and it is difficult to maximize them directly. A sequential convex programming (SCP) based on iteratively linearizing the concave objective function is then utilized to reformualte the non-convex problem to a series of convex subproblems, each of which can be optimally solved using convex programming [133]. Taking the problem in Eq. (8.55) as an example, the symbol selection in the (k+1)th iteration can be formulated based on the solution $\mathbf{z}^{(k)}$ from the kth iteration as,

$$\begin{aligned}
\max_{\mathbf{z},\mathbf{w},\rho} \ & \rho + \mu[(2\mathbf{z}^{(k)} - \mathbf{1})^T \mathbf{z} - \mathbf{z}^{(k)T}\mathbf{z}^{(k)}], \qquad (8.56) \\
\text{s.t.} \ & |\mathbf{w}^H \mathbf{c}(\theta_i) - e^{j\mu(\theta_i)}| \leq \rho, \ \theta_i \in \Theta, i = 1, \ldots, L_m, \\
& |\mathbf{w}^H \mathbf{c}(\theta_k)| \leq \epsilon, \ \theta_k \in \bar{\Theta}, k = 1, \ldots, L_s \\
& |\mathbf{J}_m \mathbf{w}| \leq z_m, m = 1, \ldots, M \\
& 0 \leq \mathbf{z} \leq 1, \mathbf{1}^T \mathbf{z} = K.
\end{aligned}$$

Note that the SCP is a local heuristic and its performance depends on the initial point $\mathbf{z}^{(0)}$. It is, therefore, feasible to construct the symbol dictionary by initializing the algorithm with different feasible points $\mathbf{z}^{(0)}$. The symbol subset selection considering both criteria can be achieved by combining the constraints in Eqs. (8.52) and (8.55). The detailed description of the symbol subset selection is illustrated in Table. 8.3. After obtaining the selection vector $\mathbf{z}^*$, the corresponding symbol can be calculated as $\tilde{\mathbf{a}} = \mathbf{a}(\mathbb{Z}(\mathbf{z}^*)) = \{e^{j2\pi(m-1)/M}, z_m^* = 1, m = 1, \ldots, M\}$.

Table 8.3 The detailed description of symbol subset selection

Initialization	Initialize symbol subset $\mathbb{D} = \{\}$, initialize $\mu = 1$
Step 1 (Outer Loop)	**WHILE:** $\lvert\mathbb{D}\rvert < L_b$
Step 2	Set $k = 0$ and maximum iteration number K_m. Randomly initialize $\mathbf{z}^{(k)}$
Step 3 (Inner Loop)	**FOR** $k < K_m$
	Solve Eq. (8.52) or (8.55) using Matlab embedded software CVX, set k=k+1;
	END OF INNER LOOP
Step 4	If the obtained selection vector $\mathbf{z}$ is boolean and not included in $\mathbb{D}$, calculate the corresponding symbol $\tilde{\mathbf{a}}$, set $\mathbb{D} = [\mathbb{D}, \tilde{\mathbf{a}}]$ and go to Step 1
Step 5	If the obtained selection vector $\mathbf{z}$ is not boolean, go to Step 2
Step 6	**END OF OUTER LOOP**

8.3.3 Combined Antenna Selection and Waveform Permutation

The MIMO radar receiver requires the knowledge of transmit waveform and transmit antenna pairing, and does not require pinning a specific waveform to a specific antenna. The flexibility of varying the selected transmit antennas across K orthogonal waveforms over different PRIs can be exploited to embed a large constellation of symbols. For each selected K-antenna sparse array, the number of symbols that can be embedded is a factorial of the number of transmit antennas, that is $K!$. Thus, combining antenna selection with permutating K independent waveforms to each selected antenna over one PRI, a data rate of megabits per second can be achieved by a moderate number of transmit antennas. Taking this notion further, we propose a hybrid selection and permutation based signaling strategy for DFRC systems in this section. Since permutations used to assign the antennas to the waveform set are known to the radar, the reordering enables restoring the coherent structure of the MIMO radar data, i.e. the primary MIMO radar operation is unaffected by the secondary communication function.

The structure of combined selection and permutation based signaling strategy remains the same as that of selection only method and is depicted in Fig. 8.20. There are M antennas installed on the platform and a specific subset of K antennas associated with the communication symbols are switched on for transmitting independent waveforms during each radar pulse. Denote the $K \times M$ selection matrix and $K \times K$ permutation matrix as $\mathbf{P}(\tau)$ and $\mathbf{Q}(\tau)$, respectively. The signal at the output of the communication receiver antenna is remodelled as,

$$x_c(t, \tau) = \alpha_{\text{ch}} \mathbf{a}^T(\theta_c) \mathbf{P}^T(\tau) \mathbf{Q}^T(\tau) \Psi(t) + n_c(t, \tau). \tag{8.57}$$

Matched filtering the received data with the set of orthogonal waveforms yields,

$$\mathbf{y}_c(\tau) = \text{vec}\left\{ \int_T x_c(t, \tau) \Psi^H(t) dt \right\} = \alpha_{\text{ch}} \mathbf{M}(\tau) \mathbf{a}(\theta_c) + \mathbf{n}_c(\tau), \tag{8.58}$$

where $\mathbf{M}(\tau) = \mathbf{Q}(\tau)\mathbf{P}(\tau)$. Thus, the communication receiver signal at the output of the matched-filter is a (scaled and noisy) selected permutation of the steering vector $\mathbf{a}(\theta_c)$, meaning that the product of selection and permutation matrices $\mathbf{M}(\tau)$ can be recovered from the received vector $\mathbf{y}_c(\tau)$ by determining the ordering of K selected transmit antennas. We propose to utilize the selected permutation of the steering vector $\mathbf{M}(\tau)\mathbf{a}(\theta_c)$, that is the ordered set of phases induced by selected antenna positions, as the codes to embed communication symbols.

To mitigate angular ambiguity and maximize communication performance, the phase rotation imposed to each transmit antenna per Eq. (8.42) can be deployed here. Thus, a dictionary of $K! \times L$ symbols is constructed as,

$$\mathbb{D} = \{\mathbf{A}_1, \ldots, \mathbf{A}_L\}, \tag{8.59}$$

where $\mathbf{A}_l = [\mathbf{Q}_1\tilde{\mathbf{a}}_l, \ldots, \mathbf{Q}_{K!}\tilde{\mathbf{a}}_l]$ with $\mathbf{Q}_k, k = 1, \ldots, K!$ denoting the permutation matrix. In addition, $\tilde{\mathbf{a}}_l = [e^{j\tilde{\phi}_{l_1}}, \ldots, e^{j\tilde{\phi}_{l_K}}]^T$ with $l_k \in \{1, \ldots, M\}$ and $\tilde{\phi}_{l_k} = 2\pi(l_k - 1)/M$. During each radar pulse, the K orthogonal waveforms $\Psi_k(t), k = 1, \ldots, K$ are transmitted through the ordered subset of antennas with positions p_k corresponding to the N_b-bit information. Assume that communication receiver has a prior knowledge of its angle θ_c relative to the joint transmit array. The ordered selected steering vector can be estimated as,

$$\hat{\mathbf{a}}(\theta_c; \tau) = (1/\alpha_{\mathrm{ch}})\mathbf{y}_c(\tau) \approx \mathbf{M}(\tau)\mathbf{a}(\theta_c). \tag{8.60}$$

The communication receiver can then compare the estimated vector $\hat{\mathbf{a}}(\theta; \tau)$ to the dictionary $\mathbb{D}$ to obtain the embedded communication symbols. As there are $K!$ different ordering for each selected subarray, the message bits that can be transmitted during each pulse are

$$N_b = \lfloor \log_2(L \times K!) \rfloor = \lfloor \log_2 L + \log_2 K! \rfloor. \tag{8.61}$$

Thus, the data rate, measured in bps, for the proposed hybrid selection and permutation based signaling scheme can be expressed as,

$$R = \lfloor \log_2 L + \log_2 K! \rfloor \times f_{\mathrm{PRF}}. \tag{8.62}$$

It is worth noting that, not only the data rate can be increased, but also the symbol error rate for the hybrid scheme can be significantly reduced compared with that of selection-only signaling scheme. The reason is that the permutation of antenna positions can be utilized to further increase the distance between the selected symbols $\tilde{\mathbf{a}}_l, l = 1, \ldots, L_b$ in $\mathbb{D}_c$. The symbol selection for the dictionary $\mathbb{D}_p$ utilizing both antenna selection and permutation can be formulated as,

$$\begin{aligned} &\max_{\mathbf{Q},\mathbf{P},\nu} \quad \nu, \\ &\text{subject to} \quad \|\mathbf{QPa} - \bar{\mathbf{a}}_k\|^2 \geq \nu, k = 1, \ldots, |\mathbb{D}_p|, \end{aligned} \tag{8.63}$$

where $\bar{\mathbf{a}}_k = \mathbf{Q}_k\mathbf{P}_k\mathbf{a} \in \mathbb{D}_p$ are already-selected symbols. As the optimization variables $\mathbf{Q}$ and $\mathbf{P}$ are required to satisfy the conditions of permutation matrix and selection matrix, respectively, the problem is highly non-convex. Moreover, enumerating all $L \times K!$ different permutations and combinations is prohibitively exhausitive. Instead, we resort to enumerate the optimum permutation for each selected symbol in $\mathbb{D}_c$ and obtain a sub-optimum dictionary $\mathbb{D}_p$. It is worth noting that the proposed signaling modulation strategy is different from the waveform shuffling scheme introduced in [315], where permutation matrix $\mathbf{Q}$ only is utilized for embedding communication symbols and symbol detection is accomplished by a complicated minimization problem in terms of the permutation matrix.

8.3.4 Regularized Selection Based Spatial Index Modulation

The two aforementioned signaling strategies implement an unrestricted antenna selection, that is an arbitrary K-antenna sparse array might be selected for waveform transmitting according to the embedded symbols. As there are only K RF front-ends installed in the platform, antenna selection network is required to be capable of connecting an arbitrary subset of K antennas with front-ends. This may put a high pressure on the hardware realization especially when the selected antennas locate far from the front-ends. In order to preserve original radar functions, the MIMO radar receiver is assumed to know the association of the orthogonal waveforms to the transmit antennas for the hybrid selection and permutation scheme. The complete transparency between the two functions may cause practical implementation issues as well. To counteract these implementation issues, we propose a regularized selection based signaling strategy to embed communication symbols into the transmit array configuration in the following.

The concept of the proposed signaling scheme is shown in Fig. 8.21. There are $M = 2K$ uniformly spaced transmit antennas with an inter-element spacing of d. The M antennas are divided into K subgroups with each subgroup consisting of two adjacent antennas. Each subgroup represents one-bit symbol, where the symbol "0" implies the first antenna selected and the second antenna discarded, and vice versa for the symbol "1". The restriction of only one selected antenna for each subgroup can guarantee a constant number of K transmit antennas. Let $\Psi_k(t)$, $m = 1, \ldots, K$ be K orthogonal waveforms corresponding to the K subgroups of antennas. The antenna selection matrix in Eq. (8.32) becomes $\mathbf{P}(\tau) \in \{0, 1\}^{K \times 2K}$, which is a rectangular diagonal selection matrix with a K-bit message matrix $\mathbf{E} \in \{0, 1\}^{2 \times K}$ populating along the diagonal. Each row $\mathbf{e}_k$, $k = 1, \ldots, K$ of the message matrix $\mathbf{E}$ is defined as follows,

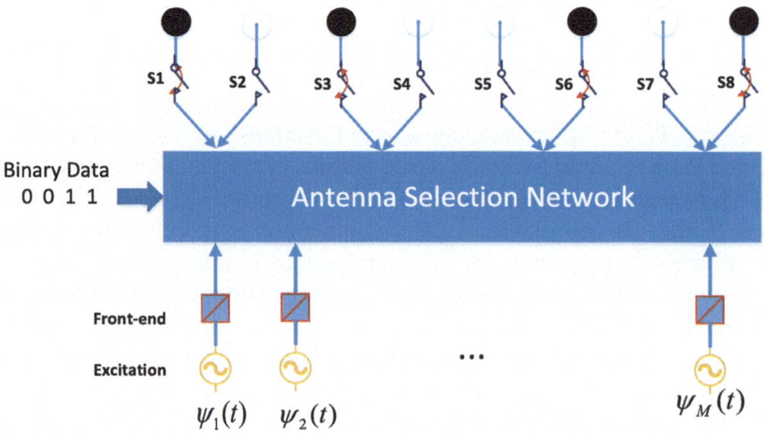

Fig. 8.21 Illustration of regularized antenna selection based modulation signaling scheme

$$\mathbf{e_k} = \begin{cases} [1, 0] & \text{if the kth message bit is } b_k = 0, \\ [0, 1] & \text{if the kth message bit is } b_k = 1. \end{cases} \tag{8.64}$$

In order to decouple the dependency of communication performance on the arrival angle θ_c, a set of phase rotations can be pre-multiplied with orthogonal waveforms before transmitting. As proved in Sect. 8.3.2.3, the phase-rotated waveforms are capable of preserving the orthogonality, and do not affect the normal radar operations. To approach the performance of BPSK scheme, the additional phase rotations φ_k, $k = 1, \ldots, K$ are calculated as,

$$\varphi_k = \begin{cases} -(2k-2)k_0 d \sin\theta_c & \text{if } b_k = 0, \\ \pi - (2k-1)k_0 d \sin\theta_c & \text{if } b_k = 1. \end{cases} \tag{8.65}$$

Denoting the phase rotation vector as $\mathbf{u} = [e^{j\varphi_1}, \ldots, e^{j\varphi_K}]^T$, the received communication signal is

$$x_c(t, \tau) = \alpha_{\text{ch}} \mathbf{a}^T(\theta_c) \mathbf{P}^T(\tau) \text{diag}(\mathbf{u}) \Psi(t) + n_c(t, \tau), \tag{8.66}$$

Matched filtering the received data with the kth waveform yields,

$$\begin{aligned} y_{c,k}(\tau) &= \int_T x_c(t, \tau) \Psi_k(t) dt, \qquad (8.67) \\ &= \alpha_{\text{ch}} e^{j\varphi_k} \left[\mathbf{a}_{2k-1}(\theta_c)(1 - b_k) + \mathbf{a}_{2k}(\theta_c) b_k\right] + n_{c,k}(\tau), k = 1, \ldots, K, \end{aligned}$$

Then, each message bit can be deciphered from the phase of the received signal, that is,

$$\begin{aligned} \hat{\phi}_k(\tau) &= \text{angle}\{y_{c,k}(\tau)\} - \text{angle}\{\alpha_{\text{ch}}\}, \qquad (8.68) \\ &\approx \begin{cases} 0 & \text{if } b_k = 1, \\ \pi & \text{if } b_k = 0, \end{cases} \end{aligned}$$

where angle$(\cdot)$ stands for the angle of a complex number. As the number of embedded bits during each radar pulse equals to the number K of selected antennas, the data rate in bps can be expressed as,

$$R = K \times f_{\text{PRF}}, \tag{8.69}$$

The bit error rate for the proposed regularized selection strategy is the same as that of BPSK, that is

$$\text{BER}_r = Q(\rho, 1). \tag{8.70}$$

As the correct detection of the K-bit communication symbol requires the accurate estimate of each bit, the symbol error rate of the regularized antenna selection scheme can be expressed in terms of the bit error rate,

$$\mathrm{SER}_r = 1 - (1 - Q(\rho, 1))^K. \tag{8.71}$$

Note that there are totally 2^K symbols in the dictionary for the regularized selection based signaling scheme and no symbol subset selection is further required. The pair of symbols with the minimum distance is obtained by switching one antenna to the other in one subgroup with maintaining others unchanged and those with the maximum distance is obtained by switching on/off the antennas in all K subgroups. It is worth noting that the association between antennas and orthogonal waveforms is fixed during the entire process and the assumption of communication operation transparency is no more necessary to MIMO radar.

8.3.5 Simulations

In our simulations, we consider a radar with $M = 16$ antennas arranged in a ULA with an inter-element spacing of 0.25 wavelength. Throughout the simulations, we assume a number of $K = 8$ antennas are selected during each PRI to simultaneously embed one communication symbol while performing the radar operation. The radar receiver array is a 10-antenna ULA. Unless otherwise stated, we evaluate the performance of the system by showing the symbol error rate as a function of SNR.

8.3.5.1 Example 1: Antenna Selection Based Spatial Index Modulation

In the first example, we assume that the main radar operation takes place within the angular sector $\Theta = [-10°, 10°]$. A single communication receiver is assumed to locate at the direction of $\theta_c = 14.4775°$. In this case, the total number of unique subarray configurations which can be obtained by antenna selection equals $C_{16}^8 = 12870$. We embed one communication symbol per PRI. The highest number of bits per symbol is $\lfloor \log_2(C_{16}^8) \rfloor = 13$. Here, we consider the cases of 1, 2, 4, and 8 bits per symbol which can be achieved by building four dictionaries of 2, 4, 16, and 256 subarrays, respectively. Symbol subset selection of the 256 configurations, drawn from the total 12870 available combinations, is performed offline for two scenarios. In the first scenario, the radar operation was given the priority by enforcing the selected sparse arrays to have the smallest peak ripples within the main radar beam. We refer to this set of configurations as $\mathbb{D}_r$. The power patterns of different sparse arrays in the dictionary $\mathbb{D}_r$ are almost the same with a small mainlobe ripple, as shown in Fig. 8.22. In the second scenario, we select the sparse arrays such that the Euclidean distance between different symbols in the dictionary is maximized. We refer to this set of configurations as $\mathbb{D}_c$. A peak sidelobe level of -20 dB is required in both scenarios.

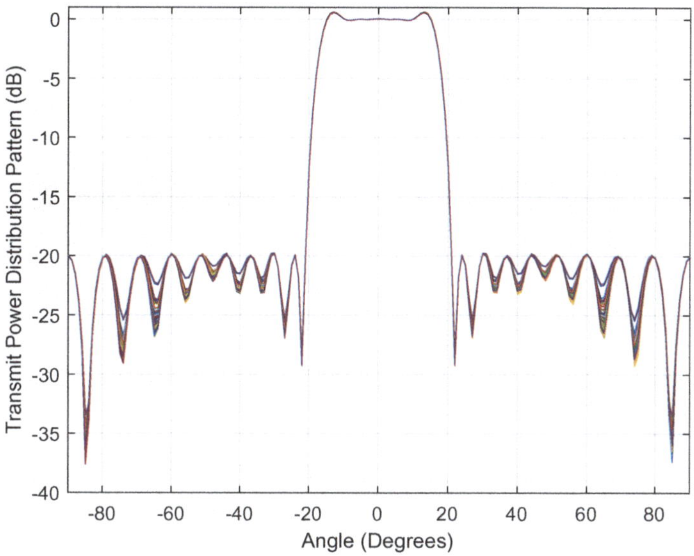

Fig. 8.22 Overall power patterns of all the 8-antenna sparse arrays in the dictionary $\mathbb{D}_r$

Unfortunately, the larger Euclidean distance between different symbols comes at the price of having larger difference between the corresponding beampatterns as shown in Fig. 8.23. For small dictionary size, the individual beampatterns have almost the same mainbeam level as the nominal value. However, as the dictionary size increases, some of the individual beampatterns exhibit noticeable deviation from the nominal beampattern within the mainbeam. This may cause some loss in radar performance which is the price paid for having an improved communication detection performance.

To test the communication performance and show the trade-off between the communications and the radar beampattern requirements, a number of 10^7 symbols are randomly generated. Information embedding is performed using the dictionaries $\mathbb{D}_r$ and $\mathbb{D}_c$, which emphasize the radar requirement and the communication performance, respectively. Figure 8.24 shows the SER versus SNR for various numbers of bits per symbol using the constellations drawn from $\mathbb{D}_r$. The figure shows that the SER curves exhibit the expected standard behavior of a communication system, with the SER increasing with decreasing SNR and with increasing number of bits per symbol. At high SNR values, a SER smaller than 10^{-5} can be achieved for all cases considered. The figure shows that for a fixed SNR the use of a dictionary of smaller size results in lower SER and vice versa. At low SNR values where noise is dominant, the communication receiver detects each symbol in the dictionary with equal probability. For example, the SER is approximately 0.5 at the SNR of -20 dB in the case of 1 bit per symbol, that is the probability of detecting the symbol correctly equals the probability of detecting it erroneously. Figure 8.25 shows the SER versus SNR for various numbers of bits per symbol in the scenario where the dictionary $\mathbb{D}_c$ is used.

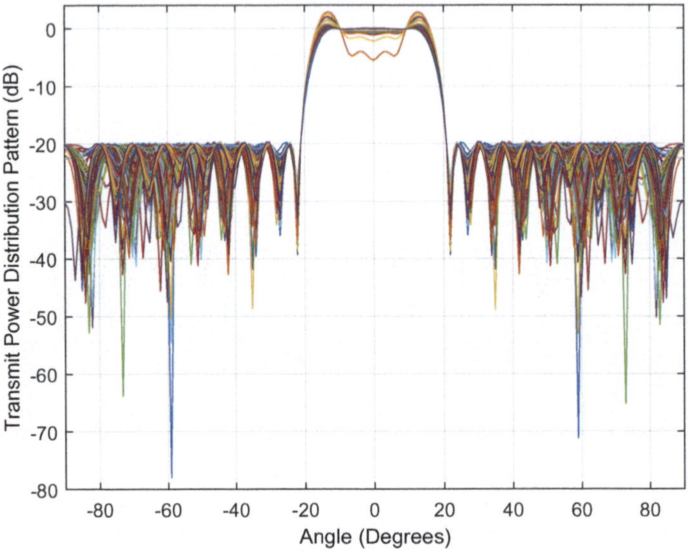

Fig. 8.23 Overall power patterns of all the 8-antenna sparse arrays in the dictionary $\mathbb{D}_c$

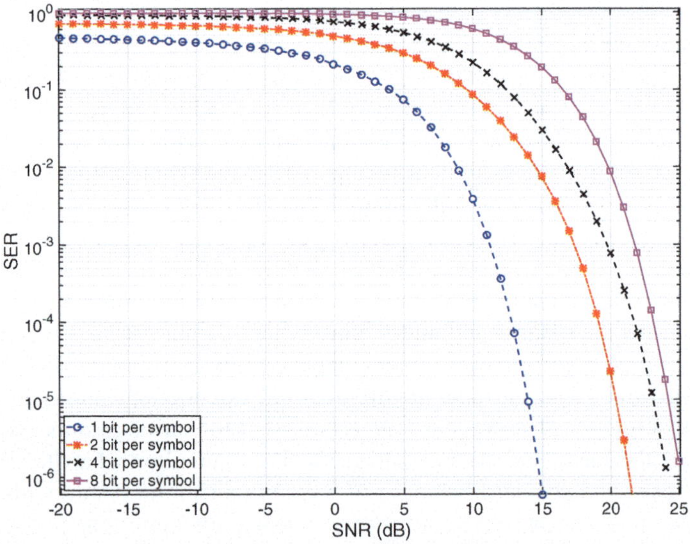

Fig. 8.24 SER versus SNR in the case where the communication receiver is at direction $\theta_c = 14.4775°$; The dictionary $\mathbb{D}_r$ is selected in favor of radar operation

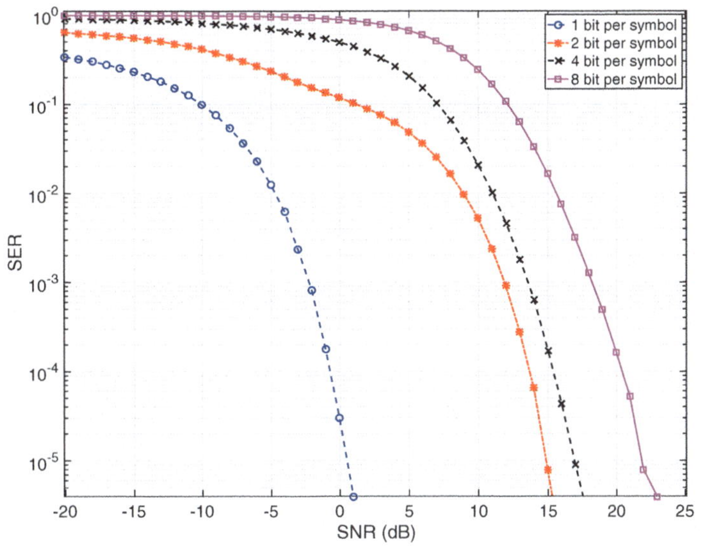

Fig. 8.25 SER versus SNR in the case where the communication receiver is at direction $\theta_c = 14.4775°$; The dictionary $\mathbb{D}_c$ is selected in favor of communications at the price of increased mainlobe ripples

The figure shows that the SER curves exhibit better SER performance as compared to that of the first scenario. This can be attributed to the fact that the dictionary $\mathbb{D}_c$ is designed to enhance the communication performance.

8.3.5.2 Example 2: Combined Antenna Selection and Waveform Permutation

We proceed to investigate the hybrid selection and permutation based signaling scheme in this example. For the dictionary $\mathbb{D}_c$ constructed based on the metric of communication performance, we enumerate all potential permutations of each symbol such that the distance between arbitrary two symbols in the dictionary is further maximized. The new dictionary is denoted as $\mathbb{D}_p$. We can calculate the minimum and maximum distances between the kth symbol and the remaining 255 symbols as follows,

$$d_{\min}^k = \begin{cases} \min\left\{\|\tilde{\mathbf{a}}_k - \tilde{\mathbf{a}}_i\|^2, i = 1, \ldots, k-1, k+1, \ldots, 256\right\}, & \text{for } \mathbb{D}_c, \mathbb{D}_r; \\ \min\left\{\|\bar{\mathbf{a}}_k - \bar{\mathbf{a}}_i\|^2, i = 1, \ldots, k-1, k+1, \ldots, 256\right\}, & \text{for } \mathbb{D}_p; \end{cases} \tag{8.72}$$

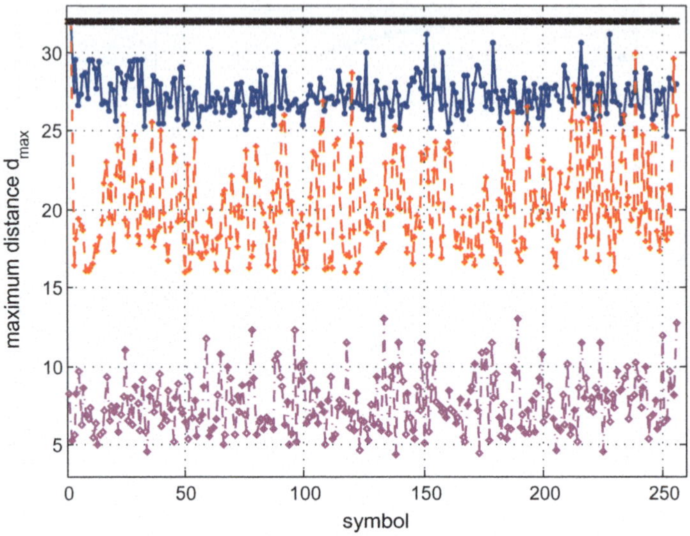

Fig. 8.26 Maximum distance d_{max}^k between the kth symbol and any other symbol in the dictionaries

and

$$d_{\text{max}}^k = \begin{cases} \max\left\{\|\tilde{\mathbf{a}}_k - \tilde{\mathbf{a}}_i\|^2, i = 1, \ldots, k-1, k+1, \ldots, 256\right\}, & \text{for } \mathbb{D}_c, \mathbb{D}_r; \\ \max\left\{\|\bar{\mathbf{a}}_k - \bar{\mathbf{a}}_i\|^2, i = 1, \ldots, k-1, k+1, \ldots, 256\right\}, & \text{for } \mathbb{D}_p; \end{cases} \tag{8.73}$$

The maximum and minimum distances of the constructed three dictionaries $\mathbb{D}_r$, $\mathbb{D}_c$, $\mathbb{D}_p$ are plotted in Figs. 8.26 and 8.27 for comparison. Clearly, the minimum distance of the dictionary $\mathbb{D}_p$ after antenna permutation is much larger than those of the two dictionaries, which directly determines the communication accuracy.

To test the communication performance, a number of 10^7 symbols are randomly generated. Figure 8.28 shows the SER versus SNR for various numbers of bits per symbol. We embed one communication symbol per PRI. The highest number of bits per symbol is $\lfloor \log_2(C_{16}^8 \times 8!) \rfloor = 28$. Similar to Example 1, we consider the cases of 1, 2, 4, and 8 bits per symbol, respectively. We can see that the communication performance is significantly improved especially for the case of 8 bits per symbol.

8.3.5.3 Example 3: Regularized Selection Based Spatial Index Modulation

We continue to investigate the regularized selection based signaling scheme. The 16-antenna ULA is divided into 8 subgroups and each subgroup consists of two antennas. During each radar pulse, one out of two antennas in each subgroup are switched on according to the communication symbol. There are totally $2^8 = 256$

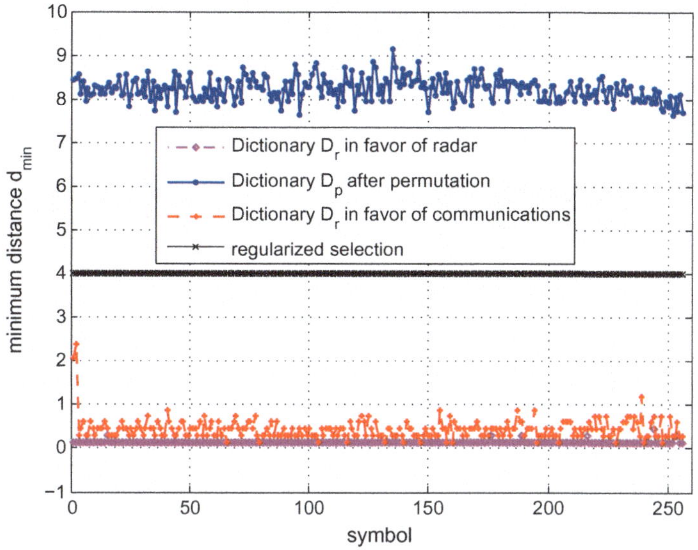

Fig. 8.27 Minimum distance d^k_{min} between the kth symbol and any other symbol in the dictionaries

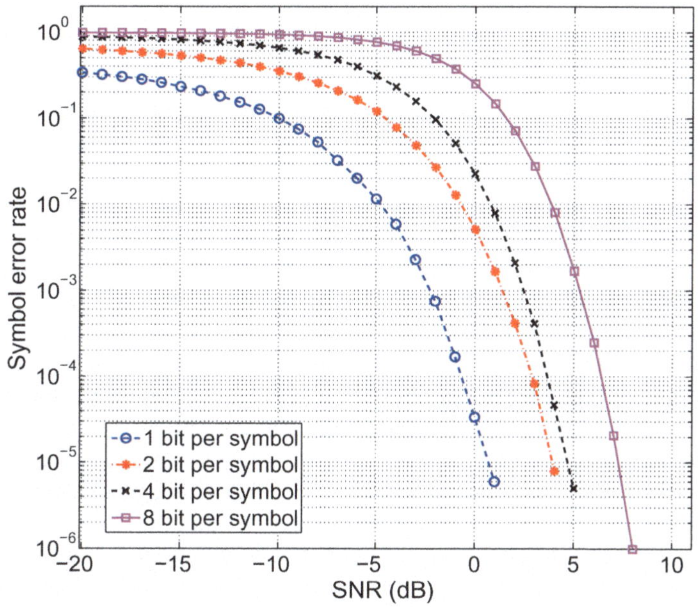

Fig. 8.28 SER versus SNR in the case where the communication receiver is at direction $\theta_c = 14.4775°$; The dictionary $\mathbb{D}_p$ is obtained by applying permutation to the dictionary $\mathbb{D}_c$ to increase the distance between arbitrary two symbols

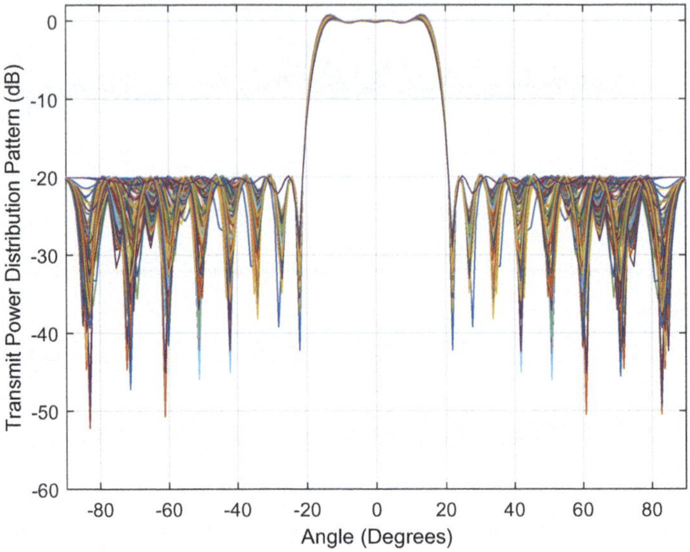

Fig. 8.29 Power patterns of the 256 different 8-antenna sparse arrays in the regularized selection scheme

symbols and no symbol subset selection is required. The maximum distance is 32, which is obtained by changing the statuses of all 8 subgroups, as shown in Fig. 8.26. The minimum distance is 4, which is achieved by changing the antenna status of one subgroup and maintaining the other subgroups unchanged, as shown in Fig. 8.27. The power patterns of the 256 sparse arrays are depicted in Fig. 8.29, although worse than those of the dictionary $\mathbb{D}_r$ constructed in favor of radar functions, but much better than those of the dictionary $\mathbb{D}_c$ constructed in favor of communication function.

To test the communication performance, we consider the cases of 1, 2, 4, and 8 bits per symbol respectively. For the case of 1 bits per symbol, all the 8 subgroups transmit the same bit information. For the case of 2 bits per symbol, the first four subgroups transmit the first bit and the last four subgroups transmit the second bit. For the case of 4 bits per symbol, each two adjacent subgroups transmit one bit information. For the case of 8 bits per symbol, every subgroup transmits one bit. The SER curve versus the SNR is plotted in Fig. 8.30. Although the communication performance is inferior to that of the hybrid selection strategy, it is much better than those of the antenna-selection scheme with both constellations $\mathbb{D}_r$ and $\mathbb{D}_c$.

Finally, we compare the robustness against the estimation error of communication receiver angle between the hybrid scheme and the regularized selection scheme. Assume that the true angle of the communication receiver is normally distributed with mean $\theta_c = 14.4775°$ and standard variance σ. The dual-function platform transmits the communication symbol towards the assumed angle $\theta_c = 14.4775°$ and calculates the phase rotation of each antenna according to that assumed angle. The communication receiver detects the symbol based on the dictionary constructed with the

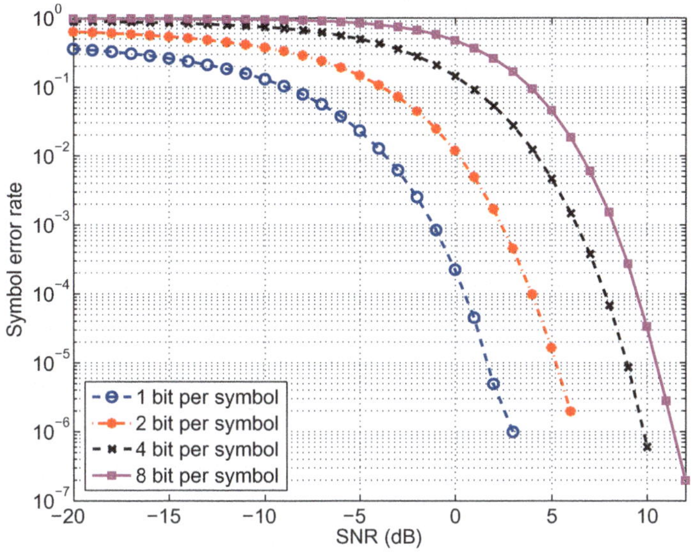

Fig. 8.30 SER versus SNR for the case where the communication receiver is at direction $\theta_c = 14.4775°$ using the regularized antenna selection scheme

assumed angle. The estimation standard variance σ is changing from 1 to 5 in steps of 1 and 500 Monte Carlo simulations are executed for each value. The SER curve versus the standard variance is plotted in Fig. 8.31 in four cases of 1, 2, 4, and 8 bits per symbol, respectively. We can observe that the regularized selection scheme is more robust against the communication angle estimation error than the hybrid scheme.

8.4 Summary

In this chapter, we investigated the sparse sensing for dual-functional radar communications.

In Sect. 8.2, we explored the optimal sparse array design for dual-function radar communications. We addressed the problem of sparse array design by antenna selection under the framework of dual functional radar communications systems. The additional spatial DoFs and configuration flexibility provided by sparse arrays were utilized to suppress the cross-interference and facilitate the cohabitation of the two functions. We solved the new problem of integrated sparse array design and transmit beampattern synthesis with additional constraints imposed by the co-design of two simultaneous functions. To increase the robustness of antenna selection algorithm against initial search points, we proposed a new selection vector updating rule. Furthermore, we proposed two new dual-function systems with a common sparse

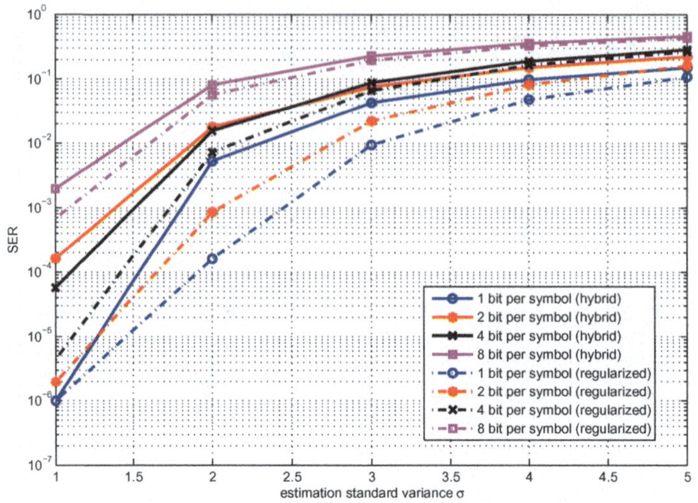

Fig. 8.31 SER versus the standard variance of communication angle estimation, the communication receiver is assumed at direction $\theta_c = 14.4775^{\circ}$, the actual angle is normally distributed around θ_c with variance σ

array associated with different beamformers and complementary sparse arrays with shared aperture. Simulation results showed that sparse arrays can synthesize a narrow mainlobe with well-controlled sidelobes, thus increasing the spatial resolution and suppressing cross-interference between the two simultaneous functions. Moreover, the comparable performance exhibited by fewer-antenna sparse arrays with large ULAs demonstrated their advantages in resource management and hardware efficiency.

In Sect. 8.3, we considered the sparse array reconfiguration for spatial index modulation. We investigated the deployment of sparse arrays by antenna selection for the design of dual functional MIMO radar communications systems. We proposed three new techniques, namely antenna selection, combined selection and permutation, and regularized selection based signaling schemes, utilizing transmit array configurations in tandem with waveform diversity for communication information embedding. The strategy of combined selection and permutation was able of achieving a megabits high data rate with low symbol error rate. The regularized selection scheme was proposed from the viewpoint of practical implementation and exhibited the best robustness against the estimation error of communication receiver angle. Simulation results validated the successful deployment of sparse arrays in dual functional MIMO radar communications systems for communication performance enhancement without impacting primary radar functions.

Part III
Sparsity Sensed Using Sparse Arrays

Chapter 9
Sparsity Sensed with Thinned Antenna Array

The performance of DOA estimation is ultimately determined by the deployed algorithm and array geometry, and it is nevertheless lower-bounded by Cramer Rao bound (CRB). The dependence of CRB on the array configuration has been examined and important formulas have been derived in terms of antenna positions in some existing works. In this chapter, we investigate the sparse array design to promote the DOA estimation accuracy. In Sect. 9.2, we study the optimal sparse array design in single-source scenario. Performance enhancement of DOA estimation is achieved by reconfiguring the multi-antenna receiver via antenna selection. We derive the Cramer-Rao Bound (CRB) in terms of the selected antenna positions for both peak sidelobe level (PSL) constrained isotropic and directional arrays. Since the configurations of directional arrays are angle dependent, a Dinklebach type algorithm and convex relaxation are introduced to maintain the optimum selection by adaptively reconfiguring the directional subarrays using semi-definite programming. In Sect. 9.3, we consider the thinned array design in the multi-source scenario. We propose the optimal sparse MIMO transceiver design via cognitive antenna selection. A reweighted l_1-norm minimization with convex relaxation is adopted to promote the binary sparsity. Simulation results validate the effectiveness of the proposed antenna selection strategies.

9.1 Introduction

Estimating the direction of arrival (DOA) using antenna arrays has been an important topic in signal processing with diverse applications. DOA estimation accuracy is dependent not only on the employed algorithm, but also on the transmitter/receiver array configuration. Extensive research has been devoted to investigate the effect of array configuration on DOA estimation performance for both near-field [316, 317] and far-field scenarios [318]. The Cramer-Rao Bound (CRB) is commonly used as a metric for characterizing the estimation performance in terms of the array

X. Wang et al., *Sparse Sensing and Sparsity Sensed in Multi-sensor Array Applications*, https://doi.org/10.1007/978-981-99-9558-5_9

configuration[319]. A compact formula of the CRB in terms of antenna positions for isotropic 2D and 3D arrays was derived in [88, 320–322], and a Bayesian CRB approach for a single source with known prior probability distribution was proposed in [323]. A study of the CRB for 2D arrays, presented in [154, 155, 324], showed that the optimum array is V-shaped under the assumptions of equal inter-element spacing and concave array geometry. The design of optimum directional arrays was also introduced in [155], but the work considered only the most favourable direction and the proposed exhaustive search strategy places limitations on the practicality of this method.

Beside the computational cost, the prohibitive hardware cost of large arrays, where a separate receiver is used for each antenna, is impractical and presents a significant limiting factor. In Sect. 9.2, we maximize the DOA estimation performance for a given reduced number of antennas (i.e. reduced hardware and computational cost) by varying the array geometry [318]. In order to realize the array reconfigurability, we thin a full array by acting on a sequence of Radio Frequency (RF) switches. The problem of sensor selection for a desired CRB with the smallest number of antennas was considered in [325], albeit for a uniform linear array. We, on the other hand, generalize in this work the antenna selection to achieve the lowest CRB with a fixed number of antennas over arbitrarily shaped arrays that are either isotropic or directional. Since the non-uniformity of selected antennas typically results in high sidelobes, the trade-off between peak sidelobe suppression and estimation accuracy is controlled via the Spatial Correlation Coefficient (SCC) introduced in Chap. 4. Continuously changing operational environments in radar, satellite communication etc. require adaptive enhanced interference localization for subsequent cancellation. It is well-known that antenna selection is essentially an NP-hard combinatorial optimization. Optimum peak sidelobe level (PSL) constrained isotropic subarrays may be found using an exhaustive search, whereas directional subarrays are angle dependent and require a polynomial-time selection algorithm to implement array thinning adaptively. Therefore, we employ an effective Dinklebach-type algorithm and convex relaxation for antenna selection through semi-definite programming. In Sect. 9.3, we proceed to examine the array thinning for enhanced DOA estimation in the multi-source scenario. We propose the optimal sparse MIMO transceiver design via cognitive antenna selection. A reweighted l_1-norm minimization with convex relaxation is adopted to promote the binary sparsity. Finally, some conclusions are made in Sect. 9.4.

9.2 Array Thinning for DOA Estimation in Single-Signal Case

Antenna array configurations play an important role in DOA estimation. Performance enhancement of DOA estimation can be achieved by reconfiguring the multi-antenna receiver via antenna selection. In this section, we consider the array thinning for enhanced accuracy of single-DOA estimation.

9.2.1 Mathematical Model

Consider a set of N antennas located in the (x, y) plane. We associate each antenna with the x and y coordinates x_n and y_n, $n = 1, \ldots, N$ respectively. A single narrow-band signal $s(t)$ with wavelength λ is impinging on the array from azimuth $\phi \in [0, 2\pi]$ and elevation angle $\theta \in [0, \pi/2]$. The steering vector of the signal is,

$$\mathbf{a} = [e^{jk_0(x_1 u_x + y_1 u_y)}, \ldots, e^{jk_0(x_N u_x + y_N u_y)}], \tag{9.1}$$

where $k_0 = 2\pi/\lambda$, $u_x = \sin\theta\cos\phi$ and $u_y = \sin\theta\sin\phi$. Assuming omni-directional antennas and far-field sources, the received signal can then be expressed as

$$\mathbf{x}(t) = \mathbf{a}s(t) + \mathbf{n}(t),\ t = 1, \ldots, T. \tag{9.2}$$

The model is referred to as deterministic if $s(t)$ is a deterministic unknown signal, and random if $s(t)$ is assumed random. Since the CRBs of both data models have the same dependence on the array structure [88], we consider the random waveform model in what follows. We assume both the estimated signal and noise to be Gaussian with zero mean, and constant variances σ_s^2 and σ_n^2, respectively. The signal-to-noise ratio (SNR) is defined as $\rho = \sigma_s^2/\sigma_n^2$.

Given the full array, we define the following vectors using the antenna positions,

$$\mathbf{x} = [x_1, \ldots, x_N]^T, \mathbf{y} = [y_1, \ldots, y_N]^T, \mathbf{x}_x = [x_1^2, \ldots, x_N^2]^T, \tag{9.3}$$

$$\mathbf{y}_y = [y_1^2, \ldots, y_N^2]^T, \mathbf{x}_y = [x_1 y_1, \ldots, x_N y_N]^T. \tag{9.4}$$

Let the Fisher Information Matrix (FIM) for the estimation of elevation angle θ and azimuth angle ϕ be $\mathbf{J}$, i.e.,

$$\mathbf{J} = \begin{bmatrix} J_{\theta\theta} & J_{\theta\phi} \\ J_{\phi\theta} & J_{\phi\phi} \end{bmatrix}. \tag{9.5}$$

We define $\mathbf{w} \in \{0, 1\}^N$ to be a binary selection vector where an entry of zero means that the corresponding antenna is discarded and one means it is selected. Suppose the number of selected antennas is K, the center of gravity of the thinned array is defined as [321],

$$x_c = \frac{1}{K}\sum_{i=1}^{N} w_i x_i = \frac{1}{K}\mathbf{w}^T\mathbf{x},\ y_c = \frac{1}{K}\sum_{i=1}^{N} w_i y_i = \frac{1}{K}\mathbf{w}^T\mathbf{y}, \tag{9.6}$$

where w_i denotes the ith entry of $\mathbf{w}$. We assume, without loss of generality, that the centre of the coordinate system is colocated with the centre of gravity of the thinned array [321],

$$x_c = \frac{1}{K}\mathbf{w}^T\mathbf{x} = 0, \quad y_c = \frac{1}{K}\mathbf{w}^T\mathbf{y} = 0. \tag{9.7}$$

As shown in [320, 321, 326], the CRB is a function of the array configuration through the following parameters involving the selected antenna positions,

$$Q_{xx} = \mathbf{w}^T\mathbf{x}_x, \, Q_{yy} = \mathbf{w}^T\mathbf{y}_y, \, Q_{xy} = \mathbf{w}^T\mathbf{x}_y. \tag{9.8}$$

The FIM, **J**, can then be expressed in terms of the selected antennas as follows [320, 321, 326],

$$J_{\theta\theta} = \mathcal{G}\cos^2\theta\left\{\cos^2\phi Q_{xx} + \sin^2\phi Q_{yy} + \sin(2\phi)Q_{xy}\right\}, \tag{9.9}$$

$$J_{\phi\phi} = \mathcal{G}\sin^2\theta\left\{\sin^2\phi Q_{xx} + \cos^2\phi Q_{yy} - \sin(2\phi)Q_{xy}\right\}, \tag{9.10}$$

$$J_{\phi\theta} = \frac{\mathcal{G}}{4}\sin(2\theta)\left\{\sin(2\phi)(Q_{xx} - Q_{yy}) - 2\cos(2\phi)Q_{xy}\right\}, \tag{9.11}$$

where $J_{\theta\phi} = J_{\phi\theta}$ and $\mathcal{G} = 2N\rho^2k_0^2/(1+\rho N)$ is angle independent.

9.2.2 *Optimum PSL Constrained Isotropic Subarray*

The single source CRB of an isotropic array is independent of the azimuth angle in [322]. The configuration of an optimum isotropic array is independent on arrival directions, i.e. both elevation and azimuth angles. Then isotropic arrays are obtained when, [321],

$$Q_{xy} = 0, \quad Q_{xx} = Q_{yy} = Q. \tag{9.12}$$

Combining the condition in Eq. (9.12) with Eqs. (9.9)–(9.11) implies that the FIM is a diagonal matrix and the CRB becomes,

$$\mathbf{C} = \mathbf{J}^{-1} = \frac{1}{\mathcal{G}}\begin{bmatrix} 1/(\cos^2\theta Q) & 0 \\ 0 & 1/(\sin^2\theta Q) \end{bmatrix}. \tag{9.13}$$

It should be noted that an isotropic subarray satisfying the symmetric condition of Eq. (9.12) does not always exist.

Now proceeding from Eq. (9.13), and minimizing the trace of the CRB for optimum array thinning [327, 328], we have that,

$$\min\left\{\operatorname{tr}(\mathbf{C}) = \frac{1}{\sin^2\theta\cos^2\theta Q}\right\} \Leftrightarrow \max\{Q\}, \tag{9.14}$$

where tr(•) denotes the trace of the matrix •. It is clear from Eq. (9.14) that the optimum isotropic thinned array includes the boundary antennas, which can guarantee the largest aperture. This observation agrees with the conclusion in [329] for linear arrays. But, the optimum isotropic subarrays that consist of the boundary antennas typically exhibit high sidelobes [275, 330, 331]. In order to solve this problem, we utilise the spatial correlation coefficients (SCC) [104, 174] to control the trade-off between the estimation variance and the synthesized beampattern. Since the SCC denotes the cross correlation between the steering vectors of two separated incoming sources, it is only dependent on electrical angle differences, i.e. $\Delta\mathbf{u} = [\Delta u_x, \Delta u_y]^T$. Let $\Delta\mathbf{u}_{i,j} = [\Delta u_x^i, \Delta u_y^j] \in [-2, 2] \times [-2, 2], i, j = 1, ..., L_1(L_2)$ be the samples in u-space. Then the correlation steering vector $\mathbf{v}_{i,j}$ is defined as, [104],

$$\mathbf{v}_{i,j} = e^{jk_0(\Delta u_x^i \mathbf{x} + \Delta u_x^j \mathbf{y})}, i, j = 1, ..., L_1(L_2). \tag{9.15}$$

The samples, $\Delta\mathbf{u}_{i,j}$, can be set to be the specified electrical angular region with the constrained PSL.

Now, we consider a set of subarrays, each with K antennas and the array center of gravity colocated with the center of the coordinate system,

$$\mathcal{S} = \{\mathbf{w} \in \{0, 1\}^N : \mathbf{w}^T\mathbf{x} = 0; \mathbf{w}^T\mathbf{y} = 0; \mathbf{1}^T\mathbf{w} = K\}, \tag{9.16}$$

where $\mathbf{1}$ is a vector with all ones. Note that the set $\mathcal{S}$ comprises the extreme points of the polyhedra,

$$\mathcal{P} = \{\mathbf{w} \in [0, 1]^N : \mathbf{w}^T\mathbf{x} = 0; \mathbf{w}^T\mathbf{y} = 0; \mathbf{1}^T\mathbf{w} = K\}. \tag{9.17}$$

The problem of determining the optimum isotropic subarray with constrained PSL is formulated as,

$$\begin{aligned} \max_{\mathbf{w}} \ & \mathbf{w}^T\mathbf{x}_x \\ \text{s.t.} \ & \mathbf{w} \in \mathcal{S}; \quad \mathbf{w}^T\mathbf{x}_y = 0; \quad \mathbf{w}^T(\mathbf{x}_x - \mathbf{y}_y) = 0; \\ & \mathbf{w}^T\mathbf{V}_{i,j}\mathbf{w} \leq \delta_{i,j}, i, j = 1, ..., L_1(L_2), \end{aligned} \tag{9.18}$$

where $\mathbf{V}_{i,j} = \text{real}(\mathbf{v}_{i,j}\mathbf{v}_{i,j}^H)$ and $\delta_{i,j} < 1$ is the desired normalised sidelobe power level with respect to the mainlobe. The problem in Eq. (9.18) is convex programming, except for the binary constraints $\mathbf{w} \in \{0, 1\}^N$. Since the isotropic array is independent of the estimated angle, the optimum solution of Eq. (9.18) may be calculated offline through an exhaustive search. In order to reduce computational load, another method of solving Eq. (9.18) is to relax the binary constraints through the difference of two convex sets (DCS), which is a polynomial-time algorithm with the detailed implementation procedure given in [104]. Here, we formulate the DCS for antenna selection in the $(k+1)^{th}$ iteration as follows,

$$\max_{\mathbf{w}} \ \mathbf{w}^T(\mathbf{x}_x + 2\mu\mathbf{w}_k - \mu\mathbf{1})$$
$$\text{s.t.} \ \mathbf{w} \in \mathcal{P}; \quad \mathbf{w}^T\mathbf{x}_y = 0; \quad \mathbf{w}^T(\mathbf{x}_x - \mathbf{y}_y) = 0;$$
$$\mathbf{w}^T\mathbf{V}_{i,j}\mathbf{w} \leq \delta_{i,j}, i, j = 1, ..., L_1(L_2), \tag{9.19}$$

where μ is a trade-off parameter which compromises between the solution sparseness and the CRB.

9.2.3 Optimum Directional Subarray

As shown in Eqs. (9.9)–(9.11), the FIM depends on the array geometry and the source DOAs. Therefore, the optimum array for one angle is not necessarily optimum for other angles. If prior information on the source DOAs is available, it would be desirable to select directional subarrays that optimize the estimation performance in a certain neighbourhood of the privileged direction, including both elevation and azimuth. For example, some prior information on the DOAs can be obtained utilising PSL constrained isotropic subarrays. Since the estimated angle is narrowed down within a small range, it becomes unnecessary to consider peak sidelobe suppression for directional subarrays.

9.2.3.1 Problem Formulation

Assume the neighbourhood of interest is around the angle $[\theta, \phi]$. After mathematical manipulations, the CRB of θ is

$$C_{\theta\theta} = \frac{1}{\mathcal{G}} \cdot \frac{1}{\cos^2\theta} \cdot \frac{\sin^2\phi Q_{xx} + \cos^2\phi Q_{yy} - \sin(2\phi)Q_{xy}}{Q_{xx}Q_{yy} - Q_{xy}^2}, \tag{9.20}$$

similarly, the CRB of ϕ is

$$C_{\phi\phi} = \frac{1}{\mathcal{G}} \cdot \frac{1}{\sin^2\theta} \cdot \frac{\cos^2\phi Q_{xx} + \sin^2\phi Q_{yy} + \sin(2\phi)Q_{xy}}{Q_{xx}Q_{yy} - Q_{xy}^2}. \tag{9.21}$$

Unlike [326], where the volume of the confidence region was introduced for azimuth-invariant cases and was defined only in terms of the determinant of the FIM, i.e., the denominator $Q_{xx}Q_{yy} - Q_{xy}^2$, we utilise the trace of the CRB as a metric,

$$\text{tr}(\mathbf{C}) = \frac{1}{\mathcal{G}} \cdot \frac{1}{Q_{xx}Q_{yy} - Q_{xy}^2} \cdot \left[\alpha Q_{xx} + \beta Q_{yy} + \varsigma Q_{xy}\right], \tag{9.22}$$

where

$$\alpha = \frac{\sin^2\phi}{\cos^2\theta} + \frac{\cos^2\phi}{\sin^2\theta}, \beta = \frac{\cos^2\phi}{\cos^2\theta} + \frac{\sin^2\phi}{\sin^2\theta}, \zeta = \frac{\sin(2\phi)}{\sin^2\theta} - \frac{\sin(2\phi)}{\cos^2\theta}.$$

The optimization problem can be formulated as,

$$\begin{aligned} \min_{\mathbf{w}} \quad & \frac{\mathbf{w}^T(\tilde{\alpha}\mathbf{x}_x\mathbf{1}^T + \tilde{\beta}\mathbf{y}_y\mathbf{1}^T + \tilde{\zeta}\mathbf{x}_y\mathbf{1}^T)\mathbf{w}}{\mathbf{w}^T(\mathbf{x}_x\mathbf{y}_y^T - \mathbf{x}_y\mathbf{x}_y^T)\mathbf{w}} \\ \text{s.t.} \quad & \mathbf{w} \in \mathcal{S}; \end{aligned} \tag{9.23}$$

where $\tilde{\alpha} = \alpha/K$, $\tilde{\beta} = \beta/K$ and $\tilde{\zeta} = \zeta/K$. Contrary to isotropic subarrays, an optimum K-antenna directional subarray always exists. The objective function in Eq. (9.23) is a quadratic fractional, which makes Eq. (9.23) difficult to solve. In order to alleviate the problem, we introduce another variable $\mathbf{W}$ along with the rank-one constraint $\mathbf{W} = \mathbf{w}\mathbf{w}^T$. Then, the binary constraint can be reformulated as,

$$\text{tr}(\mathbf{W}\mathbf{E}_i) - \mathbf{e}_i^T\mathbf{w} = 0, i = 1, \ldots, N \tag{9.24}$$

where $\mathbf{e}_i$ is the ith unit vector with the ith entry being one and all others being zero and $\mathbf{E}_i = \mathbf{e}_i\mathbf{e}_i^T$. Correspondingly, the set $\mathcal{S}$ is rewritten as,

$$\begin{aligned} \mathcal{S} = \{\mathbf{w}, \mathbf{W} : \mathbf{w}^T\mathbf{x} = 0; \mathbf{w}^T\mathbf{y} = 0; \mathbf{1}^T\mathbf{w} = K; \\ \text{trace}(\mathbf{W}\mathbf{E}_i) - \mathbf{e}_i^T\mathbf{w} = 0, i = 1, \ldots, N\}. \end{aligned}$$

We rewrite Eq.(9.23) by relaxing the rank-one constraint as follows,

$$\begin{aligned} \min_{\mathbf{w},\mathbf{W}} \quad & \frac{\text{tr}(\mathbf{W}\mathbf{N}_u)}{\text{tr}(\mathbf{W}\mathbf{D}_e)} \\ \text{s.t.} \quad & \{\mathbf{w}, \mathbf{W}\} \in \mathcal{S}, \quad \mathbf{W} \succeq \mathbf{w}\mathbf{w}^T, \end{aligned} \tag{9.25}$$

where $\mathbf{N}_u = \tilde{\alpha}\mathbf{x}_x\mathbf{1}^T + \tilde{\beta}\mathbf{y}_y\mathbf{1}^T + \tilde{\zeta}\mathbf{x}_y\mathbf{1}^T$ and $D_e = \mathbf{x}_x\mathbf{y}_y^T - \mathbf{x}_y\mathbf{x}_y^T$. An exhaustive search within $\mathcal{S}$ can be conducted for small arrays to find the optimum directional subarray, while for large arrays, we propose a Dinklebach-type algorithm for adaptive directional subarray selection.

9.2.3.2 Dinkelbach-Type Algorithm

The Dinkelbach-type algorithm is based on a theorem concerning the relationship between factional and parametric programming[332]. The parametric objective function is transformed from the fraction in Eq. (9.25),

$$F(\eta) = \text{tr}(\mathbf{W}\mathbf{N}_u) - \eta\text{tr}(\mathbf{W}\mathbf{D}_e). \tag{9.26}$$

The detailed derivation and performance analysis can be found in [332]. Here we give an outline of the procedure as follows.

Step 0: Initialize η_1 and the termination threshold $\varepsilon = 0.01$;

Step 1: Solve the following minimization problem to obtain global solutions $\mathbf{w}_k$, $\mathbf{W}_k$ and the optimum value $F(\eta_k)$:

$$\begin{aligned} \min_{\mathbf{w},\mathbf{W}} \quad & F(\eta_k) \\ \text{s.t.} \quad & \{\mathbf{w}, \mathbf{W}\} \in \mathcal{S}; \quad \mathbf{W} \geq \mathbf{w}\mathbf{w}^T. \end{aligned} \tag{9.27}$$

Step 2: If $F(\eta_k) \leq \varepsilon$, then terminate. Otherwise, let

$$\eta_{k+1} = \frac{\mathrm{tr}(\mathbf{W}_k \mathbf{N}_u)}{\mathrm{tr}(\mathbf{W}_k \mathbf{D}_e)}. \tag{9.28}$$

and return to Step 1.

The selection vector $\mathbf{w}$ is generated by setting the K largest entries to be one. The initial η_1 can take the corresponding value of the isotropic subarray to accelerate convergence rate.

9.2.4 Simulations

Firstly, we select a 10-antenna subarray from a 20-antenna uniform linear array (ULA). Since the effect of the array geometry on the CRB for a linear array can be separated from the arrival angle, the optimum linear subarray is always isotropic. The two optimum subarrays with and without constrained PSL are shown in Fig. 9.1. The subarray without constrained PSL comprises two clusters of antennas, one at each end of the linear array. For the subarray with constrained PSL, the squared SCC value is set to be $\delta = 0.5$, which implies the PSL is -6 dB as shown in Fig. 9.2. Finally, the estimation variance versus SNR for the two subarrays and the 20-antenna ULA is shown in Fig. 9.3. The subarray with constrained PSL has 2.1 dB performance loss

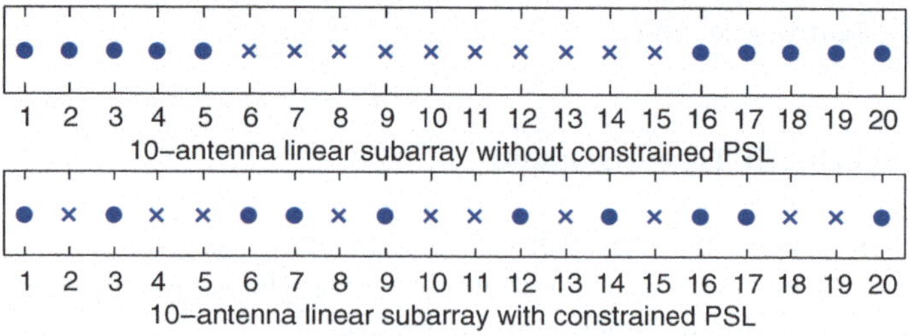

Fig. 9.1 Optimum 10-antenna linear subarrays

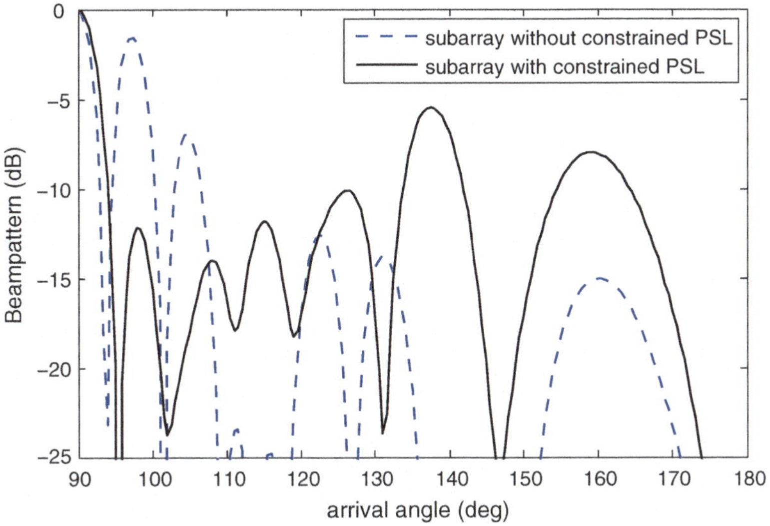

Fig. 9.2 Beampatterns of two linear subarrays

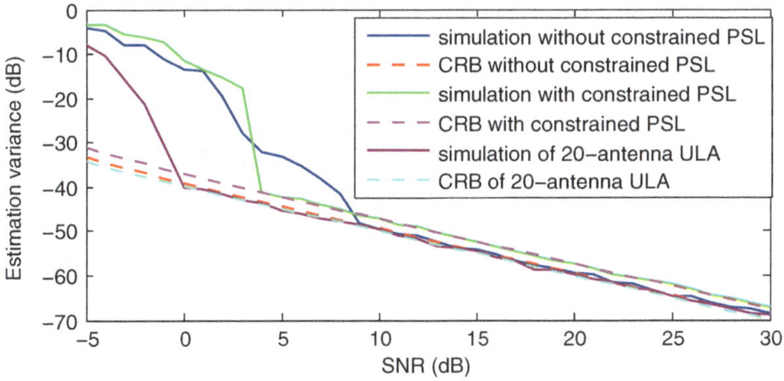

Fig. 9.3 Estimation variance of two subarrays and ULA versus SNR. Each point is averaged by 500 Monte-Carlo runs

compared to the other subarray, however it exhibits a 5dB smaller threshold value due to the well synthesized beampattern. The ULA exhibits 3.14dB and 1.04dB smaller estimation variance than the two subarrays with and without PSL constraints respectively, although with 10 more antennas.

Next, we select a 10-antenna subarray from a 6×4 rectangular planar array. Due to the symmetric requirement, it is impossible to select an isotropic 10-antenna subarray from this rectangular array. Thus we assume another 5×5 square array for comparison. The desired signal is impinging on the array from an azimuth of $10°$ and elevation of $175°$. The isotropic and two directional subarrays are shown in Fig. 9.4.

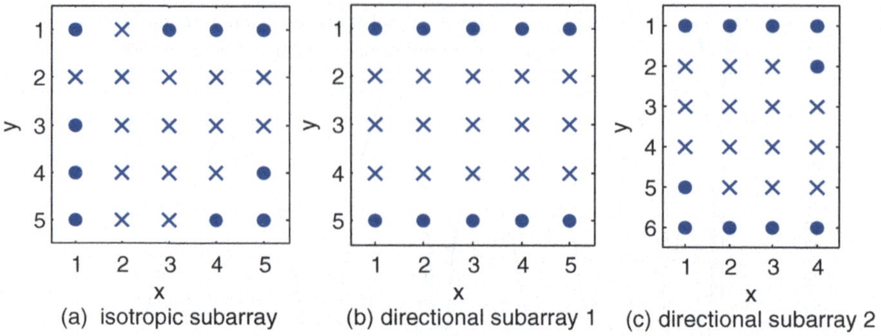

Fig. 9.4 Three 10-antenna planar subarrays

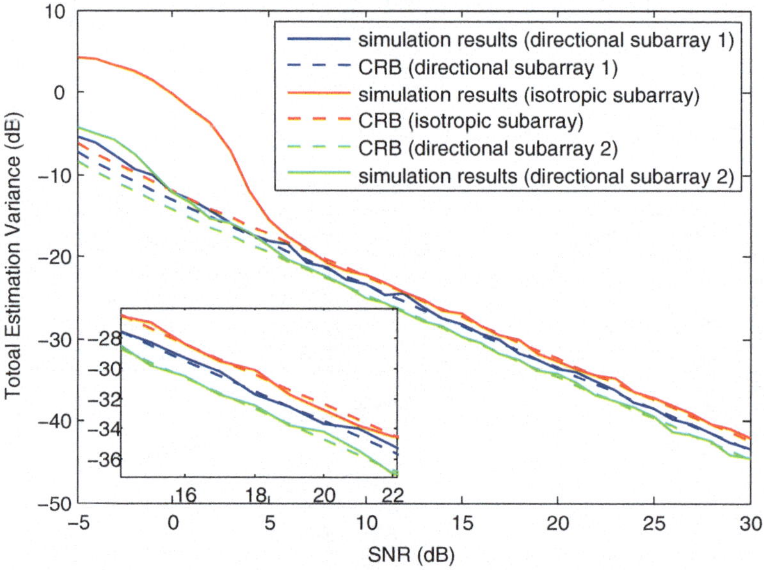

Fig. 9.5 Total estimation variance for three subarrays versus SNR. Each point is averaged by 500 Monte-Carlo runs

Note that the directional subarrays may have grating lobes, which does not affect the DOA estimation performance with some prior knowledge of the arrival angle. The total estimation variance, given by the sum of the elevation and azimuth, for the three subarrays are shown in Fig. 9.5. The first directional subarray has 1.11dB better performance than the isotropic subarray, while the second one exhibits the best estimation performance with 2.26 dB smaller variance compared to the isotropic one. It should be noted that the low threshold value for both directional subarrays results from the prior information.

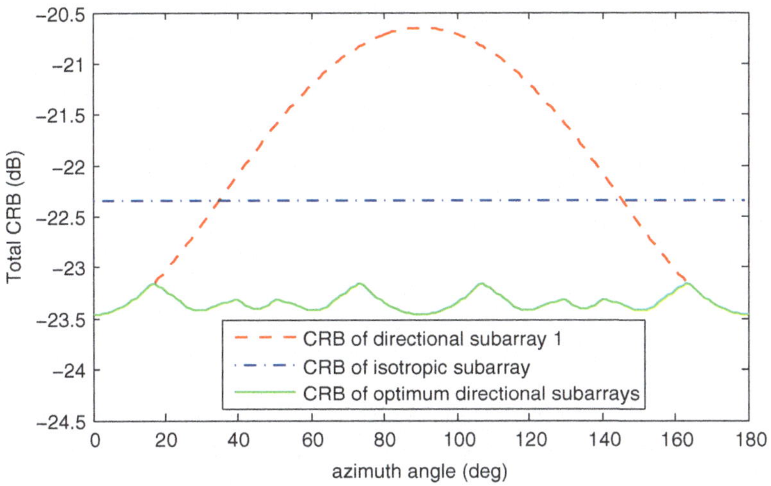

Fig. 9.6 Total CRB versus the azimuth angle. The SNR is 10dB

Finally, we investigate the relationship between the optimum directional and isotropic subarrays. It is clear from Eq. (9.22) that when ζ is close to zero and α is close to β, the optimum directional subarray is essentially the isotropic one. This occurs when the estimated elevation angle is in the neighbourhood of 45°. Now we fix the elevation angle to be 10° and sweep the azimuth angle from 0° to 180°. The total CRB, given by the sum of the CRBs for the elevation and azimuth, is shown in Fig. 9.6 for both the isotropic and the first directional subarray in Fig. 9.4(b). The CRB of optimum directional subarrays corresponding to each azimuth angle is also shown for comparison. We can see that reconfiguring optimum directional subarrays adaptively can achieve almost the same estimation performance regardless of the azimuth angle. In other words, the dependence of the estimation variance on arrival angles can be compensated by reconfiguring array geometry, which enables directional subarrays to mimic the angle-independent performance as isotropic subarrays while offering a better estimation performance.

9.3 Array Thinning for DOA Estimation in Multi-Signal Case

In this section, we consider MIMO transceiver array thinning for enhanced DOA estimation in the multi-source scenario. We first derive the formula of CRB as the optimality criterion for antenna selection. To solve the underlying NP-hard problem, we apply convex relaxation strategies to transform the original problem into a convex form. The cardinality constraints are relaxed into boxing constraints with binary sparsity promoted via a reweighted l_1-norm minimization.

9.3.1 Mathematical Model

Consider a co-located MIMO array consisting of a M-antenna transmitter and a N-antenna receiver. Both the transmit and receive arrays are uniform linear arrays (ULAs). To achieve a large virtual array, the inter-element spacing of receive array is set as $d_R = Md_T$, where d_T denotes the inter-element spacing of the transmit array. As such, the sum co-array of the MIMO array is a large ULA with an inter-element spacing of d_T. Suppose there are Q targets in the MIMO radar field of view, the received target echo can be expressed as,

$$\mathbf{y}(l, \tau) = \mathbf{A}_R \text{diag}\{\eta(\tau)\}\mathbf{A}_T^T \mathbf{s}(l) + \mathbf{v}(l, \tau). \tag{9.29}$$

where $\mathbf{s}(l) \in \mathbb{C}^{M\times 1}$ is the orthogonal radar waveform vector transmitted by the M antennas at the time instant l, l denotes the fast-time sample index within one radar pulse, τ represents the slow-time index (i.e., pulse number), $\eta(\tau) = [\eta_1(\tau), ..., \eta_Q(\tau)]^T$ is the vector consisting of the reflection coefficients of all Q targets which are assumed to satisfy the Swerling II model, $\mathbf{v}(l, \tau) = [v_1(l, \tau), ..., v_N(l, \tau)]^T$ is modeled as a zero-mean complex white Gaussian process with power of σ_n^2, $\mathbf{A}_T = [\mathbf{a}_T(\theta_1), ..., \mathbf{a}_T(\theta_Q)]$ and $\mathbf{A}_R = [\mathbf{a}_R(\theta_1), ..., \mathbf{a}_R(\theta_Q)]$. Here, $\mathbf{a}_T(\theta_l)$ and $\mathbf{a}_R(\theta_l)$ are the steering vectors towards direction θ_l which are defined by,

$$\mathbf{a}_{T/R}(\theta_l) = [1, e^{j\frac{2\pi}{\lambda}d_{T/R}\sin\theta_l}, ..., e^{j\frac{2\pi}{\lambda}(N-1)d_{T/R}\sin\theta_l}]^T. \tag{9.30}$$

Performing the matched-filtering to the received data vector with the transmit orthogonal radar waveforms, we can obtain a $N \times M$-dimensional matrix as follows,

$$\begin{aligned}\mathbf{Y}(\tau) &= \frac{1}{L}\sum_{l=1}^{L}\mathbf{y}(l, \tau)\mathbf{s}^H(l), \\ &= \mathbf{A}_R \text{diag}\{\eta(\tau)\}\mathbf{A}_T^T + \mathbf{N}(\tau),\end{aligned} \tag{9.31}$$

where L is the number of samples within one pulse and $\mathbf{N}(\tau) = \frac{1}{L}\sum_{l=1}^{L}\mathbf{v}(l, \tau)\mathbf{s}^H(l)$ is still additive Gaussian noise.

Vectorizing $\mathbf{Y}(\tau)$, we have that,

$$\mathbf{y}(\tau) = \text{vec}\{\mathbf{Y}^T(\tau)\} = \mathbf{A}\eta(\tau) + \mathbf{n}(\tau), \tag{9.32}$$

where $\mathbf{n}(\tau) = \text{vec}\{\mathbf{N}^T(\tau)\}$ is the extended additive Gaussian noise term with zero mean and a covariance $\sigma_n^2\mathbf{I}$, $\mathbf{A} = [\mathbf{a}(\theta_1), ..., \mathbf{a}(\theta_Q)]$ with the vector $\mathbf{a}(\theta_q)$ is defined as $\mathbf{a}(\theta_q) = \mathbf{a}_R(\theta_q) \otimes \mathbf{a}_T(\theta_q)$. As mentioned above, since $d_R = Nd_T$, $\mathbf{a}(\theta_q)$ is the steering vector corresponding to a ULA consisting of MN virtual antennas with an interval of d_R.

The Cramer-Rao Bound (CRB) represents the best performance of any unbiased estimator and is widely applied to assess the sensing performance. As shown in [87], the CRB-minimization beamformers outperform other methods based on beampattern matching for target sensing. Based on the signal model given in Eq. (9.32), the CRB of the multi-source DOA estimation can be expressed by [87],

$$\text{CRB} = \frac{\sigma_n^2}{2J} \left\{ \text{Re}\left[\{\mathbf{D}^H(\mathbf{I} - \mathbf{A}(\mathbf{A}^H\mathbf{A})^{-1}\mathbf{A}^H)\mathbf{D}\} \odot \mathbf{P}^T \right] \right\}^{-1}, \tag{9.33}$$

where J is the total number of processed snapshots, and $\mathbf{D} = [\mathbf{d}_1, \ldots, \mathbf{d}_Q]$ with the vector $\mathbf{d}_q$ being defined as the first order derivative of $\mathbf{a}(\theta_q)$ with respect to ω_q, which is defined as $\omega_q = \frac{2\pi}{\lambda} d_T \sin\theta_q$. That is,

$$\mathbf{d}_q = \frac{\partial \mathbf{a}(\theta_q)}{\partial \omega_q} = [0, je^{j\omega_q}, \ldots, j(MN-1)e^{j(MN-1)\omega_q}]^T. \tag{9.34}$$

Matrix $\mathbf{P}$ is the echo covariance matrix, which is given by,

$$\mathbf{P} = \mathbb{E}\{\eta(\tau)\eta^H(\tau)\}. \tag{9.35}$$

When the reflection coefficients of different targets are uncorrelated, the matrix $\mathbf{P}$ becomes diagonal with the reflection power $P_l, l = 1, \ldots, Q$ populating the diagonal.

9.3.2 *Sparse Transceiver Design for Enhanced DOA Estimation*

In this subsection, we design the sparse MIMO array transceiver via antenna selection to optimize the DOA estimation accuracy. Suppose that we select M_0 out of M transmit antennas to construct the sparse transmitter and N_0 out of N receive antennas for the receiver. CRB is chosen as the performance metric to optimize both configurations of the transmitter and receiver. In the multi-target detection scenario, the cross-correlation terms in the CRB matrix in Eq. (9.33) are all zeros under the assumption of uncorrelated signals, that is, $\mathbf{P}$ is a diagonal matrix. Accordingly, the CRB in Eq. (9.33) is diagonal, with its diagonal elements denoting the lower bounds for the DOA estimation variance of different signals. To improve the estimation accuracy, the upper bound of CRBs of all estimated DOAs is chosen as the objective function to lower the estimation variance for each signal, that is,

$$f = \max_{1 \le i \le Q} \left\{ P_i \mathbf{d}_i^H (\mathbf{I} - \mathbf{A}(\mathbf{A}^H\mathbf{A})^{-1}\mathbf{A}^H)\mathbf{d}_i \right\}^{-1}. \tag{9.36}$$

We proceed to incorporate antenna selection into the objective function to design sparse MIMO array transceiver. Define an antenna selection vector $\mathbf{z} \in \{0, 1\}^{MN}$, where "1" denotes the corresponding virtual antenna being selected and "0" discarded. Note that the selection vector $\mathbf{z}$ exhibits a special group sparsity structure since the pairing of each transmit antenna and receive antenna is associated with a certain corresponding virtual antenna under the MIMO framework. Combining the antenna selection vector $\mathbf{z}$ with Eq. (9.36), we can derive the cost function of the sparse MIMO transceiver as follows,

$$f(\mathbf{z}) = \max_{1 \leq i \leq Q} \{P_i[\mathbf{d}_i^H \text{diag}(\mathbf{z})\mathbf{d}_i - \mathbf{d}_i^H \text{diag}(\mathbf{z})\mathbf{A} \\ (\mathbf{A}^H \text{diag}(\mathbf{z})\mathbf{A})^{-1} \mathbf{A}^H \text{diag}(\mathbf{z})\mathbf{d}_i]\}^{-1}, \tag{9.37}$$

Capitalizing upon the analysis above, the optimization problem of sparse array design for enhanced DOA estimation can be mathematically formulated as,

$$\min_{\mathbf{z}} f(\mathbf{z}), \tag{9.38}$$

$$\text{s.t. } \mathbf{z} \in \{0, 1\}^{MN}, \tag{9.38a}$$

$$(1/N_0)\mathbf{p}_i^T \mathbf{z} \in \{0, 1\},\ i = 1, ..., M, \tag{9.38b}$$

$$(1/M_0)\mathbf{q}_i^T \mathbf{z} \in \{0, 1\},\ i = 1, ..., N, \tag{9.38c}$$

$$||\mathbf{P}\mathbf{z}||_0 = M_0, \tag{9.38d}$$

$$||\mathbf{Q}\mathbf{z}||_0 = N_0, \tag{9.38e}$$

where the constraints (9.38b) and (9.38d) mean that M_0 transmit antennas are selected, while constraints (9.38c) and (9.38e) imply that N_0 receive antennas are selected. Here, $\mathbf{p}_i \in \mathbb{R}^{MN \times 1}$ and $\mathbf{q}_i \in \mathbb{R}^{MN \times 1}$ are the transmit and receive selection vectors for automotive sensing respectively, which are defined as,

$$\mathbf{p}_i = [\underbrace{\overbrace{0, ..., 0}^{(i-1)\ 0s}, 1, 0, ..., 0}_{M \text{ elements}}, ..., \underbrace{\overbrace{0, ..., 0}^{(i-1)\ 0s}, 1, 0, ..., 0}_{M \text{ elements}}]^T, \tag{9.39}$$

and

$$\mathbf{q}_i = [\underbrace{0, ..., 0}_{(i-1)M\ 0s}, \underbrace{1, ..., 1}_{M\ 1s}, 0, ..., 0]^T. \tag{9.40}$$

Accordingly, $\mathbf{P}$ and $\mathbf{Q}$ are the transmit and receive selection matrices, which are defined as $\mathbf{P} = [\mathbf{p}_1, ..., \mathbf{p}_M]^T$ and $\mathbf{Q} = [\mathbf{q}_1, ..., \mathbf{q}_N]^T$ respectively.

Both the objective function and the cardinality constraints in problem (9.38) are non-convex, which makes it difficult to solve. Next, we emloy a sequential convex relaxation method, which simplifies the original problem by applying a series of convex relaxation to transform the problem into a convex form and solve it iteratively.

Convex Relaxation of Objective: Utilizing the Schur complement of a block matrix, we can transform the maximum value function in Eq. (9.36) into the following equivalent linear matrix inequality (LMI) constraint optimization problem,

$$\min_{\mathbf{z},\delta} \ -\delta, \tag{9.41}$$

$$\text{s.t.} \begin{bmatrix} \mathbf{A}^H\text{diag}(\mathbf{z})\mathbf{A} & \mathbf{A}^H\text{diag}(\mathbf{z})\mathbf{d}_i \\ \mathbf{d}_i^H\text{diag}(\mathbf{z})\mathbf{A} & \mathbf{d}_i^H\text{diag}(\mathbf{z})\mathbf{d}_i - \frac{\delta}{P_i} \end{bmatrix} \succeq 0, \quad 1 \le i \le Q. \tag{9.41a}$$

The LMI constraint in Eq. (9.41a) implies that,

$$P_i^{-1}\{\mathbf{d}_i^H\text{diag}(\mathbf{z})\mathbf{d}_i - \mathbf{d}_i^H\text{diag}(\mathbf{z})\mathbf{A}(\mathbf{A}^H\text{diag}(\mathbf{z})\mathbf{A})^{-1} \mathbf{A}^H\text{diag}(\mathbf{z})\mathbf{d}_i\}^{-1} \le \frac{1}{\delta}, i \in \{1, 2, \ldots, Q\} \tag{9.42}$$

Note that the signal power $P_i, i = 1, \ldots, Q$ may not need to be exactly known and it plays a role of imposing different emphasis on the Q estimated signals.

Convex Relaxation of Binary Constraint: As for the binary constraints (9.38a), (9.38b) and (9.38c), we first relax them into box constraints and then utilize a modified iterative reweighted l_1-norm to promote the binary sparsity via a set of judiciously designed reweights [40]. In the jth iteration, combining Eq. (9.41), we can transform the original problem (9.38) into,

$$\min_{\mathbf{z},\delta} \ -\delta + \mu_1\mathbf{m}_1^{(j)T}\mathbf{z}_1 + \mu_2\mathbf{m}_2^{(j)T}\mathbf{z}_2, \tag{9.43}$$

$$\text{s.t. } 0 \le \mathbf{z} \le 1, \tag{9.43a}$$

$$\mathbf{z}_1 = (1/N_0)\mathbf{P}\mathbf{z}, \ 0 \le \mathbf{z}_1 \le 1, \ \mathbf{1}^T\mathbf{z}_1 = M_0, \tag{9.43b}$$

$$\mathbf{z}_2 = (1/M_0)\mathbf{Q}\mathbf{z}, \ 0 \le \mathbf{z}_2 \le 1, \ \mathbf{1}^T\mathbf{z}_2 = N_0, \tag{9.43c}$$

$$\begin{bmatrix} \mathbf{A}^H\text{diag}(\mathbf{z})\mathbf{A} & \mathbf{A}^H\text{diag}(\mathbf{z})\mathbf{d}_i \\ \mathbf{d}_i^H\text{diag}(\mathbf{z})\mathbf{A} & \mathbf{d}_i^H\text{diag}(\mathbf{z})\mathbf{d}_i - \frac{\delta}{P_i} \end{bmatrix} \succeq 0, \quad 1 \le i \le Q, \tag{9.43d}$$

where μ_1 and μ_2 are artificially-set trade-off parameters between the sensing performance and the boolean sparsity. Note that we relax the binary constraint in Eq. (9.38a) on $\mathbf{z}$ to a box constraint $0 \le \mathbf{z} \le 1$ in Eq. (9.43a). Further, we adopt a modified reweighted l_1-norm to promote the boolean sparsity of the selection vector. Here, $\mathbf{m}_1^{(j)}$ and $\mathbf{m}_2^{(j)}$ are reweighted coefficient vectors for the transmit antennas and receive antennas, respectively. The reweighted l_1-norm is a commonly-used technique to promote sparsity [35], however, the reweighted coefficients have to be judiciously defined in order to promote a more strict boolean sparsity. For the jth iteration, the reweighted coefficient of the ith transmit antenna, that is the ith element of $\mathbf{m}_1$, is updated by,

$$(\mathbf{m}_1^{(j)})_i = \frac{1 - (\mathbf{z}_1^{(j-1)})_i}{1 - e^{-\beta_0(\mathbf{z}_1^{(j-1)})_i} + \epsilon} - (\frac{1}{\epsilon})((\mathbf{z}_1^{(j-1)})_i)^{\alpha_0}, \tag{9.44}$$

where $\mathbf{z}_1^{(j-1)}$ is calculated by the optimal solution of the $(j-1)$th iteration. Similarly, the reweighted coefficient of the ith receive antenna, that is the ith element of $\mathbf{m}_2$, is updated by,

$$(\mathbf{m}_2^{(j)})_i = \frac{1-(\mathbf{z}_2^{(j-1)})_i}{1-e^{-\beta_0(\mathbf{z}_2^{(j-1)})_i}+\epsilon} - (\frac{1}{\epsilon})((\mathbf{z}_2^{(j-1)})_i)^{\alpha_0}. \tag{9.45}$$

According to the characteristics of the reweighted iterative strategy [40], (9.43) can converge to a local optimal solution of (9.38).

9.3.3 Simulations

In this subsection, we verify the proposed sparse MIMO transceiver design for enhanced DOA estimation.

First, we consider an MIMO array comprising a 10-antenna ULA transmitter with an inter-element spacing of half wavelength and a 10-antenna ULA receiver with an inter-element spacing of 5 wavelengths. We select 5 antennas from the transmit array and another 5 antennas from the receive array to construct the sparse transceiver. There are 5 targets in the radar field of view which are located at $-70°$, $-30°$, $10°$, $40°$ and $70°$ respectively, with SNR all begin -20 dB. In the considered scenario, we apply the iterative reweighted algorithm proposed to optimize the configuration of the sparse MIMO transceiver in terms of minimizing the DOA estimation CRB and the optimal sparse MIMO transceiver is shown in Fig. 9.7, where we can see that both the transmit and receiver arrays select the 1st, 2nd, 3rd, 9th and 10th antennas to assure the largest virtual array aperture. Accordingly, the generated virtual array is dipicted in Fig. 9.8, where virtual antennas of the same color are associated with one transmit antenna and virtual antennas in one row are associated with the same receive antenna.

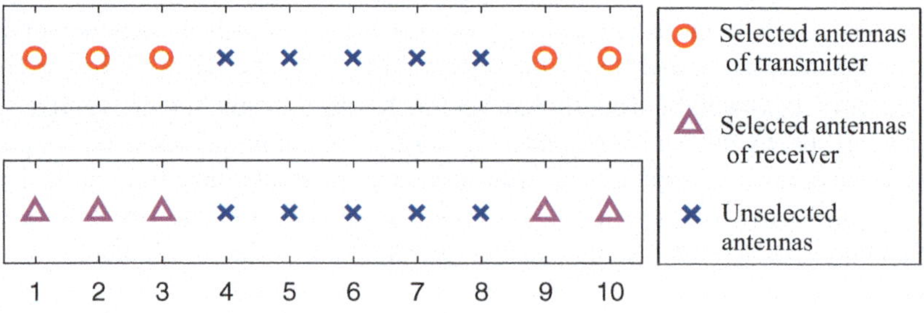

Fig. 9.7 The configurations of selected transmit array (upper figure) and receive array (lower figure) corresponding to the optimal sparse MIMO transceiver

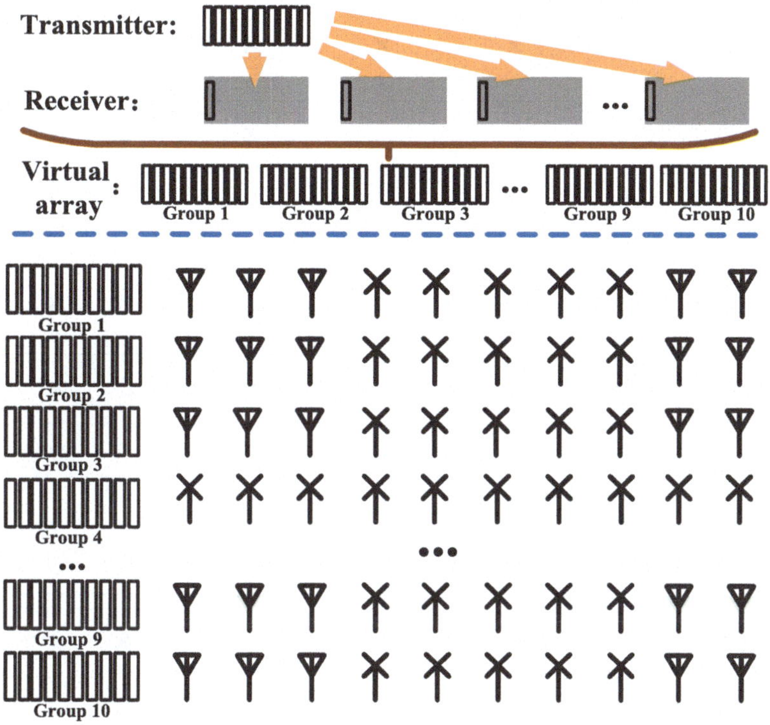

Fig. 9.8 Virtual array generated by the optimally co-designed sparse MIMO transceiver. Here, "down-triangle" denotes the corresponding virtual antenna selected and "cross" unselected

Second, we verify the enhanced DOA estimation accuracy of the optimal transceiver. We choose three sparse transceivers, as shown in Fig. 9.9, to compare with the optimal MIMO transceiver obtained by the proposed algorithm. The high-resolution MUSIC algorithm is employed to estimate DOAs of radar targets and 10000 Monte-Carlo runs are carried out for statistical performance analysis. The curves of mean square estimation errors (MSEs) versus SNRs using different MIMO transceivers are plotted in Fig. 9.10. In Fig. 9.10a, we calculate the MSEs of each target and average them to obtain the averaged estimation accuracy. In Fig. 9.10b–f, we depict the MSE versus SNR curve of each target separately. In Fig. 9.10, the red and pink lines represent the CRB and MSE of the optimal transceiver respectively. We can observe that with the increase of SNR, the MSE curves gradually approach the CRB. The blue line indicates the performance of the case where the virtual array is a ULA and its corresponding transceiver configuration is given in Fig. 9.9a. The sapphire and green lines represent two randomly selected sparse transceivers, whose configurations are depicted in Fig. 9.9b, c, respectively. From Fig. 9.10, we can see that the DOA estimation accuracy of the optimal transceiver is obviously better than that of other arrays, which validates the DOA estimation enhancement enabled by the constructed optimal transceiver. To further illustrate the DOA estimation accuracy

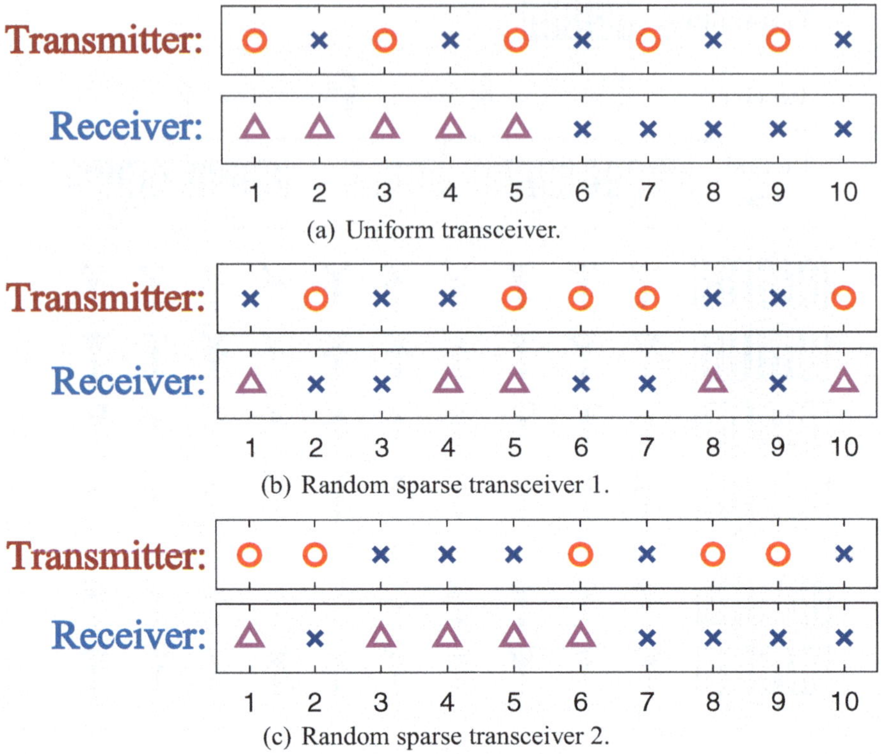

Fig. 9.9 The configurations of the selected transmit and receive antennas of uniform, random 1 and random 2

of the optimal transceiver, we fix the SNR of the target echoes at 10dB and depict the MUSIC spectrum obtained by the optimal transceiver, as shown in Fig. 9.11. It can be seen from the result that the peak values at the target's directions are more than 20 dB higher than the sidelobe. Therefore, the target's DOAs can be easily identified which indicates that the optimal transceiver shows good DOA estimation accuracy.

Finally, to verify the tightness of problem (9.43), we depict the performance of all possible configurations of the transceiver by exhaustive method, as shown in Fig. 9.12. For the sparse transceiver for DOA estimation, there are a total of $C_{10}^5 \times C_{10}^5 = 63504$ different cases of configurations. We fix the SNR of the target echoes to be 10dB and calculate the MSEs under all these configurations, as shown in Fig. 9.12. It evident from the results that the optimal transceiver selected by (9.43) can achieve the optimal DOA estimation accuracy. The applied relaxation strategies exhibit good tightness.

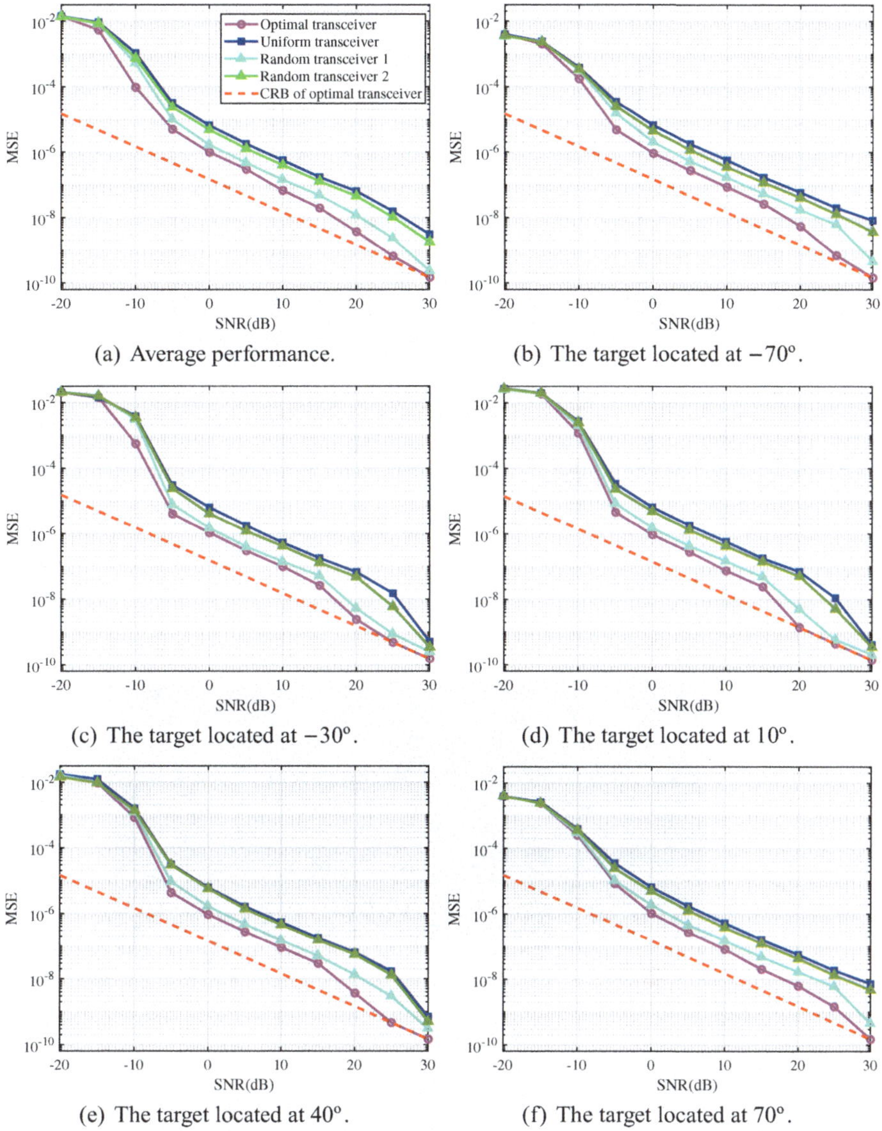

(a) Average performance. (b) The target located at -70°.

(c) The target located at -30°. (d) The target located at 10°.

(e) The target located at 40°. (f) The target located at 70°.

Fig. 9.10 The curves of MSE versus SNR for different sparse transceivers **a** the average of five signals, and **b–f** from the first to the fifth signal

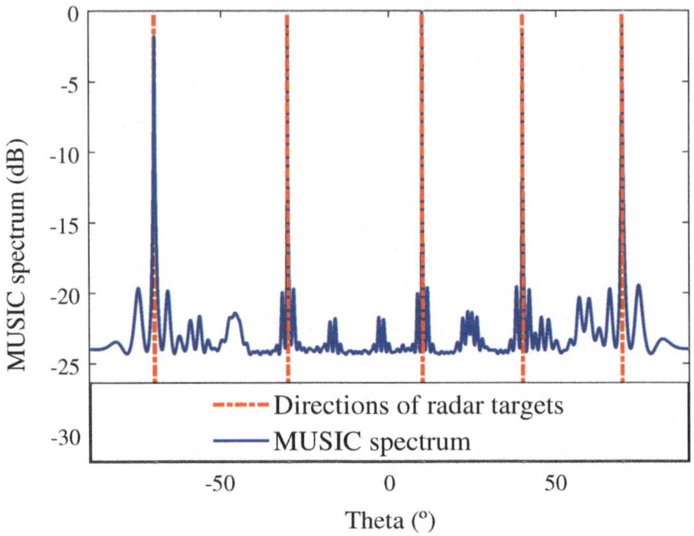

Fig. 9.11 The MUSIC spectrum of the optimal transceiver

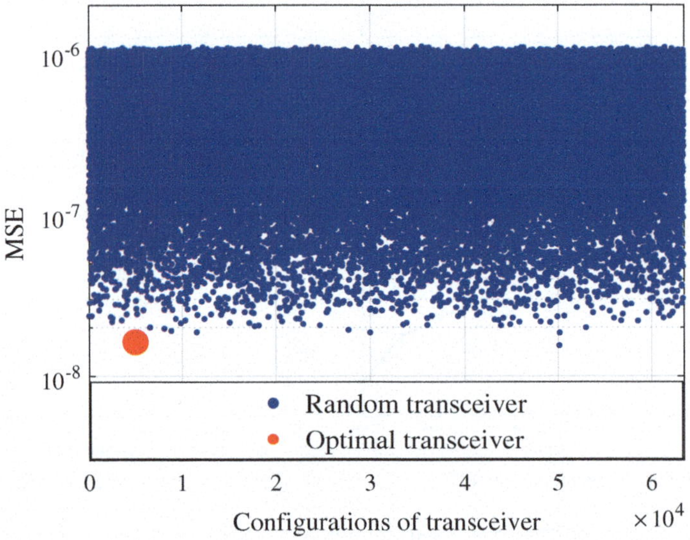

Fig. 9.12 The averaged MSE of DOA estimation for all configurations of transceiver

9.4 Summary

In this chapter, we proposed the array thinning via antenna selection for enhanced DOA estimation in both single-source and multi-source scenarios. In Sect. 9.2, we considered the single-source case. Problem formulation and optimal antenna selection in terms of minimizing the CRB for both isotropic and directional arrays were provided. We proposed a Dinklebach-type algorithm and convex relaxation to solve the combinational optimization in polynomial time. In Sect. 9.3, we investigated the optimal sparse MIMO transceiver design for enhanced DOA estimation in the multi-source scenario. A reweighted l_1-norm minimization with convex relaxation was adopted to promote the binary sparsity. Simulation results validated the effectiveness of the proposed strategy.

Chapter 10
Sparsity Sensed for Enhanced DOA Estimation

In this chapter, we investigate the DOA estimation by exploiting the sparsity sensed in the coarray domain using two types of sparse arrays, those are fully augmentable arrays and partially augmentable arrays. This enables estimation of more sources than the number of physical antennas. We adopt the covariance vectorization technique to construct the received signal vectors of coarrays for both fully and partially augmentable arrays. The difference is that the coarray is filled for fully augmentable arrays and exhibits holes for partially augmentable arrays. In Sect. 10.2, the DOAs of incoming signals are estimated utilizing the Minimum Output Power (MOP) method applied to the receiver coarray. Two different approaches of spectrum sensing and polynomial rooting are proposed based on MOP. Supporting simulation results demonstrate that DOA estimation based on polynomial rooting outperforms spectrum sensing. In Sect. 10.3, we utilize Bayesian Compressive Sensing (BCS) for DOA estimation, which formulates the problem from a probabilistic perspective and solve it with the relevance vector machine (RVM) concept. Supporting simulation results for both sparse linear arrays and circular arrays demonstrate the effectiveness of the proposed approach in terms of high resolution and estimation accuracy compared to the MUSIC and sparse signal reconstruction based methods.

10.1 Introduction

Estimating the direction-of-arrival (DOA) using antenna arrays has been an important topic in signal processing with diverse applications, such as radar, satellite navigation and telecommunication to list a few [1]. There has been extensive research on high-resolution DOA estimation techniques, among which the ones evolving around Capon's methods and MUSIC algorithms are commonly used [2]. Another kind of effective DOA estimation techniques based on sparse signal reconstruction (SSR) has emerged in recent years, including the l_1-SVD method proposed in [3]. These SSR methods fully leverage the sparsity sensed for enhanced DOA estimation. In the

X. Wang et al., *Sparse Sensing and Sparsity Sensed in Multi-sensor Array Applications*, https://doi.org/10.1007/978-981-99-9558-5_10

case of a single measurement vector (SMV), l_1 optimization approach is considered attractive to sparse signal recovery due to its guaranteed recovery accuracy. A major issue encountered in l_1 optimization, however, is that reliable recovery is guaranteed only when the restricted isometry property is satisfied [4]. It is worth noting that, for all aforementioned DOA estimation methods, the number of estimated sources cannot exceed the number of physical antennas.

When the number of estimated sources is larger than the number of physical sensors, high-resolution DOA estimation can be accomplished based on two approaches, neither requires increasing the number of physical antennas: (1) Different spatial lags of the covariance matrix of the sparse array are used to form an augmented Toeplitz matrix, which is equivalent to the true covariance matrix of an equivalent filled uniform array [5–7]; (2) The covariance matrix of the sparse array is vectorized to emulate observations at the corresponding difference coarray [8–10], which is defined as the set of points at which the spatial covariance function can be sampled with the physical array [11, 12]. The former technique requires positive definite Toeplitz completion for partially augmentable arrays [13], which is difficult to implement, especially when there are multiple holes corresponding to missing autocorrelation lags in the coarray points. In the second approach, the sources are replaced by their powers, casting them as mutually coherent. Spatial smoothing must then be applied to decorrelate signals and restore the full rank of the resulting covariance matrix before high-resolution DOA estimation can be performed [8, 14]. Spatial smoothing, however, requires availability of a set of contiguous coarray points without any holes which limits its applicability to partially augmentable arrays.

In Sect. 10.2, we study the DOA estimation using fully augmentable arrays. We propose a novel DOA estimation method utilizing the Minimum Output Power (MOP) based on the covariance vectorization technique. The reciprocal of the MOP nulling spectrum is the MOP signal spectrum where the spectral peaks indicate the signal presence, constituting the first proposed method for interference direction finding. Motivated by the efficient search-free root-MUSIC algorithm [15, 16], we propose the second method for DOA estimation based on polynomial rooting. The MOP is amenable to fast implementation [17], and thus favorable to practical engineering.

In Sect. 10.3, we study the DOA estimation using partially augmentable arrays. In order to better utilize the coarray aperture and increase the number of DoFs without the requirement of contiguous spatial lags, an SSR method for DOA estimation has been adopted based on the second approach of covariance matrix vectorization [18, 19]. There are inevitably spurious peaks in the sensing spectrum for SSR based methods due to the coherency of ill-conditioned measurement dictionary. Thus, a more reliable estimation approach is required, especially for partially augmentable arrays. To this end, we utilize the Bayesian Compressive Sensing (BCS) [20] which formulates the problem from a probabilistic perspective and solve it with the relevance vector machine (RVM) concept [21]. The sparse solution is obtained by assuming a Laplace prior for the sources of interest. It has been shown that the BCS-based spectrum sensing method is an effective and robust DOA estimation technique [22–24].

10.2 DOA Estimation Using Fully Augmentable Arrays

In this section, we propose a novel DOA estimation method utilizing the Minimum Output Power (MOP) based on the covariance vectorization technique. The reciprocal of the MOP nulling spectrum is the MOP signal spectrum where the spectral peaks indicate the signal presence, constituting the first proposed method for direction finding. Motivated by the efficient search-free root-MUSIC algorithm, we propose the second method for DOA estimation based on polynomial rooting.

10.2.1 DOA Estimation Based on Coarray Using MOP

In this subsection, we present two methods to estimate the DOAs of all incoming signals by utilizing the MOP based on the covariance vectorization. We assume the number of estimated signals is either known or can be estimated, for example using an information theoretic measure [25].

10.2.1.1 Problem Formulation

We consider K uncorrelated narrowband signals and L source signals impinging on an M-antenna nonuniform linear array ($K > M - 1$) with antenna positions $\mathbf{d} = [d_1, \ldots, d_M]^T$, which are spatially distributed in angle ϕ_k, $k = 1, \ldots, K$, and $\hat{\phi}_l$, $l = 1, \ldots, L$, respectively. Then, the received signal vector at time t can be written as,

$$\mathbf{x}(t) = \sum_{l=1}^{L} s_l(t)\hat{\mathbf{v}}_l + \sum_{k=1}^{K} q_k(t)\mathbf{v}_k + \mathbf{n}(t), t = 1, \ldots, \hat{T} \tag{10.1}$$

where $s_l(t)$ and $q_k(t)$ are complex amplitudes of the lth satellite and the kth jammer, respectively, $\mathbf{n}(t)$ is the noise vector and $\hat{T}$ is the total number of snapshots. The steering vectors $\hat{\mathbf{v}}_l$, $l = 1, \ldots, L$ and $\mathbf{v}_k$, $k = 1, \ldots, K$ are defined as,

$$\hat{\mathbf{v}}_l = [e^{jk_0 d_1 \cos(\hat{\phi}_l)}, \ldots, e^{jk_0 d_M \cos(\hat{\phi}_l)}]^T, l = 1, \ldots, L, \tag{10.2}$$

and

$$\mathbf{v}_k = [e^{jk_0 d_1 \cos(\phi_k)}, \ldots, e^{jk_0 d_M \cos(\phi_k)}]^T, k = 1, \ldots, K, \tag{10.3}$$

where $k_0 = 2\pi/\lambda$ is the wavenumber, $d_m = i_m \frac{\lambda}{2}$, $m = 1, \ldots, M$ is the position of the mth antenna, which is typically an integer multiple of half-wavelengths, and the superscript "T" denotes matrix transpose. Then, the covariance matrix $\mathbf{R}_x$ of the received signal can be written as,

$$\mathbf{R}_x = \mathbf{V}\mathbf{R}_i\mathbf{V}^H + \tilde{\mathbf{V}}\mathbf{R}_s\tilde{\mathbf{V}}^H + \sigma_n^2\mathbf{I}, \tag{10.4}$$

where the two steering matrices are $\mathbf{V} = [\mathbf{v}_1, \ldots, \mathbf{v}_K]$ and $\tilde{\mathbf{V}} = [\tilde{\mathbf{v}}_1, \ldots, \tilde{\mathbf{v}}_L]$, σ_n^2 is the noise power, and $\mathbf{R}_s$ and $\mathbf{R}_i$, respectively, represent source and interference diagonal correlation matrices. The corresponding diagonal elements are the source powers $\tilde{\sigma}_1^2, \ldots, \tilde{\sigma}_L^2$ and jammer powers $\sigma_1^2, \ldots, \sigma_K^2$. The ijth element of $\mathbf{R}_x$ is expressed as,

$$\begin{aligned}\mathbf{R}_x(i, j) = &\sum_{k=1}^{K} \sigma_k^2 e^{jk_0(d_i - d_j)\cos(\phi_k)} + \sigma_n^2 \delta(i, j) \\ &+ \sum_{l=1}^{L} \hat{\sigma}_l^2 e^{jk_0(d_i - d_j)\cos(\hat{\phi}_l)},\end{aligned} \tag{10.5}$$

where $\delta(i, j)$ is the Kronecker Delta function. It is clear that $\mathbf{R}_x(i, j)$ can be treated as the data received by the coarray element position $(d_i - d_j)$. Thus, for a linear array with antenna positions $\mathbf{d} = [d_1, \ldots, d_L]$, the corresponding difference coarray has positions,

$$\tilde{\mathbf{d}} = \{d_i - d_j : i, j = 1, \ldots, M\}. \tag{10.6}$$

That is, the difference coarray is the set of pairwise differences of the array element positions. The received signal correlation can be calculated at all lags comprising the difference coarray. Hence, by suitable configuration of the physical array $\mathbf{d}$, the number of spatial lags can be substantially increased for a given number, M, of physical antennas. In order to briefly explain the concept of difference coarray, we consider, as an example, the four-antenna minimum redundancy array (MRA) with element positions $\mathbf{d} = [0, 1, 4, 6]\frac{\lambda}{2}$ [11], as shown in the top plot of Fig. 10.1. The corresponding difference coarray is a ULA with 13 antennas as shown in the bottom plot of Fig. 10.1.

Vectorizing $\mathbf{R}_x$, we obtain

$$\tilde{\mathbf{x}} = \text{vec}(\mathbf{R}_x) = \tilde{\mathbf{V}}\mathbf{b} + \sigma_n^2\mathbf{i}, \tag{10.7}$$

where $\tilde{\mathbf{V}} = [\mathbf{V}, \hat{\mathbf{V}}]^* \odot [\mathbf{V}, \hat{\mathbf{V}}]$ with $\odot$ and $*$ denoting Khatri-Rao product and complex conjugate, respectively, $\mathbf{b} = [\sigma_1^2, \ldots, \sigma_K^2, \hat{\sigma}_1^2, \ldots, \hat{\sigma}_L^2]^T$ and $\mathbf{i} = \text{vec}(\mathbf{I})$. The vector $\tilde{\mathbf{x}}$ can be viewed as a single snapshot received by the difference coarray. The equivalent source signal $\mathbf{b}$ consists of the source powers and the noise becomes a deterministic vector. Therefore, the rank of the covariance matrix of the coarray measurement $\tilde{\mathbf{x}}$ is one and spatial smoothing can be utilized to restore the rank of the covariance matrix. Note that the sample covariance $\hat{\mathbf{R}}_x = \frac{1}{\hat{T}} \sum_{t=1}^{\hat{T}} \mathbf{x}(t)\mathbf{x}^H(t)$ is used instead of $\mathbf{R}_x$ when there are $\hat{T}$ samples.

10.2.1.2 The MOP Signal Spectrum

The optimization formulation of the MOP criterion is as follows [26],

$$\min_{\mathbf{w}} \mathbf{w}^H \mathbf{R}_x \mathbf{w} \text{ subject to } \mathbf{w}^H \mathbf{f} = 1, \tag{10.8}$$

where $\mathbf{w}$ is the weight vector and the superscript "H" denotes the Hermitian operator, the constraint vector $\mathbf{f}$ equals $[1, 0, \ldots, 0]^T$, i.e., only the first entry is one and others are zero, so as to avoid all the weights becoming zero.

Implementing the Lagrange Multiplier to Eq. (10.8) yields the optimal MOP weight vector,

$$\mathbf{w}_{\mathrm{MOP}} = \mu \mathbf{R}_x^{-1} \mathbf{f}, \tag{10.9}$$

where μ is a scaling factor which does not affect the output signal-to-interference-plus-noise ratio (SINR). The MOP interference spectrum based on the MOP weight vector can be expressed as,

$$P(\theta) = \frac{1}{\mathbf{w}_{\mathrm{MOP}}^H \mathbf{v}(\theta) \mathbf{v}^H(\theta) \mathbf{w}_{\mathrm{MOP}}}, \tag{10.10}$$

where $\mathbf{v}(\theta)$ is the steering vector evaluated at angle θ following the definition in (10.3).

It is well-known that applying the MOP weights in (10.9) directly can identify maximally $M - 1$ signals. The number of estimated sources can be increased by ulitizing the vectorized data covariance matrix, which are the measurements corresponding to a virtual ULA with M_a elements. The ULA with $(M_a + 1)/2$ virtual antennas after implementing spatial smoothing, which is coarray equivalent to the physical array, is capable of identifying DOAs of $(M_a - 1)/2$ jammers, which is higher than the number M of physical sensors. In order to briefly explain the main concept behind the proposed scheme, we consider, as an example, the four-antenna MRA [11], shown in the upper plot of Fig. 10.1. The coarray measurements are calculated as follows,

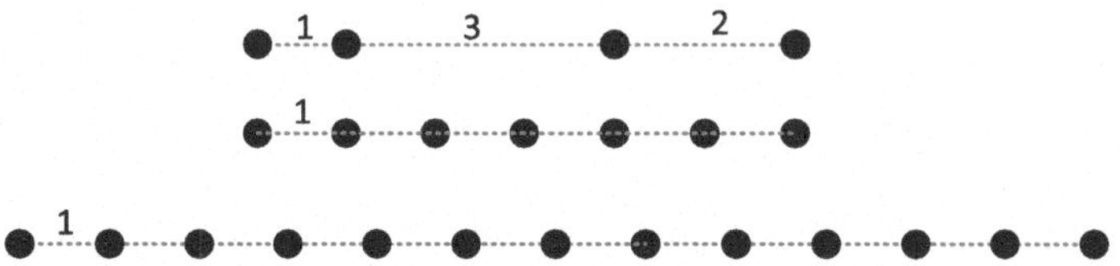

Fig. 10.1 Four-antenna MRA (top), virtual six-antenna ULA (center), and the corresponding difference coarray (bottom)

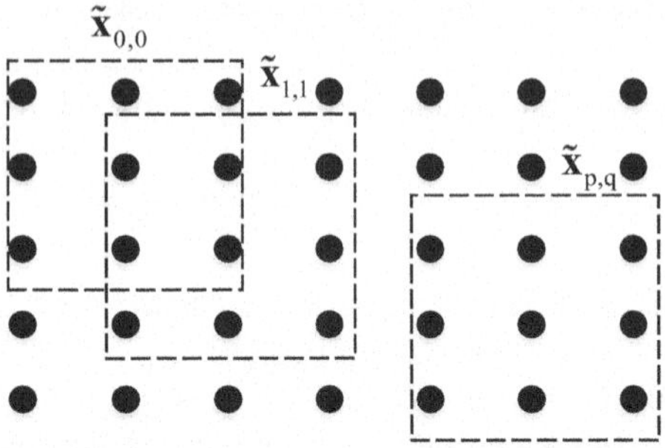

Fig. 10.2 Two-dimensional spatial smoothing

$$\tilde{\mathbf{x}}_i = \begin{cases} \mathbf{R}_x(l, m), & l - m = i, -6 \leq i \leq 6, i \neq 0 \\ \frac{1}{M} \sum_{k=1}^{M} \mathbf{R}_x(k, k) & i = 0 \end{cases}$$

where $\mathbf{R}_x(l, m)$ is the correlation between the signals received by the lth and mth sensors. As the number of estimated sources K is assumed known, a virtual ULA with $K + 1$-antennas is constructed in order to circumvent the sysmtematic peaks generated by the MOP weight vector. A forward spatial smoothing [27] is applied to the coarray measurement vector $\tilde{\mathbf{x}}$, i.e.,

$$\tilde{\mathbf{R}}_x = \frac{1}{N} \sum_{i=1}^{N} \tilde{\mathbf{x}}_{i:K+i} \tilde{\mathbf{x}}_{i:K+i}^H, \tag{10.11}$$

where $N = M_a - K + 1$ and $\tilde{\mathbf{x}}_{i:K+i}$ denotes the subvector of $\tilde{\mathbf{x}}$ containing entries from i to $K + i$. Then, $\tilde{\mathbf{R}}_x$ is the covariance matrix of the coarray equivalent virtual six-antenna ULA, shown in the center plot of Fig. 10.1, which is of full rank and positive definite. For a uniform rectangular array (URA), the spatial smoothing can be applied as shown in Fig. 10.2, i.e.,

$$\tilde{\mathbf{R}}_x = \frac{1}{N_1 N_2} \sum_{i=1}^{N_1} \sum_{j=1}^{N_2} \tilde{\mathbf{x}}_{i,j} \tilde{\mathbf{x}}_{i,j}^H, i = 0, \ldots, N_1, j = 0, \ldots, N_2, \tag{10.12}$$

where $\tilde{\mathbf{x}}_{i,j}$ is the received signal by the ijth coarray (see Fig. 10.2) and $\tilde{\mathbf{R}}_x$ is the spatially smoothed covariance matrix of a coarray equivalent virtual URA with dimensions $N_1 \times N_2$.

Now, the optimum MOP weight vector in the coarray framework can be expressed in terms of the spatial smoothed covariance matrix $\tilde{\mathbf{R}}_x$ as,

$$\tilde{\mathbf{w}}_{\text{MOP}} = \mu \tilde{\mathbf{R}}_x^{-1} \tilde{\mathbf{f}}, \tag{10.13}$$

Here, $\tilde{\mathbf{f}}$ is a vector of length $M_a = 7$ with one element being one and others being zero. Since the number of DoFs is increased for the vectorized covariance matrix, the maximum number of estimated signals is increased from 3 to 6. In essence, for fully augmentable arrays, the maximum number of detected signals is limited by the number of the distinct non-negative spatial lags. It should be noted that the increased size of the coarray covariance matrix results in increased variance, on the order of twice the variance of a physical ULA [28]. In order to compensate for performance degradation, the number of samples in estimating the data covariance matrix should be sufficiently large.

10.2.1.3 The MOP-Based Polynomial Rooting

The weight $\tilde{\mathbf{w}}_{\text{MOP}}$ produces deep nulls corresponding to the jammers' directions, i.e.,

$$\tilde{\mathbf{w}}_{\text{MOP}}^H \tilde{\mathbf{v}}(\phi_k) = 0, k = 1, \ldots, K \tag{10.14}$$

where $\tilde{\mathbf{v}}(\phi_k)$ is the steering vector of the difference coarray. For an inter-element spacing of half-wavelength, we obtain

$$\tilde{\mathbf{v}}(\phi_k) = [1, e^{j\pi \cos(\phi_k)}, \ldots, e^{j\pi \cos(\phi_k)(M_a - 1)}]^T. \tag{10.15}$$

Defining $z = e^{j\pi \cos(\phi_k)}$ permits (10.14) to be expressed as a polynomial,

$$\tilde{\mathbf{w}}_{\text{MOP}}^H \tilde{\mathbf{v}}(\phi_k) = \sum_{i=1}^{M_a} \tilde{w}_{\text{MOP},i}^* z^{i-1} = 0. \tag{10.16}$$

where $\tilde{w}_{\text{MOP},i}$ is the ith element of $\tilde{\mathbf{w}}_{\text{MOP}}$. This property enables DOA estimation to be performed by calculating the roots of the polynomial with coefficients $\tilde{\mathbf{w}}_{\text{MOP}}$. Unlike signal spectrum based DOA estimation approaches, where the estimation accuracy depends on the density of searching grid points, the search-free polynomial rooting method exhibits two main advantages, namely, reduced computational complexity and off-grid DOA estimation.

10.2.2 Simulations

In this section, we present simulation results to validate the performance of the proposed DOA estimation approaches.

10.2.2.1 Example 1

First, we compare the DOA estimation performance based on the coarray and the physical array for the two proposed methods, namely MOP-based polynomial rooting and MOP interference spectrum, and the well-known MUSIC algorithm. We employ the 4-antenna MRA as shown in Fig. 10.1. We consider an increasing number of signals from 3 to 6 with signal-to-noise ratio (SNR) of 20 dB and arrival angles uniformly distributed within an angular sector starting from 30° with an angular interval of 20°. The number of snapshots is set as $\hat{T} = 2046$, which corresponds to 2 ms integration time. The MOP sensing spectra based on both the physical array and coarray, as well as the MUSIC spectra based on coarray are depicted in Fig. 10.3. The MOP based polynomial roots are plotted in Fig. 10.4. We can see that the physical array cannot distinguish more signals than the number of physical antennas due to its limited DoFs and the estimated DOAs exhibit large bias. The MOP signal spectrum and the MOP polynomial rooting can identify all signals accurately.

In order to further investigate the estimation accuracy of the proposed methods, we calculate the sum estimation variance of the six signals with respect to different numbers of snapshots and different SNR values. The sum estimation variance is defined as

$$\text{Var} = \frac{1}{Q}\sum_{i=1}^{K}\sum_{n=1}^{Q}(\hat{\phi}_{n,k} - \phi_k)^2, \tag{10.17}$$

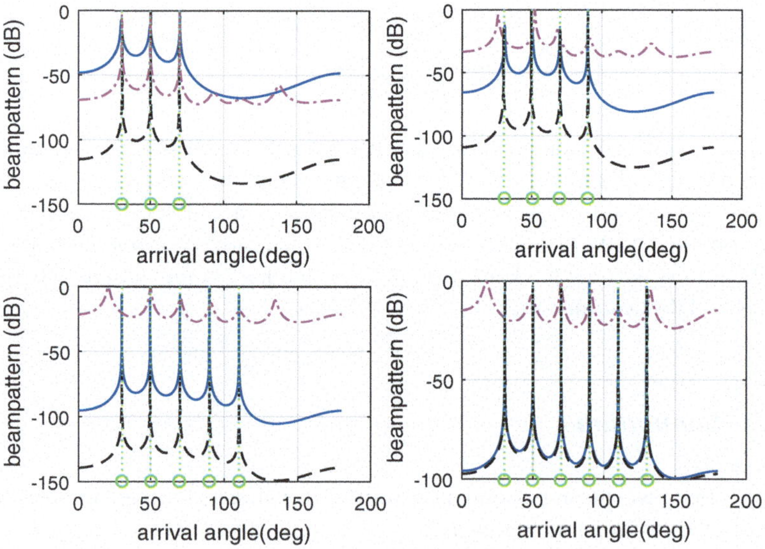

Fig. 10.3 MOP sensing spectra based on both physical array (dash-dotted) and coarray (solid) and the MUSIC spectrum based on coarray (dashed) for number of jammers varying from 3 to 6. The dotted vertical lines in each subfigure indicate the jammers' true arrival directions

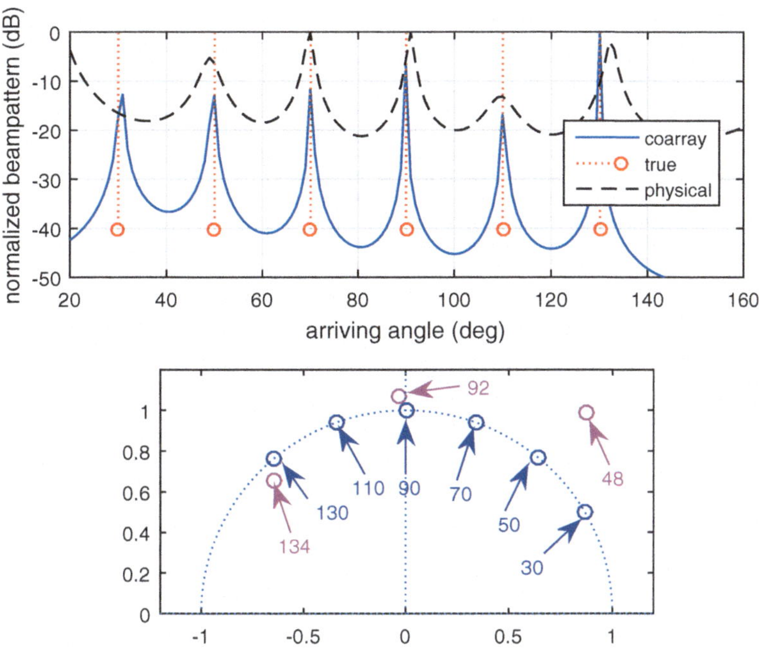

Fig. 10.4 The MOP polynomial rooting based on both physical array and coarray. The number of jammers is 6. The numbers on the arrows indicate the estimated arrival angles, blue is for coarray and magenta is for the physical array

where $\hat{\phi}_{n,k}$ denotes the estimated angle of the kth signal in the nth Monte-Carlo test and Q is the total number of Monte-Carlo runs. We set $Q = 500$ in the simulation and vary the SNR value from 0 dB to 30 dB in 1 dB increments. The number of snapshots is varied from 100 to 10000 in steps of 100. Figure 10.5 compares the sum estimation variance of the MOP sensing spectrum with the MUSIC algorithm based on coarray. The sum estimation variance of the MOP sensing spectrum applied to a 7-element ULA is alo plotted as a benchmark in the same figure. Clearly, the proposed method exhibits almost the same accuracy as the well-known MUSIC algorithm based on coarray in terms of detecting more sources than the number of physical sensors. Note that both methods in general perform worse than the MOP sensing spectrum applied to a 7-element ULA. This is because the coarray based method employs fewer antennas and a limited number of snapshots for estimating the covariance matrix [8].

10.2.2.2 Example 2

The same 4-antenna MRA is employed for off-grid DOA estimation. Consider six signals impinging on the array from directions [30.5°,50.6°,70.7°,90.8°,110.4°, 130.3°]. All the six signals are of equal power and the SNR value is varying from 0dB to 30dB.

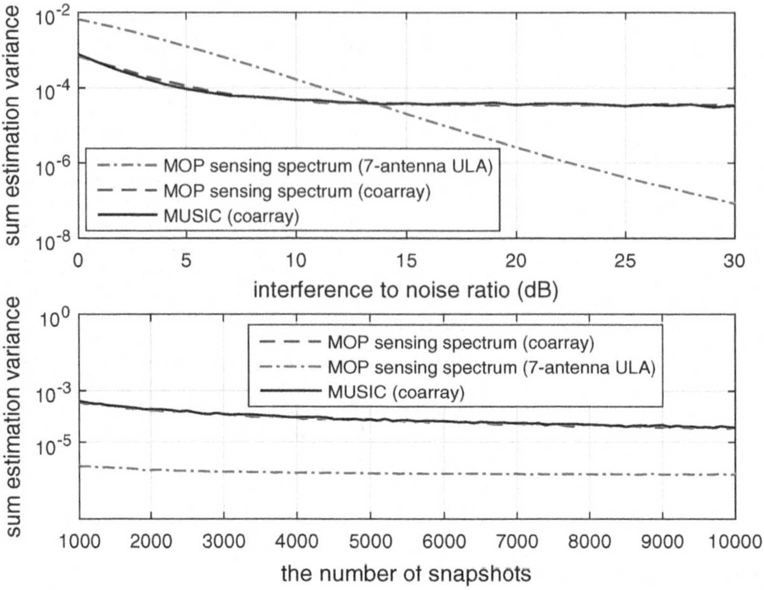

Fig. 10.5 Sum estimation variance versus SNR (top plot) with the number of snapshots set to 2046; Sum estimation variance versus numbers of snapshots (bottom plot) with INR set to 20 dB

The number of snapshots is set to $\hat{T} = 2046$ and the sampling grid interval is chosen to be 1°. The estimation accuracy of the MOP spectrum sensing method is highly dependent on the density of sampling grid points. On the other hand, since MOP based polynomial rooting is search-free, the off-grid DOAs can be estimated both accurately and efficiently. Similar to Example 1, we implement 500 Monte-Carlo trials for each SNR value and plot the sum estimation variance in the top plot of Fig. 10.6. The sum estimation variance versus number of snapshots for a fixed 20 dB SNR is depicted in the bottom plot of Fig. 10.6. We observe that the spectrum sensing method rounds off the DOAs to the closest grid points and presents a limited estimation accuracy, whereas the MOP based polynomial rooting exhibits superior estimation performance due to its search-free property.

10.2.2.3 Example 3

In the third example, there are three pairs of closely spaced signals with SNR values being [30,30,10,10,30,10] dB and arriving from the directions [30°,35°,70°,90°,120°, 130°] respectively. The searching grid interval is set to 1° and there are in total 2046 snapshots, i.e. the integration time is 2 ms. The beampattern of the MOP signal spectrum based on the coarray is shown in the upper plot of Fig. 10.7 and the MOP polynomial roots based on the coarray are depicted in the lower plot. It is clear that the MOP signal spectrum cannot distinguish the third pair of signals due to their

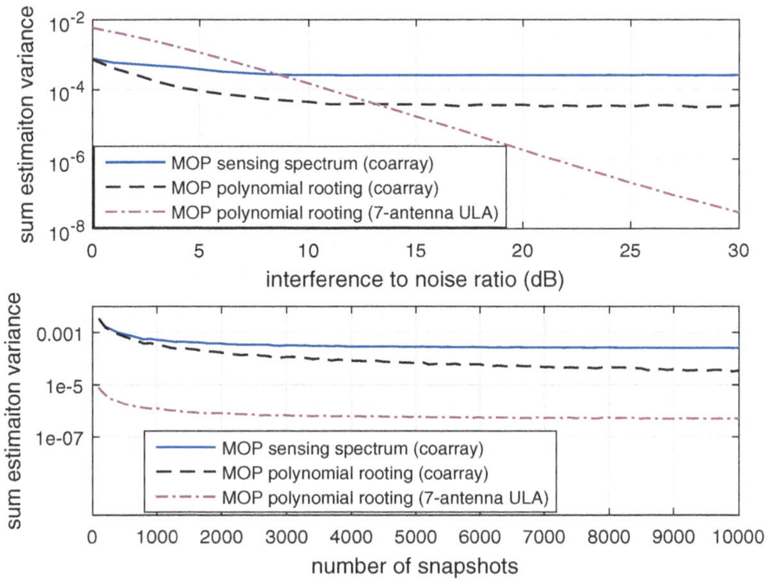

Fig. 10.6 Sum estimation variance versus SNR (top plot) with the number of snapshots set to 2046; Sum estimation variance versus numbers of snapshots (bottom plot) with INR set to 20 dB

large power difference, and also exhibits estimation biases for the other two pairs of signals. The MOP based polynomial rooting exhibits the best estimation performance and clearly shows three pairs of roots.

10.2.2.4 Example 4

Next, the proposed MOP signal spectrum sensing method based on the coarray is verified for two-dimensional (2-D) DOA estimation. Note that the MOP polynomial rooting method is not applicable to 2-D arrays. Figure 10.8 depicts a 12-antenna sparse rectangular array (filled dots) and the corresponding difference coarray (circles). The latter is a uniform rectangular array (URA) with 91 elements. A 2-D spatial smoothing is utilized to recover the full-rank covariance matrix of the URA with dimensions 7×4, which implies the maximum number of estimated signals is 27. Suppose there are 15 signals uniformly distributed in the azimuth sector $[20^\circ, 300^\circ]$ with 20° increment and arriving from five equi-spaced directions in the elevation sector $[10^\circ, 50^\circ]$. Each signal has a 20 dB SNR. The 2-D DOA estimation results are shown in Fig. 10.9, where the radial direction denotes the elevation angle and the circumference direction is the azimuth. We observe that the estimated 2-D angles almost coincide with the true angles, which further validates the effectiveness of the proposed method.

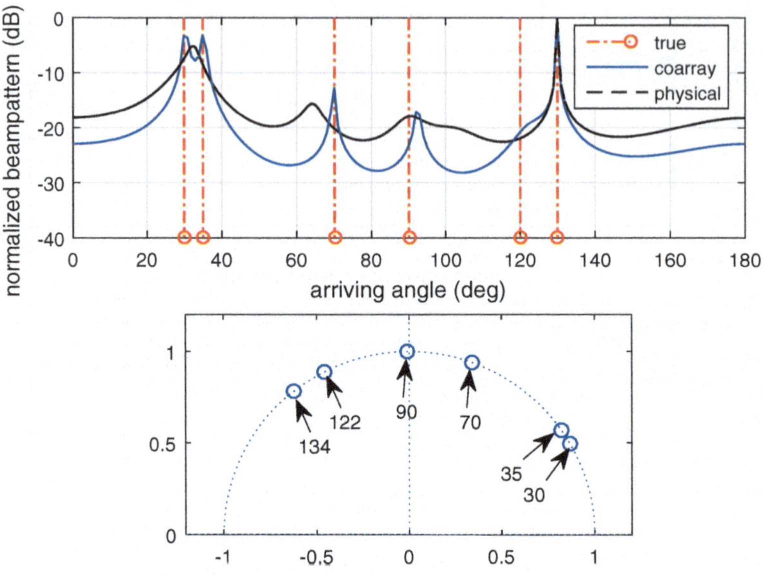

Fig. 10.7 MOP sensing spectra based on both physical array and coarray (top) and the polynomial roots based on coarray (bottom). The numbers on the arrows in the bottom plot indicate the estimated arrival angles

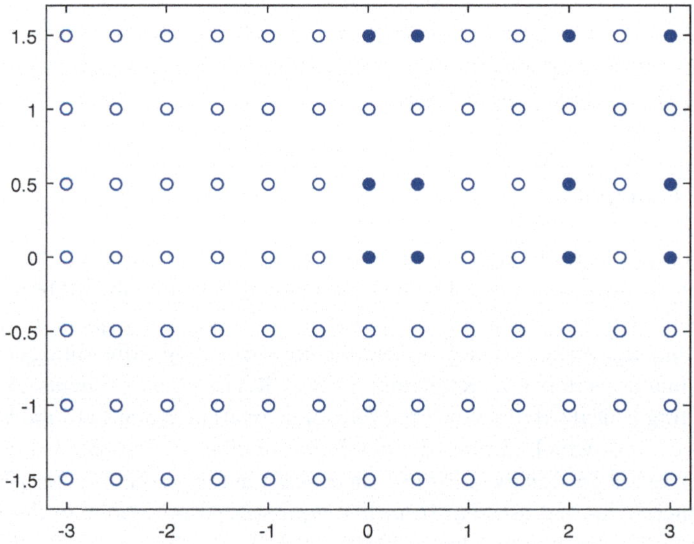

Fig. 10.8 The 12-antenna sparse rectangular array (filled dots) and the corresponding difference coarray (circles)

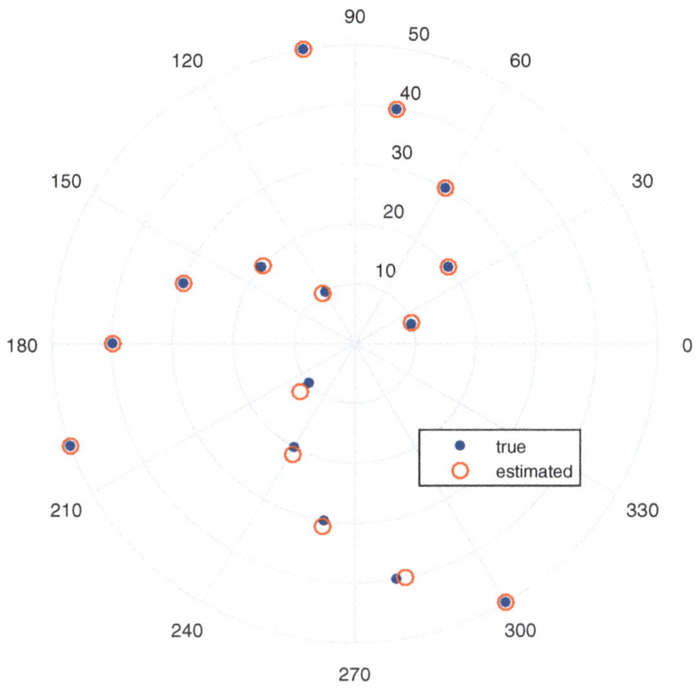

Fig. 10.9 2-D DOA estimation: the filled dots and circles indicate the true and estimated directions, respectively. The number of snapshots is set to 10230, i.e., the integration time is 10 ms

10.3 DOA Estimation Using Partially Augmentable Arrays

In this section, we introduce an SSR based DOA estimation using partially augmentable arrays. There are inevitably spurious peaks in the sensing spectrum for SSR based methods due to the coherency of ill-conditioned measurement dictionary. Thus, a more reliable estimation approach is required, especially for partially augmentable arrays. To this end, we utilize the Bayesian Compressive Sensing (BCS) [20] which formulates the problem from a probabilistic perspective and solve it with the relevance vector machine (RVM) concept [21]. The sparse solution is obtained by assuming a Laplace prior for the sources of interest.

10.3.1 DOA Estimation Based on Difference Coarray

Assuming the positions of the array elements form the set,

$$\mathcal{S} = \{(x_i, y_i), i = 1, \ldots, M\}, \tag{10.18}$$

where $x_i = n_i d_0$, $y_i = m_i d_0$ with d_0 being the unit inter-element spacing and n_i, m_i being integer numbers. The corresponding difference coarray has positions,

$$\mathcal{S}_d = \{(x_i - x_j, y_i - y_j), i, j = 1, \dots, M\}, \tag{10.19}$$

i.e., the difference coarray is the set of pairwise differences of the array element positions and the received signal correlation can be calculated at all "lags" comprising the difference coarray [12]. Minimum Redundancy Arrays (MRAs) and Minimum Hole Arrays (MHAs) are the common classes of sparse arrays [11, 29]. MRAs are those configurations of M elements that satisfy minimum (R | H = 0; M = constant), where R and H denote the number of redundancies and holes in the coarray, respectively. MRAs are also referred to as fully augmentable arrays in [30]. MHAs are sparse arrays of M elements that satisfy minimum (H | R = 0; M = constant), which belong to the class of partially augmentable arrays.

Consider K narrowband far-field uncorrelated sources $s_k(t), k = 1, \dots, K$, impinging on an array of M omnidirectional sensors from directions $\theta_k, k = 1, \dots, K$ in elevation and $\phi_k, k = 1, \dots, K$ in azimuth. The array output can be expressed as,

$$\mathbf{y}(t) = \mathbf{A}\mathbf{s}(t) + \mathbf{e}(t), t = 1, \dots, \tilde{T}, \tag{10.20}$$

where $\mathbf{y}(t) = [y_1(t), \dots, y_M(t)]^T$, $\mathbf{s}(t) = [s_1(t), \dots, s_K(t)]^T$ and $\mathbf{e}(t) = [e_1(t), \dots, e_M(t)]^T$ is the noise vector. The matrix $\mathbf{A} = [\mathbf{a}(\mathbf{u}_1), \dots, \mathbf{a}(\mathbf{u}_K)]$ is the array manifold and $\mathbf{a}(\mathbf{u}_k)$ is the steering vector of the kth source, which is defined as,

$$\mathbf{a}(\mathbf{u}_k) = [1, e^{jk_0(u_2^x x_2 + u_2^y y_2)}, \dots, e^{jk_0(u_M^x x_M + u_M^y y_M)}]^T, \tag{10.21}$$

where $k_0 = 2\pi/\lambda$ is the wavenumber and $\mathbf{u}_k = [u_k^x, u_k^y]^T = [\cos\theta_k \cos\phi_k, \cos\theta_k \sin\phi_k]^T$.

The correlation matrix $\mathbf{R}$ of the received signal is given by,

$$\mathbf{R} = \sum_{t=1}^{\tilde{T}} \mathbf{y}(t)\mathbf{y}^H(t) = \mathbf{A}\mathbf{R}_s\mathbf{A}^H + \sigma_0^2\mathbf{I}, \tag{10.22}$$

where $\mathbf{R}_s$ represents the source correlation matrix, which is diagonal with the source powers $\sigma_1^2, \dots, \sigma_K^2$ populating its main diagonal, $\mathbf{I}$ is the identity matrix of corresponding rank, σ_0^2 is the noise variance and the superscript 'H' denotes conjugate transpose. The ijth element of $\mathbf{R}$ is,

$$(\mathbf{R})_{ij} = \sum_{k=1}^{K} \sigma_k^2 e^{jk_0(u_k^x(x_i - x_j) + u_k^y(y_i - y_j))} + \sigma_0^2\delta(i - j), \tag{10.23}$$

where $\delta(i - j)$ is the Kronecker Delta function. It is clear that $(\mathbf{R})_{ij}$ can be treated as the data received by the coarray element position $(x_i - x_j, y_i - y_j)$.

10.3.2 SMV-BCS Based on Covariance Vectorization

We elaborate on the SMV-BCS method based on covariance vectorization for DOA estimation in the case where the sources are more than the physical antennas.

10.3.2.1 Covariance Vectorization

Vectorizing $\mathbf{R}$, we obtain

$$\tilde{\mathbf{y}} = \text{vec}(\mathbf{R}) = \tilde{\mathbf{A}}\mathbf{b} + \sigma_0^2\tilde{\mathbf{i}}, \tag{10.24}$$

where $\tilde{\mathbf{A}} = [\tilde{\mathbf{a}}(\mathbf{u}_1), \ldots, \tilde{\mathbf{a}}(\mathbf{u}_K)]$ with $\tilde{\mathbf{a}}(\mathbf{u}_k) = \mathbf{a}(\mathbf{u}_k) \otimes \mathbf{a}^*(\mathbf{u}_k)$. Here $\otimes$ denotes the Kronecker product and '*' is the conjugate operation. The two vectors $\tilde{\mathbf{i}} = \text{vec}(\mathbf{I})$ and $\mathbf{b} = [\sigma_1^2, \ldots, \sigma_K^2]^T$. The vector $\tilde{\mathbf{y}}$ can be viewed as a single snapshot received by a much larger virtual array, whose element positions are given by the difference coarray. Utilizing the coarray measurement vector $\tilde{\mathbf{y}}$ for DOA estimation permits handling of a greater number of sources than the number of physical antennas. The equivalent source signal $\mathbf{b}$ consists of the powers of the estimated sources and the noise becomes a deterministic vector. Therefore, the rank of the covariance matrix of $\tilde{\mathbf{y}}$ is one and subspace-based DOA estimation techniques, such as MUSIC, would fail. It should be noted that if the sparse array is fully augmentable, spatial smoothing can be utilized to restore the rank of the covariance matrix.

10.3.2.2 SMV-BCS

Let $\hat{\theta} = \{\hat{\theta}_1, \ldots, \hat{\theta}_{N_e}\}$ and $\hat{\phi} = \{\hat{\phi}_1, \ldots, \hat{\phi}_{N_a}\}$ be two fixed sampling grid sets covering all possible DOAs, where N_e and N_a are the number of grid points in elevation and azimuth, respectively. Let $\tilde{\Phi} = [\tilde{\mathbf{a}}(\hat{\mathbf{u}}_1), \ldots, \tilde{\mathbf{a}}(\hat{\mathbf{u}}_{N_e N_a}), \tilde{\mathbf{i}}]$. The observation model in Eq. (10.24) can be rewritten as,

$$\tilde{\mathbf{y}} = \tilde{\Phi}\mathbf{x} + \mathbf{e}, \tag{10.25}$$

where $\mathbf{x}$ represents the source signals, with the entry corresponding to the angles $[\theta_k, \phi_k]$ equals σ_k^2 if $\theta_k \in \{\hat{\theta}_1, \ldots, \hat{\theta}_{N_e}\}$ and $\phi_k \in \{\hat{\phi}_1, \ldots, \hat{\phi}_{N_a}\}$, $k = 1, \ldots, K$. The signal $\mathbf{x}$ is sparse since $N_e N_a \gg K$. Here, $\mathbf{e}$ denotes the estimation noise which, without loss of generality, is circularly symmetric Gaussian distributed with probability density function,

$$p(\mathbf{e}|\alpha_0) = \mathcal{CN}(\mathbf{e}|\mathbf{0}, \alpha_0^{-1}\mathbf{I}). \tag{10.26}$$

The complex Normal distribution $\mathcal{CN}(\mathbf{u}|\mu, \Sigma)$ is defined as,

$$\mathcal{CN}(\mathbf{u}|\mu, \Sigma) = \frac{1}{\pi^{N_e N_a}|\Sigma|}\exp\{-(\mathbf{u}-\mu)^H\Sigma^{-1}(\mathbf{u}-\mu)\}. \tag{10.27}$$

Therefore, we have

$$p(\tilde{\mathbf{y}}|\mathbf{x}, \alpha_0) = \mathcal{CN}(\tilde{\mathbf{y}}|\tilde{\Phi}\mathbf{x}, \alpha_0^{-1}\mathbf{I}). \tag{10.28}$$

A two-stage hierarchical prior is adopted for $\mathbf{x}$ to introduce a Laplace prior for both real part $\mathcal{R}(\mathbf{x})$ and imaginary part $\mathcal{I}(\mathbf{x})$. First, we define a zero mean complex Gaussian prior,

$$p(\mathbf{x}|\alpha) = \mathcal{CN}(\mathbf{x}|\mathbf{0}, \Lambda), \tag{10.29}$$

with $\Lambda = \text{diag}(\alpha)$ and $\alpha = [\alpha_1, \ldots, \alpha_{N_e N_a}]^T \in \mathbb{R}_+^{N_e N_a}$. Further, a Gamma hyperprior is considered over α,

$$p(\alpha|\rho) = \Pi_{n=1}^{N_e N_a}\Gamma(\alpha_n|1, \rho), \tag{10.30}$$

with $\rho \in \mathbb{R}_+$ being fixed a priori. Similarly, a Gamma prior is introduced on α_0,

$$p(\alpha_0|c, d) = \Gamma(\alpha_0, c, d) = \alpha_0^{c-1}e^{-\alpha_0/d}d^{-c}\Gamma(c)^{-1}, \tag{10.31}$$

with $c, d \in \mathbb{R}_+$ being fixed a priori and $\Gamma(c)$ being the Gamma function evaluated at c. To make the Gamma prior non-informative, $c, d \to 0$ are adopted [22].

By combining the stages of the hierarchical model, the joint distribution is obtained as

$$p(\mathbf{x}, \tilde{\mathbf{y}}, \alpha_0, \alpha) = p(\tilde{\mathbf{y}}|\mathbf{x}, \alpha_0)p(\mathbf{x}|\alpha)p(\alpha_0)p(\alpha). \tag{10.32}$$

A type-II ML approach is exploited to perform the Bayesian inference since the posterior distribution $p(\mathbf{x}, \alpha_0, \alpha|\tilde{\mathbf{y}})$ cannot be expressed explicitly. First, it is easy to show that the posterior for $\mathbf{x}$ is a complex Gaussian distribution, [22]

$$p(\mathbf{x}|\mathbf{y}, \alpha_0, \alpha) = \frac{p(\mathbf{y}|\mathbf{x}, \alpha_0)p(\mathbf{x}|\alpha)}{p(\mathbf{y}|\alpha_0, \alpha)} = \mathcal{CN}(\mathbf{x}|\mu, \Sigma), \tag{10.33}$$

with mean and covariance,

$$\mu = \alpha_0\Sigma\tilde{\Phi}^H\tilde{\mathbf{y}}, \;\; \Sigma = (\Lambda^{-1} + \alpha_0\tilde{\Phi}^H\tilde{\Phi})^{-1}. \tag{10.34}$$

Then, the hyperparameters α_0 and α are estimated by the maxima of the posterior $p(\alpha_0, \alpha|\tilde{\mathbf{y}})$, or equivalently, the maxima of the joint distribution $p(\tilde{\mathbf{y}}, \alpha_0, \alpha) \propto p(\alpha_0, \alpha|\tilde{\mathbf{y}})$, which can be expressed as

$$p(\tilde{\mathbf{y}}, \alpha_0, \alpha) = p(\tilde{\mathbf{y}}|\alpha_0, \alpha)p(\alpha_0)p(\alpha). \tag{10.35}$$

We can readily show that $p(\tilde{\mathbf{y}}|\alpha_0, \alpha)$ is the convolution of two Gaussian distributions, i.e.,

$$p(\tilde{\mathbf{y}}|\alpha_0, \alpha) = \int p(\tilde{\mathbf{y}}|\mathbf{x}, \alpha_0)p(\mathbf{x}|\alpha)\mathrm{d}\mathbf{x} = \mathcal{CN}(\tilde{\mathbf{y}}|\mathbf{0}, \mathbf{C}), \tag{10.36}$$

with

$$\mathbf{C} = \alpha_0^{-1}\mathbf{I} + \tilde{\Phi}\Lambda\tilde{\Phi}^H. \tag{10.37}$$

To alleviate the computational complexity of Eq.(10.34), we use the Woodbury matrix identity to obtain

$$\Sigma = \Lambda - \Lambda\tilde{\Phi}^H\mathbf{C}^{-1}\tilde{\Phi}\Lambda. \tag{10.38}$$

The Maximum Likelihood (ML) function is the logarithm of the joint PDF, $\mathcal{L}(\alpha_0, \alpha) = \log p(\tilde{\mathbf{y}}, \alpha_0, \alpha)$. An expectation maximization (EM) algorithm is implemented to maximize the ML function. Hence, the update of α_n is given as

$$\alpha_n^{\text{new}} = \frac{\sqrt{1 + 4\rho(|\mu_n|^2 + \Sigma_{nn})} - 1}{2\rho}, \tag{10.39}$$

where μ_n and Σ_{nn} denote the n_{th} entry of the mean vector μ and the diagonal of the covariance matrix Σ respectively. Similarly, for α_0 we have

$$\alpha_0^{\text{new}} = \frac{M_a + c - 1}{\|\tilde{\mathbf{y}} - \tilde{\Phi}\mu\|_2^2 + \alpha_0^{-1}\sum_{n=1}^{N_e N_a}\gamma_n + d}, \tag{10.40}$$

with $\gamma_n = 1 - \alpha_n^{-1}\Sigma_{nn}$ and c, d are defined in Eq.(10.31). Here, M_a is the number of virtual antennas in the coarray. Most entries of μ and Σ converge to very small values, which implies that the posterior for these x_n becomes strongly peaked at zero. As a result, these x_n are zeros and, hence, sparsity is realized. The source power in the direction $[\hat{\theta}_{n_1}, \hat{\phi}_{n_2}]$ with $n = N_a * (n_1 - 1) + n_2$ is estimated by $\sigma_n^2 = |\mu(n)|$.

10.3.3 Simulations

Extensive simulation results are presented to validate the proposed SMV-BCS method.

10.3.3.1 Sparse Linear Array

We use two sparse linear arrays with 10 antennas with configurations $[0, 1, 3, 6, 13, 20, 27, 31, 35, 36]\lambda/2$ for MRA and $[0, 1, 6, 10, 23, 26, 34, 41, 53, 55]\lambda/2$ for MHA respectively [11, 31]. We set the BCS parameters as $\rho = c = d = 1e^{-4}$ and initialize $\alpha_0 = 100/\mathrm{Var}(\tilde{\mathbf{y}})$, $\alpha = |\tilde{\Phi}^H\tilde{\mathbf{y}}|$. The number of time snapshots is 2000 for covariance matrix estimation. The performance comparison of the different techniques reported in this section also holds when using higher and lower number

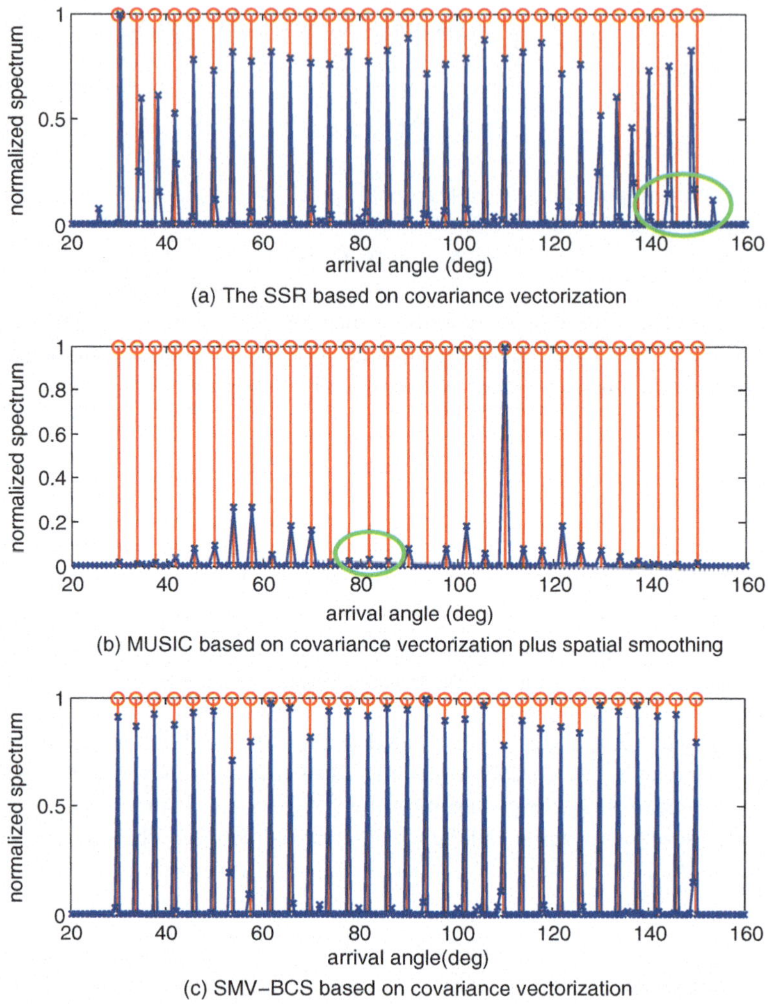

Fig. 10.10 Sensing spectra of three DOA estimation algorithms for fully augmentable arrays

of snapshots. The MRA is fully augmentable with the coarray being a 73-antenna uniform linear array (ULA). Therefore, in addition to the SSR and SMV-BCS, the MUSIC equipped with spatial smoothing can also be utilized for DOA estimation. Consider 31 sources uniformly distributed within the range $[30^\circ, 150^\circ]$ with an angular interval 4° and signal-to-noise ratio (SNR) 20dB. The sensing spectra of the three methods are shown in Fig. 10.10. We can observe that there are spurious and biased peaks in the spectrum of the SSR method. Some spectral lines are too weak to identify for the MUSIC pseudo-spectrum. The SMV-BCS based on the covariance vectoriza-

Table 10.1 DOA estimation MSE of MUSIC, SSR and BCS for both MRA and MHA

Arrays	MUSIC(°)	SSR(°)	SMV-BCS(°)
MRA	2.35	1.07	0.012
MHA	–	4.78	0.035

tion exhibits the best performance, and clearly shows 31 peaks with almost the same power level.

In order to examine the estimation accuracy, we utilize the mean squared error (MSE) as the metric defined as,

$$\mathrm{MSE} = \frac{1}{K}\sum_{k=1}^{K}(\theta_k - \tilde{\theta}_k)^2, \tag{10.41}$$

where $\tilde{\theta}_k$ denotes the estimated angle of the kth source. The MSE values of the three considered methods, MUSIC, the SSR and the SMV-BCS, are listed in the first row of Table 10.1 with 200 Monte-Carlo runs. It is evident that the MUSIC approach exhibits the worst estimation performance, whereas the proposed SMV-BCS approach based on covariance vectorization demonstrates much higher estimation accuracy than the other two methods.

The coarray for the 10-antenna MHA is a 91-antenna linear array with 10 holes, thus only the SSR and the SMV-BCS can be used for DOA estimation. Consider 41 sources uniformly distributed within the range [30°, 150°] with an angular interval 3° and 20dB SNR. The sensed spectra of the two methods are shown in Fig. 10.11. Again, the proposed method is superior to the SSR and enjoys high resolution DOA estimation for this case of closely spaced sources. Similar to the MRA, we also utilize 200 Monte-Carlo runs to calculate the estimation MSE, which is provided in the second row of Table 10.1. It is clear that the proposed method presents high estimation accuracy for partially augmentable arrays as well. Note that we do not present the DOA estimation performance based on the physical array with its limited degrees of freedom, as it cannot deal with more sources than the number of antennas.

10.3.3.2 Circular Arrays

A circular array with a single antenna in the center and 7 antennas uniformly distributed along the circumference, as indicated by filled dots in Fig. 10.12, is typically employed in satellite navigation applications. The corresponding coarray is a four-circle concentric array, also shown in Fig. 10.12. Suppose there are 10 jammers arriving from [10°, 20°, 30°, 10°, 20°, 30°, 10°, 20°, 30°, 38°] in elevation and uniformly distributed in azimuth sector [30°, 300°] with an angular interval 30° and 20dB interference to noise ratio. The two-dimensional (2-D) DOA estimation in both elevation and azimuth is shown in Fig. 10.13, where the radial direction denotes the

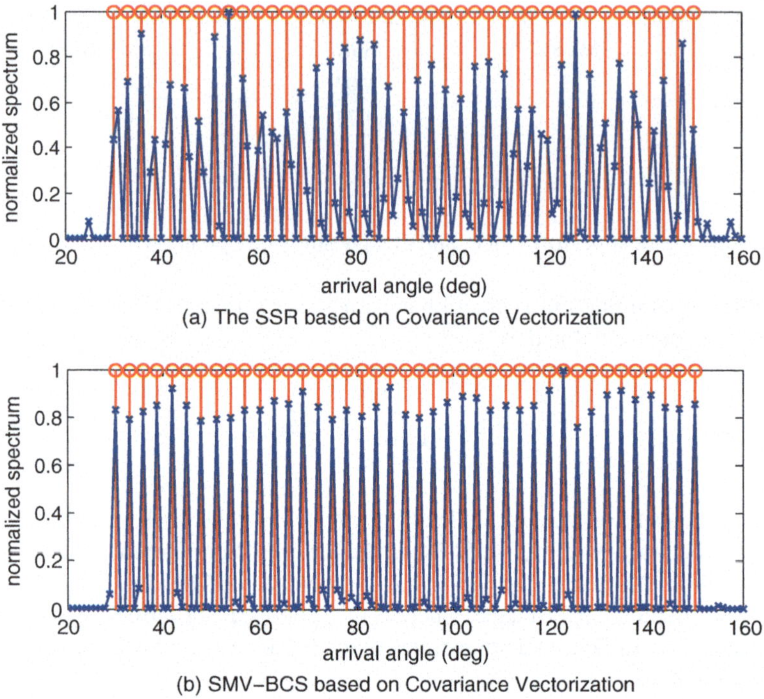

Fig. 10.11 Sensing spectra of two DOA estimation algorithms for partially augmentable arrays

elevation angle and the circumference direction is the azimuth. We can observe that the estimated 2-D angles coincide with the true angles which further validates the effectiveness of the proposed approach.

10.4 Summary

In this chapter, we studied the sparsity sensed for enhanced DOA estimation in the coarray domain employing both fully augmentable arrays and partially augmentable arrays. The utility of difference coarray could increase the number of estimated signals, which was far more than the number of physical antennas. In Sect. 10.2, a novel DOA estimation approach utilizing the MOP based on covariance vectorization was proposed. Although the MUSIC algorithm and sparse reconstruction had been employed previously for DOA estimation based on the coarray, the proposed MOP signal spectrum and polynomial rooting were preferable due to their simplicity and comparable performance. Furthermore, the MOP based polynomial rooting was superior in estimating off-grid DOAs accurately. In Sect. 10.3, we utilized a probabilistic Bayesian inference method for DOA estimation based on the difference

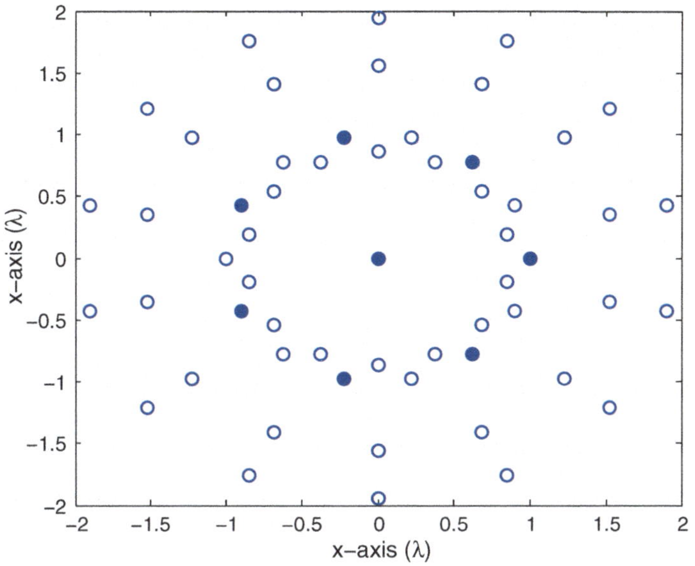

Fig. 10.12 The 8-antenna circular array and its difference coarray

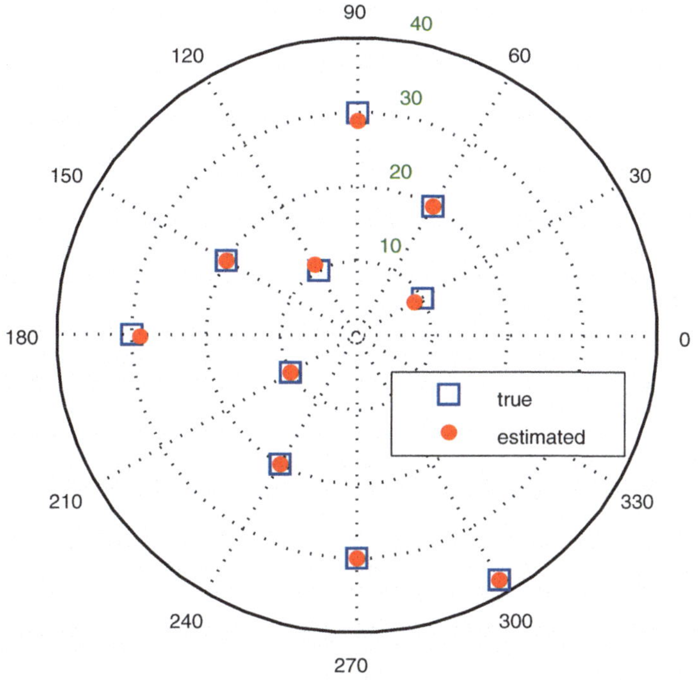

Fig. 10.13 2-D DOA estimation: the square and circle indicate the true and estimated directions respectively

coarray. We combined the SMV-BCS approach with covariance vectorization for both fully and partially augmentable arrays. Simulation results showed that the proposed method can overcome the shortcomings of the SSR method and the MUSIC, such as spurious peaks and inaccurate power estimation. However, this performance superiority has to be weighted against slow convergence of BCS for high dimension dictionary matrix.

References

1. H. Krim and M. Viberg. Two decades of array signal processing research: the parametric approach. *Signal Processing Magazine, IEEE*, 13(4):67–94, 1996.
2. P. Stoica and A. Nehorai. Performance study of conditional and unconditional direction-of-arrival estimation. *Acoustics, Speech and Signal Processing, IEEE Transactions on*, 38(10):1783–1795, Oct 1990.
3. D. Malioutov, M. Cetin, and A.S. Willsky. A sparse signal reconstruction perspective for source localization with sensor arrays. *Signal Processing, IEEE Transactions on*, 53(8):3010–3022, 2005.
4. E. J. Candes, J. K. Romberg, and T. Tao. Stable signal recovery from incomplete and inaccurate measurements. *Communications on pure and applied mathematics*, 59(8):1207–1223, 2006.
5. S.U. Pillai, Y. Bar-Ness, and F. Haber. A new approach to array geometry for improved spatial spectrum estimation. *Proceedings of the IEEE*, 73(10):1522–1524, Oct 1985.
6. Y.I. Abramovich, D.A. Gray, A.Y. Gorokhov, and N.K. Spencer. Positive-definite Toeplitz completion in DOA estimation for nonuniform linear antenna arrays. I. Fully augmentable arrays. *Signal Processing, IEEE Transactions on*, 46(9):2458–2471, Sep 1998.
7. Y.I. Abramovich, N.K. Spencer, and A.Y. Gorokhov. Positive-definite Toeplitz completion in DOA estimation for nonuniform linear antenna arrays. II. Partially augmentable arrays. *Signal Processing, IEEE Transactions on*, 47(6):1502–1521, Jun 1999.
8. P. Pal and P.P. Vaidyanathan. Nested arrays: A novel approach to array processing with enhanced degrees of freedom. *Signal Processing, IEEE Transactions on*, 58(8):4167–4181, 2010.
9. P. Pal and P.P. Vaidyanathan. Correlation-aware techniques for sparse support recovery. In *Statistical Signal Processing Workshop (SSP), 2012 IEEE*, pages 53–56, Aug 2012.
10. P. Pal and P.P. Vaidyanathan. On application of LASSO for sparse support recovery with imperfect correlation awareness. In *Signals, Systems and Computers (ASILOMAR), 2012 Conference Record of the Forty Sixth Asilomar Conference on*, pages 958–962, Nov 2012.
11. A. Moffet. Minimum-redundancy linear arrays. *Antennas and Propagation, IEEE Transactions on*, 16(2):172–175, 1968.
12. R.T. Hoctor and S.A. Kassam. The unifying role of the coarray in aperture synthesis for coherent and incoherent imaging. *Proceedings of the IEEE*, 78(4):735–752, Apr 1990.
13. Y.I. Abramovich, N.K. Spencer, and A.Y. Gorokhov. Detection-estimation of more uncorrelated gaussian sources than sensors in nonuniform linear antenna arrays. II. Partially augmentable arrays. *Signal Processing, IEEE Transactions on*, 51(6):1492–1507, June 2003.
14. P.P. Vaidyanathan and P. Pal. Sparse sensing with co-prime samplers and arrays. *Signal Processing, IEEE Transactions on*, 59(2):573–586, Feb 2011.
15. A Barabell. Improving the resolution performance of eigenstructure-based direction-finding algorithms. In *Acoustics, Speech, and Signal Processing, IEEE International Conference on (ICASSP)*, volume 8, pages 336–339, Apr 1983.
16. A.H. Tewfik and W. Hong. On the application of uniform linear array bearing estimation techniques to uniform circular arrays. *Signal Processing, IEEE Transactions on*, 40(4):1008–1011, 1992.

17. R.L. Fante and JJ Vaccaro. Wideband cancellation of interference in a GPS receive array. *Aerospace and Electronic Systems, IEEE Transactions on*, 36(2):549–564, 2000.
18. Y.D. Zhang, M.G. Amin, and B. Himed. Sparsity-based DOA estimation using co-prime arrays. In *Acoustics, Speech and Signal Processing (ICASSP), 2013 IEEE International Conference on*, pages 3967–3971, May 2013.
19. X. Wang, E. Aboutanios, M. Trinkle, and M.G. Amin. Reconfigurable adaptive array beamforming by antenna selection. *Signal Processing, IEEE Transactions on*, 62(9):2385–2396, May 2014.
20. S. Ji, Y. Xue, and L. Carin. Bayesian compressive sensing. *Signal Processing, IEEE Transactions on*, 56(6):2346–2356, 2008.
21. M.E. Tipping. Sparse Bayesian learning and the relevance vector machine. *The Journal of Machine Learning Research*, 1:211–244, 2001.
22. K. Yang, Z. Zhao, and Q. H. Liu. Fast pencil beam pattern synthesis of large unequally spaced antenna arrays. *Antennas and Propagation, IEEE Transactions on*, 61(2):627 –634. 2013.
23. M. Carlin, P. Rocca, G. Oliveri, F. Viani, and A. Massa. Directions-of-arrival estimation through Bayesian compressive sensing strategies. *Antennas and Propagation, IEEE Transactions on*, 61(7):3828–3838, 2013.
24. Q. Wu, Y.D. Zhang, M.G. Amin, and B. Himed. Complex multitask Bayesian compressive sensing. In *Acoustics, Speech and Signal Processing (ICASSP), 2014 IEEE International Conference on*, pages 3375–3379, May 2014.
25. J. Li and P. Stoica. Efficient mixed-spectrum estimation with applications to target feature extraction. *IEEE Transactions on Signal Processing*, 44(2):281–295, 1996.
26. R.T. Compton. The power-inversion adaptive array: Concept and performance. *Aerospace and Electronic Systems, IEEE Transactions on*, AES-15(6):803–814, Nov 1979.
27. S. U. Pillai and B. H. Kwon. Forward/backward spatial smoothing techniques for coherent signal identification. *Acoustics, Speech and Signal Processing, IEEE Transactions on*, 1989.
28. S.U. Pillai and F. Haber. Statistical analysis of a high resolution spatial spectrum estimator utilizing an augmented covariance matrix. *Acoustics, Speech and Signal Processing, IEEE Transactions on*, 35(11):1517–1523, Nov 1987.
29. G.S. Bloom and S.W. Golomb. Application of numbered undirected graphs. *Proceedings of the IEEE*, 65(4):562–570, 1977.
30. Y.I. Abramovich, D.A. Gray, A.Y. Gorokhov, and N.K. Spencer. Comparison of DOA estimation performance for various types of sparse antenna array geometries. In *Proc. EUSIPCO*, volume 96, pages 915–918, 1996.
31. E. Vertatschitsch and S. Haykin. Nonredundant arrays. *Proceedings of the IEEE*, 74(1):217–217, Jan 1986.

Appendix A
Linear and Matrix Algebranning

Important results from linear and algebra theory are reviewed in Appendix A. In the discussions to follow, it is assumed that the reader already has some familiarity with these topics. The specific concepts to be described are used heavily throughout the book.

A.1 Definitions

Consider an $m \times n$ matrix $\mathbf{A}$ with elements $a_{i,j}, i = 1, 2, \ldots, m; j = 1, 2, \ldots, n$. A shorthand notation for describing $\mathbf{A}$ is,

$$[\mathbf{A}]_{ij} = a_{ij}. \tag{A.1}$$

When $n = 1$, the matrix $\mathbf{A}$ becomes an $m \times 1$ column vector $\mathbf{a}$; when $m = 1$, the matrix $\mathbf{A}$ then becomes a $1 \times n$ row vector $\tilde{\mathbf{a}}$. The *transpose* of $\mathbf{A}$, which is denoted by $\mathbf{A}^T$, is defined as the $n \times m$ matrix with element $a_{j,i}$ or

$$\left[\mathbf{A}^T\right]_{ij} = a_{ji}. \tag{A.2}$$

A *square* matrix is one for which $m = n$. A square matrix is *symmetric*, that is $\mathbf{A}^T = \mathbf{A}$.

The *rank* of a matrix is the number of linearly independent rows or columns, whichever is less. The *inverse* of a square $n \times n$ matrix is the square $n \times n$ matrix $\mathbf{A}^{-1}$ for which,

$$\mathbf{A}^{-1}\mathbf{A} = \mathbf{A}\mathbf{A}^{-1} = \mathbf{I}, \tag{A.3}$$

where $\mathbf{I}$ is the $n \times n$ identity matrix. The inverse will exist if and only if the rank of $\mathbf{A}$ is n. If the inverse does not exist, then $\mathbf{A}$ is *singular*.

X. Wang et al., *Sparse Sensing and Sparsity Sensed in Multi-sensor Array Applications*, https://doi.org/10.1007/978-981-99-9558-5

The *determinant* of a square $n \times n$ matrix $\mathbf{A}$ is denoted by $\det(\mathbf{A})$. It is computed as,

$$\det(\mathbf{A}) = \sum_{j=1}^{n} a_{ij} C_{ij}, \tag{A.4}$$

where

$$C_{ij} = (-1)^{i+j} M_{ij}, \tag{A.5}$$

and M_{ij} is the determinant of the submatrix of $\mathbf{A}$ obtained by deleting the ith row and jth column and is termed as the *minor* of a_{ij}. C_{ij} is the *cofactor* of a_{ij}. Note that any choice of i for $i = 1, 2, \ldots, n$ will yield the same value for $\det(A)$.

A *quadratic form* Q is defined as,

$$Q = \sum_{i=1}^{n} \sum_{j=1}^{n} a_{ij} x_i x_j. \tag{A.6}$$

In defining the quadratic form, it is assumed that $a_{ji} = a_{ij}$. Q may also be expressed as matrix-vector from as,

$$Q = \mathbf{x}^T \mathbf{A} \mathbf{x}, \tag{A.7}$$

where $\mathbf{x} = [x_1, x_2, \ldots, x_n]^T$ and $\mathbf{A}$ is a square $n \times n$ matrix with $a_{ji} = a_{ij}$ or $\mathbf{A}$ is a symmetric matrix.

A square $n \times n$ matrix $\mathbf{A}$ is *positive semi-definite matrix* if $\mathbf{A}$ is symmetric and

$$\mathbf{x}^T \mathbf{A} \mathbf{x} \geq 0, \tag{A.8}$$

for all real $\mathbf{x} \neq \mathbf{0}$. If the quadratic form is strictly positive, then $\mathbf{A}$ is *positive definite*. When referring to a matrix as positive matrix or positive semi-definite, it is always assumed that the matrix is symmetric.

The *trace* of a square $n \times n$ matrix is the sum of its diagonal elements or

$$\mathrm{tr}(\mathbf{A}) = \sum_{i=1}^{n} a_{ii}. \tag{A.9}$$

A partitioned $n \times n$ matrix $\mathbf{A}$ is the one that expressed in terms of its submatrices. An example is the 2×2 partitioning,

$$\mathbf{A} = \begin{bmatrix} \mathbf{A}_{11} & \mathbf{A}_{12} \\ \mathbf{A}_{21} & \mathbf{A}_{22} \end{bmatrix}. \tag{A.10}$$

Each "element" $\mathbf{A}_{ij}$ is a submatrix of $\mathbf{A}$. The dimensions of the partitions are given as

$$\begin{bmatrix} k \times l & k \times (n-l) \\ (m-k) \times l & (m-k) \times (n-l) \end{bmatrix}. \tag{A.11}$$

A.2 Special Matrices

A *diagonal* matrix is a square $n \times n$ matrix with $a_{ij} = 0$ for $i \neq j$, or all elements off the principle diagonal are zero. A diagonal matrix appears as

$$\mathbf{A} = \begin{bmatrix} a_{11} & 0 & \cdots & 0 \\ 0 & a_{22} & \cdots & 0 \\ \vdots & \vdots & \ddots & \vdots \\ 0 & 0 & \cdots & a_{nn} \end{bmatrix}. \tag{A.12}$$

A diagonal matrix will sometimes be denoted by $\text{diag}(a_{11}, a_{22}, \ldots, a_{nn})$. The inverse of a diagonal matrix is found by simple inverting each element on the principal diagonal.

A generalization of the diagonal matrix is the square $n \times n$ *block-diagonal* matrix

$$\mathbf{A} = \begin{bmatrix} \mathbf{A}_{11} & 0 & \cdots & 0 \\ 0 & \mathbf{A}_{22} & \cdots & 0 \\ \vdots & \vdots & \ddots & \vdots \\ 0 & 0 & \cdots & \mathbf{A}_{kk} \end{bmatrix}, \tag{A.13}$$

in which all submatrices $\mathbf{A}_{ii}$ are square and the other submatrices are identically zero. The dimensions of the submatrices need not be identical. For instance, if $k = 2$, $\mathbf{A}_{11}$ might have dimension 2×2 while $\mathbf{A}_{22}$ might be a scalar. If all $\mathbf{A}_{ii}$ are nonsingular, then the inverse is easily found as

$$\mathbf{A}^{-1} = \begin{bmatrix} \mathbf{A}_{11}^{-1} & 0 & \cdots & 0 \\ 0 & \mathbf{A}_{22}^{-1} & \cdots & 0 \\ \vdots & \vdots & \ddots & \vdots \\ 0 & 0 & \cdots & \mathbf{A}_{kk}^{-1} \end{bmatrix}. \tag{A.14}$$

Also, the determinant is

$$\det(A) = \prod_{i=1}^{k} \det(\mathbf{A}_{ii}). \tag{A.15}$$

A square $n \times n$ matrix is *orthogonal* if

$$\mathbf{A}^{-1} = \mathbf{A}^{T}. \tag{A.16}$$

For a matrix to be orthogonal the columns (and rows) must be orthonormal or if

$$\mathbf{A} = [\mathbf{a}_1, \mathbf{a}_2, \ldots, \mathbf{a}_n], \tag{A.17}$$

where $\mathbf{a}_i$ denotes the ith column, the conditions

$$\mathbf{a}_i^T \mathbf{a}_j = \begin{cases} 0, & \text{for } i \neq j \\ 1, & \text{for } i = j, \end{cases} \tag{A.18}$$

must be satisfied. An important example of an orthogonal matrix in modeling of data by a sum of harmonically related sinusoids or by a discrete Fourier series. As an example, for n even

$$\mathbf{A} = \frac{1}{\sqrt{\frac{n}{2}}} \begin{bmatrix} \frac{1}{\sqrt{2}} & 1 & \cdots & \frac{1}{\sqrt{2}} & 0 & \cdots & 0 \\ \frac{1}{\sqrt{2}} & \cos\frac{2\pi}{n} & \cdots & \frac{1}{\sqrt{2}}\cos\frac{2\pi(\frac{n}{2})}{n} & \sin\frac{2\pi}{n} & \cdots & \sin\frac{2\pi(\frac{n}{2}-1)}{n} \\ \vdots & \vdots & \vdots & \vdots & \vdots & \vdots & \vdots \\ \frac{1}{\sqrt{2}} & \cos\frac{2\pi(n-1)}{n} & \cdots & \frac{1}{\sqrt{2}}\cos\frac{2\pi\frac{n}{2}(n-1)}{n} & \sin\frac{2\pi(n-1)}{n} & \cdots & \sin\frac{2\pi(\frac{n}{2}-1)(n-1)}{n} \end{bmatrix} \tag{A.19}$$

is an orthogonal matrix. This follows from the orthogonality relationship for $i, j = 0, 1, \ldots, n/2$,

$$\sum_{k=0}^{n-1} \cos\frac{2\pi ki}{n} \cos\frac{2\pi kj}{n} = \begin{cases} 0, \ i \neq j \\ \frac{n}{2}, \ i = j = 1, 2, \ldots, \frac{n}{2} - 1 \\ n, \ i = j = 0, \frac{n}{2} \end{cases} \tag{A.20}$$

and for $i, j = 1, 2, \ldots, n/2 - 1$,

$$\sum_{k=0}^{n-1} \sin\frac{2\pi ki}{n} \sin\frac{2\pi kj}{n} = \frac{n}{2}\delta_{ij}, \tag{A.21}$$

and finally for $i = 0, 1, \ldots, n/2$; $j = 1, 2, \ldots, n/2 - 1$,

$$\sum_{k=0}^{n-1} \cos\frac{2\pi ki}{n} \sin\frac{2\pi kj}{n} = 0. \tag{A.22}$$

These orthogonality relationships may be proven by expressing the sines and cosines in terms of complex exponentials and using the result,

$$\sum_{k=0}^{n-1} \exp\left(j\frac{2\pi}{n}kl\right) = n\delta_l, \tag{A.23}$$

for $l = 0, 1, \ldots, n-1$ and

$$\delta_l = \begin{cases} 1 & l = 0, \\ 0 & l = 1, \ldots, n-1. \end{cases} \tag{A.24}$$

An *idempotent* matrix implies is a square $n \times n$ matrix that satisfies,

$$\mathbf{A}^2 = \mathbf{A}. \tag{A.25}$$

This condition implies that $\mathbf{A}^l = \mathbf{A}$ for $l \geq 1$. An example is the *projection matrix*,

$$\mathbf{A} = \mathbf{H}\left(\mathbf{H}^T\mathbf{H}\right)^{-1}\mathbf{H}^T \tag{A.26}$$

where $\mathbf{H}$ is an $m \times n$ full rank matrix with $m > n$.

A square $n \times n$ *Toeplitz* matrix is defined as

$$[\mathbf{A}]_{ij} = a_{i-j}, \tag{A.27}$$

or

$$\mathbf{A} = \begin{bmatrix} a_0 & a_{-1} & a_{-2} & \cdots & a_{-(n-1)} \\ a_1 & a_0 & a_{-1} & \cdots & a_{-(n-2)} \\ \vdots & \vdots & \vdots & \ddots & \vdots \\ a_{n-1} & a_{n-2} & a_{n-3} & \cdots & a_0 \end{bmatrix}. \tag{A.28}$$

Each element along a northwest-southeast diagonal is the same. If in addition, $a_{-k} = a_k$, then $\mathbf{A}$ is *symmetric Toeplitz*.

A.3 Matrix Manipulation and Formulas

Some useful formulas for the algebraic manipulation of the matrices are summarized in Appendix B. For $n \times n$ matrices $\mathbf{A}$ and $\mathbf{B}$, the following relationships are useful.

$$\begin{aligned}
(\mathbf{AB})^T &= \mathbf{B}^T\mathbf{A}^T, \\
\left(\mathbf{A}^T\right)^{-1} &= \left(\mathbf{A}^{-1}\right)^T, \\
(\mathbf{AB})^{-1} &= \mathbf{B}^{-1}\mathbf{A}^{-1}, \\
\det(\mathbf{A}^T) &= \det(\mathbf{A}), \\
\det(c\mathbf{A}) &= c^n\det(\mathbf{A}), \quad c \text{ is a scalar}, \\
\det(\mathbf{AB}) &= \det(\mathbf{A})\det(\mathbf{B}), \\
\det(\mathbf{A}^{-1}) &= \frac{1}{\det(\mathbf{A})}, \\
\mathrm{tr}(\mathbf{AB}) &= \mathrm{tr}(\mathbf{BA}), \\
\mathrm{tr}(\mathbf{A}^T\mathbf{B}) &= \sum_{i=1}^{n}\sum_{j=1}^{n}[\mathbf{A}]_{ij}[\mathbf{B}]_{ij}.
\end{aligned}$$

In differentiating linear and quadratic forms, the following formulas for the gradient are useful.

$$\frac{\partial \mathbf{b}^T\mathbf{x}}{\partial \mathbf{x}} = \mathbf{b},$$
$$\frac{\partial \mathbf{x}^T\mathbf{A}\mathbf{x}}{\partial \mathbf{x}} = 2\mathbf{A}\mathbf{x},$$

where it is assumed that $\mathbf{A}$ is a symmetric matrix. Also, for vectors $\mathbf{x}$ and $\mathbf{y}$ we have,

$$\mathbf{y}^T\mathbf{x} = \text{tr}(\mathbf{x}\mathbf{y}^T). \tag{A.29}$$

It is frequently necessary to determine the inverse of a matrix analytically. To do so, one can make use of the following formula. The inverse of a square $n \times n$ matrix is

$$\mathbf{A}^{-1} = \frac{\mathbf{C}^T}{\det(\mathbf{A})}, \tag{A.30}$$

where $\mathbf{C}$ is the square $n \times n$ matrix of cofactors of $\mathbf{A}$. The cofactor matrix is defined by

$$[\mathbf{C}]_{ij} = (-1)^{i+j} M_{ij}, \tag{A.31}$$

where M_{ij} is the minor of a_{ij} obtained by deleting the ith row and jth column of $\mathbf{A}$.

Another formula that is quite useful is the *matrix inversion lemma*,

$$(\mathbf{A}+\mathbf{BCD})^{-1} = \mathbf{A}^{-1} - \mathbf{A}^{-1}\mathbf{B}\left(\mathbf{D}\mathbf{A}^{-1}\mathbf{B}+\mathbf{C}^{-1}\right)^{-1}\mathbf{D}\mathbf{A}^{-1} \tag{A.32}$$

where it is assumed that $\mathbf{A}$ is $n \times n$, $\mathbf{B}$ is $n \times m$, $\mathbf{C}$ is $m \times m$, and $\mathbf{D}$ is $m \times n$, and that the indicated inverse exist. A special case, known as *Woodbury's identity* results, for $\mathbf{B}$ an $n \times 1$ column vector $\mathbf{u}$, $\mathbf{C}$ a scalar of unity, and $\mathbf{D}$ a $1 \times n$ row vector $\mathbf{u}^T$. Then,

$$\left(\mathbf{A}+\mathbf{u}\mathbf{u}^T\right)^{-1} = \mathbf{A}^{-1} - \frac{\mathbf{A}^{-1}\mathbf{u}\mathbf{u}^T\mathbf{A}^{-1}}{1+\mathbf{u}^T\mathbf{A}^{-1}\mathbf{u}}. \tag{A.33}$$

Partitioned matrices may be manipulated according to the usual rule of matrix algebra by considering each submatrix as an element. For multiplication of partitioned matrices the submatrices that are multiplied together must be conformable. As an illustration, for 2×2 partitioned matrices,

$$\mathbf{AB} = \begin{bmatrix}\mathbf{A}_{11} & \mathbf{A}_{12}\\ \mathbf{A}_{21} & \mathbf{A}_{22}\end{bmatrix}\begin{bmatrix}\mathbf{B}_{11} & \mathbf{B}_{12}\\ \mathbf{B}_{21} & \mathbf{B}_{22}\end{bmatrix}, \tag{A.34}$$

$$= \begin{bmatrix}\mathbf{A}_{11}\mathbf{B}_{11}+\mathbf{A}_{12}\mathbf{B}_{21} & \mathbf{A}_{11}\mathbf{B}_{12}+\mathbf{A}_{12}\mathbf{B}_{22}\\ \mathbf{A}_{21}\mathbf{B}_{11}+\mathbf{A}_{22}\mathbf{B}_{21} & \mathbf{A}_{21}\mathbf{B}_{12}+\mathbf{A}_{22}\mathbf{B}_{22}\end{bmatrix}. \tag{A.35}$$

The transposition of a partitioned matrix is formed by transposing the submatrices of the matrix and applying T to each submatrix. For a 2×2 partitioned matrix,

$$\begin{bmatrix} \mathbf{A}_{11} & \mathbf{A}_{12} \\ \mathbf{A}_{21} & \mathbf{A}_{22} \end{bmatrix}^T = \begin{bmatrix} \mathbf{A}_{11}^T & \mathbf{A}_{21}^T \\ \mathbf{A}_{12}^T & \mathbf{A}_{22}^T \end{bmatrix}. \tag{A.36}$$

The extension of these properties to arbitrary partitioning is straightforward. Determination of the inverses and determinants of partitioned matrices is facilitated by employing the following formulas. Let $\mathbf{A}$ is a square $n \times n$ matrix partitioned as

$$\mathbf{A} = \begin{bmatrix} \mathbf{A}_{11} & \mathbf{A}_{12} \\ \mathbf{A}_{21} & \mathbf{A}_{22} \end{bmatrix} = \begin{bmatrix} k \times k & k \times (n-k) \\ (n-k) \times k & (n-k) \times (n-k) \end{bmatrix}. \tag{A.37}$$

Then,

$$\mathbf{A}^{-1} = \begin{bmatrix} \left(\mathbf{A}_{11} - \mathbf{A}_{12}\mathbf{A}_{22}^{-1}\mathbf{A}_{21}\right)^{-1} & -\left(\mathbf{A}_{11} - \mathbf{A}_{12}\mathbf{A}_{22}^{-1}\mathbf{A}_{21}\right)^{-1}\mathbf{A}_{12}\mathbf{A}_{22}^{-1} \\ -\left(\mathbf{A}_{22} - \mathbf{A}_{21}\mathbf{A}_{11}^{-1}\mathbf{A}_{12}\right)^{-1}\mathbf{A}_{21}\mathbf{A}_{11}^{-1} & \left(\mathbf{A}_{22} - \mathbf{A}_{21}\mathbf{A}_{11}^{-1}\mathbf{A}_{12}\right)^{-1} \end{bmatrix}, \tag{A.38}$$

where the inverses of $\mathbf{A}_{11}$ and $\mathbf{A}_{22}$ are assumed to exist.

A.4 Theorems

Some important theorems used throughout the book are summarized in Appendix C.

1. A square $n \times n$ matrix $\mathbf{A}$ is invertible (nonsingular) if and only if its columns (or rows) are linearly independent or, equivalently, if its determinant is nonzero. In such a case, $\mathbf{A}$ is *full rank*. Otherwise, it is singular.
2. A square $n \times n$ matrix $\mathbf{A}$ is positive definite if and only if

   a. it can be written as

   $$\mathbf{A} = \mathbf{C}\mathbf{C}^T \tag{A.39}$$

   where $\mathbf{C}$ is $n \times n$ full rank and hence invertible, or

   b. the principal minors are all positive. (The ith principle minor is the determinant of the submatrix formed by deleting all rows and columns with an index greater than i.) If $\mathbf{A}$ can be written as in (A.39), but $\mathbf{C}$ is not full rank or the principle minors are only nonnegative, the $\mathbf{A}$ is positive semidefinite.

3. If $\mathbf{A}$ is positive definite, then the inverse exists and may be found from (A.39) as $\mathbf{A}^{-1} = (\mathbf{C}^{-1})^T(\mathbf{C}^{-1})$.
4. Let $\mathbf{A}$ be positive definite. If $\mathbf{B}$ is an $m \times n$ matrix of full rank with $m \le n$, the $\mathbf{BAB}^T$ is also positive definite.

5. If **A** is positive definite (positive semidefinite), then
   a. the diagonal elements are positive (nonnegative),
   b. the determinant of **A**, which is a principal minor, is positive (nonnegative).

A.5 Eigendecomposition of Matrices

An *eigenvector* of a square $n \times n$ matrix $\mathbf{A}$ is an $n \times 1$ vector $\mathbf{v}$ satisfying,

$$\mathbf{A}\mathbf{v} = \lambda \mathbf{v}, \tag{A.40}$$

for some scalar λ, which may be complex. λ is the *eigenvalue* of $\mathbf{A}$ corresponding to the eigenvector $\mathbf{v}$. It is assumed that the eigenvector is normalized to have unit length or $\mathbf{v}^T\mathbf{v} = 1$. If $\mathbf{A}$ is symmetric, then one can always find n linearly independent eigenvectors, although they will not in general be unique. An example is the identity matrix for which any vector is an eigenvector with eigenvalue 1. If $\mathbf{A}$ is symmetric, then the eigenvectors corresponding to the distinct eigenvalues are orthogonal or $\mathbf{v}_i^T\mathbf{v}_j = \delta_{ij}$, and the eigenvalues are real. If, furthermore, the matrix is positive definite (positive semidefinite), then the eigenvalues are positive (nonnegative). For a positive semidefinite matrix, the rank is equal to the number of nonzero eigenvalues.

The defining relation of (A.40) can also be written as

$$\mathbf{A}\left[\, \mathbf{v}_1 \; \mathbf{v}_2 \; \ldots \; \mathbf{v}_n \,\right] = \left[\, \lambda_1\mathbf{v}_1 \; \lambda_2\mathbf{v}_2 \; \ldots \; \lambda_n\mathbf{v}_n \,\right], \tag{A.41}$$

or

$$\mathbf{A}\mathbf{V} = \mathbf{V}\Lambda, \tag{A.42}$$

where

$$\begin{aligned} \mathbf{V} &= [\mathbf{v}_1, \mathbf{v}_2, \ldots, \mathbf{v}_n], \\ \Lambda &= \mathrm{diag}(\lambda_1, \lambda_2, \ldots, \lambda_n). \end{aligned} \tag{A.43}$$

If $\mathbf{A}$ is symmetric so that the eigenvectors corresponding to the distinct eigenvalues are orthogonal and the remaining eigenvectors are chosen to yield an orthonormal eigenvector set, then $\mathbf{V}$ is an orthogonal matrix. The matrix $\mathbf{V}$ is termed the *modal matrix*. As such, its inverse is $\mathbf{V}^T$, so that (A.42) becomes

$$\mathbf{A} = \mathbf{V}\Lambda\mathbf{V}^T = \sum_{i=1}^{n} \lambda_i \mathbf{v}_i \mathbf{v}_i^T. \tag{A.44}$$

Also, the inverse is easily determined as

$$\begin{aligned}\mathbf{A}^{-1} &= \mathbf{V}^{T-1}\lambda^{-1}\mathbf{V}^{-1}, \\ &= \mathbf{V}\lambda^{-1}\mathbf{V}^{T}, \\ &= \sum_{i=1}^{n} \frac{1}{\lambda_i}\mathbf{v}_i\mathbf{v}_i^T .\end{aligned} \tag{A.45}$$

A final useful relationship follows form (A.42) as

$$\begin{aligned}\det(A) &= \det(\mathbf{V})\det(\Lambda)\det(\mathbf{V}^{-1}), \\ &= \det(\Lambda) = \prod_{i=1}^{n} \lambda_i .\end{aligned} \tag{A.46}$$

It should be noted that the modal matrix $\mathbf{V}$, which diagonalizes a symmetric matrix, can be employed to decorrelate a set of random variables. Assume that $\mathbf{x}$ is a vector random variable with zero mean and covariance matrix $\mathbf{C}_x$. Then $\mathbf{y} = \mathbf{V}^T\mathbf{x}$ for $\mathbf{V}$ the modal matrix of $\mathbf{C}_x$ is a vector random variable with zero mean and diagonal covariance matrix. The follows for (A.42) since the covariance matrix of $\mathbf{y}$ is,

$$\begin{aligned}\mathbf{C}_y &= \mathbb{E}\{\mathbf{y}\mathbf{y}^T\} = \mathbb{E}\{\mathbf{V}^T\mathbf{x}\mathbf{x}^T\mathbf{V}\}, \\ &= \mathbf{V}^T\mathbb{E}\{\mathbf{x}\mathbf{x}^T\} = \mathbf{V}^T\mathbf{C}_x\mathbf{V}, \\ &= \Lambda,\end{aligned} \tag{A.47}$$

where Λ is a diagonal matrix.

A.6 Inequalities

The Cauchy-Schwarz inequality can be used to simplify maximization problems and provide explicit solution. For two vectors $\mathbf{x}$ and $\mathbf{y}$ it asserts that

$$\left(\mathbf{y}^T\mathbf{x}\right)^2 \leq \left(\mathbf{y}^T\mathbf{y}\right)\left(\mathbf{x}^T\mathbf{x}\right), \tag{A.48}$$

with equality if and only if $\mathbf{y} = c\mathbf{x}$ for c an arbitrary constant. As applied to function $g(x)$ and $h(x)$, which we assume to be complex functions of a real variable for generality, it takes the form,

$$\left|\int g(x)h(x)dx\right|^2 \leq \int |g(x)|^2 dx \int |h(x)|^2 dx, \tag{A.49}$$

with equality if only if $g(x) = ch^*(x)$ for c an arbitrary complex constant.

Appendix B
Random Processes and Power Spectrum Estimation

The signals that we are dealing with can be divided into deterministic signal and random signal, where the former type of signal can be uniquely described by an explicit mathematical expression and the latter type of signal evolves in time in an unpredictable manner. In this appendix, we will review some fundamentals of random signals, starting from random variables and random vectors to random process.

B.1 Random Variables and Vectors

B.1.1 Random Variables

A *random variable (RV)*, usually denoted as upper case X, is a variable whose possible values are numerical outcomes of a random phenomenon. The numerical outcome is referred to as a realization of the random variable, denoted as lower case x. You may think about X as a place-holder for the unknown outcome, while x carries around our prior expectations about the likelihood of each possible outcome. There are two types of random variables, discrete and continuous. A discrete random variable takes on only a countable number of distinct values such as $\{x_1, x_2, \ldots\}$. A RV X is continuous if there exists a function f_X such that $f_X(x) \geq 0, \forall x$, and $\int_{-\infty}^{\infty} f_X(x)dx = 1$. Here, $f_X(x)$ is called *Probability Density Function (PDF)* of X, defined by,

$$f_X(x) = \lim_{\|\delta(x)\| \to 0} \frac{P(X \in \delta(x))}{\|\delta(x)\|}, \tag{B.1}$$

where $\delta(x)$ is a small interval which contains x, and $\|\delta(x)\|$ is its length.

The *cumulative distribution function, (CDF)* is the function $F_X : \mathbb{R} \to [0, 1]$ defined by,

$$F_X(x) = P(X \leq x). \tag{B.2}$$

X. Wang et al., *Sparse Sensing and Sparsity Sensed in Multi-sensor Array Applications*, https://doi.org/10.1007/978-981-99-9558-5

Clearly, we have that,

$$F_X(x) = \int_{-\infty}^{x} f_X(x)dt, \tag{B.3}$$

and $f_X(x) = F'_X(x)$ at all points x at which F_X is differentiable.

Next, we introduce two important moments of RV, as follows:

(1) *The mean (or expectation, or first moment) of* X, μ_X is defined by

$$\mu_X = E[X] = \int x f_X(x)dx. \tag{B.4}$$

It follows that expectation is a linear operator, with

$$E[\alpha X] = \alpha E[X], \quad \alpha \text{ is a deterministic constant,} \tag{B.5}$$

and

$$E\left[g_1(X) + g_2(X)\right] = E\left[g_1(X)\right] + E\left[g_2(Y)\right], \tag{B.6}$$

More generally, given any two random variables, X and Y,

$$E[\alpha X + \beta Y] = \alpha E[X] + \beta E[Y]. \tag{B.7}$$

(2) *The variance of* X is defined by,

$$\sigma_X^2 = \text{var}(X) = E\left[(X - \mu_X)^2\right] = E\left[X^2\right] - \mu_X^2 = \left(\int x^2 f_X(x)dx\right) - \mu_X^2. \tag{B.8}$$

More generally, if $g(x)$ is any function which maps x to y according to

$$y = g(x), \tag{B.9}$$

we may view the y values as outcomes of another random variable, Y. We write $Y = g(X)$ and we can easily show that the expectation of Y satisfies,

$$\mu_Y = E[Y] = \int y f_Y(y)dy = \int g(x) f_X(x)dx. \tag{B.10}$$

Thus, the variance of X is nothing other than the mean of another random variable, Y, whose outcome is $y = (x - \mu_X)^2$, whenever x is the outcome of X.

B.1.2 Independence and Correlation

Two random variables, X and Y, are said to be statistically independent if their distributions are separable, i.e.,

$$f_{X,Y}(x, y) = f_X(x) f_Y(y). \tag{B.11}$$

This definition agrees with our intuition concerning independent events. Specifically if two events, X and Y, have independent outcomes and we want to know the probability that $X \in \delta(x)$ and $Y \in \delta(y)$ simultaneously. We expect that this should be the product of the probability that $X \in \delta(x)$ and the probability that $Y \in \delta(y)$. That is, we expect independent events to satisfy,

$$\begin{aligned} f_{X,Y}(x, y) &\triangleq \lim_{\|\delta(x)\|, \|\delta(y)\| \to 0} \frac{P(X \in \delta(x) \text{ and } Y \in \delta(y))}{\|\delta(x)\| \cdot \|\delta(y)\|}, \\ &= \lim_{\|\delta(x)\|, \|\delta(y)\| \to 0} \frac{P(X \in \delta(x)) \cdot P(Y \in \delta(y))}{\|\delta(x)\| \cdot \|\delta(y)\|}, \\ &= f_X(x) \cdot f_Y(y). \end{aligned} \tag{B.12}$$

The notion of independence is perhaps even more intuitive if we consider conditional probability.

We write $f_{X|Y}(x|y)$ for the conditional probability distribution of X given that $Y = y$, where $f_{X|Y}(x|y)$ is defined by,

$$f_{X|Y}(x|y) \triangleq \frac{f_{X,Y}(x, y)}{f_Y(y)} = \frac{f_{X,Y}(x, y)}{\int f_{X,Y}(x, y)dx}. \tag{B.13}$$

It is easy to see that $f_{X|Y}(x|y) = f_X(x)$ if and only if X and Y are statistically independent. That is, when X and Y are independent, the outcome of Y is irrelevant to the distribution of X.

The correlation of two random variables, X and Y, is defined by

$$E[XY] = \iint xy f_{X,Y}(x, y)dxdy, \tag{B.14}$$

where $f_{X,Y}(\cdot, \cdot)$ is the joint distribution function (PDF) of X and Y.

The covariance of X and Y is defined by,

$$\text{cov}(X, Y) = E\left[(X - \mu_X)(Y - \mu_Y)\right] = E[XY] - \mu_X \mu_Y. \tag{B.15}$$

Two random variables whose covariance is 0 *are said to be "uncorrelated".*

The relationship between "being independent" and " being uncorrelated" is stated as follows: If X and Y are statistically independent, then they must be uncorrelated. To see this, observe that,

$$
\begin{aligned}
E[XY] &= \iint xy f_{X,Y}(x,y)dxdy, \\
&= \int x f_X(x)dx \cdot \int y f_Y(y)dy, \\
&= \mu_X \mu_Y.
\end{aligned}
\tag{B.16}
$$

B.1.3 Random Vectors

An $m-$dimensional random vector $\mathbf{X}$ is nothing other than a collection of m random variables, X_1 through X_m, which we may write as,

$$
\mathbf{X} = \begin{pmatrix} X_1 \\ X_2 \\ \vdots \\ X_m \end{pmatrix}. \tag{B.17}
$$

We write $f_{\mathbf{X}}(\mathbf{x})$ for the PDF of $\mathbf{X}$, which is the joint probability density function of the m random variables. We sometimes write this as $f_{X_1,X_2\ldots,X_m}(x_1, x_2 \ldots, x_m)$ to emphasize the individual random variables whose joint probability distribution is being considered. The joint PDF is defined in the natural way as,

$$
f_{X_1,\ldots,X_m}(x_1,\ldots,x_m) \triangleq_{\|\delta(x_1)\|,\ldots,\|\delta(x_m)\|\to 0} \frac{P\left(X_1 \in \delta x_1 \text{ and } \ldots \text{ and } X_m \in \delta\left(x_m\right)\right)}{\|\delta\left(x_1\right)\| \cdot \ldots \cdot \|\delta\left(x_m\right)\|}. \tag{B.18}
$$

We refer to the individual distributions of the random variables, $f_{X_k}(x_k)$, as their "marginal" distributions or marginal PDF's. The kth marginal distribution is obtained by integrating the joint PDF over all variables except the kth one x_k.

It is easy to show that *the mean of* $\mathbf{X}$ is the vector formed from the means of the individual random variables. That is,

$$
\mu_{\mathbf{X}} = E[\mathbf{X}] = \begin{pmatrix} E\left[X_1\right] \\ E\left[X_2\right] \\ \vdots \\ E\left[X_m\right] \end{pmatrix} = \begin{pmatrix} \mu_{X_1} \\ \mu_{X_2} \\ \vdots \\ \mu_{X_m} \end{pmatrix}. \tag{B.19}
$$

The correlation matrix, $R_{\mathbf{X}}$, *of a random vector*, $\mathbf{X}$, is defined by,

$$
R_{\mathbf{X}} = E\left[\mathbf{X}\mathbf{X}^T\right] = \begin{pmatrix} E\left[X_1X_1\right] & E\left[X_1X_2\right] & \cdots & E\left[X_1X_m\right] \\ E\left[X_2X_1\right] & E\left[X_2X_2\right] & \cdots & E\left[X_2X_m\right] \\ \vdots & \vdots & \ddots & \vdots \\ E\left[X_mX_1\right] & E\left[X_mX_2\right] & \cdots & E\left[X_mX_m\right] \end{pmatrix}. \tag{B.20}
$$

The covariance matrix, $C_{\mathbf{X}}$, of a random vector, $\mathbf{X}$, is defined by,

$$C_{\mathbf{X}} = E\left[(\mathbf{X} - \mu_{\mathbf{X}}) \cdot (\mathbf{X} - \mu_{\mathbf{X}})^T\right] = R_{\mathbf{X}} - \mu_{\mathbf{X}}\mu_{\mathbf{X}}^T. \tag{B.21}$$

The entry at the i'th row and j'th column of the covariance matrix is $\operatorname{cov}\left(X_i X_j\right) = E\left[X_i X_j\right] - \mu_{X_i}\mu_{X_j}$.

Evidently, the correlation and covariance matrices are both symmetric, those are

$$R_{\mathbf{X}} = R_{\mathbf{X}}^T, \quad C_{\mathbf{X}} = C_{\mathbf{X}}^T. \tag{B.22}$$

Similar to the case of two RVs, if the m random variables, X_1 through X_m are mutually independent, then they must be mutually uncorrelated, meaning that $C_{\mathbf{X}}$ is a diagonal matrix,

$$\begin{aligned} C_{\mathbf{X}} &= \begin{pmatrix} \operatorname{cov}(X_1, X_1) & \operatorname{cov}(X_1, X_2) & \cdots & \operatorname{cov}(X_1, X_m) \\ \operatorname{cov}(X_2, X_1) & \operatorname{cov}(X_2, X_2) & \cdots & \operatorname{cov}(X_2, X_m) \\ \vdots & \vdots & \ddots & \vdots \\ \operatorname{cov}(X_m, X_1) & \operatorname{cov}(X_m, X_2) & \cdots & \operatorname{cov}(X_m, X_m) \end{pmatrix}, \\ &= \begin{pmatrix} \sigma_1^2 & 0 & \cdots & 0 \\ 0 & \sigma_2^2 & \cdots & 0 \\ \vdots & \vdots & \ddots & \vdots \\ 0 & 0 & \cdots & \sigma_m^2 \end{pmatrix}. \end{aligned} \tag{B.23}$$

Unfortunately, the converse is not generally true. That is, uncorrelated random variables are not generally statistically independent, despite the fact that we often use the word "uncorrelated" informally to suggest statistical independence. One important exception occurs when $\mathbf{X}$ has a Gaussian distribution. In this case, mutually uncorrelated random variables are always independent, and vice-versa.

We say that $\mathbf{X}$ has a Gaussian distribution, or equivalently that X_1 through X_m are mutually Gaussian, if the joint PDF, $f_{\mathbf{X}}(\mathbf{x})$ has the form:

$$f_{\mathbf{X}}(\mathbf{x}) = \frac{1}{\sqrt{(2\pi)^m \det(C_{\mathbf{X}})}} e^{-\frac{1}{2}(\mathbf{x}-\mu_{\mathbf{X}})^T C_{\mathbf{X}}^{-1}(\mathbf{x}-\mu_{\mathbf{X}})} \tag{B.24}$$

Here, $\det(C_{\mathbf{X}})$ *identifies the determinant of the covariance matrix, $C_{\mathbf{X}}$. It can be shown that the determinant of a covariance matrix must be non-negative.*

The Gaussian distribution evidently depends only upon means and covariances, i.e. the expectations of polynomials of degree 1 or 2 in the m random variables in $\mathbf{X}$. We call these the *first and second order statistics*. The distribution of a Gaussian random vector is determined entirely by its first and second order statistics. This is one of the reasons why so much of statistical signal processing is restricted to consideration of the first and second order statistics.

If **X** is Gaussian and its random variables are all mutually uncorrelated, $C_{\mathbf{X}}$ is diagonal and the Gaussian PDF expression separates into,

$$f_{\mathbf{X}}(\mathbf{x}) = \prod_{i=1}^{m} \frac{1}{\sqrt{2\pi\sigma_{X_i}^2}} e^{-\frac{1}{2}\frac{x_i^2}{\sigma_{X_i}}}. \tag{B.25}$$

Gaussian distributions are the only distributions for which statistical independence is equivalent to zero covariance.

B.2 Fundamental PDF and Properties

Scalar Gaussian PDF: The Gaussian PDF (also referred to as the normal PDF) for a scalar random variable X is defined as,

$$p(x) = \frac{1}{\sqrt{2\pi\sigma^2}} \exp\left[-\frac{1}{2\sigma^2}(x-\mu)^2\right], -\infty < x < \infty, \tag{B.26}$$

where μ is the mean and σ^2 is the variance of X. It is denoted by $\mathcal{N}(\mu, \sigma^2)$ and we say that $X \sim \mathcal{N}(\mu, \sigma^2)$, where "$\sim$" means "is distributed according to." If $\mu = 0$, its nth moments are,

$$E(x^n) = \begin{cases} 1 \cdot 3 \cdot 5 \cdots (n-1)\sigma^n & n \text{ even}, \\ 0 & n \text{ odd}, \end{cases} \tag{B.27}$$

Otherwise, we use,

$$E[(x+\mu)^n] = \sum_{k=0}^{n} \binom{n}{k} E(x^k)\mu^{n-k}. \tag{B.28}$$

The CDF for $\mu = 0$ and $\sigma^2 = 1$, for which the PDF is termed as a standard normal PDF, is defined as,

$$\Phi(x) = \int_{-\infty}^{x} \frac{1}{\sqrt{2\pi}} \exp(-\frac{1}{2}t^2)dt. \tag{B.29}$$

A convenient description, which is termed the *right-tail probability* and is the probability of exceeding a given value, is defined as $Q(x) = 1 - \Phi(x)$, where

$$Q(x) = \int_{x}^{\infty} \frac{1}{\sqrt{2\pi}} \exp(-\frac{1}{2}t^2)dt. \tag{B.30}$$

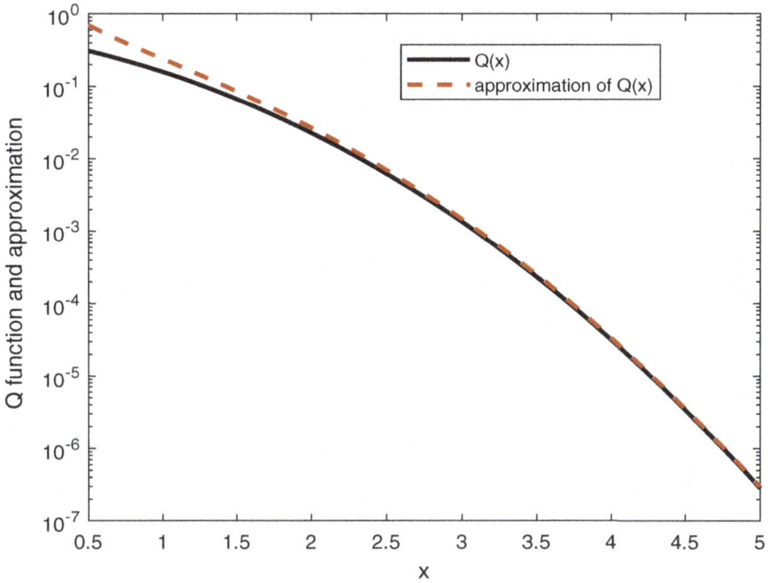

Fig. B.1 Approximation to Q function

The function $Q(x)$ is also referred to as the *complementary cumulative distribution function*. This cannot be evaluated in a closed-form. We can use the MATLAB command "y = Q(x) = 0.5 * erfc(x/sqrt(2))" to calculate the complementary cumulative distribution function. Here, erfc(·) is the MATLAB command for complementary error function. An approximation that is sometimes useful is,

$$Q(x) \approx \frac{1}{\sqrt{2\pi}x} \exp\left(-\frac{1}{2}x^2\right). \tag{B.31}$$

It is shown in Fig. B.1 along with the exact value of $Q(x)$. The approximation is quite accurate for $x > 4$.

If it is known that a probability is given by $P = Q(\gamma)$, then we can determine γ for a given P. Symbolically, we have $\gamma = Q^{-1}(P)$, where Q^{-1} is the inverse function. The latter must exist since $Q(x)$ is strictly monotonically decreasing. We can use the MATLAB command "y = Q^{-1}(x) = sqrt(2) * erfinv(1–2*x)" to calculate inverse function. Here, erfinv(·) is the MATLAB command for inverse error function.

Rayleigh PDF: The Rayleigh PDF is obtained as the PDF of $x = \sqrt{x_1^2 + x_2^2}$, where $x_1 \sim \mathcal{N}(0, \sigma^2)$, $x_2 \sim \mathcal{N}(0, \sigma^2)$, and x_1, x_2 are independent. Its PDF is,

$$p(x) = \begin{cases} \frac{x}{\sigma^2} \exp\left(-\frac{1}{2\sigma^2}x^2\right), & x > 0, \\ 0, & x < 0. \end{cases} \tag{B.32}$$

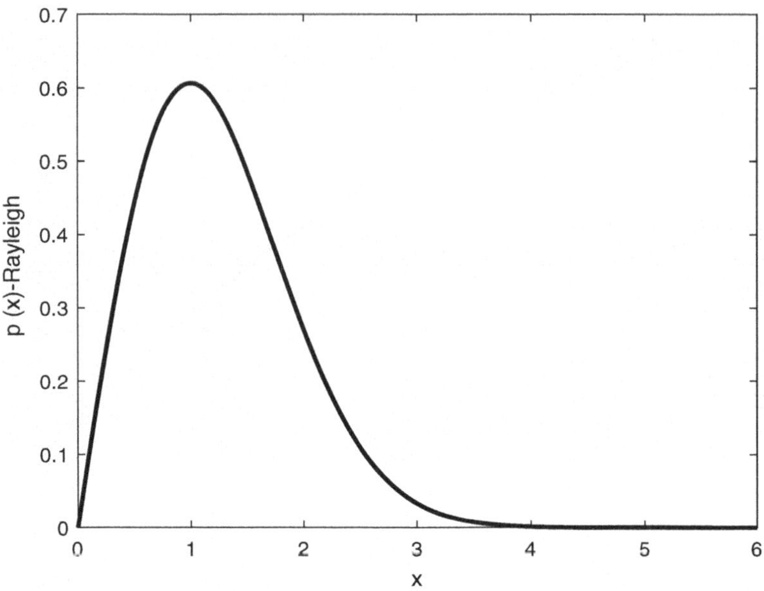

Fig. B.2 PDF for Rayleigh random variable ($\sigma^2 = 1$)

The mean and variance are given by,

$$E(x) = \sqrt{\frac{\pi\sigma^2}{2}},$$
$$var(x) = (2 - \frac{\pi}{2})\sigma^2.$$

The PDF of Rayleigh distribution with σ^2 is given in Fig. B.2.

Chi-Squared PDF: A chi-squared (central) PDF with ν degrees of freedom is defined as,

$$p(x) = \begin{cases} \frac{1}{2^{\nu/2}\Gamma(\nu/2)} x^{\nu/2-1} \exp(-\frac{1}{2}x), & x > 0, \\ 0, & x < 0, \end{cases} \tag{B.33}$$

and is denoted by χ^2_ν. The degrees of freedom ν is assumed to be an integer with $\nu \geq 1$. The function $\Gamma(u)$ is the Gamma function, which is defined as,

$$\Gamma(u) = \int_0^\infty t^{u-1} \exp(-t)dt. \tag{B.34}$$

The relations $\Gamma(u) = (u-1)\Gamma(u-1)$ for any u, $\Gamma(\frac{1}{2}) = \sqrt{\pi}$ and $\Gamma(n) = (n-1)!$ for n an integer can be used to evaluate it.

The mean and variance are

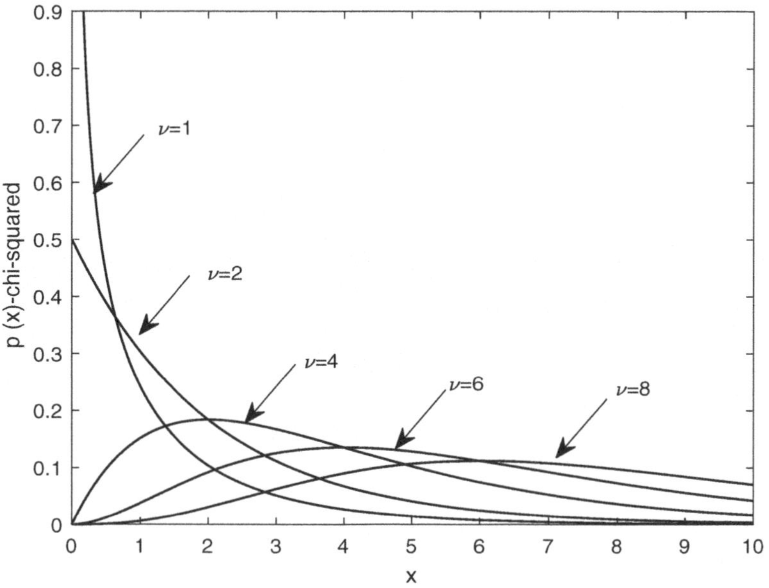

Fig. B.3 PDF for chi-squared random variable

$$E(x) = \nu,$$
$$var(x) = 2\nu.$$

The chi-squared PDF arises as the PDF of x where $x = \sum_{i=1}^{\nu} x_i^2$ if $x_i \sim \mathcal{N}(0, 1)$ and the $x_i's$ are independent and identically distributed (IID). Some examples of the chi-squared PDF are given in Fig. B.3.

B.3 Random Processes

B.3.1 Definition and Examples

Random processes are a generalization of random vectors, to infinitely long sequences of random variables. The time sequence, $x[n]$, is understood as a realiztion of some underlying random process, $X[n]$.

Example 2.1 Consider the addition of a deterministic signal, $x_0[n]$, to a random noise process, $v[n]$, yielding

$$x[n] = x_0[n] + v[n]. \qquad \text{(B.35)}$$

Since the noise is random, so is $x[n]$. The corresponding random processes satisfy

$$X[n] = x_0[n] + V[n]. \tag{B.36}$$

Note that any random component, such as noise, in the time sequence renders it a random process. ***Consequently, almost all signals which we deal with in signal processing are realizations of random processes.***

Having defined a random process $X[n]$ as an ensemble of random variables, the statistics of a random process can be characterized by their joint PDF denoted as $f_{X[1],X[2],\ldots}(x)$. In general, we cannot gain access to the statistics of a random process by monitoring it over time, because the entire time signal itself is only a single realization of the process. We will need to restrict the random process to a narrower class of "ergodic" random processes for this to be possible.

Suppose we have k samples of the random process $X[n]$ at $n \in \mathcal{I} = \{i_1, i_2, \ldots, i_k\}$ and another set of k samples displaced in time from the first set by an amount of m. *We say that a random process is stationary in the strict sense if the statistics are invariant to shifts. In other words,* $X[n]$ *is stationary if for all finite index sets,* $\mathcal{I} = \{i_1, i_2, \ldots, i_k\}$, *and all delays,* m, *the joint distribution of* $\mathbf{X}_{\mathcal{I}}$ *is identical to the joint distribution of the random vector,* $\mathbf{X}_{\mathcal{I}+m}$, *where* $\mathcal{I} + m = \{i_1 + m, \ldots, i_k + m\}$.

It is obvious that a stationary random process has a constant mean. That is,

$$\mu_{X[n]} = E[X[n]] = \int x f_{X[n]}(x) dx = \int x f_{X[0]}(x) dx = \mu_{X[0]}. \tag{B.37}$$

We then consider two random variables $X[k]$ and $X[k-m]$ corresponding to samples of $X[n]$ taken at $n = k$ and $n = k - m$. The statistical correlation between $X[k]$ and $X[k-m]$ is called the *autocorrelation function of the random process.*

$$R_{XX}[k, k-m] = E[X[k] \cdot X[k-m]]. \tag{B.38}$$

If the random process $X[n]$ *is stationary, we may simplify this to*

$$R_{XX}[k, k-m] = E[X[0]X[-m]] = R_{XX}[m], \tag{B.39}$$

since that the autocorrelation function depends only upon the lag, m.

In general, stationary means a great deal more than having mean and autocorrelation properties which are independent of time. Such a process is called Wide-Sense Stationary (WSS). A stationary process is certainly WSS, but a WSS process may not be truly stationary in the sense that all of its statistical properties are invariant to time shifts. Nevertheless, at least the mean and autocorrelation functions of a WSS process are invariant to time shifts. Wide sense stationary is often all we need, since we largely restrict our attention to second order statistics. Note that Gaussian processes are described entirely by their second order statistics and so for Gaussian processes wide sense stationary is equivalent to stationary. A WSS Gaussian process is described completely by its mean, μ_X, and autocorrelation, $R_{XX}[m]$.

Related to the autocorrelation function is the autocovariance function, defined as,

$$
\begin{aligned}
C_{XX}[m] &= \text{cov}(X[k], X[k-m]), \\
&= E\{[X[k] - \mu_{X[k]}][X[k-m] - \mu_{X[k-m]}]\}, \\
&= R_{XX}[k, k-m] - \mu_{X[k]}\mu_{X[k-m]},
\end{aligned} \tag{B.40}
$$

where $\mu_{X[k]} = E\{X[k]\}$ and $\mu_{X[k-m]} = E\{X[k-m]\}$ are the mean values of $X[k]$ and $X[k-m]$. When the process is stationary,

$$
C_{XX}[m] = R_{XX}[m] - \mu_X^2, \tag{B.41}
$$

only relevant to the lag, m.

B.3.2 Time Averages

In this section, we restrict our attention to WSS random processes. Recall that the mean is defined by $\mu_{X[n]} = E[X[n]]$, which for WSS processes is independent of the index, n. We can also define the time average of a specific realization,

$$
m_x = \lim_{N\to\infty} \frac{1}{2N+1} \sum_{n=-N}^{N} x[n]. \tag{B.42}
$$

The random process is said to be *mean-ergodic* if the time average, m_x of the realization $x[n]$, is equal to the *ensemble* (or statistical) average, μ_X, with probability 1. That is, regardless of which realization we observe, the time average turns out to be identical to μ_X with probability 1.

It is clear that $E[m_X] = \mu_X$, that is, the ensemble average of time average must equal to the ensemble average. This is because,

$$
\begin{aligned}
E[m_X] &= E\left[\lim_{N\to\infty} \frac{1}{2N+1} \sum_{n=-N}^{N} X[n]\right], \\
&= \lim_{N\to\infty} \frac{1}{2N+1} \sum_{n=-N}^{N} E[X[n]], \\
&= \mu_X.
\end{aligned} \tag{B.43}
$$

However, it is not clear that the variance of the time averages, m_X, should necessarily be zero so that only one value has any likelihood of occurring. In fact, it is possible to construct examples of WSS and even stationary random processes for which the time average has a non-zero variance.

We then define a time-autocorrelation for any specific realization, $x[n]$, of the random process, $X[n]$, according to

$$r_{xx}[m] = \lim_{N\to\infty} \frac{1}{2N+1} \sum_{n=-N}^{N} x[n]x[n-m]. \tag{B.44}$$

Again, in the case of a *correlation ergodic* random process, this time average always yields the same autocorrelation sequence, $r_{xx}[m]$, for all realizations of the underlying random process, with probability 1. And thus we can equate $r_{xx}[m] = R_{XX}[m]$. Note carefully that $r_{xx}[m]$ is calculated from a single realization, $x[n]$, while $R_{XX}[m]$ is a statistical property of the underlying random process.

The cross-correlation between two WSS random processes, $X[n]$ and $Y[n]$, is defined by

$$R_{XY}[m] = E[X[k], Y[k-m]]. \tag{B.45}$$

The WSS property ensures that this definition is independent of k.

When the process is appropriately ergodic, the cross-correlation is equal (with probability 1) to the time average,

$$r_{xy}[m] = \lim_{N\to\infty} \frac{1}{2N+1} \sum_{n=-N}^{N} x[n]y[n-m], \tag{B.46}$$

for all realizations, $x[n]$ and $y[n]$, of $X[n]$ and $Y[n]$. Note that $R_{XY}[m] = R_{YX}[-m]$.

We generally restrict our attention to mean ergodic and correlation ergodic random processes, because this allows us to access the statistical properties of the random process by taking time averages of a single realization.

B.3.3 Power Density Spectrum

In this section, we restrict our attention to WSS ergodic random processes.

A stationary random process is an infinite-energy signal and hence its Fourier transform does not exit. The spectral characteristic of a random process is obtained according to the Wiener-Khintchine theorem, by computing the Fourier transform of the autocorrelation function. That is,

$$S_{XX}(\omega) = \sum_{m} R_{XX}[m]e^{-j\omega m}. \tag{B.47}$$

$S_{XX}(\omega)$ estimated the distribution of power with frequency, and thus is called (Power Density Spectrum, PDS) for the random process, $X[n]$. Note that $S_{XX}(\omega)$ is entirely real-valued, since the autocorrelation sequence is symmetric, i.e., $R_{XX}[m] = R_{XX}[-m]$.

The above relationships also suggest the autocorrelation function can be obtained by the inverse Fourier transform of the PDF,

$$R_{XX}[m] = \frac{1}{2\pi}\int_{-\pi}^{\pi} S_{XX}(\omega)e^{j\omega m}d\omega. \tag{B.48}$$

Evaluating this at $m = 0$, we find that

$$E\left[X[n]^2\right] = R_{XX}[0] = \frac{1}{2\pi}\int_{-\pi}^{\pi} S_{XX}(\omega)d\omega. \tag{B.49}$$

Thus, integrating $S_{XX}(\omega)$ over frequency, gives us the "power" of the random process, $X[n]$.

Now suppose we design a band-pass filter which suppresses all but a narrow range of frequencies in the neighbourhood of ω_0, i.e.,

$$\hat{h}(\omega) = \begin{cases} 1 & \text{if } |\omega - \omega_0| < \delta, \\ 0 & \text{otherwise} , \end{cases} \tag{B.50}$$

and we measure the power of the output random process from this filter. Then we will obtain

$$\begin{aligned} E\left[Y[n]^2\right] &= \frac{1}{2\pi}\int_{-\pi}^{\pi} S_{YY}(\omega)d\omega, \\ &= \frac{1}{2\pi}\int_{\omega_0-\delta}^{\omega_0+\delta} S_{XX}(\omega)d\omega. \end{aligned} \tag{B.51}$$

So we see that the power around frequency, ω_0, is directly related to the value of the PDS, $S_{XX}(\omega)$, in the neighbourhood of ω_0.

The cross-power density spectrum $S_{XY}(\omega)$ is defined as the Fourier transform of $R_{XY}[m]$. We can interpret the cross PDS from the viewpoint of LTI system.

Now suppose that $Y[n]$ is the output of an LTI system with impulse response, $h[n]$, which is excited by the random process, $X[n]$. Then, the following relationships can be shown to hold:

- The cross-correlation sequence, $R_{YX}[m]$, is given by the convolution of the auto-correlation sequence, $R_{XX}[m]$, with the impulse response, $h[m]$, i.e.,

  $$R_{YX}[m] = \sum_k h[k]R_{XX}[m-k]. \tag{B.52}$$

  In the Fourier domain this becomes

  $$S_{YX}(\omega) = S_{XX}(\omega)\hat{h}(\omega), \tag{B.53}$$

  where x^*denotes the complex conjugate of x and $\hat{h}(\omega)$ is the spectrum of the LTI system.

- Since $R_{XY}[m] = R_{YX}[-m]$, we also have

$$R_{XY}[m] = \sum_k h[-k] R_{XX}[m-k], \tag{B.54}$$

$$S_{XY}(\omega) = S_{XX}(\omega)\hat{h}^*(\omega). \tag{B.55}$$

- The autocorrelation of the output random process, $R_{YY}[m]$, is obtained by convolving $R_{XX}[m]$ by the sequence,

$$h_2[m] = (h * \tilde{h})[m] = \sum_k h[k]h[k-m]. \tag{B.56}$$

  In the spectral domain, this is easily expressed as

$$\begin{aligned} S_{YY}(\omega) &= S_{XX}(\omega)\hat{h}(\omega)\hat{h}^*(\omega), \\ &= S_{XX}(\omega) \cdot |\hat{h}(\omega)|^2. \end{aligned} \tag{B.57}$$

The above relationships suggest a practical method for measuring the impulse response, $h[n]$, of a system. Suppose we excite the system by white noise, i.e. a random process for which $R_{XX}[m] = \delta[m]$ and hence $S_{XX}(\omega) = 1$. Suppose then that we use time averages to estimate the cross-correlation between the input and output sequences, $R_{YX}[m]$. We find that,

$$S_{YX}(\omega) = S_{XX}(\omega)\hat{h}(\omega) = \hat{h}(\omega), \tag{B.58}$$

and so the cross-correlation sequence is the system impulse response which we seek.

References

1. H. L. Van Trees. *Detection, Estimation, and Modulation Theory, Optimum Array Processing*. John Wiley & Sons, 2004.
2. C. A. Balanis. *Antenna theory: analysis and design*. John Wiley & Sons, 2012.
3. Y. Mitra, S. K.and Kuo. *Digital signal processing: a computer-based approach*, volume 2. McGraw-Hill New York, 2006.
4. S. Applebaum. Adaptive arrays. *Antennas and Propagation, IEEE Transactions on*, 24(5):585–598, 1976.
5. R. L. Riegler and R. T. Compton. An adaptive array for interference rejection. *Proceedings of the IEEE*, 61(6):748–758, 1973.
6. B. Widrow, P.E. Mantey, L. Griffiths, and B. Goode. Adaptive antenna systems. *The Journal of the Acoustical Society of America*, 42(5):1175–1176, 1967.
7. O. L. Frost III. An algorithm for linearly constrained adaptive array processing. *Proceedings of the IEEE*, 60(8):926–935, 1972.
8. B. Widrow and S. D. Stearns. Adaptive signal processing. *Englewood Cliffs, NJ, Prentice-Hall, Inc., 1985, 491 p.*, 1, 1985.
9. R. A. Monzingo and T. W. Miller. *Introduction to adaptive arrays*. SciTech Publishing, 1980.
10. J. E. Hudson. *Adaptive array principles*, volume 11. IET, 1981.
11. S. Haykin. *Array signal processing*. Englewood Cliffs, NJ, Prentice-Hall, 1985.
12. B. D. Steinberg. Principles of aperture and array system design: Including random and adaptive arrays. *New York, Wiley-Interscience, 1976. 374 p.*, 1, 1976.
13. B. Friedlander and B. Porat. Performance analysis of a null-steering algorithm based on direction-of-arrival estimation. *Acoustics, Speech and Signal Processing, IEEE Transactions on*, 37(4):461–466, April 1989.
14. C. L. Dolph. A current distribution for broadside arrays which optimizes the relationship between beamwidth and sidelobe level. *Proceedings of IRE*, 34:335–348, June 1945.
15. T. T. Taylor. One parameter family of line sources producing modified SinU/U patterns. Technical report, 324, Hughs Aircraft Co., Culver City, California, 1953.
16. T. T. Taylor. Design of line-source antennas for narrow beamwidth and low sidelobes. *Antennas and Propagation, IRE Transactions on*, AP-3:16–28, January 1955.
17. H. Unz. Linear arrays with arbitrarily distributed elements. *Antennas and Propagation, IRE Transactions on*, 8(2):222–223, 1960.
18. R. Harrington. Sidelobe reduction by nonuniform element spacing. *Antennas and Propagation, IRE Transactions on*, 9(2):187–192, 1961.

X. Wang et al., *Sparse Sensing and Sparsity Sensed in Multi-sensor Array Applications*, https://doi.org/10.1007/978-981-99-9558-5

19. M. Andreasen. Linear arrays with variable interelement spacings. *Antennas and Propagation, IRE Transactions on*, 10(2):137–143, March 1962.
20. B.P. Kumar and G.R. Branner. Design of unequally spaced arrays for performance improvement. *Antennas and Propagation, IEEE Transactions on*, 47(3):511–523, 1999.
21. B.P. Kumar and G.R. Branner. Generalized analytical technique for the synthesis of unequally spaced arrays with linear, planar, cylindrical or spherical geometry. *Antennas and Propagation, IEEE Transactions on*, 53(2):621–634, 2005.
22. P.Y. Zhou and M.A. Ingram. Pattern synthesis for arbitrary arrays using an adaptive array method. *Antennas and Propagation, IEEE Transactions on*, 47(5):862–869, 1999.
23. T.H. Ismail and M.M. Dawoud. Null steering in phased arrays by controlling the element positions. *Antennas and Propagation, IEEE Transactions on*, 39(11):1561–1566, 1991.
24. H. Lebret and S. Boyd. Antenna array pattern synthesis via convex optimization. *Signal Processing, IEEE Transactions on*, 45(3):526–532, 1997.
25. F. Wang, V. Balakrishnan, P.Y. Zhou, J.J. Chen, R. Yang, and C. Frank. Optimal array pattern synthesis using semidefinite programming. *Signal Processing, IEEE Transactions on*, 51(5):1172–1183, 2003.
26. Z.L. Yu, M.H. Er, and W. Ser. A novel adaptive beamformer based on semidefinite programming (SDP) with magnitude response constraints. *Antennas and Propagation, IEEE Transactions on*, 56(5):1297–1307, 2008.
27. B. Fuchs and J.J. Fuchs. Optimal narrow beam low sidelobe synthesis for arbitrary arrays. *Antennas and Propagation, IEEE Transactions on*, 58(6):2130–2135, 2010.
28. M. Comisso and R. Vescovo. Fast iterative method of power synthesis for antenna arrays. *Antennas and Propagation, IEEE Transactions on*, 57(7):1952–1962, 2009.
29. K. Yang, Z. Zhao, and Q. H. Liu. Fast pencil beam pattern synthesis of large unequally spaced antenna arrays. *Antennas and Propagation, IEEE Transactions on*, 61(2):627 –634, feb 2013.
30. K.K. Yan and Y. Lu. Sidelobe reduction in array-pattern synthesis using genetic algorithm. *Antennas and Propagation, IEEE Transactions on*, 45(7):1117–1122, 1997.
31. L. Cen, Z.L. Yu, W. Ser, and W. Cen. Linear aperiodic array synthesis using an improved genetic algorithm. *Antennas and Propagation, IEEE Transactions on*, 60(2):895–902, 2012.
32. M.M. Khodier and C.G. Christodoulou. Linear array geometry synthesis with minimum sidelobe level and null control using particle swarm optimization. *Antennas and Propagation, IEEE Transactions on*, 53(8):2674–2679, 2005.
33. C. Lin, A. Qing, and Q. Feng. Synthesis of unequally spaced antenna arrays by using differential evolution. *Antennas and Propagation, IEEE Transactions on*, 58(8):2553–2561, 2010.
34. T. Isernia, F.J.A. Pena, O.M. Bucci, M. D'Urso, J.F. Gomez, and J.A. Rodriguez. A hybrid approach for the optimal synthesis of pencil beams through array antennas. *Antennas and Propagation, IEEE Transactions on*, 52(11):2912–2918, Nov 2004.
35. S.E. Nai, W. Ser, Z.L. Yu, and H. Chen. Beampattern synthesis for linear and planar arrays with antenna selection by convex optimization. *Antennas and Propagation, IEEE Transactions on*, 58(12):3923–3930, 2010.
36. A. Massa and G. Oliveri. Bayesian compressive sampling for pattern synthesis with maximally sparse non-uniform linear arrays. *Antennas and Propagation, IEEE Transactions on*, 59(2):467–481, 2011.
37. X. Wang, E. Aboutanios, and M.G. Amin. Thinned array beampattern synthesis by iterative soft-thresholding-based optimization algorithms. *Antennas and Propagation, IEEE Transactions on*, 62(12):6102–6113, Dec 2014.
38. R.C. Nongpiur and D.J. Shpak. Synthesis of linear and planar arrays with minimum element selection. *IEEE Transactions on Signal Processing*, 62(20):5398–5410, 2014.
39. X. Wang, M. Amin, and E. Aboutanios. Interferometric array design under regularized antenna placements for interference suppression. In *2018 IEEE 10th Sensor Array and Multichannel Signal Processing Workshop (SAM)*, pages 657–661, 2018.
40. X. Wang and E. Aboutanios. Sparse array design for multiple switched beams using iterative antenna selection method. *Digital Signal Processing*, 105:102684, 2020. Special Issue on Optimum Sparse Arrays and Sensor Placement for Environmental Sensing.

41. A. Garcia-Rodriguez, C. Masouros, and P. Rulikowski. Efficient large scale antenna selection by partial switching connectivity. In *2017 IEEE International Conference on Acoustics, Speech and Signal Processing (ICASSP)*, pages 6269–6273, 2017.
42. H. Nosrati, E. Aboutanios, and D.B. Smith. Spatial array thinning for interference cancellation under connectivity constraints. In *2018 IEEE International Conference on Acoustics, Speech and Signal Processing (ICASSP)*, pages 3325–3329, 2018.
43. P. Pal and P.P. Vaidyanathan. Nested arrays: A novel approach to array processing with enhanced degrees of freedom. *Signal Processing, IEEE Transactions on*, 58(8):4167–4181, Aug 2010.
44. M. Guo, Y. D. Zhang, and T. Chen. Doa estimation using compressed sparse array. *IEEE Transactions on Signal Processing*, 66(15):4133–4146, 2018.
45. G. Qin, M. G. Amin, and Y. D. Zhang. Doa estimation exploiting sparse array motions. *IEEE Transactions on Signal Processing*, 67(11):3013–3027, 2019.
46. X. Liu, T. Huang, N. Shlezinger, Y. Liu, J. Zhou, and Y. C. Eldar. Joint transmit beamforming for multiuser mimo communications and mimo radar. *IEEE Transactions on Signal Processing*, 68:3929–3944, 2020.
47. M. S. Davis, G. A. Showman, and A. D. Lanterman. Coherent mimo radar: The phased array and orthogonal waveforms. *IEEE Aerospace and Electronic Systems Magazine*, 29(8):76–91, 2014.
48. A. M. Haimovich, R. S. Blum, and L. J. Cimini. Mimo radar with widely separated antennas. *IEEE Signal Processing Magazine*, 25(1):116–129, 2008.
49. C. Gao, K. C. Teh, and A. Liu. Orthogonal frequency diversity waveform with range-doppler optimization for mimo radar. *IEEE Signal Processing Letters*, 21(10):1201–1205, 2014.
50. X. Hu, C. Feng, Y. Wang, and Y. Guo. Adaptive waveform optimization for mimo radar imaging based on sparse recovery. *IEEE Transactions on Geoscience and Remote Sensing*, 58(4):2898–2914, 2020.
51. L. Wu, P. Babu, and D. P. Palomar. Transmit waveform/receive filter design for mimo radar with multiple waveform constraints. *IEEE Transactions on Signal Processing*, 66(6):1526–1540, 2018.
52. J. Zhang, H. Wang, and X. Zhu. Adaptive waveform design for separated transmit/receive ula-mimo radar. *IEEE Transactions on Signal Processing*, 58(9):4936–4942, 2010.
53. P. W. Moo and Z. Ding. Tracking performance of mimo radar for accelerating targets. *IEEE Transactions on Signal Processing*, 61(21):5205–5216, 2013.
54. A. Hassanien and S. A. Vorobyov. Phased-mimo radar: A tradeoff between phased-array and mimo radars. *IEEE Transactions on Signal Processing*, 58(6):3137–3151, 2010.
55. D. E. Hack, L. K. Patton, B. Himed, and M. A. Saville. Detection in passive mimo radar networks. *IEEE Transactions on Signal Processing*, 62(11):2999–3012, 2014.
56. Y. Wang and S. Zhu. Range ambiguous clutter suppression for fda-mimo forward looking airborne radar based on main lobe correction. *IEEE Transactions on Vehicular Technology*, 70(3):2032–2046, 2021.
57. K. Wang, G. Liao, J. Xu, Y. Zhang, and L. Huang. Clutter rank analysis in airborne fda-mimo radar with range ambiguity. *IEEE Transactions on Aerospace and Electronic Systems*, 58(2):1416–1430, 2022.
58. P. Wang, H. Li, and B. Himed. Moving target detection using distributed mimo radar in clutter with nonhomogeneous power. *IEEE Transactions on Signal Processing*, 59(10):4809–4820, 2011.
59. J. Li and P. Stoica. Mimo radar with colocated antennas. *IEEE Signal Processing Magazine*, 24(5):106–114, 2007.
60. S. D. Blunt and E. L. Mokole. Overview of radar waveform diversity. *IEEE Aerospace and Electronic Systems Magazine*, 31(11):2–42, 2016.
61. R. Ibernon-Fernandez, J. Molina-Garcia-Pardo, and L. Juan-Llacer. Comparison between measurements and simulations of conventional and distributed mimo system. *IEEE Antennas and Wireless Propagation Letters*, 7:546–549, 2008.

62. Y. Yuan, W. Yi, R. Hoseinnezhad, and P. K. Varshney. Robust power allocation for resource-aware multi-target tracking with colocated mimo radars. *IEEE Transactions on Signal Processing*, 69:443–458, 2021.
63. P. Chen, L. Zheng, X. Wang, H. Li, and L. Wu. Moving target detection using colocated mimo radar on multiple distributed moving platforms. *IEEE Transactions on Signal Processing*, 65(17):4670–4683, 2017.
64. J. R. Guerci. *Space-time adaptive processing for radar*. Artech House, 2003.
65. R. Klemm. *Principles of space-time adaptive processing*. Number 159. Inspec/Iee, 2002.
66. J. Capon. High-resolution frequency-wavenumber spectrum analysis. *Proceedings of the IEEE*, 57(8):1408–1418, Aug 1969.
67. J. Zhuang, B. Shi, X. Zuo, and A. H. Ali. Robust adaptive beamforming with minimum sensitivity to correlated random errors. *Signal Processing*, 131:92–98, 2017.
68. H. Ruan and R. C. de Lamare. Robust adaptive beamforming based on low-rank and cross-correlation techniques. *IEEE Transactions on Signal Processing*, 64(15):3919–3932, 2016.
69. S. D. Somasundaram. Linearly constrained robust capon beamforming. *IEEE Transactions on Signal Processing*, 60(11):5845–5856, 2012.
70. D. Chapman. Partial adaptivity for the large array. *IEEE Transactions on Antennas and Propagation*, 24(5):685–696, 1976.
71. A. Farina. Single sidelobe canceller: Theory and evaluation. *IEEE Transactions on Aerospace and Electronic Systems*, AES-13(6):690–699, 1977.
72. G. Rombouts, A. Spriet, and M. Moonen. Generalized sidelobe canceller based combined acoustic feedback- and noise cancellation. *Signal Processing*, 88(3):571–581, 2008.
73. R. Levanda and A. Leshem. Adaptive selective sidelobe canceller beamformer with applications in radio astronomy. In *2010 IEEE 26-th Convention of Electrical and Electronics Engineers in Israel*, pages 000839–000843, 2010.
74. H. Shin. The mitigation of multiple jammers using a sidelobe canceller designed with independently configured loops. In *2018 IEEE Radar Conference (RadarConf18)*, pages 0969–0974, 2018.
75. J. Ward and R.T. Compton. Sidelobe level performance of adaptive sidelobe canceller arrays with element reuse. *IEEE Transactions on Antennas and Propagation*, 38(10):1684–1693, 1990.
76. A. Farina, P. Lombardo, and L. Ortenzi. A unified approach to adaptive radar processing with general antenna array configuration. *Signal Processing*, 84(9):1593–1623, 2004. Special Section on New Trends and Findings in Antenna Array Processing for Radar.
77. S. Yu and J. Lee. Design of two-dimensional rectangular array beamformers with partial adaptivity. *IEEE Transactions on Antennas and Propagation*, 45(1):157–167, 1997.
78. D. Morgan. Partially adaptive array techniques. *IEEE Transactions on Antennas and Propagation*, 26(6):823–833, 1978.
79. S. M. Kim and H. K. Kim. Multi-microphone target signal enhancement using generalized sidelobe canceller controlled by phase error filter. *IEEE Sensors Journal*, 16(21):7566–7567, 2016.
80. J. P. Townsend and K. D. Donohue. Stability analysis for the generalized sidelobe canceller. *IEEE Signal Processing Letters*, 17(6):603–606, 2010.
81. K. Elkhalil, A. Kammoun, T. Y. Al-Naffouri, and M. Alouini. Fluctuations of the snr at the output of the mvdr with regularized tyler estimators. *Signal Processing*, 135:1–8, 2017.
82. Y. Huang, M. Zhou, and S. A. Vorobyov. New designs on mvdr robust adaptive beamforming based on optimal steering vector estimation. *IEEE Transactions on Signal Processing*, 67(14):3624–3638, 2019.
83. J. Lee and C. Cho. Gsc-based adaptive beamforming with multiple-beam constraints under random array position errors. *Signal Processing*, 84(2):341–350, 2004. Special Section on Independent Component Analysis and Beyond.
84. R.O. Schmidt. Multiple emitter location and signal parameter estimation. *Antennas and Propagation, IEEE Transactions on*, 34(3):276–280, 1986.

85. R. Schmidt and R. Franks. Multiple source DF signal processing: an experimental system. *Antennas and Propagation, IEEE Transactions on*, 34(3):281–290, 1986.
86. Y. Bresler and A. Macovski. Exact maximum likelihood parameter estimation of superimposed exponential signals in noise. *Acoustics, Speech and Signal Processing, IEEE Transactions on*, 34(5):1081–1089, Oct 1986.
87. P. Stoica and Nehorai A. MUSIC, maximum likelihood, and Cramer-Rao bound. *Acoustics, Speech and Signal Processing, IEEE Transactions on*, 37(5):720–741, May 1989.
88. Y. Hua and T.K. Sarkar. A note on the Cramer-Rao bound for 2-D direction finding based on 2-D array. *Signal Processing, IEEE Transactions on*, 39(5):1215–1218, 1991.
89. G. Bienvenu and L. Kopp. Adaptivity to background noise spatial coherence for high resolution passive methods. In *Acoustics, Speech, and Signal Processing, IEEE International Conference on (ICASSP)*, volume 5, pages 307–310, Apr 1980.
90. A Barabell. Improving the resolution performance of eigenstructure-based direction-finding algorithms. In *Acoustics, Speech, and Signal Processing, IEEE International Conference on (ICASSP)*, volume 8, pages 336–339, Apr 1983.
91. A. Paulraj, R. Roy, and T. Kailath. A subspace rotation approach to signal parameter estimation. *Proceedings of the IEEE*, 74(7):1044–1046, 1986.
92. R. Roy and T. Kailath. Esprit-estimation of signal parameters via rotational invariance techniques. *IEEE Transactions on Acoustics, Speech, and Signal Processing*, 37(7):984–995, 1989.
93. F. Gao and A.B. Gershman. A generalized esprit approach to direction-of-arrival estimation. *IEEE Signal Processing Letters*, 12(3):254–257, 2005.
94. L. Qu, Q. Sun, T. Yang, L. Zhang, and Y. Sun. Time-delay estimation for ground penetrating radar using esprit with improved spatial smoothingtechnique. *IEEE Geoscience and Remote Sensing Letters*, 11(8):1315–1319, 2014.
95. J. Steinwandt, F. Roemer, and M. Haardt. Generalized least squares for esprit-type direction of arrival estimation. *IEEE Signal Processing Letters*, 24(11):1681–1685, 2017.
96. I. S. Merrill. *Introduction to radar systems*. Tata McGraw Hill, 2003.
97. S. M. Kay. Fundamentals of statistical signal processing: Detection theory, vol. 2, 1998.
98. F.C. Robey, D.R. Fuhrmann, E.J. Kelly, and R. Nitzberg. A CFAR adaptive matched filter detector. *Aerospace and Electronic Systems, IEEE Transactions on*, 28(1):208–216, Jan 1992.
99. E.J. Kelly. An adaptive detection algorithm. *Aerospace and Electronic Systems, IEEE Transactions on*, AES-22(2):115–127, March 1986.
100. M. Rangaswamy, J. H. Michels, and B. Himed. Statistical analysis of the non-homogeneity detector for STAP applications. *Digital Signal Processing*, 14(3):253–267, 2004.
101. K. Buckley and L. Griffiths. An adaptive generalized sidelobe canceller with derivative constraints. *IEEE Transactions on Antennas and Propagation*, 34(3):311–319, 1986.
102. B.D. Van Veen and K.M. Buckley. Beamforming: a versatile approach to spatial filtering. *IEEE ASSP Magazine*, 5(2):4–24, 1988.
103. H.C. Lin. Spatial correlations in adaptive arrays. *Antennas and Propagation, IEEE Transactions on*, 30(2):212–223, 1982.
104. X. Wang, E. Aboutanios, M. Trinkle, and M.G. Amin. Reconfigurable adaptive array beamforming by antenna selection. *Signal Processing, IEEE Transactions on*, 62(9):2385–2396, May 2014.
105. X. Wang, E. Aboutanios, and M. G. Amin. Slow radar target detection in heterogeneous clutter using thinned space-time adaptive processing. *IET Radar, Sonar & Navigation*, 10(4):726–734, 2016.
106. O.M. Bucci, G. D'Elia, G. Mazzarella, and G. Panariello. Antenna pattern synthesis: a new general approach. *Proceedings of the IEEE*, 82(3):358–371, 1994.
107. C.-Y. Tseng and L.J. Griffiths. A simple algorithm to achieve desired patterns for arbitrary arrays. *IEEE Transactions on Signal Processing*, 40(11):2737–2746, 1992.
108. D. J. Bekers, S. Jacobs, S. Monni, R. J. Bolt, D. Fortini, P. Capece, and G. Toso. A ka-band spaceborne synthetic aperture radar instrument: A modular sparse array antenna design. *IEEE Antennas and Propagation Magazine*, 61(5):97–104, 2019.

109. M.G. Amin. Concurrent nulling and locations of multiple interferences in adaptive antenna arrays. *IEEE Transactions on Signal Processing*, 40(11):2658–2668, 1992.
110. B. Friedlander and B. Porat. Performance analysis of a null-steering algorithm based on direction-of-arrival estimation. *IEEE Transactions on Acoustics, Speech, and Signal Processing*, 37(4):461–466, 1989.
111. S. Nai, W. Ser, Z. Yu, and H. Chen. Beampattern synthesis for linear and planar arrays with antenna selection by convex optimization. *IEEE Transactions on Antennas and Propagation*, 58(12):3923–3930, 2010.
112. P. Rocca, R. L. Haupt, and A. Massa. Interference suppression in uniform linear arrays through a dynamic thinning strategy. *IEEE Transactions on Antennas and Propagation*, 59(12):4525–4533, 2011.
113. L. Poli, P. Rocca, M. Salucci, and A. Massa. Reconfigurable thinning for the adaptive control of linear arrays. *IEEE Transactions on Antennas and Propagation*, 61(10):5068–5077, 2013.
114. L. Cen, Z. Yu, W. Ser, and C. Linear aperiodic array synthesis using an improved genetic algorithm. *IEEE Transactions on Antennas and Propagation*, 60(2):895–902, 2012.
115. B. Fuchs. Synthesis of sparse arrays with focused or shaped beampattern via sequential convex optimizations. *IEEE Transactions on Antennas and Propagation*, 60(7):3499–3503, 2012.
116. G. Oliveri, P. Rocca, and A. Massa. Reliable diagnosis of large linear arrays–a bayesian compressive sensing approach. *IEEE Transactions on Antennas and Propagation*, 60(10):4627–4636, 2012.
117. Y. Liu, Z. Nie, and Q. Liu. Reducing the number of elements in a linear antenna array by the matrix pencil method. *IEEE Transactions on Antennas and Propagation*, 56(9):2955–2962, 2008.
118. K. Yang, Z. Zhao, and Q. Liu. Fast pencil beam pattern synthesis of large unequally spaced antenna arrays. *IEEE Transactions on Antennas and Propagation*, 61(2):627–634, 2013.
119. W. P. du Plessis. Weighted thinned linear array design with the iterative fft technique. *IEEE Transactions on Antennas and Propagation*, 59(9):3473–3477, 2011.
120. W. P. M. N. Keizer. Linear array thinning using iterative fft techniques. *IEEE Transactions on Antennas and Propagation*, 56(8):2757–2760, 2008.
121. O.M. Bucci, L. Caccavale, and T. Isernia. Optimal far-field focusing of uniformly spaced arrays subject to arbitrary upper bounds in nontarget directions. *IEEE Transactions on Antennas and Propagation*, 50(11):1539–1554, 2002.
122. A. Capozzoli, C. Curcio, E. Iavazzo, A. Liseno, M. Migliorelli, and G. Toso. Phase-only synthesis of a-periodic reflectarrays. In *Proceedings of the 5th European Conference on Antennas and Propagation (EUCAP)*, pages 987–991, 2011.
123. E. Treister and I. Yavneh. A multilevel iterated-shrinkage approach to l1 penalized least-squares minimization. *Signal Processing, IEEE Transactions on*, 60(12):6319–6329, 2012.
124. Z. Shi and Z. Feng. A new array pattern synthesis algorithm using the two-step least-squares method. *Signal Processing Letters, IEEE*, 12(3):250–253, 2005.
125. B. Fuchs. Synthesis of sparse arrays with focused or shaped beampattern via sequential convex optimizations. *Antennas and Propagation, IEEE Transactions on*, 60(7):3499–3503, 2012.
126. W. Deng, W. Yin, and Y. Zhang. Group sparse optimization by alternating direction method. *TR11-06, Department of Computational and Applied Mathematics, Rice University*, 2011.
127. Z. Han, H. Li, and W. Yin. *Compressive Sensing for Wireless Networks*. Cambridge University Press, 2013.
128. G. Oliveri, M. Carlin, and A. Massa. Complex-weight sparse linear array synthesis by bayesian compressive sampling. *Antennas and Propagation, IEEE Transactions on*, 60(5):2309–2326, 2012.
129. A. Beck and M. Teboulle. A fast iterative shrinkage-thresholding algorithm for linear inverse problems. *SIAM Journal on Imaging Sciences*, 2(1):183–202, 2009.
130. E.T. Hale, W. Yin, and Y. Zhang. A fixed-point continuation method for l1-regularized minimization with applications to compressed sensing. *CAAM TR07-07, Rice University*, 2007.
131. H. Tuy. *Convex analysis and global optimization*, volume 22. Springer, 1998.
132. R. A. Horn and C. R. Johnson. *Matrix Analysis*. Cambridge University Press, 1985.

133. S. Boyd and L. Vandenberghe. *Convex Optimization*. Cambridge University Press, 2004.
134. J. Kovacevic and A. Chebira. Life beyond bases: The advent of frames (part i). *IEEE Signal Processing Magazine*, 24(4):86–104, 2007.
135. E. Prugovecki. Chapter iii theory of linear operators in hilbert spaces. volume 41, pages 172–256. 1971.
136. T. Isernia, A. Massa, A. F. Morabito, and P. Rocca. On the optimal synthesis of phase-only reconfigurable antenna arrays. In *Proceedings of the 5th European Conference on Antennas and Propagation (EUCAP)*, pages 2074–2077, 2011.
137. R. Horst. *Introduction to global optimization*. Springer, 2000.
138. S. Zionts. Programming with linear fractional functionals. *Naval Research Logistics Quarterly*, 15:449–451, 1968.
139. H. Crowder, E. L. Johnson, and M. Padberg. Solving large-scale zero-one linear programming problems. *Oper. Res.*, 31(5):803–834, oct 1983.
140. J. Mattingley and S. Boyd. Real-time convex optimization in signal processing. *IEEE Signal Processing Magazine*, 27(3):50–61, 2010.
141. B. Fuchs. Application of convex relaxation to array synthesis problems. *Antennas and Propagation, IEEE Transactions on*, 62(2):634–640, Feb 2014.
142. M. Grant, S. Boyd, and Y. Ye. *CVX: Matlab software for disciplined convex programming[Online]*. Available: http://www.stanford.edu/ boyd/cvx, 2008.
143. L.E. Brennan and L.S. Reed. Theory of adaptive radar. *IEEE Transactions on Aerospace and Electronic Systems*, AES-9(2):237–252, 1973.
144. R.L. Fante and JJ Vaccaro. Wideband cancellation of interference in a GPS receive array. *Aerospace and Electronic Systems, IEEE Transactions on*, 36(2):549–564, 2000.
145. A.-J. van der Veen, Amir Leshem, and A.-J. Boonstra. Signal processing for radio astronomical arrays. In *Processing Workshop Proceedings, 2004 Sensor Array and Multichannel Signal*, pages 1–10, 2004.
146. R.T. Compton. An adaptive array in a spread-spectrum communication system. *Proceedings of the IEEE*, 66(3):289–298, 1978.
147. D. Johnson and S. DeGraaf. Improving the resolution of bearing in passive sonar arrays by eigenvalue analysis. *IEEE Transactions on Acoustics, Speech, and Signal Processing*, 30(4):638–647, 1982.
148. Y. Selén, H. Tullberg, and J. Kronander. Sensor selection for cooperative spectrum sensing. In *New Frontiers in Dynamic Spectrum Access Networks, The 3rd IEEE Symposium on*, pages 1–11. IEEE, 2008.
149. X. Shen and P.K. Varshney. Sensor selection based on generalized information gain for target tracking in large sensor networks. *Signal Processing, IEEE Transactions on*, 62(2):363–375, Jan 2014.
150. M. B. Hawes and W. Liu. Sparse array design for wideband beamforming with reduced complexity in tapped delay-lines. *IEEE/ACM Transactions on Audio, Speech, and Language Processing*, 22(8):1236–1247, 2014.
151. P.P. Vaidyanathan and P. Pal. Sparse sensing with co-prime samplers and arrays. *Signal Processing, IEEE Transactions on*, 59(2):573–586, Feb 2011.
152. P. Pal and P.P. Vaidyanathan. Correlation-aware techniques for sparse support recovery. In *Statistical Signal Processing Workshop (SSP), 2012 IEEE*, pages 53–56, Aug 2012.
153. X. Wang, M.G. Ahmad, F. Amin, and E. Aboutanios. Interference DOA estimation for GNSS receivers using fully augmentable arrays. *The Jounral of Franklin Institute (accepted)*, 2015.
154. H. Gazzah and S. Marcos. Cramer-Rao bounds for antenna array design. *Signal Processing, IEEE Transactions on*, 54(1):336–345, 2006.
155. H. Gazzah and K. Abed-Meraim. Optimum ambiguity-free directional and omnidirectional planar antenna arrays for DOA estimation. *Signal Processing, IEEE Transactions on*, 57(10):3942–3953, Oct 2009.
156. S. Joshi and S. Boyd. Sensor selection via convex optimization. *Signal Processing, IEEE Transactions on*, 57(2):451–462, 2009.

157. X. Wang, M.G. Amin, F. Ahmad, and E. Aboutanios. Coarray based optimum geometries for DOA estimation with multiple CRPA GPS arrays. In *Proceedings of the 27th International Technical Meeting of The Satellite Division of the Institute of Navigation (ION GNSS+ 2014)*, Sep 2014.
158. M. Wax and Y. Anu. Performance analysis of the minimum variance beamformer in the presence of steering vector errors. *IEEE Transactions on Signal Processing*, 44(4):938–947, 1996.
159. A. Bertrand, J. Szurley, P. Ruckebusch, I. Moerman, and M. Moonen. Efficient calculation of sensor utility and sensor removal in wireless sensor networks for adaptive signal estimation and beamforming. *Signal Processing, IEEE Transactions on*, 60(11):5857–5869, 2012.
160. O. Mehanna, N.D. Sidiropoulos, and G.B. Giannakis. Joint multicast beamforming and antenna selection. *Signal Processing, IEEE Transactions on*, 61(10):2660–2674, 2013.
161. N. K. Dhingra, M. R. Jovanovic, and Z. Luo. An admm algorithm for optimal sensor and actuator selection. In *53rd IEEE Conference on Decision and Control*, pages 4039–4044, 2014.
162. J. Ranieri, A. Chebira, and M. Vetterli. Near-optimal sensor placement for linear inverse problems. *IEEE Transactions on Signal Processing*, 62(5):1135–1146, 2014.
163. N. B. Mehta, S. Kashyap, and A. F. Molisch. Antenna selection in lte: from motivation to specification. *IEEE Communications Magazine*, 50(10):144–150, 2012.
164. Q. Yuan, Q. Chen, and K. Sawaya. Performance of adaptive array antenna with arbitrary geometry in the presence of mutual coupling. *IEEE Transactions on Antennas and Propagation*, 54(7):1991–1996, 2006.
165. A. Garcia-Rodriguez, C. Masouros, and P. Rulikowski. Efficient large scale antenna selection by partial switching connectivity. In *2017 IEEE International Conference on Acoustics, Speech and Signal Processing (ICASSP)*, pages 6269–6273, 2017.
166. G. van Keuk and S.S. Blackman. On phased-array radar tracking and parameter control. *IEEE Transactions on Aerospace and Electronic Systems*, 29(1):186–194, 1993.
167. W.D. White. Low-angle radar tracking in the presence of multipath. *IEEE Transactions on Aerospace and Electronic Systems*, AES-10(6):835–852, 1974.
168. D. Rahamim, J. Tabrikian, and R. Shavit. Source localization using vector sensor array in a multipath environment. *IEEE Transactions on Signal Processing*, 52(11):3096–3103, 2004.
169. P. Setlur, M. Amin, and F. Ahmad. Multipath model and exploitation in through-the-wall and urban radar sensing. *IEEE Transactions on Geoscience and Remote Sensing*, 49(10):4021–4034, 2011.
170. J.W. Betz. Effect of narrowband interference on GPS code tracking accuracy. In *Proceedings of the 2000 National Technical Meeting of The Institute of Navigation*, pages 16–27, 2000.
171. J.W. Betz. Effect of partial-band interference on receiver estimation of C/N0: Theory. Technical report, DTIC Document, 2001.
172. A.T. Balaei and E. Aboutanios. Characterization of interference effects in multiple antenna GNSS receivers. In *Image and Signal Processing (CISP), 2010 3rd International Congress on*, volume 8, pages 3930–3934. IEEE, 2010.
173. M. Li, A.G. Dempster, A.T. Balaei, C. Rizos, and F. Wang. Switchable beam steering/null steering algorithm for CW interference mitigation in GPS C/A code receivers. *Aerospace and Electronic Systems, IEEE Transactions on*, 47(3):1564–1579, 2011.
174. X. Wang and E. Aboutanios. Reconfigurable adaptive linear array signal processing in GNSS applications. In *Acoustics, Speech and Signal Processing (ICASSP), 2013 IEEE International Conference on*, pages 4154–4158, May 2013.
175. A.M. Haimovich and Y. Bar-Ness. An eigenanalysis interference canceler. *Signal Processing, IEEE Transactions on*, 39(1):76–84, 1991. Jan
176. X. Wang, E. Aboutanios, and M.G. Amin. Adaptive array thinning for enhanced DOA estimation. *Signal Processing Letters, IEEE*, 22(7):799–803, July 2015.
177. P. Wedin. On angles between subspaces of a finite dimensional inner product space. *Lecture Notes in Mathematics*, 973:263–285, 1983.

178. H. Gunawan, O. Neswan, and W. Setya-Budhi. A formula for angles between subspaces of inner product spaces. 2005.
179. E. Hitzer. Angles Between Subspaces Computed in Clifford Algebra. *AIP Conference Proceedings*, 1281(1):1476–1479, 09 2010.
180. J. Miao and A. Ben-Israel. On principal angles between subspaces in rn. *Linear Algebra and its Applications*, 171:81–98, 1992.
181. R. A. Horn and C. R. Johnson. *Matrix Analysis*. Cambridge University Press, 2 edition, 2012.
182. M. Fazel, H. Hindi, and S.P. Boyd. Log-det heuristic for matrix rank minimization with applications to hankel and euclidean distance matrices. In *Proceedings of the 2003 American Control Conference, 2003.*, volume 3, pages 2156–2162 vol.3, 2003.
183. S. G. Nash and A. Sofer. Linear and nonlinear programming. 1987.
184. Z. Luo, W. Ma, A. M. So, Y. Ye, and S. Zhang. Semidefinite relaxation of quadratic optimization problems. *IEEE Signal Processing Magazine*, 27:20–34, 2010.
185. S. Shahbazpanahi, A.B. Gershman, Z. Luo, and K. Wong. Robust adaptive beamforming for general-rank signal models. *IEEE Transactions on Signal Processing*, 51(9):2257–2269, 2003.
186. H. Deng, Z. Geng, and B. Himed. Mimo radar waveform design for transmit beamforming and orthogonality. *IEEE Transactions on Aerospace and Electronic Systems*, 52(3):1421–1433, 2016.
187. X. Wang, E. Aboutanios, M. Trinkle, and M. G. Amin. Reconfigurable adaptive array beamforming by antenna selection. *IEEE Transactions on Signal Processing*, 62(9):2385–2396, 2014.
188. B. Liao and S. Chan. Adaptive beamforming for uniform linear arrays with unknown mutual coupling. *IEEE Antennas and Wireless Propagation Letters*, 11:464–467, 2012.
189. X. Wu, Y. Cai, M. Zhao, R. C. de Lamare, and B. Champagne. Adaptive widely linear constrained constant modulus reduced-rank beamforming. *IEEE Transactions on Aerospace and Electronic Systems*, 53(1):477–492, 2017.
190. S. He, Y. Huang, J. Wang, L. Yang, and W. Hong. Joint antenna selection and energy-efficient beamforming design. *IEEE Signal Processing Letters*, 23(9):1165–1169, 2016.
191. L. Lei, J. P. Lie, A. B. Gershman, and C. See. Robust adaptive beamforming in partly calibrated sparse sensor arrays. *IEEE Transactions on Signal Processing*, 58(3):1661–1667, 2010.
192. X. Wang, C. Pin Tan, Y. Wang, and Z. Zhang. Active fault tolerant control based on adaptive interval observer for uncertain systems with sensor faults. *International Journal of Robust and Nonlinear Control*, 31(8):2857–2881, 2021.
193. X. Wang, W. Li, and V. C. Chen. Hand gesture recognition using radial and transversal dual micromotion features. *IEEE Transactions on Aerospace and Electronic Systems*, 58(6):5963–5973, 2022.
194. W. Zhang, Z. He, H. Li, J. Li, and X. Duan. Beam-space reduced-dimension space-time adaptive processing for airborne radar in sample starved heterogeneous environments. *IET Radar, Sonar & Navigation*, 10(9):1627–1634, 2016.
195. J. Yan, H. Liu, W. Pu, S. Zhou, Z. Liu, and Z. Bao. Joint beam selection and power allocation for multiple target tracking in netted colocated mimo radar system. *IEEE Transactions on Signal Processing*, 64(24):6417–6427, 2016.
196. X. Wang, M. S. Greco, and F. Gini. Adaptive sparse array beamformer design by regularized complementary antenna switching. *IEEE Transactions on Signal Processing*, 69:2302–2315, 2021.
197. X. Wang, M. Amin, X. Wang, and X. Cao. Sparse array quiescent beamformer design combining adaptive and deterministic constraints. *IEEE Transactions on Antennas and Propagation*, 65(11):5808–5818, 2017.
198. Y. Liu, L. Zhang, L. Ye, Z. Nie, and Q. Liu. Synthesis of sparse arrays with frequency-invariant-focused beam patterns under accurate sidelobe control by iterative second-order cone programming. *IEEE Transactions on Antennas and Propagation*, 63(12):5826–5832, 2015.

199. H. Shen and B. Wang. An effective method for synthesizing multiple-pattern linear arrays with a reduced number of antenna elements. *IEEE Transactions on Antennas and Propagation*, 65(5):2358–2366, 2017.
200. H. Nosrati, E. Aboutanios, and D. Smith. Multi-stage antenna selection for adaptive beamforming in mimo radar. *IEEE Transactions on Signal Processing*, 68:1374–1389, 2020.
201. J. Wang, W. Sheng, Y. Han, and X. Ma. Adaptive beamforming with compressed sensing for sparse receiving array. *IEEE Transactions on Aerospace and Electronic Systems*, 50(2):823–833, 2014.
202. S. A. Hamza and M. G. Amin. Hybrid sparse array beamforming design for general rank signal models. *IEEE Transactions on Signal Processing*, 67(24):6215–6226, 2019.
203. S. A. Hamza and M. G. Amin. Sparse array design for maximizing the signal-to-interference-plus-noise-ratio by matrix completion. *Digital Signal Processing*, 105:102678, 2020. Special Issue on Optimum Sparse Arrays and Sensor Placement for Environmental Sensing.
204. S. A. Hamza and M. G. Amin. Learning sparse array capon beamformer design using deep learning approach. In *2020 IEEE Radar Conference (RadarConf20)*, pages 1–5, 2020.
205. A. M. Elbir and K. V. Mishra. Joint antenna selection and hybrid beamformer design using unquantized and quantized deep learning networks. *IEEE Transactions on Wireless Communications*, 19(3):1677–1688, 2020.
206. S. Haykin, Y. Xue, and P. Setoodeh. Cognitive radar: Step toward bridging the gap between neuroscience and engineering. *Proceedings of the IEEE*, 100(11):3102–3130, 2012.
207. M. S. Greco, F. Gini, P. Stinco, and K. Bell. Cognitive radars: On the road to reality: Progress thus far and possibilities for the future. *IEEE Signal Processing Magazine*, 35(4):112–125, 2018.
208. G. Hakobyan, K. Armanious, and B. Yang. Interference-aware cognitive radar: A remedy to the automotive interference problem. *IEEE Transactions on Aerospace and Electronic Systems*, 56(3):2326–2339, 2020.
209. X. Wang, Y. Jiao, and Y. Tan. Synthesis of large thinned planar arrays using a modified iterative fourier technique. *Antennas and Propagation, IEEE Transactions on*, 62(4):1564–1571, April 2014.
210. H. Krim and M. Viberg. Two decades of array signal processing research: the parametric approach. *Signal Processing Magazine, IEEE*, 13(4):67–94, 1996.
211. Q. Tran-Dinh and M. Diehl. *Local Convergence of Sequential Convex Programming for Nonconvex Optimization*, pages 93–102. 01 2010.
212. T. Lipp and S. Boyd. Variations and extension of the convex–concave procedure. *Optimization and Engineering*, 17, 06 2016.
213. E.J. Candes, M.B. Wakin, and S.P. Boyd. Enhancing sparsity by reweighted l1 minimization. *Journal of Fourier Analysis and Applications*, 14(5):877–905, 2008.
214. W. Zhai, X. Wang, M. S. Greco, and F. Gini. Joint optimization of sparse fdas for time invariant transmit beampattern synthesis. *IEEE Signal Processing Letters*, 29:110–114, 2022.
215. P. Stoica, Z. Wang, and J. Li. Robust capon beamforming. *IEEE Signal Processing Letters*, 10(6):172–175, 2003.
216. J. Li, P. Stoica, and Z. Wang. On robust capon beamforming and diagonal loading. *IEEE Transactions on Signal Processing*, 51(7):1702–1715, 2003.
217. Z. Tian, K.L. Bell, and H.L. Van Trees. A recursive least squares implementation for lcmp beamforming under quadratic constraint. *IEEE Transactions on Signal Processing*, 49(6):1138–1145, 2001.
218. X. Wang, M.s Amin, and X. Wang. Robust sparse array design for adaptive beamforming against doa mismatch. *Signal Processing*, 146:41–49, 2018.
219. X. Wang and E. Aboutanios. Sparse array design for multiple switched beams using iterative antenna selection method. *Digital Signal Processing*, 105:102684, 2020. Special Issue on Optimum Sparse Arrays and Sensor Placement for Environmental Sensing.
220. D. Wang, X. Ma, A. Chen, and Y. Su. High-resolution imaging using a wideband mimo radar system with two distributed arrays. *IEEE Transactions on Image Processing*, 19(5):1280–1289, 2010.

221. S. G. Dontamsetti and R. V. R. Kumar. A distributed mimo radar with joint optimal transmit and receive signal combining. *IEEE Transactions on Aerospace and Electronic Systems*, 57(1):623–635, 2021.
222. J. Qian, M. Lops, L. Zheng, X. Wang, and Z. He. Joint system design for coexistence of mimo radar and mimo communication. *IEEE Transactions on Signal Processing*, 66(13):3504–3519, 2018.
223. Y. I. Abramovich and G. J. Frazer. Bounds on the volume and height distributions for the mimo radar ambiguity function. *IEEE Signal Processing Letters*, 15:505–508, 2008.
224. C. Liu and P. P. Vaidyanathan. Correlation subspaces: Generalizations and connection to difference coarrays. *IEEE Transactions on Signal Processing*, 65(19):5006–5020, 2017.
225. A. Hassanien, M. G. Amin, Y. D. Zhang, and F. Ahmad. High-resolution single-snapshot doa estimation in mimo radar with colocated antennas. In *2015 IEEE Radar Conference (RadarCon)*, pages 1134–1138, 2015.
226. A. Hassanien and S. A. Vorobyov. Transmit energy focusing for doa estimation in mimo radar with colocated antennas. *IEEE Transactions on Signal Processing*, 59(6):2669–2682, 2011.
227. S. Gogineni and A. Nehorai. Target estimation using sparse modeling for distributed mimo radar. *IEEE Transactions on Signal Processing*, 59(11):5315–5325, 2011.
228. X. Wang, A. Hassanien, and M. G. Amin. Dual-function mimo radar communications system design via sparse array optimization. *IEEE Transactions on Aerospace and Electronic Systems*, 55(3):1213–1226, 2019.
229. M. Masood, L. H. Afify, and T. Y. Al-Naffouri. Efficient coordinated recovery of sparse channels in massive mimo. *IEEE Transactions on Signal Processing*, 63(1):104–118, 2015.
230. X. Wang, C. Tan, F. Wu, and J. Wang. Fault-tolerant attitude control for rigid spacecraft without angular velocity measurements. *IEEE Transactions on Cybernetics*, 51(3):1216–1229, 2021.
231. A. Moffet. Minimum-redundancy linear arrays. *IEEE Transactions on Antennas and Propagation*, 16(2):172–175, 1968.
232. P. Pal and P. P. Vaidyanathan. Nested arrays: A novel approach to array processing with enhanced degrees of freedom. *IEEE Transactions on Signal Processing*, 58(8):4167–4181, 2010.
233. S. Qin, Y. D. Zhang, and M. G. Amin. Generalized coprime array configurations for direction-of-arrival estimation. *IEEE Transactions on Signal Processing*, 63(6):1377–1390, 2015.
234. A. Ahmed, Y. D. Zhang, and J. Zhang. Coprime array design with minimum lag redundancy. In *ICASSP 2019 - 2019 IEEE International Conference on Acoustics, Speech and Signal Processing (ICASSP)*, pages 4125–4129, 2019.
235. S. A. Hamza and M. G. Amin. Sparse array beamforming design for wideband signal models. *IEEE Transactions on Aerospace and Electronic Systems*, 57(2):1211–1226, 2021.
236. X. Wang, M. G. Amin, and X. Cao. Optimum adaptive beamformer design with controlled quiescent pattern by antenna selection. In *2017 IEEE Radar Conference (RadarConf)*, pages 0749–0754, 2017.
237. S. A. Hamza and M. G. Amin. Optimum sparse array receive beamforming for wideband signal model. In *2018 52nd Asilomar Conference on Signals, Systems, and Computers*, pages 89–93, 2018.
238. X. Wang, M. Amin, and X. Cao. Analysis and design of optimum sparse array configurations for adaptive beamforming. *IEEE Transactions on Signal Processing*, 66(2):340–351, 2018.
239. S. A. Hamza and M. G. Amin. Hybrid sparse array design for under-determined models. In *ICASSP 2019 - 2019 IEEE International Conference on Acoustics, Speech and Signal Processing (ICASSP)*, pages 4180–4184, 2019.
240. W. Huleihel, J. Tabrikian, and R. Shavit. Optimal adaptive waveform design for cognitive mimo radar. *IEEE Transactions on Signal Processing*, 61(20):5075–5089, 2013.
241. J. S. Bergin, J. R. Guerci, R. M. Guerci, and M. Rangaswamy. Mimo clutter discrete probing for cognitive radar. In *2015 IEEE Radar Conference (RadarCon)*, pages 1666–1670, 2015.
242. N. Sharaga, J. Tabrikian, and H. Messer. Optimal cognitive beamforming for target tracking in mimo radar/sonar. *IEEE Journal of Selected Topics in Signal Processing*, 9(8):1440–1450, 2015.

243. P.D. Karamalis, N.D. Skentos, and A.G. Kanatas. Selecting array configurations for mimo systems: an evolutionary computation approach. *IEEE Transactions on Wireless Communications*, 3(6):1994–1998, 2004.
244. M. G. Amin, X. Wang, Y. D. Zhang, F. Ahmad, and E. Aboutanios. Sparse arrays and sampling for interference mitigation and doa estimation in gnss. *Proceedings of the IEEE*, 104(6):1302–1317, 2016.
245. W. Roberts, L. Xu, J. Li, and P. Stoica. Sparse antenna array design for mimo active sensing applications. *IEEE Transactions on Antennas and Propagation*, 59(3):846–858, 2011.
246. Z. Cheng, Y. Lu, Z. He, Yufengli, J. Li, and X. Luo. Joint optimization of covariance matrix and antenna position for mimo radar transmit beampattern matching design. In *2018 IEEE Radar Conference (RadarConf18)*, pages 1073–1077, 2018.
247. S. A. Hamza and M. G. Amin. Sparse array design for transmit beamforming. In *2020 IEEE International Radar Conference (RADAR)*, pages 560–565, 2020.
248. O. Mehanna, N. D. Sidiropoulos, and G. B. Giannakis. Joint multicast beamforming and antenna selection. *IEEE Transactions on Signal Processing*, 61(10):2660–2674, 2013.
249. S. A. Hamza, M. G. Amin, and G. Fabrizio. Optimum sparse array beamforming for general rank signal models. In *2018 IEEE Radar Conference (RadarConf18)*, pages 1343–1347, 2018.
250. N. Kwak. Principal component analysis based on l1-norm maximization. *IEEE Transactions on Pattern Analysis and Machine Intelligence*, 30(9):1672–1680, 2008.
251. J. Ward. Space-time adaptive processing for airborne radar. *Lincoln Laboratory Tech. Rept. ESC-TR-94-109*, Dec 1994.
252. W.L. Melvin. A STAP overview. *IEEE Aerospace and Electronic Systems Magazine*, 19(1):19–35, 2004.
253. J.R. Guerci, J.S. Goldstein, and I.S. Reed. Optimal and adaptive reduced-rank STAP. *IEEE Transactions on Aerospace and Electronic Systems*, 36(2):647–663, 2000.
254. C.D. Richmond. Performance of a class of adaptive detection algorithms in nonhomogeneous environments. *Signal Processing, IEEE Transactions on*, 48(5):1248–1262, May 2000.
255. L. Scott and B. Mulgrew. Sparse lcmv beamformer design for suppression of ground clutter in airborne radar. *IEEE Transactions on Signal Processing*, 43(12):2843–2851, 1995.
256. M. Rangaswamy, J. H. Michels, and B. Himed. Statistical analysis of the non-homogeneity detector for stap applications. *Digital Signal Processing*, 14(3):253–267, 2004.
257. E. Aboutanios and B. Mulgrew. Hybrid detection approach for STAP in heterogeneous clutter. *IEEE Transactions on Aerospace and Electronic Systems*, 46(3):1021–1033, 2010.
258. D. J. Rabideau and A. Steinhardt. Improved adaptive clutter cancellation through data-adaptive training. *Aerospace and Electronic Systems, IEEE Transactions on*, 35(3):879–891, 1999.
259. J. F. Degurse, L. Savy, and S. Marcos. Reduced-rank STAP for target detection in heterogeneous environments. *Aerospace and Electronic Systems, IEEE Transactions on*, 50(2):1153–1162, April 2014.
260. E. Aboutanios and B. Mulgrew. Heterogeneity detection for hybrid stap algorithm. In *2008 IEEE Radar Conference*, pages 1–4, 2008.
261. W.L. Melvin and J.R. Guerci. Knowledge-aided signal processing: a new paradigm for radar and other advanced sensors. *IEEE Transactions on Aerospace and Electronic Systems*, 42(3):983–996, 2006.
262. S. Bidon, O. Besson, and J. Tourneret. Knowledge-aided STAP in heterogeneous clutter using a hierarchical Bayesian algorithm. *IEEE Transactions on Aerospace and Electronic Systems*, 47(3):1863–1879, July 2011.
263. M. Steiner and K. Gerlach. Fast converging adaptive processor or a structured covariance matrix. *IEEE Transactions on Aerospace and Electronic Systems*, 36(4):1115–1126, 2000.
264. A. Aubry, A. De Maio, L. Pallotta, and A. Farina. Maximum likelihood estimation of a structured covariance matrix with a condition number constraint. *IEEE Transactions on Signal Processing*, 60(6):3004–3021, 2012.
265. T.K. Sarkar, H. Wang, S. Park, R. Adve, J. Koh, K. Kim, Y. Zhang, M.C. Wucjs, and R.D. Brown. A Deterministic Least-Squares Approach to Space-Time Adaptive Processing (STAP). *IEEE Transactions on Antennas and Propagation*, 49(1):91–103, January 2001.

266. E. Aboutanios and B. Mulgrew. A STAP algorithm for radar target detection in heterogeneous environments. In *Proc. IEEE/SP 13th Workshop on Statistical Signal Processing*, pages 966–971, 2005.
267. H. Deng, B. Himed, and M.C. Wicks. Image feature-based space-time processing for ground moving target detection. *Signal Processing Letters, IEEE*, 13(4):216–219, April 2006.
268. C. H. Gierull. Statistical analysis of the eigenvector projection method for adaptive spatial filtering of interference. *Radar, Sonar and Navigation, IEE Proceedings*, 144(2):57–63, Apr 1997.
269. S. Sen. Ofdm radar space-time adaptive processing by exploiting spatio-temporal sparsity. *IEEE Transactions on Signal Processing*, 61(1):118–130, 2013.
270. K. Sun, H. Meng, Y. Wang, and X. Wang. Direct data domain stap using sparse representation of clutter spectrum. *Signal Processing*, 91(9):2222–2236, 2011.
271. Z. Yang, R. C. de Lamare, and X. Li. L1 regularized stap algorithm with a generalized sidelobe canceler architecture for airborne radar. In *2011 IEEE Statistical Signal Processing Workshop (SSP)*, pages 329–332, 2011.
272. E.J. Baranoski. Sparse network array processing. In *Seventh IEEE Workshop on Statistical Signal and Array Processing*, pages 145–148, 1994.
273. F.M. Brennan, L.E.and Staudaher. Subclutter visibility demonstration. *Adaptive Sensors, Tech. Rep., RL-TR-92-21*, 1992.
274. Y. Wu, J. Tang, and Y. Peng. On the essence of knowledge-aided clutter covariance estimate and its convergence. *IEEE Transactions on Aerospace and Electronic Systems*, 47(1):569–585, 2011.
275. M. Gavish and A.J. Weiss. Array geometry for ambiguity resolution in direction finding. *Antennas and Propagation, IEEE Transactions on*, 44(6):889–895, 1996.
276. M. Elad. *Sparse and redundant representations: from theory to applications in signal and image processing*. Springer, 2010.
277. R. Horst and N. V. Thoai. DC programming: overview. *Journal of Optimization Theory and Applications*, 103(1):1–43, 1999.
278. G. Gasso, A. Rakotomamonjy, and S. Canu. Recovering sparse signals with a certain family of nonconvex penalties and DC programming. *Signal Processing, IEEE Transactions on*, 57(12):4686–4698, 2009.
279. A. Khabbazibasmenj and S.A. Vorobyov. Robust adaptive beamforming for general-rank signal model with positive semi-definite constraint via POTDC. *Signal Processing, IEEE Transactions on*, 61(23):6103–6117, 2013.
280. A. Beck, A. Ben-Tal, and L. Tetruashvili. A sequential parametric convex approximation method with applications to nonconvex truss topology design problems. *Journal of Global Optimization*, 47(1):29–51, 2010.
281. Z. Luo, W. Ma, A. M. So, Y. Ye, and S. Zhang. Semidefinite relaxation of quadratic optimization problems. *IEEE Signal Processing Magazine*, 27(3):20–34, 2010.
282. T. H. Cormen, C. E. Leiserson, R. L. Rivest, and C. Stein. *Introduction to algorithms*. MIT press, 2001.
283. B. Himed. MCARM/STAP data analysis. volume II. Technical report, DTIC Document, 1999.
284. R.S. Adve and M.C. Wicks. Joint domain localized processing using measured spatial steering vectors. In *Radar Conference, 1998. RADARCON 98. Proceedings of the 1998 IEEE*, pages 165–170, May 1998.
285. H. Griffiths, S. Blunt, L. Cohen, and L. Savy. Challenge problems in spectrum engineering and waveform diversity. In *2013 IEEE Radar Conference (RadarCon13)*, pages 1–5, 2013.
286. H. Deng and B. Himed. Interference mitigation processing for spectrum-sharing between radar and wireless communications systems. *IEEE Transactions on Aerospace and Electronic Systems*, 49(3):1911–1919, 2013.
287. D. W. Bliss. Cooperative radar and communications signaling: The estimation and information theory odd couple. In *2014 IEEE Radar Conference*, pages 0050–0055, 2014.
288. H. T. Hayvaci and B. Tavli. Spectrum sharing in radar and wireless communication systems: A review. In *2014 International Conference on Electromagnetics in Advanced Applications (ICEAA)*, pages 810–813, 2014.

289. C. Baylis, M. Fellows, L. Cohen, and R. Marks II. Solving the spectrum crisis: Intelligent, reconfigurable microwave transmitter amplifiers for cognitive radar. *IEEE Microwave Magazine*, 15(5):94–107, 2014.
290. K. Huang, M. Bica, U. Mitra, and V. Koivunen. Radar waveform design in spectrum sharing environment: Coexistence and cognition. In *2015 IEEE Radar Conference (RadarCon)*, pages 1698–1703, 2015.
291. H. Griffiths, L. Cohen, S. Watts, E. Mokole, C. Baker, M. Wicks, and S. Blunt. Radar spectrum engineering and management: Technical and regulatory issues. *Proceedings of the IEEE*, 103(1):85–102, 2015.
292. R. M. Mealey. A method for calculating error probabilities in a radar communication system. *IEEE Transactions on Space Electronics and Telemetry*, 9(2):37–42, 1963.
293. S. D. Blunt, P. Yatham, and J. Stiles. Intrapulse radar-embedded communications. *IEEE Transactions on Aerospace and Electronic Systems*, 46(3):1185–1200, 2010.
294. D. Ciuonzo, A. De Maio, G. Foglia, and M. Piezzo. Intrapulse radar-embedded communications via multiobjective optimization. *IEEE Transactions on Aerospace and Electronic Systems*, 51(4):2960–2974, 2015.
295. J. R. Guerci, R. M. Guerci, A. Lackpour, and D. Moskowitz. Joint design and operation of shared spectrum access for radar and communications. In *2015 IEEE Radar Conference (RadarCon)*, pages 0761–0766, 2015.
296. A. Aubry, A. De Maio, M. Piezzo, and A. Farina. Radar waveform design in a spectrally crowded environment via nonconvex quadratic optimization. *IEEE Transactions on Aerospace and Electronic Systems*, 50(2):1138–1152, 2014.
297. A. Aubry, A. De Maio, Y. Huang, M. Piezzo, and A. Farina. A new radar waveform design algorithm with improved feasibility for spectral coexistence. *IEEE Transactions on Aerospace and Electronic Systems*, 51(2):1029–1038, 2015.
298. B. Li, A. P. Petropulu, and W. Trappe. Optimum co-design for spectrum sharing between matrix completion based mimo radars and a mimo communication system. *IEEE Transactions on Signal Processing*, 64(17):4562–4575, 2016.
299. A. Hassanien, M. G. Amin, Y. D. Zhang, and F. Ahmad. A dual function radar-communications system using sidelobe control and waveform diversity. In *2015 IEEE Radar Conference (RadarCon)*, pages 1260–1263, 2015.
300. A. Hassanien, M. G. Amin, Y. D. Zhang, and F. Ahmad. Dual-function radar-communications using phase-rotational invariance. In *2015 23rd European Signal Processing Conference (EUSIPCO)*, pages 1346–1350, 2015.
301. A. Hassanien, M. G. Amin, Y. D. Zhang, and F. Ahmad. Phase-modulation based dual-function radar-communications. *IET Radar, Sonar & Navigation*, 10(8):1411–1421, 2016.
302. A. Hassanien, M. G. Amin, Y. D. Zhang, and F. Ahmad. Dual-function radar-communications: Information embedding using sidelobe control and waveform diversity. *IEEE Transactions on Signal Processing*, 64(8):2168–2181, 2016.
303. P. M. McCormick, S. D. Blunt, and J. G. Metcalf. Simultaneous radar and communications emissions from a common aperture, part i: Theory. In *2017 IEEE Radar Conference (RadarConf)*, pages 1685–1690, 2017.
304. P. M. McCormick, B. Ravenscroft, S. D. Blunt, A. J. Duly, and J. G. Metcalf. Simultaneous radar and communication emissions from a common aperture, part ii: Experimentation. In *2017 IEEE Radar Conference (RadarConf)*, pages 1697–1702, 2017.
305. F. Liu, C. Masouros, A. Li, and T. Ratnarajah. Robust mimo beamforming for cellular and radar coexistence. *IEEE Wireless Communications Letters*, 6(3):374–377, 2017.
306. S. E. Nai, W. Ser, Z. L. Yu, and S. Rahardja. A robust adaptive beamforming framework with beampattern shaping constraints. *IEEE Transactions on Antennas and Propagation*, 57(7):2198–2203, 2009.
307. S. Sanayei and A. Nosratinia. Antenna selection in mimo systems. *IEEE Communications Magazine*, 42(10):68–73, 2004.
308. Z. Xiao, P. Xia, and X. Xia. Full-duplex millimeter-wave communication. *IEEE Wireless Communications*, 24(6):136–143, 2017.

309. Z. Xiao, C. Yin, P. Xia, and X. Xia. Codebook design for millimeter-wave channel estimation with hybrid precoding structure. In *2016 IEEE International Conference on Communication Systems (ICCS)*, pages 1–6, 2016.
310. A.F. Molisch. Mimo systems with antenna selection - an overview. In *Radio and Wireless Conference, 2003. RAWCON '03. Proceedings*, pages 167–170, 2003.
311. A. Domahidi, E. Chu, and S. Boyd. Ecos: An socp solver for embedded systems. In *2013 European Control Conference (ECC)*, pages 3071–3076, 2013.
312. S. P. Boyd. Real-time embedded convex optimization. *IFAC Proceedings Volumes*, 42(11):9, 2009. 7th IFAC Symposium on Advanced Control of Chemical Processes.
313. S. S. Haykin. Communication systems. 1978.
314. S. D. Blunt, M. R. Cook, and J. Stiles. Embedding information into radar emissions via waveform implementation. In *2010 International Waveform Diversity and Design Conference*, pages 000195–000199, 2010.
315. E. BouDaher, A. Hassanien, E. Aboutanios, and M. G. Amin. Towards a dual-function mimo radar-communication system. In *2016 IEEE Radar Conference (RadarConf)*, pages 1–6, 2016.
316. M. N. El Korso, R. Boyer, A. Renaux, and Sy. Marcos. Conditional and unconditional Cramer-Rao bounds for near-field source localization. *Signal Processing, IEEE Transactions on*, 58(5):2901–2907, 2010.
317. H. Gazzah and J.-P. Delmas. CRB-based design of linear antenna arrays for near-field source localization. *Antennas and Propagation, IEEE Transactions on*, 62(4):1965–1974, April 2014.
318. E. J. Vertatschitsch and S. Haykin. Impact of linear array geometry on direction-of-arrival estimation for a single source. *Antennas and Propagation, IEEE Transactions on*, 39(5):576–584, May 1991.
319. P. Stoica and A. Nehorai. Performance study of conditional and unconditional direction-of-arrival estimation. *Acoustics, Speech and Signal Processing, IEEE Transactions on*, 38(10):1783–1795, Oct 1990.
320. A.N. Mirkin and L.H. Sibul. Cramer-Rao bounds on angle estimation with a two-dimensional array. *Signal Processing, IEEE Transactions on*, 39(2):515–517, 1991.
321. R.O. Nielsen. Azimuth and elevation angle estimation with a three-dimensional array. *Oceanic Engineering, IEEE Journal of*, 19(1):84–86, Jan 1994.
322. U. Baysal and R.L. Moses. On the geometry of isotropic arrays. *Signal Processing, IEEE Transactions on*, 51(6):1469–1478, 2003.
323. U. Oktel and R.L. Moses. A bayesian approach to array geometry design. *Signal Processing, IEEE Transactions on*, 53(5):1919–1923, 2005.
324. H. Gazzah and J.-P. Delmas. Direction finding antenna arrays for the randomly located source. *Signal Processing, IEEE Transactions on*, 60(11):6063 –6068, 2012.
325. V. Roy, S.P. Chepuri, and G. Leus. Sparsity-enforcing sensor selection for DOA estimation. In *Computational Advances in Multi-Sensor Adaptive Processing (CAMSAP), 2013 IEEE 5th International Workshop on*, pages 340–343, Dec 2013.
326. A. Dogandzic and A. Nehorai. Cramer-Rao bounds for estimating range, velocity, and direction with an active array. *Signal Processing, IEEE Transactions on*, 49(6):1122–1137, Jun 2001.
327. B. Yang and J. Scheuing. Cramer-Rao bound and optimum sensor array for source localization from time differences of arrival. In *Acoustics, Speech, and Signal Processing, 2005. Proceedings. (ICASSP '05). IEEE International Conference on*, volume 4, pages iv/961–iv/964 Vol. 4, March 2005.
328. K. W. Lui and H.C. So. A study of two-dimensional sensor placement using time-difference-of-arrival measurements. *Digital Signal Processing*, 19(4):650–659, 2009.
329. A.B. Gershman and J. F. Bohme. A note on most favorable array geometries for DOA estimation and array interpolation. *Signal Processing Letters, IEEE*, 4(8):232–235, 1997.
330. H. Messer. Source localization performance and the array beampattern. *Signal processing*, 28(2):163–181, 1992.
331. T. Birinci and Y.I. TanIk. Optimization of nonuniform array geometry for DOA estimation with the constraint on gross error probability. *Signal Processing*, 87(10):2360–2369, 2007.

332. W. Dinkelbach. On nonlinear fractional programming. *Management Science*, 13(7):492–498, 1967.
333. D. Malioutov, M. Cetin, and A.S. Willsky. A sparse signal reconstruction perspective for source localization with sensor arrays. *Signal Processing, IEEE Transactions on*, 53(8):3010–3022, 2005.
334. E. J. Candes, J. K. Romberg, and T. Tao. Stable signal recovery from incomplete and inaccurate measurements. *Communications on pure and applied mathematics*, 59(8):1207–1223, 2006.
335. S.U. Pillai, Y. Bar-Ness, and F. Haber. A new approach to array geometry for improved spatial spectrum estimation. *Proceedings of the IEEE*, 73(10):1522–1524, Oct 1985.
336. Y.I. Abramovich, D.A. Gray, A.Y. Gorokhov, and N.K. Spencer. Positive-definite Toeplitz completion in DOA estimation for nonuniform linear antenna arrays. I. Fully augmentable arrays. *Signal Processing, IEEE Transactions on*, 46(9):2458–2471, Sep 1998.
337. Y.I. Abramovich, N.K. Spencer, and A.Y. Gorokhov. Positive-definite Toeplitz completion in DOA estimation for nonuniform linear antenna arrays. II. Partially augmentable arrays. *Signal Processing, IEEE Transactions on*, 47(6):1502–1521, Jun 1999.
338. P. Pal and P.P. Vaidyanathan. On application of LASSO for sparse support recovery with imperfect correlation awareness. In *Signals, Systems and Computers (ASILOMAR), 2012 Conference Record of the Forty Sixth Asilomar Conference on*, pages 958–962, Nov 2012.
339. A. Moffet. Minimum-redundancy linear arrays. *Antennas and Propagation, IEEE Transactions on*, 16(2):172–175, Mar 1968.
340. R.T. Hoctor and S.A. Kassam. The unifying role of the coarray in aperture synthesis for coherent and incoherent imaging. *Proceedings of the IEEE*, 78(4):735–752, Apr 1990.
341. Y.I. Abramovich, N.K. Spencer, and A.Y. Gorokhov. Detection-estimation of more uncorrelated gaussian sources than sensors in nonuniform linear antenna arrays. II. Partially augmentable arrays. *Signal Processing, IEEE Transactions on*, 51(6):1492–1507, June 2003.
342. A.H. Tewfik and W. Hong. On the application of uniform linear array bearing estimation techniques to uniform circular arrays. *Signal Processing, IEEE Transactions on*, 40(4):1008–1011, Apr 1992.
343. Y.D. Zhang, M.G. Amin, and B. Himed. Sparsity-based DOA estimation using co-prime arrays. In *Acoustics, Speech and Signal Processing (ICASSP), 2013 IEEE International Conference on*, pages 3967–3971, May 2013.
344. S. Ji, Y. Xue, and L. Carin. Bayesian compressive sensing. *Signal Processing, IEEE Transactions on*, 56(6):2346–2356, 2008.
345. M.E. Tipping. Sparse Bayesian learning and the relevance vector machine. *The Journal of Machine Learning Research*, 1:211–244, 2001.
346. M. Carlin, P. Rocca, G. Oliveri, F. Viani, and A. Massa. Directions-of-arrival estimation through Bayesian compressive sensing strategies. *Antennas and Propagation, IEEE Transactions on*, 61(7):3828–3838, 2013.
347. Q. Wu, Y.D. Zhang, M.G. Amin, and B. Himed. Complex multitask Bayesian compressive sensing. In *Acoustics, Speech and Signal Processing (ICASSP), 2014 IEEE International Conference on*, pages 3375–3379, May 2014.
348. J. Li and P. Stoica. Efficient mixed-spectrum estimation with applications to target feature extraction. *IEEE Transactions on Signal Processing*, 44(2):281–295, 1996.
349. R.T. Compton. The power-inversion adaptive array: Concept and performance. *Aerospace and Electronic Systems, IEEE Transactions on*, AES-15(6):803–814, Nov 1979.
350. S. U. Pillai and B. H. Kwon. Forward/backward spatial smoothing techniques for coherent signal identification. *Acoustics, Speech and Signal Processing, IEEE Transactions on*, 1989.
351. S.U. Pillai and F. Haber. Statistical analysis of a high resolution spatial spectrum estimator utilizing an augmented covariance matrix. *Acoustics, Speech and Signal Processing, IEEE Transactions on*, 35(11):1517–1523, Nov 1987.

352. G.S. Bloom and S.W. Golomb. Application of numbered undirected graphs. *Proceedings of the IEEE*, 65(4):562–570, 1977.
353. Y.I. Abramovich, D.A. Gray, A.Y. Gorokhov, and N.K. Spencer. Comparison of DOA estimation performance for various types of sparse antenna array geometries. In *Proc. EUSIPCO*, volume 96, pages 915–918, 1996.
354. E. Vertatschitsch and S. Haykin. Nonredundant arrays. *Proceedings of the IEEE*, 74(1):217–217, Jan 1986.

The manufacturer's authorised representative in the EU is Springer
Nature Customer Service Centre GmbH, Europaplatz 3, 69115 Heidelberg,
Germany. If you have any concerns regarding our products, please
contact ProductSafety@springernature.com

Printed and bound by CPI Group (UK) Ltd, Croydon, CR0 4YY
01/10/2026
02225957-0012